Carol Jacoby and Peter Loth

Abelian Groups

De Gruyter Studies in Mathematics

Volume 73

Carol Jacoby and Peter Loth

Abelian Groups

Structures and Classifications

DE GRUYTER

Mathematics Subject Classification 2010
Primary: 20-2, 20Kxx, 03C52, 13C05; Secondary: 03C60, 03E10, 22B05, 22D35

Authors
Dr. Carol Jacoby
Jacoby Consulting
Long Beach, CA
USA
cjacoby@jacobyconsulting.com

Prof. Dr. Peter Loth
Department of Mathematics
Sacred Heart University
5151 Park Avenue
Fairfield, CT 06825
USA
lothp@sacredheart.edu

ISBN 978-3-11-043211-4
e-ISBN (PDF) 978-3-11-042768-4
e-ISBN (EPUB) 978-3-11-042786-8
ISSN 0179-0986

Library of Congress Control Number: 2019938386

Bibliographic information published by the Deutsche Nationalbibliothek
The Deutsche Nationalbibliothek lists this publication in the Deutsche Nationalbibliografie; detailed bibliographic data are available on the Internet at http://dnb.dnb.de.

Typesetting: VTeX UAB, Lithuania
Printing and binding: CPI books GmbH, Leck

MIX
Papier aus verantwortungsvollen Quellen
FSC www.fsc.org FSC® C083411

www.degruyter.com

For our spouses Jerry and Alice
In memoriam Rüdiger Göbel

Preface

How do you get your arms around an infinite abelian group? Such groups are all around us—integers, rationals, reals, and complex numbers. Yet these examples do not even hint at the richness and complexities of abelian groups, as we will see. To help in getting through the complexity we want to focus in on some set of characteristics that completely determines the group. Specifically, we seek some invariants of a group such that if two groups share the same invariants, then they are of necessity isomorphic.

As it turns out, this problem cannot be solved in general. There is a trade-off between the size of the class of groups classified and the complexity of the invariants. This means that a key part of the classification task is finding classes of groups that admit to an insightful classification. Thus, the classification problem has three parts:

(1) Identify a class of groups to be classified. The broader the class the better. Much of the work in this area involves expanding the class. For example, Ulm was able to classify countable torsion groups, and later work expanded the classification to other torsion groups, and then to certain mixed groups (that is, those that are neither torsion nor torsion-free) with additional invariants. The ideal class will be closed under key operations to aid studying it. In particular, the class should be closed under direct sums and summands, and ideally G should be in the class if and only if both $p^{\alpha}G$ and $G/p^{\alpha}G$ are in the class.

(2) Identify an invariant, usually a cardinal, that can be determined for each group. It is somewhat misleading to say "an invariant" since it may actually consist of infinitely many values. Prove that it is indeed an invariant by proving that it is unchanged under isomorphism.

(3) Prove that the invariant classifies the groups in the class. Specifically, prove that if two groups in the class have the same invariants, then they are isomorphic. In some cases, we can expand the class by demanding something less than full isomorphism.

This exploration leads to the discovery of interesting classes of groups, and hence to the structure problem. The goal is to describe the structure or other defining characteristics of a class of groups. For example, we will see that groups in certain classes may be written and thought of as direct sums of certain cyclic groups. Often this structure suggests invariants to use for classification. Structure and classification are often combined into uniqueness theorems that present the class as the unique one that contains certain groups, is closed under certain operations and classified by certain invariants.

It can be argued that Gauss gave us the first classification result in 1801, the finite version of the primary decomposition theorem. This was even before groups and abelian groups were defined by Kronecker in 1870. The first classification result of infinite groups was due to Ulm in 1933. He developed cardinal invariants that com-

https://doi.org/10.1515/9783110427684-201

pletely classified countable torsion groups. Soon after, Baer (1937) developed invariants called types that classify torsion-free groups of rank 1. This was the first work that systematically addressed groups that were not necessarily countable. Interest in abelian groups reignited in the mid-20th century with the publication of books on the topic [78, 24] by Kaplansky (1954) and Fuchs (1958), the latter containing 86 open problems. Hill extended Ulm's classification to a larger class of torsion groups, the totally projective groups. These torsion and torsion-free results seemed to be unrelated until Warfield developed a classification theory of a class of groups that includes both, but also some groups that do not split into a direct sum of torsion and torsion-free subgroups. An apparent way to look at these mixed groups is as extensions of their torsion subgroup by a torsion-free group. Warfield flipped this around and viewed these mixed groups in terms of a torsion-free basis, characterized by the Warfield invariants, with the quotient over this basis torsion and totally projective, and thus characterized by the Ulm invariants.

We extend this work in two ways. One is to weaken the requirement for isomorphism. Barwise and Eklof extended Ulm's theorem to all torsion groups up to partial isomorphism. Since partial isomorphism implies isomorphism on countable groups, this is a true generalization. In Chapter 5, we define a class of groups that includes the Warfield groups and classify it up to partial isomorphism. Partial isomorphism has an interesting model-theoretic counterpart that we explore.

The other extension looks at topological groups. In Chapters 6 and 7, we use Pontrjagin duality to determine the dual counterparts of various constructs in topological groups. Every group result has a corresponding result in compact abelian groups. In particular, this gives us classifications for various classes of compact abelian groups.

The classes of groups all have a range of characterizations. We explore each class in terms of presentations, homology, increasing sequences of subgroups, decomposition bases, and many others. Many of the chapters culminate in a list of varied characterizations that are proved equivalent.

We assume the reader is familiar with the definition of group, and key constructs, such as subgroups, factor groups, homomorphisms, and isomorphisms. When we use the word "group", we will mean additive abelian group. At times, we will consider groups as modules over the integers or over $\mathbb{Z}_p$, a ring that will be useful for looking at groups one prime at a time. Necessary background in the key characteristics of infinite abelian groups is provided in Chapter 1.

Acknowledgments

We would like to express our gratitude to the memory of the late Professor Rüdiger Göbel of the University of Duisburg–Essen. He greatly impacted the direction of Peter Loth's research and initiated a research project that led to our subsequent collaboration. Carol Jacoby would like to thank him for his generous help and encouragement in incorporating and extending her previous research. She gratefully acknowledges Professor Emeritus Paul Eklof of the University of California, Irvine, for recognizing the synergy possible between this research and her previous work, and providing suggestions and connections to make it happen.

This project was partially supported by a University Research/Creativity Grant (URCG) from Sacred Heart University. We wish to thank Walter de Gruyter Verlag for the publication of this book.

https://doi.org/10.1515/9783110427684-202

Contents

1 Basic characteristics of abelian groups

This introductory chapter presents the background and key characteristics needed to classify abelian groups. Key distinguishing characteristics of groups and their subgroups include reduced, divisible, torsion, and torsion-free. The key characteristic of a group element is its height, and we also often focus on elements in the p-socle for various primes p. Key techniques that will be used throughout the development are introduced here: localization, breaking the group into direct summands, extension theorems, commutative diagrams, short exact sequences, and basic homology. We introduce the two basic classical classification results, Ulm's theorem for countable torsion groups and Baer's theorem for completely decomposable torsion-free groups. This chapter may be used alone as a short introduction to the study of abelian groups.

1.1 Preliminaries

In this section, we introduce some basic notations, definitions, and facts, mostly without proof. For background and more details, we refer the reader to the book by Fuchs, "Abelian Groups" [29]. Throughout, we accept the ZFC axioms of set theory (Zermelo–Fraenkel with the Axiom of Choice). Let X be a **partially ordered set**, that is, a set with a binary relation $\leq$ such that for all $x, y, z \in X$, we have:

1. $x \leq x$;
2. $x \leq y$ and $y \leq x$ imply $x = y$;
3. $x \leq y$ and $y \leq z$ imply $x \leq z$.

We write $x < y$ whenever $x \leq y$ and $x \neq y$. An element $m \in X$ is called **minimal** if $x \leq m$ with $x \in X$ implies $m = x$. Likewise, an element $m' \in X$ is said to be **maximal** if $m' \leq x$ with $x \in X$ implies $m' = x$. The set X satisfies the **minimum** (or **maximum**) **condition** if every nonempty subset of X contains a minimal (or maximal) element. An element $a \in X$ is called the **least** element (or **smallest** element) of X if $a \leq x$ for all $x \in X$, and $b \in X$ is said to be the **greatest** element (or **largest** element) of X if $x \leq b$ for all $x \in X$. A subset C of X is called a **totally ordered set** or a **chain** if $x \leq y$ or $y \leq x$ whenever $x, y \in C$. An element $u \in X$ is called an **upper bound** of C in X if $x \leq u$ for all $x \in C$.

We will make frequent use of the following lemma, which is equivalent to the Axiom of Choice.

Zorn's lemma. *Let X be a nonempty partially ordered set, in which every chain has an upper bound in X. Then X contains at least one maximal element.*

A partially ordered set is called **well-ordered** if every nonempty subset has a least element. Note that a well-ordered set is automatically totally ordered. If C is a

https://doi.org/10.1515/9783110427684-001

nonempty subset of a well-ordered set X such that C has an upper bound in X, then C possesses a *least* upper bound, called the **supremum** of C. It is denoted by $\sup C$.

Many of the invariants that are used to classify groups are based on ordinals. A set S is called **transitive** if for all $x \in S$ and $y \in x$ we have $y \in S$. Then **ordinals** are defined to be transitive sets whose elements are transitive sets. Since elements of an ordinal α are ordinals, it will be convenient to write $\beta < \alpha$ instead of $\beta \in \alpha$, and we write $\alpha = \{\beta : \beta < \alpha\}$. Notice that α is well-ordered by $\leq$, thus every set of ordinals is well-ordered. In particular, any decreasing sequence of ordinals becomes stationary. Finite ordinals are denoted by $0, 1, 2, 3, \ldots$. Specifically, we define $0 = \emptyset$ and $n + 1$, the successor of the finite ordinal n, is defined as $\{n, \{n\}\}$. Then

$$\omega = \{0, 1, 2, 3, \ldots\}$$

is the smallest infinite ordinal, and $n < \omega$ or $n \in \omega$ simply means that n is a nonnegative integer. The set of all positive integers (also called **natural numbers**) is denoted by $\mathbb{N}$. We assume that the reader is familiar with ordinal arithmetic. Recall the following properties of ordinals. If α, β, and γ are ordinals, then

1. $\alpha + (\beta + \gamma) = (\alpha + \beta) + \gamma$
2. $\alpha \leq \beta$ if and only if there is an ordinal δ such that $\alpha + \delta = \beta$
3. $\beta \leq \alpha + \beta$
4. Left cancelation: If $\alpha + \beta \leq \alpha + \gamma$, then $\beta \leq \gamma$
5. Division algorithm: If $\beta \neq 0$, then there are unique ordinals ψ and ν such that $\alpha = \beta\psi + \nu$ and $\nu < \beta$.

In particular, every ordinal can be written as $\alpha = \omega\delta + n$ with an ordinal δ and $n < \omega$. Note that ordinal arithmetic is not commutative. For example, $1 + \omega = \omega$, whereas $\omega + 1 > \omega$. Note also that right cancelation is invalid. For example, $1 + \omega = 2 + \omega$.

Let Ord denote the class of all ordinals, and let $\mathrm{Ord}_\infty = \mathrm{Ord} \cup \{\infty\}$. An ordinal of the form $\alpha = \sigma + 1$ for some ordinal σ is called a **successor ordinal**; in this case, we say that $\alpha - 1$ **exists**. An ordinal > 0 is called a **limit ordinal** if it is not a successor ordinal. A subset A of a limit ordinal α is called **cofinal in** α if $\sup A = \alpha$. We adopt the convention that $\alpha < \infty$ for all $\alpha \in \mathrm{Ord}_\infty$. For any $\alpha, \beta \in \mathrm{Ord}_\infty$, we write $\alpha \lneqq \beta$ whenever $\alpha < \beta$ and $\alpha \neq \beta$, i. e., if $\infty \neq \alpha < \beta$. Throughout this book we will use the notation

$$\alpha_1 \wedge \cdots \wedge \alpha_n \quad \text{or} \quad \min\{\alpha_1, \ldots, \alpha_n\}$$

to denote the minimum of the ordinals $\alpha_1, \ldots, \alpha_n$. The choice will depend on which is clearer in the context. In particular, we will usually use the first notation whenever only two ordinals are involved. Collections of sets are frequently indexed by ordinals; for instance, by $\{S_\alpha\}_{\alpha<\lambda}$, where λ is an ordinal, we mean the collection of all sets S_α with $\alpha < \lambda$.

The principle of **transfinite induction** will be used frequently:

Theorem 1.1.1. *For every ordinal α, let $S(\alpha)$ be a statement. Suppose that the following conditions hold:*
1. *$S(0)$ is true;*
2. *for each $\alpha > 0$, $S(\alpha)$ is true whenever $S(\beta)$ is true for all $\beta < \alpha$.*

Then $S(\alpha)$ is true for all ordinals α.

Proof. Assume that $S(\gamma)$ is false for some ordinal γ. Then the set X of all ordinals $\beta \leq \gamma$, for which $S(\beta)$ is false is nonempty; so X has a least element α. Since $\alpha > 0$ by (1) and $S(\beta)$ is true for all $\beta < \alpha$, it follows from (2) that $S(\alpha)$ is true, a contradiction. □

We will be interested in cardinality to compare the sizes of infinite groups. Two sets S and T are called **equinumerous**, written $|S| = |T|$, if there is a bijection from S onto T. Then **cardinals** can be viewed as ordinals α such that $|\beta| \neq |\alpha|$ for all $\beta < \alpha$. If S is a set and κ is a cardinal with $|S| = |\kappa|$, we simply write $|S| = \kappa$ and call κ the **cardinality** of S. The first infinite cardinal is $\aleph_0 = \omega$. Note that for cardinals $\kappa \leq \lambda$, we have

$$\kappa + \lambda = \lambda = \kappa\lambda$$

whenever λ is infinite.

In this book, all groups are assumed to be abelian, so the word "abelian" is mostly omitted. A subgroup A of a group G is called a **proper** subgroup if $A \neq G$. By $\mathbb{Z}$, we mean the group or ring of **integers**; $\mathbb{P}$ is the set of primes in $\mathbb{Z}$, and $\mathbb{Q}$ is the group or ring of **rational numbers.** For a prime p, $\mathbb{Z}_p$ denotes the group or ring of **integers localized at** p, that is,

$$\mathbb{Z}_p = \left\{ \frac{m}{n} : m, n \in \mathbb{Z} \text{ and } p \nmid n \right\}.$$

By R, we will always mean an arbitrary principal ideal domain with 1, unless stated otherwise. Recall that any elements $a, b \in R$ have a greatest common divisor (gcd) and that there exist $k, l \in R$ such that $ka + lb = \gcd(a, b)$.

An R-**module** is an abelian group G, together with a map $R \times G \to G$, written $(r, x) \mapsto rx$, such that

$$(r+s)x = rx + sx \quad (rs)x = r(sx)$$
$$r(x+y) = rx + ry \quad 1x = x$$

for all $r, s \in R$ and $x, y \in G$. If G and H are R-modules, then an R-**module homomorphism** of G to H is a map $\alpha : G \to H$, satisfying

$$\alpha(x+y) = \alpha(x) + \alpha(y) \quad \text{and} \quad \alpha(rx) = r(\alpha(x))$$

for all $x, y \in G$ and $r \in R$. We often refer to α simply as a **homomorphism.**

We will find it useful to consider groups as special cases of R-modules. In particular, groups are $\mathbb{Z}$-modules. We will also find it useful to focus on a particular prime p by considering $\mathbb{Z}_p$-modules. Then a $\mathbb{Z}_p$-module is also called a ***p*-local group**. Notice that a p-group can be regarded as a p-local group. A group homomorphism $G \to H$ is automatically a $\mathbb{Z}_p$-module homomorphism if G and H are p-local groups.

Let G and H be R-modules. For a homomorphism $\alpha : G \to H$, the **image** (also called **range**) of α is denoted by $\operatorname{im}\alpha$ or $\alpha(G)$. The **kernel** of α is $\ker\alpha = \{x \in G : \alpha(x) = 0\}$, and by the **cokernel** of α, we mean $\operatorname{coker}\alpha = H/\operatorname{im}\alpha$. If α is injective, then we sometimes call α a **monomorphism**. A surjective homomorphism is called an **epimorphism**. If $\alpha : G \to H$ is a bijective homomorphism, it is said to be an **isomorphism,** and we write $G \cong H$. A homomorphism $\nu : G \to G$ is called an **endomorphism**. For example, the **zero map** $0_G : x \mapsto 0$, and the **identity map** $1_G : x \mapsto x$ are endomorphisms. In the case where the endomorphism ν is an isomorphism, it is called an **automorphism.** The **composite** of two homomorphisms $\alpha : G \to H$ and $\beta : H \to K$ is written $\beta\alpha : G \to K$.

For a prime $p \in R$ and a nonnegative integer n, we define

$$p^n G = \{p^n x : x \in G\} \quad \text{and} \quad G[p^n] = \{x \in G : p^n x = 0\}.$$

Then $G[p] = \{x \in G : px = 0\}$ is called the ***p*-socle of** G, and G is called ***p*-bounded** if $G = G[p]$. Notice that $G[p]$ can be regarded as a vector space over the field $R/(p)$.

For a subset S of G, the submodule of G generated by S is written $\langle S\rangle$. Specifically, $\langle S\rangle$ consists of all linear combinations of elements of S. We will use this same notation for submodules generated by combinations of subsets and elements, such as $\langle S, x\rangle$ or $\langle x_1, \dots, x_n\rangle$. Of particular interest is the cyclic submodule of G generated by an element $x \in G$, written $\langle x\rangle$. The trivial R-module is written $\{0\}$ or 0.

If A and B are submodules of G, the submodule C of G generated by A and B is written $C = A + B$. If in addition $A \cap B = 0$, we write $C = A \oplus B$ and call C the **(internal) direct sum** of its submodules A and B. Then A is said to be a **direct summand** of C.

Likewise, the submodule of G generated by a family of submodules A_i $(i \in I)$ is denoted by $A = \sum_{i\in I} A_i$. If the submodules A_i satisfy $A_i \cap \sum_{j\neq i} A_j = 0$ for all $i \in I$, then we call A the **(internal) direct sum** of submodules A_i and write $A = \bigoplus_{i\in I} A_i$. If I is finite, say $I = \{1, \dots, n\}$, then we often write $A = A_1 \oplus \cdots \oplus A_n$. In case A_i are cyclic submodules generated by elements x_i $(i \in I)$, we sometimes use the notation $A = \sum_{x\in X}\langle x\rangle$ ($A = \bigoplus_{x\in X}\langle x\rangle$ if the sum is direct), where $X = \{x_i : i \in I\}$.

Now suppose that A_i are any given R-modules $(i \in I)$. Then the **direct product** $\prod_{i\in I} A_i$ becomes an R-module via $(x_i)_{i\in I} + (y_i)_{i\in I} = (x_i + y_i)_{i\in I}$ and $r(x_i)_{i\in I} = (rx_i)_{i\in I}$ for $r \in R$. Its submodule $A = \{(x_i)_{i\in I} : x_i = 0 \text{ for almost all } i \in I\}$ is called the **(external) direct sum** of modules A_i. Notice that there is a natural isomorphism between internal and external direct sums. Thus, we simply write $A = \bigoplus_{i\in I} A_i$ as before, and again, we may write $A = A_1 \oplus \cdots \oplus A_n$ whenever $I = \{1, \dots, n\}$.

We make frequent use of the easily verified **modular law:**

$$(A+B)\cap C = A+(B\cap C)$$

whenever A, B and C are submodules of G with $A \subseteq C$.

For a submodule A of G, the factor group G/A becomes an R-module by defining $r(x+A) = rx + A$ for $r \in R$. Recall **Noether's isomorphism theorems** for R-modules:
1. If $\alpha : G \to H$ is an epimorphism, then $G/\ker\alpha \cong H$.
2. If A and B are submodules of G, then $(A+B)/B \cong A/(A\cap B)$.
3. If A and B are submodules of G with $A \subseteq B$, then $G/B \cong (G/A)/(B/A)$.

A sequence of R-modules and homomorphisms

$$A_1 \xrightarrow{\alpha_1} A_2 \xrightarrow{\alpha_2} \cdots \to A_n \xrightarrow{\alpha_n} A_{n+1} \quad (n \geq 2)$$

is called **exact at** A_i (with $i \in \{2,\ldots,n\}$) if $\operatorname{im}\alpha_{i-1} = \ker\alpha_i$. The sequence is called **exact** if it is exact at A_i for all $i = 2,\ldots,n$. By a **short exact sequence**, we mean an exact sequence of the form

$$0 \to A \xrightarrow{\alpha} B \xrightarrow{\beta} C \to 0.$$

In this case, we have $A \cong \alpha(A)$ and $C \cong B/\alpha(A)$.

Exercises.
1. Let X be a partially ordered set, and let $a \in X$. If a is a least member of X, then a is minimal. The converse holds whenever X is totally ordered.
2. Show that $\mathbb{Z}_p/p^n\mathbb{Z}_p \cong \mathbb{Z}/p^n\mathbb{Z}$.
3. Show that $|\bigoplus_{i\in I} G_i| = \sum_{i\in I} |G_i|$ whenever $|I| \geq \aleph_0$.
4. (a) $0 \to A \xrightarrow{\alpha} B$ is exact if and only if α is a monomorphism.
   (b) $B \xrightarrow{\beta} C \to 0$ is exact if and only if β is an epimorphism.
   (c) $0 \to A \xrightarrow{\alpha} B \to 0$ is exact if and only if α is an isomorphism.
5. If $0 \to A_i \to B_i \to C_i \to 0$ are exact sequences ($i \in I$), then the induced sequences $0 \to \bigoplus_{i\in I} A_i \to \bigoplus_{i\in I} B_i \to \bigoplus_{i\in I} C_i \to 0$ and $0 \to \prod_{i\in I} A_i \to \prod_{i\in I} B_i \to \prod_{i\in I} C_i \to 0$ are exact as well.

1.2 Commutative diagrams

A diagram of R-modules and homomorphisms is called **commutative**, and we say the **diagram commutes** if different paths along directed arrows from one module to another module result in the same composite homomorphisms. For example, the diagram

$$\begin{array}{ccccccccc} 0 & \to & A_1 & \xrightarrow{\alpha_1} & A_2 & \xrightarrow{\alpha_2} & A_3 & \to & 0 \\ & & \downarrow \phi_1 & & \downarrow \phi_2 & & \downarrow \phi_3 & & \\ 0 & \to & B_1 & \xrightarrow{\beta_1} & B_2 & \xrightarrow{\beta_2} & B_3 & \to & 0 \end{array}$$

is commutative exactly if the left square commutes (that is, if $\beta_1\phi_1 = \phi_2\alpha_1$) and the second square commutes (that is, if $\beta_2\phi_2 = \phi_3\alpha_2$).

Lemma 1.2.1. *Let $0 \to A \xrightarrow{\alpha} B \xrightarrow{\beta} C$ be an exact sequence and $\phi : G \to B$ a homomorphism. Then there is a homomorphism $\nu : G \to A$ such that the diagram*

$$\begin{array}{ccccccc} & & & & G & & \\ & & & \nu\swarrow & \downarrow\phi & & \\ 0 & \to & A & \xrightarrow{\alpha} & B & \xrightarrow{\beta} & C \end{array}$$

is commutative if and only if $\beta\phi = 0$.

Proof. If such a map ν exists, then $\beta\phi = \beta\alpha\nu = 0$. To prove the converse, assume that $\beta\phi = 0$, and define a map $\nu : G \to A$ by $g \mapsto a$ if $\alpha(a) = \phi(g)$. Such an a exists since $\beta\phi(g) = 0$ implies $\phi(g) \in \ker\beta = \operatorname{im}\alpha$. This map is well-defined since α is injective. Clearly, ν is a homomorphism and $\alpha\nu = \phi$. □

Lemma 1.2.2. *If $A \xrightarrow{\alpha} B \xrightarrow{\beta} C \to 0$ is exact and $\phi : B \to G$ is a homomorphism, then there is a homomorphism $\nu : C \to G$ such that the diagram*

$$\begin{array}{ccccccc} A & \xrightarrow{\alpha} & B & \xrightarrow{\beta} & C & \to & 0 \\ & & \phi\downarrow & \swarrow\nu & & & \\ & & G & & & & \end{array}$$

is commutative if and only if $\phi\alpha = 0$.

Proof. If there is such a map ν, then $\phi\alpha = \nu\beta\alpha = 0$. Conversely, suppose that $\phi\alpha = 0$, and define $\nu : C \to G$ by $c \mapsto \phi(b)$ if $\beta(b) = c$ with $b \in B$, which can always be done since $\operatorname{im}\beta = C$. To verify that ν is well-defined, assume that $\beta(b') = c$ with $b' \in B$. Then $b - b' \in \ker\beta = \operatorname{im}\alpha \subseteq \ker\phi$ since $\phi\alpha = 0$, thus $\phi(b) = \phi(b')$. The map ν is a homomorphism, and we have $\nu\beta = \phi$, as required. □

The next result is known as the **Five lemma:**

Lemma 1.2.3. *If the diagram*

$$\begin{array}{ccccccccc} A_1 & \xrightarrow{\alpha_1} & A_2 & \xrightarrow{\alpha_2} & A_3 & \xrightarrow{\alpha_3} & A_4 & \xrightarrow{\alpha_4} & A_5 \\ \downarrow\phi_1 & & \downarrow\phi_2 & & \downarrow\phi_3 & & \downarrow\phi_4 & & \downarrow\phi_5 \\ B_1 & \xrightarrow{\beta_1} & B_2 & \xrightarrow{\beta_2} & B_3 & \xrightarrow{\beta_3} & B_4 & \xrightarrow{\beta_4} & B_5 \end{array}$$

with exact rows is commutative, then we have:

1. *Suppose ϕ_1 is an epimorphism and ϕ_5 is a monomorphism. If ϕ_2 and ϕ_4 are isomorphisms, then so is ϕ_3.*
2. *Suppose that ϕ_1 is an epimorphism. If ϕ_2 and ϕ_4 are monomorphisms, then so is ϕ_3.*
3. *Suppose that ϕ_5 is a monomorphism. If ϕ_2 and ϕ_4 are epimorphisms, then so is ϕ_3.*

Proof. (1) follows from (2) and (3). Now assume that ϕ_1, ϕ_2, and ϕ_4 are maps as in (2). To prove that ϕ_3 is injective, let $a \in \ker \phi_3$. Then $\phi_4\alpha_3(a) = \beta_3\phi_3(a) = 0$, thus injectivity of ϕ_4 yields $\alpha_3(a) = 0$. Since $\operatorname{im} \alpha_2 = \ker \alpha_3$, we have $a = \alpha_2(a_2)$ for some $a_2 \in A_2$; hence, $\beta_2\phi_2(a_2) = \phi_3\alpha_2(a_2) = \phi_3(a) = 0$. Then $\operatorname{im} \beta_1 = \ker \beta_2$ shows that $\phi_2(a_2) = \beta_1(b_1)$ for some $b_1 \in B_1$. Now ϕ_1 is surjective, thus there is an $a_1 \in A_1$ such that $b_1 = \phi_1(a_1)$. Then $\phi_2\alpha_1(a_1) = \beta_1\phi_1(a_1) = \beta_1(b_1) = \phi_2(a_2)$, and we obtain $\alpha_1(a_1) = a_2$ since ϕ_2 is injective. But then $a = \alpha_2(a_2) = \alpha_2\alpha_1(a_1) = 0$. Therefore, ϕ_3 is injective.

Next we assume that ϕ_2, ϕ_4, and ϕ_5 are maps as in (3). We let $b \in B_3$ and show that $b \in \operatorname{im}\phi_3$. Since ϕ_4 is surjective, there is an $a_4 \in A_4$ such that $\beta_3(b) = \phi_4(a_4)$, thus $\phi_5\alpha_4(a_4) = \beta_4\phi_4(a_4) = \beta_4\beta_3(b) = 0$. It follows that $\alpha_4(a_4) = 0$ since ϕ_5 is injective. Since $\ker \alpha_4 = \operatorname{im} \alpha_3$, there is an $a_3 \in A_3$ such that $\alpha_3(a_3) = a_4$; hence, $\beta_3\phi_3(a_3) = \phi_4\alpha_3(a_3) = \phi_4(a_4) = \beta_3(b)$. Then $\ker \beta_3 = \operatorname{im} \beta_2$ shows that there is a $b_2 \in B_2$, satisfying $\phi_3(a_3) - b = \beta_2(b_2)$. By assumption, ϕ_2 is surjective; so $b_2 = \phi_2(a_2)$ for some $a_2 \in A_2$. Thus, $\phi_3\alpha_2(a_2) = \beta_2\phi_2(a_2) = \beta_2(b_2) = \phi_3(a_3) - b$, and we obtain $b \in \operatorname{im}\phi_3$, as desired. □

The 3 × 3-**Lemma** will be needed:

Lemma 1.2.4. *Suppose that the following diagram is commutative and all columns are exact:*

$$\begin{array}{ccccccccc}
 & & 0 & & 0 & & 0 & & \\
 & & \downarrow & & \downarrow & & \downarrow & & \\
0 & \to & A_1 & \xrightarrow{\alpha_1} & B_1 & \xrightarrow{\beta_1} & C_1 & \to & 0 \\
 & & \downarrow \phi_1 & & \downarrow \psi_1 & & \downarrow \chi_1 & & \\
0 & \to & A_2 & \xrightarrow{\alpha_2} & B_2 & \xrightarrow{\beta_2} & C_2 & \to & 0 \\
 & & \downarrow \phi_2 & & \downarrow \psi_2 & & \downarrow \chi_2 & & \\
0 & \to & A_3 & \xrightarrow{\alpha_3} & B_3 & \xrightarrow{\beta_3} & C_3 & \to & 0 \\
 & & \downarrow & & \downarrow & & \downarrow & & \\
 & & 0 & & 0 & & 0 & &
\end{array}$$

If the first two or the last two rows are exact, then all three rows are exact.

Proof. Assuming exactness of the first two rows, we show that the third row is exact. To prove that α_3 is injective, let $a \in \ker \alpha_3$. Then $\phi_2(a_2) = a$ for some $a_2 \in A_2$; hence, $\psi_2\alpha_2(a_2) = \alpha_3\phi_2(a_2) = 0$. Since $\ker \psi_2 = \operatorname{im} \psi_1$, there is a $b_1 \in B_1$ such that $\alpha_2(a_2) = \psi_1(b_1)$, thus $\chi_1\beta_1(b_1) = \beta_2\psi_1(b_1) = \beta_2\alpha_2(a_2) = 0$. By injectivity of χ_1, we obtain $\beta_1(b_1) = 0$. Then $\operatorname{im} \alpha_1 = \ker \beta_1$ shows that $b_1 = \alpha_1(a_1)$ for some $a_1 \in A_1$, which implies that $\alpha_2(a_2) = \psi_1(b_1) = \psi_1\alpha_1(a_1) = \alpha_2\phi_1(a_1)$. But then $a_2 - \phi_1(a_1) \in \ker \alpha_2$, thus $a_2 = \phi_1(a_1)$. It follows that $a = \phi_2(a_2) = \phi_2\phi_1(a_1) = 0$; hence, α_3 is injective.

Next we show that $\operatorname{im} \alpha_3 = \ker \beta_3$. Letting $a \in A_3$, we have $\phi_2(a_2) = a$ for some $a_2 \in A_2$. Then $\beta_3\alpha_3(a) = \beta_3\alpha_3\phi_2(a_2) = \beta_3\psi_2\alpha_2(a_2) = \chi_2\beta_2\alpha_2(a_2) = 0$ shows that $\operatorname{im} \alpha_3 \subseteq \ker \beta_3$. Now let $b \in \ker \beta_3$. Then $b = \psi_2(b_2)$ for some $b_2 \in B_2$, hence, $\chi_2\beta_2(b_2) =$

$\beta_3\psi_2(b_2) = 0$. Since $\ker\chi_2 = \operatorname{im}\chi_1$, we have $\beta_2(b_2) = \chi_1(c_1)$; for some $c_1 \in C_1$. Then there is an element $b_1 \in B_1$ such that $\beta_1(b_1) = c_1$, and we obtain $\beta_2(b_2) - \beta_2\psi_1(b_1) = \beta_2(b_2) - \chi_1\beta_1(b_1) = \beta_2(b_2) - \beta_2(b_2) = 0$, thus $b_2 - \psi_1(b_1) \in \ker\beta_2 = \operatorname{im}\alpha_2$. Consequently, we have $b_2 - \psi_1(b_1) = \alpha_2(a_2)$ for some $a_2 \in A_2$, and it follows that $\alpha_3\phi_2(a_2) = \psi_2\alpha_2(a_2) = \psi_2(b_2 - \psi_1(b_1)) = \psi_2(b_2) = b$. Thus, $b \in \operatorname{im}\alpha_3$.

To prove surjectivity of β_3, let $c \in C_3$. Then there is a $b_2 \in B_2$ such that $c = \chi_2\beta_2(b_2) = \beta_3\psi_2(b_2)$, therefore $c \in \operatorname{im}\beta_3$.

The proof of the second assertion is very similar. □

Exercises.

1. Show that the map $\nu : G \to A$ in Lemma 1.2.1 is uniquely determined.
2. Show that the map $\nu : C \to G$ in Lemma 1.2.2 is uniquely determined.
3. A homomorphism $\phi : A_1 \to A_2$ induces an exact sequence $0 \to \ker\phi \to A_1 \xrightarrow{\phi} A_2 \to \operatorname{coker}\phi \to 0$.
4. Prove the **Snake lemma**: Let

$$\begin{array}{ccccccc} & & A_1 & \to & B_1 & \to & C_1 & \to & 0 \\ & & \downarrow\phi & & \downarrow\psi & & \downarrow\chi & & \\ 0 & \to & A_2 & \to & B_2 & \to & C_2 & & \end{array}$$

be a commutative diagram with exact rows. Then there is an exact sequence $\ker\phi \to \ker\psi \to \ker\chi \to \operatorname{coker}\phi \to \operatorname{coker}\psi \to \operatorname{coker}\chi$.

1.3 The torsion part of a group

Let G be a group. By the **torsion part of** G, we mean the set

$$tG = \{x \in G : nx = 0 \text{ for some nonzero integer } n\}.$$

If $G = tG$, we call G **torsion**, and if $tG = 0$, then we say that G is **torsion-free**. If A is a subgroup of G such that G/A is torsion, we sometimes say that G **is torsion over** A. The group G is called **mixed** if G is neither torsion nor torsion-free; that is, if G has both nonzero elements of finite order and elements of infinite order.

Theorem 1.3.1. *Let G be a group. Then its torsion part tG is a subgroup of G, and G/tG is torsion-free.*

Proof. Clearly, tG is nonempty since it contains 0. If $x, y \in tG$, then $rx = 0 = sy$ for some nonzero integers r, s, which yields $rs(x - y) = 0$, thus $x - y \in tG$. Therefore, tG is a subgroup of G. To show that G/tG is torsion-free, let $g \in G$ and assume that $ng \in tG$ for some nonzero integer n. Then $mng = 0$ for some nonzero integer m, thus $g \in tG$. It follows that G/tG is torsion-free. □

For a given prime p, the **p-torsion part** of a group G is defined to be the set

$$t_pG = \{x \in G : p^n x = 0 \text{ for some nonnegative integer } n\}.$$

In the case $t_pG = 0$, we say that G **has no p-torsion.**

Theorem 1.3.2. *Let G be a group. Then t_pG is a subgroup of G for all primes p, and we have $tG = \bigoplus_{p\in\mathbb{P}} t_pG$.*

Proof. Each set t_pG is nonempty since it contains 0. If $x, y \in t_pG$, then $p^r x = 0 = p^s y$ for some $r, s < \omega$; hence, $p^{r+s}(x - y) = 0$, and therefore, $x - y \in t_pG$. Thus, t_pG is a subgroup of G.

Let $g \in tG$. Then $ng = 0$ for some positive integer n, and we write $n = p_1^{n_1} \cdots p_k^{n_k}$, where $p_1, \dots, p_k$ are distinct primes and $n_1, \dots, n_k < \omega$. For any $i \in \{1, \dots, k\}$, let $n_i' = n/p_i^{n_i}$. Then $n_i' g \in t_{p_i}G$. Since $n_1', \dots, n_k'$ are relatively prime, there are integers $s_1, \dots, s_k$ such that $s_1 n_1' + \cdots + s_k n_k' = 1$, thus

$$g = s_1 n_1' g + \cdots + s_k n_k' g \in \sum_{i=1}^{k} t_{p_i} G,$$

and we obtain $tG = \sum_{p\in\mathbb{P}} t_pG$. If p is a prime and $x \in t_pG \cap \sum_{q\neq p} t_qG$, then $p^k x = 0 = mx$ for some $k < \omega$ and integer m with $p \nmid m$, which shows that $x = 0$. Therefore, we have $tG = \bigoplus_{p\in\mathbb{P}} t_pG$. □

By the theorem above, the study of torsion groups is reduced to that of p-groups.

Exercises.

1. A group G is called **elementary** if the order of each element of G is a square-free integer. Show:
   (a) A group is elementary if and only if it is a direct sum of cyclic groups of prime orders.
   (b) Every subgroup of an elementary group is a direct summand.
2. A group is said to **split** if its torsion part is a direct summand. Show the following: If P is an infinite set of primes, then $\prod_{p\in P} \mathbb{Z}/p\mathbb{Z}$ does not split.

1.4 Independence and rank

A set X of nonzero elements of a group G is called **independent** if for all $x_1, \dots, x_n \in X$ and integers $a_1, \dots, a_n$,

$$a_1 x_1 + \cdots + a_n x_n = 0$$

implies that

$$a_1 x_1 = \cdots = a_n x_n = 0.$$

Note that—in this case—we have $a_i = 0$ provided that the order of x_i is infinite. The following fact is immediate:

Lemma 1.4.1. *Let G be a group and X a set of independent elements of G. Then X is independent if and only if the sum $\sum_{x\in X}\langle x\rangle$ is direct.*

We let I_G denote the set of all elements of G of infinite order. By Zorn's lemma, any independent subset of I_G is contained in some maximal independent subset of I_G. Likewise, for any given prime p, any independent subset of the p-torsion part t_pG is contained in a maximal independent subset of t_pG. We have the following:

Theorem 1.4.2. *Let G be a group. Then any two maximal independent subsets of I_G have the same cardinality. Likewise, for a given prime p, any two maximal independent subsets of t_pG have the same cardinality.*

Proof. Notice that a set $\{x_1,\dots,x_k\}\subseteq I_G$ is independent in G exactly if $\{x_1+tG,\dots,x_k+tG\}$ is independent in the torsion-free group G/tG. Therefore, for the first claim, we may assume that G is torsion-free. Let $X=\{x_i : i\in I\}$ be a maximal independent set in G. Letting g be a nonzero element of G, we have $rg = a_1x_{i_1}+\dots+a_tx_{i_t}$ for some $r\in\mathbb{N}$ and $0\neq a_k\in\mathbb{Z}$, $i_k\in I$ $(k=1,\dots,t)$; so we can associate with g the tuple $(r,a_1,\dots,a_t,i_1,\dots,i_t)$. Note that if $0\neq g'\in G$ is associated with the same tuple, then $r(g-g')=0$ implies that $g=g'$ since G is torsion-free. Thus, $|G|\le|X|\cdot\aleph_0$; so since $|X|\le|G|$, we obtain $|X|=|G|$ whenever X is infinite.

If X is finite, we write $X=\{x_1,\dots,x_n\}$. Then if $Z=\{z_1,\dots,z_m\}$ is any independent set in G, we have $k_iz_i\in\langle X\rangle$ for some $k_i\in\mathbb{N}$ $(i=1,\dots,m)$ by the maximality of X. Assume that $m>n$ and write $k_iz_i=\sum_{j=1}^n\lambda_{ij}x_j$ $(i=1,\dots,m)$. Then there is a nonzero $(\mu_1,\dots,\mu_m)\in\mathbb{Z}^m$ such that

$$\mu_1\begin{pmatrix}\lambda_{11}\\ \vdots\\ \lambda_{1n}\end{pmatrix}+\dots+\mu_m\begin{pmatrix}\lambda_{m1}\\ \vdots\\ \lambda_{mn}\end{pmatrix}=\begin{pmatrix}0\\ \vdots\\ 0\end{pmatrix},$$

and we have $\sum_i\mu_ik_iz_i=\sum_{i,j}\mu_i\lambda_{ij}x_j=\sum_j(\sum_i\mu_i\lambda_{ij})x_j=0$, contradicting independence of $\{z_1,\dots,z_m\}$. Thus, $|Z|\le|X|$.

Now let Y be any maximal independent subset of G. In the case where X is finite, we obtain $|Y|\le|X|$, and by symmetry, we have $|X|\le|Y|$; hence, $|X|=|Y|$. In the case where X is infinite, Y cannot be finite by the same argument we just used, thus $|X|=|G|=|Y|$.

It remains to show that any two maximal independent subsets of t_pG have the same cardinality, so we can assume that G is a p-group. Then a subset $\{x_1,\dots,x_k\}$ of G is independent if and only if $\{p^{n_1-1}x_1,\dots,p^{n_k-1}x_k\}$ is independent, where p^{n_i} is the order of x_i $(i=1,\dots,k)$; thus, it suffices to consider maximal independent subsets X of the p-socle $G[p]$. Now $G[p]$ can be considered a $\mathbb{Z}/p\mathbb{Z}$-vector space, so X is independent exactly if it is linearly independent in the vector space $G[p]$. It follows that the cardinality of any maximal independent subset of G coincides with the dimension of $G[p]$. □

The cardinality of a maximal independent subset of I_G is called the **torsion-free rank** of G, written $r_0(G)$. Likewise, if we restrict ourselves to subsets of the p-torsion part t_pG for any given prime p, the cardinality $r_p(G)$ of a maximal independent subset of t_pG is called the ***p*-rank** of G. We define $r_{p,0}(G) = r_p(G)+r_0(G)$. The **rank** of G, written $r(G)$, is defined as the sum of its torsion-free and p-ranks for all primes p. Note that the rank of a torsion-free group is the torsion-free rank, and the rank of a p-group is the p-rank. By Theorem 1.4.2 and its proof, we have the following:

Corollary 1.4.3. *Let G be a group and p a prime. Then the cardinals $r_0(G), r_p(G), r_{p,0}(G)$, and $r(G)$ are invariants of G. We have $r_0(G) = r_0(G/tG)$ and $r_p(G) = r_p(G[p])$.*

The torsion-free rank of a group can be described as follows:

Theorem 1.4.4. *Let G be a group and $\mathfrak{m}$ a cardinal. Then $r_0(G) = \mathfrak{m}$ if and only if G contains a subgroup $A \cong \bigoplus_{\mathfrak{m}} \mathbb{Z}$ such that G/A is torsion.*

Proof. If $r_0(G) = \mathfrak{m}$, then G has a maximal independent subset X of I_G with $|X| = \mathfrak{m}$. By Lemma 1.4.1, $A = \langle X\rangle$ is a direct sum of infinite cyclic groups. Assume that $g \in G \setminus X$ is an element of infinite order. Then $\langle g\rangle \cap \langle X\rangle \neq 0$ by the maximality of X, thus $ng \in \langle X\rangle$ for some positive integer n. Therefore, G/A is torsion.

Conversely, suppose that G/A is torsion for some subgroup $A \cong \bigoplus_{\mathfrak{m}} \mathbb{Z}$ of G. Then we can write $A = \bigoplus_{x\in X} \langle x\rangle$ for a subset X of A such that each element of X is nonzero and has infinite order. Since $G/\langle X\rangle$ is torsion, it follows that X is a maximal independent set in I_G. □

Exercises.
1. Let A be a subgroup of G. Then:
   (a) $r(G) \leq r(A) + r(G/A)$.
   (b) $r_0(G) = r_0(A) + r_0(G/A)$.
2. If $G = \sum_{i\in I} A_i$, then $r(G) \leq \sum_{i\in I} r(A_i)$. Equality holds whenever the sum is direct.
3. A subgroup E of a group G is said to be **essential** if $E \cap A \neq 0$ for each subgroup $A \neq 0$ of G. Show that an independent set X of elements of G is maximal if and only if $\langle X\rangle$ is an essential subgroup of G.

1.5 Free abelian groups

A group is called **free** if it is a direct sum of infinite cyclic groups. Thus, a group F is free exactly if it has a subset X such that

$$F = \bigoplus_{x\in X} \langle x\rangle$$

with $\langle x\rangle \cong \mathbb{Z}$ for all $x \in X$. It is clear that the elements of F can be written uniquely as expressions

$$z = a_1x_1 + \cdots + a_nx_n \qquad (*)$$

with coefficients $a_1,\dots,a_n \in \mathbb{Z}\setminus\{0\}$ and distinct $x_1,\dots,x_n \in X$, and that addition of two such expressions is carried out by adding coefficients belonging to the same elements of X. Alternatively, given any set X, the group F could have been defined to be the set of all formal expressions $(*)$, together with the aforementioned addition. We call F the **free group on the set** X, and say that X is a **set of free generators.** The following basic fact will be useful:

Theorem 1.5.1. *Every group is an epimorphic image of a free group.*

Proof. Let G be a group, and let X be the underlying set of elements of G. Define F_X to be the free group on X. Then the correspondence $x \mapsto x$ $(x \in X)$ uniquely defines an epimorphism $F_X \to G$. □

Lemma 1.5.2. *Let G be a group and A a subgroup of G such that $G/A = \bigoplus_{i\in I}(G_i/A)$. If $G_i = A \oplus B_i$ $(i \in I)$, then $G = A \oplus \bigoplus_{i\in I} B_i$.*

Proof. It is clear that we have $G = A + \sum_{i\in I} B_i$. To show that the sum is direct, suppose that $a + b_1 + \cdots + b_n = 0$ with elements $a \in A$ and $b_k \in B_{i_k}$ $(i_k \in I, k = 1,\dots,n)$. Then $(b_1 + A) + \cdots + (b_n + A) = 0$, which yields $b_k \in A \cap B_{i_k} = 0$ for all $k = 1,\dots,n$. It follows that $b_1 = \cdots = b_n = 0$, which implies $a = 0$. □

Theorem 1.5.3. *Let G be a group and A a subgroup of G. If G/A is free, then $G = A \oplus F$ for some free group F.*

Proof. We write $G/A = \bigoplus_{i\in I}(G_i/A)$ with $G_i/A \cong \mathbb{Z}$ $(i \in I)$, and let $i \in I$. Then $G_i/A = \langle x_i + A\rangle$ for some $x_i \in G_i$. Since this group is torsion-free, x_i has infinite order, and $A \cap \langle x_i\rangle = 0$; thus, $G_i = A \oplus \langle x_i\rangle$. Letting $F = \bigoplus_{i\in I}\langle x_i\rangle$, we obtain $G = A \oplus F$ by Lemma 1.5.2. □

The following important result will be needed:

Theorem 1.5.4. *Every subgroup of a free group is free.*

Proof. Let F be a free group and A a subgroup of F. We write $F = \bigoplus_{\alpha<\lambda}\langle x_\alpha\rangle$ with $\lambda = |F|$ and define $F_\alpha = \bigoplus_{\beta<\alpha}\langle x_\beta\rangle$ and $A_\alpha = A \cap F_\alpha$ for all $\alpha \le \lambda$. Then

$$0 = A_0 \subseteq \cdots \subseteq A_\alpha \subseteq \cdots \subseteq A_\lambda = A.$$

Letting $\alpha < \lambda$, we have $A_\alpha = A \cap F_\alpha = A_{\alpha+1} \cap A \cap F_\alpha = A_{\alpha+1} \cap F_\alpha$; thus,

$$A_{\alpha+1}/A_\alpha = A_{\alpha+1}/(A_{\alpha+1} \cap F_\alpha) \cong (A_{\alpha+1} + F_\alpha)/F_\alpha \subseteq F_{\alpha+1}/F_\alpha \cong \langle x_\alpha\rangle.$$

It follows that $A_{\alpha+1}/A_\alpha \cong \mathbb{Z}$ or $= 0$, so we have $A_{\alpha+1} = A_\alpha \oplus \langle a_\alpha \rangle$ for some $a_\alpha \in A$ (Theorem 1.5.3). Then the sum $\sum_{\alpha<\lambda} \langle a_\alpha \rangle$ is direct, and we obtain $A = \bigcup_{\alpha<\lambda} A_\alpha = \bigoplus_{\alpha<\lambda} \langle a_\alpha \rangle$. □

Exercises.

1. Let F_X and F_Y be free groups on sets X and Y, respectively. Then $F_X \cong F_Y$ if and only if $|X| = |Y|$.
2. Let F be a group and X any set of generators of F. Then F is a free group with X a free set of generators if and only if every map from X into a group G can be uniquely extended to a homomorphism $F \to G$.
3. (Mac Lane [111]) A group G is called **projective** if G has the **projective property** relative to all short exact sequences; that is, if for each exact sequence $0 \to A \to B \xrightarrow{\beta} C \to 0$ and homomorphism $\phi : G \to C$, there exists a homomorphism $\tau : G \to B$ such that the diagram

$$\begin{array}{ccccccccc} & & & & & & G & & \\ & & & & & \tau \swarrow & \downarrow \phi & & \\ 0 & \to & A & \to & B & \xrightarrow{\beta} & C & \to & 0 \end{array}$$

is commutative. Prove that G is projective if and only if G is free.

1.6 Divisible groups

Let D be a group and p a prime. We call D **p-divisible** if $D = pD$, that is, if for every $x \in D$ there is a $y \in D$ such that $x = py$. The group D is called **divisible** if for every $x \in D$ and $n \in \mathbb{N}$ there is a $y \in D$ such that $x = ny$. We have the following:

(a) D is divisible exactly if D is p-divisible for all primes p.
(b) A p-local group is divisible if and only if it is p-divisible.
(c) A torsion group with no p-torsion is p-divisible.
(d) If D_i are divisible subgroups of a group G ($i \in I$), then $\sum_{i\in I} D_i$ is divisible.
(e) The torsion part of a divisible group is divisible.
(f) The epimorphic image of a divisible group is divisible.
(g) If D_i are groups ($i \in I$), then $\bigoplus_{i\in I} D_i$ is divisible if and only if each group D_i is divisible.

Basic examples of divisible groups are the **group of rationals** $\mathbb{Q}$, the **group of real numbers** $\mathbb{R}$, the group $\mathbb{R}/\mathbb{Z}$, and the **quasicyclic p-group** $\mathbb{Z}(p^\infty)$, which is defined to be the p-torsion part of $\mathbb{R}/\mathbb{Z}$. Note that $\mathbb{Z}(p^\infty)$ can be viewed as the multiplicative group of p^n-th complex roots of unity ($n < \omega$). We include the following useful fact:

Proposition 1.6.1. *For any prime p, the only proper subgroups of $G = \mathbb{Z}(p^\infty)$ are $G[p^n] \cong \mathbb{Z}/p^n\mathbb{Z}$ with $n < \omega$.*

Proof. It is clear that for $n < \omega$, we have $G[p^n] \cong \mathbb{Z}/p^n\mathbb{Z}$; so if $x \in G[p^{n+1}] \setminus G[p^n]$ then $\langle x \rangle = G[p^{n+1}]$. Since G is a union of the ascending sequence

$$0 = G[p^0] \subseteq G[p] \subseteq \cdots \subseteq G[p^n] \subseteq \cdots,$$

it follows that a proper subgroup A of G must be finitely generated. If $A \neq 0$ and $m = \max\{n < \omega : a \in G[p^{n+1}] \setminus G[p^n]$ for some nonzero $a \in A\}$, then $A = G[p^m]$. □

This proposition and its proof highlight some interesting properties of the group $\mathbb{Z}(p^\infty)$:
(h) $\mathbb{Z}(p^\infty)$ is infinite, but all proper subgroups are finite.
(i) Any finite set of elements of $\mathbb{Z}(p^\infty)$ generates a cyclic subgroup.
(j) The subgroups of $\mathbb{Z}(p^\infty)$ are totally ordered by inclusion.

A group G is called **injective** if G has the **injective property** relative to all short exact sequences; that is, if for each exact sequence $0 \to A \xrightarrow{\alpha} B \to C \to 0$ and homomorphism $\phi : A \to G$, there exists a homomorphism $\beta : B \to G$ such that the diagram

$$\begin{array}{ccccccccc} 0 & \to & A & \xrightarrow{\alpha} & B & \to & C & \to & 0 \\ & & \phi \downarrow & \swarrow \beta & & & & & \\ & & G & & & & & & \end{array}$$

is commutative.

Theorem 1.6.2 (Baer [7]). *Divisible groups are injective.*

Proof. Let D be a divisible group, $0 \to A \xrightarrow{\alpha} B \to C \to 0$ a short exact sequence, and $\phi : A \to D$ a homomorphism. Consider the set S of all pairs (H, ν) such that H is a subgroup of B containing $\alpha(A)$, $\nu : H \to D$ and $\nu\alpha = \phi$. We partially order S by declaring that $(H, \nu) \leq (H', \nu')$ if $H \subseteq H'$ and $\nu = \nu' \restriction H$. Then every chain $\{(H_i, \nu_i)\}_{i\in I}$ in S has an upper bound, namely $(\bigcup_{i\in I} H_i, \nu)$ with $\nu_i = \nu \restriction H_i$ for all $i \in I$. By Zorn's lemma, S has a maximal element (H_0, ν_0). Assuming that $H_0 \neq B$, we let $b \in B \setminus H_0$. In the case where $H_0 \cap \langle b \rangle = 0$, the correspondence $x + rb \mapsto \nu_0(x)$ $(x \in H_0, r \in \mathbb{Z})$ yields a homomorphism

$$\nu' : \langle H_0, b \rangle \to D.$$

If $H_0 \cap \langle b \rangle \neq 0$, let n be the smallest positive integer such that $nb \in H_0$. Since D is divisible, there is a $d \in D$ such that $nd = \nu_0(nb)$. Then $x + rb \mapsto \nu_0(x) + rd$ (with $x \in H_0, 0 \leq r < n$) yields a homomorphism $\nu' : \langle H_0, b \rangle \to D$. In either case, ν' extends ν_0, contradicting the maximality of (H_0, ν_0). Thus, we have $H_0 = B$ and $\nu_0\alpha = \phi$. □

The following theorem of Baer presents a key property of divisible subgroups.

Theorem 1.6.3 (Baer [7]). *A divisible subgroup D of a group G is a direct summand of G. Moreover, if C is any subgroup of G with $C \cap D = 0$, then $G = D \oplus A$ for some subgroup A of G containing C.*

Proof. For a subgroup C of G with $C \cap D = 0$, let $\alpha : C \oplus D \to G$ be the inclusion map and $\pi : C \oplus D \to D$ the projection map. Then Theorem 1.6.2 yields a homomorphism $\beta : G \to D$ such that the diagram

$$\begin{array}{ccc} C \oplus D & \xrightarrow{\alpha} & G \\ \pi \downarrow & \swarrow \beta & \\ D & & \end{array}$$

commutes. Notice that $\beta \restriction D$ is the identity map, thus $\beta^2 = \beta$. If $x \in G$, then $x = \beta(x) + (x - \beta(x)) \in D + \ker \beta$, and we have $D \cap \ker \beta = 0$. It follows that $G = D \oplus \ker \beta$. It is clear that $C \subseteq \ker \beta$, so the proof is complete. □

For any given group G, let $\{D_i : i \in I\}$ be the set of all divisible subgroups of G. By (d), $D = \sum_{i \in I} D_i$ is divisible; thus, it is the unique **maximal divisible subgroup of** G. If 0 is the only divisible subgroup of G, then G is said to be **reduced**.

Theorem 1.6.4. *Every group G can be written as $G = D \oplus A$, where D is a divisible group and A is a reduced group. The group D is the uniquely determined maximal divisible subgroup of G, and A is unique up to isomorphism.*

Proof. Letting D be the maximal divisible subgroup of G, Theorem 1.6.3 shows that $G = D \oplus A$ for some subgroup A of G. By the maximality of D, A is reduced. If $G = D' \oplus A'$ with D' divisible and A' reduced, then $D = D' \oplus X$ for some group X (Theorem 1.6.3). Since X is divisible, $A' \cong X \oplus A$ yields $X = 0$. Thus, $D' = D$ and $A' \cong A$. □

Next we have a key extension result that will be used for building the isomorphisms for classification results. Here we will apply it to determine the structure of divisible groups.

Lemma 1.6.5. *Let G and H be groups, p a prime, and $f : S \to T$ a homomorphism, where S and T are subgroups of G and H, respectively. Let $a \in G \setminus S$ and $b \in H$ such that $pa \in S$, $pb \in T$ and $f(pa) = pb$. Then f extends to a homomorphism $g : \langle S, a\rangle \to \langle T, b\rangle$ by sending a onto b. If f is an isomorphism and $b \notin T$, then g is an isomorphism.*

Proof. Define $g : \langle S, a\rangle \to \langle T, b\rangle$ by $s + ra \mapsto f(s) + rb$ $(s \in S, r \in \mathbb{Z})$. To verify that g is well-defined, suppose $s + ra = s' + r'a$, where $s, s' \in S$ and $r, r' \in \mathbb{Z}$, and let $n = r' - r$. Then $na = s - s' \in S$. If $p \nmid n$, then there are $k, l \in \mathbb{Z}$ such that $kp + ln = 1$; hence, $a = kpa + lna \in S$, a contradiction. Therefore, we have $p | n$, say $n = tp$ for $t \in \mathbb{Z}$. It follows that $f(s) - f(s') = f(na) = tf(pa) = tpb = nb = r'b - rb$; thus, $f(s) + rb = f(s') + r'b$, and the map g is well-defined. Clearly, g is a homomorphism, and g is surjective whenever f is.

Suppose that f is injective and that $b \notin T$. To verify that g is injective, suppose $f(s) + rb = 0$ for some $s \in S$ and $r \in \mathbb{Z}$. Then $rb \in T$; so by the same argument as before, we have $p | r$, say $r = t'p$ for $t' \in \mathbb{Z}$. Then $0 = f(s) + rb = f(s) + t'f(pa) = f(s + ra)$; hence, $s + ra = 0$. Consequently, g is injective. □

Proposition 1.6.6. *Let D be a divisible group, $x \in D$, and p a prime. If x has order p, then $x \in A$ for some subgroup $A \cong \mathbb{Z}(p^\infty)$ of D. If x has infinite order, then $x \in B$ for some subgroup $B \cong \mathbb{Q}$ of D.*

Proof. Suppose that x has order p. Then there exists an infinite sequence $\{a_i : i < \omega\}$ in D such that $a_0 = x$ and $pa_{i+1} = a_i$ $(i < \omega)$. On the other hand, the group $\mathbb{Z}(p^\infty)$ is generated by elements $b_0, b_1, \ldots$ such that $pb_0 = 0$ and $pb_{i+1} = b_i$ $(i < \omega)$. Since the correspondence $a_0 \mapsto b_0$ defines an isomorphism $\langle a_0 \rangle \to \langle b_0 \rangle$, repeated application of Lemma 1.6.5 yields an isomorphism between a subgroup A of D containing x and $\mathbb{Z}(p^\infty)$.

The proof of the second claim is similar. □

Now we are able to prove the following structure theorem for divisible groups.

Theorem 1.6.7. *A group D is divisible if and only if $D \cong \bigoplus_{\mathfrak{m}} \mathbb{Q} \oplus \bigoplus_{p\in\mathbb{P}} \bigoplus_{\mathfrak{n}_p} \mathbb{Z}(p^\infty)$ with $\mathfrak{m} = r_0(D)$ and $\mathfrak{n}_p = r_p(D)$ $(p \in \mathbb{P})$. These cardinals form a complete set of isomorphism invariants for D.*

Proof. Let D be a divisible group. Then its torsion part tD is divisible, and we have $D = tD \oplus F$ for some torsion-free divisible group F (Theorem 1.6.3). We write $tD = \bigoplus_{p\in\mathbb{P}} t_pD$ (Theorem 1.3.2) and fix $p \in \mathbb{P}$. Let X be a maximal independent set in $D[p]$. By Proposition 1.6.6, each $x \in X$ can be embedded in a subgroup A_x of t_pD with $A_x \cong \mathbb{Z}(p^\infty)$. Since X is independent, the sum $\sum_{x\in X} A_x$ is direct. This group is a direct summand of t_pG (Theorem 1.6.3); thus, we have $t_pG = \bigoplus_{x\in X} A_x$ by the maximality of X.

Now let Y be a maximal independent set in F and pick $y \in Y$. Since F is divisible and torsion-free, y can be embedded in a subgroup $B_y \cong \mathbb{Q}$ of F (Proposition 1.6.6). Using the same argument as in the first part of this proof, we conclude that $F = \bigoplus_{y\in Y} B_y$. It is clear that $\mathfrak{m} = |Y|$ is the torsion-free rank and $\mathfrak{n}_p = |X|$ is the p-rank of D, both of which are invariants of D (Corollary 1.4.3). Thus, the cardinals $\mathfrak{m}$ and $\mathfrak{n}_p$ $(p \in \mathbb{P})$ determine D uniquely up to isomorphism. □

Notice that by the theorem above, and since $|\mathbb{R}| = 2^{\aleph_0} > |\mathbb{Q}|$, we have

$$\mathbb{R} \cong \bigoplus_{2^{\aleph_0}} \mathbb{Q} \quad \text{and} \quad \mathbb{R}/\mathbb{Z} \cong \bigoplus_{2^{\aleph_0}} \mathbb{Q} \oplus \bigoplus_{p\in\mathbb{P}} \mathbb{Z}(p^\infty).$$

Theorem 1.6.8. *Every group can be embedded in a divisible group.*

Proof. Let G be any group. Then G is an epimorphic image of a free group, say $F = \bigoplus_{\mathfrak{m}} \mathbb{Z}$; thus, $G \cong F/K$ for some subgroup K of F. There is an embedding $\phi : F \to \bigoplus_{\mathfrak{m}} \mathbb{Q}$, so we can identify G with $\phi(F)/\phi(K)$, which is a subgroup of the divisible group $(\bigoplus_{\mathfrak{m}} \mathbb{Q})/\phi(K)$. □

Let G be a group and suppose that G is a subgroup of the divisible group E. Then E is called a **divisible hull of** G if E is minimal in the sense that there is no proper divisible subgroup of E containing G.

Theorem 1.6.9. *Suppose that G is embedded in the divisible group D (see Theorem 1.6.8). Then any such group D contains a divisible hull E of G. If G is torsion-free (or torsion), then E is torsion-free (or torsion). We have $r_0(G) = r_0(E)$ and $r_p(G) = r_p(E)$ for all primes p.*

Proof. The set of all divisible subgroups C of D satisfying $G \cap C = 0$ is partially ordered by inclusion, so it has a maximal element C' by Zorn's lemma. Then $D = C' \oplus E$ with $G \subseteq E$ (Theorem 1.6.3); thus, E is divisible. To show that E is a divisible hull of G, assume that D' is a divisible subgroup of E containing G. If $D' \neq E$, then $E = D' \oplus A'$ for some group $A' \neq 0$ (Theorem 1.6.3); thus, A' is divisible. Since the sum $D' + (A' \oplus C')$ is direct, we have $G \cap (A' \oplus C') = 0$, contradicting the maximality of C'. Thus, E is a divisible hull of G.

Now suppose that G is torsion-free. Assuming that E has a subgroup $H \cong \mathbb{Z}(p^\infty)$ for some prime p, we have $H \cap G = 0$. Thus, $E = H \oplus X$ for some subgroup X of E containing G (Theorem 1.6.3). But then X is divisible, contradicting the minimality of E. Therefore, E is torsion-free. For the torsion case, simply replace $\mathbb{Z}(p^\infty)$ by $\mathbb{Q}$.

To prove that $r_0(G) = r_0(E)$, assume that E has an element x of infinite order such that $G \cap \langle x\rangle = 0$. Then $E = tE \oplus F$ for some torsion-free group F with $x \in F$ (Theorem 1.6.3). Since F is divisible, x can be embedded in a subgroup $B_x \cong \mathbb{Q}$ of F (Proposition 1.6.6), and we have $G \cap B_x = 0$. It follows that $E = E' \oplus B_x$ for some subgroup E' of E containing G (Theorem 1.6.3), contradicting the minimality of E. Thus, $r_0(G) = r_0(E)$. A similar argument shows that $r_p(G) = r_p(E)$ for all primes p. □

Exercises.
1. Show that if E_1 and E_2 are divisible hulls of a group G, then there is an isomorphism between E_1 and E_2 extending the identity map on G.
2. Prove the converse of Theorem 1.6.2.
3. A group G is called **cocyclic** if there exists an $x \in G$ such that every homomorphism $\phi : G \to H$ with $x \notin \ker\phi$ is injective. Show that a group G is cocyclic if and only if there is a prime p such that $G \cong \mathbb{Z}/p^n\mathbb{Z}$ with $n < \omega$ or $G \cong \mathbb{Z}(p^\infty)$.

1.7 Purity, p-independence and p-basic subgroups

Let G be an R-module and $p \in R$ a prime. A submodule A of G is called **p-pure** if

$$p^nG \cap A = p^nA$$

for all $n < \omega$. If G is a group, then for a positive integer n, we define

$$nG = \{nx : x \in G\} \quad \text{and} \quad G[n] = \{x \in G : nx = 0\}.$$

It is clear that $m(nG) = (mn)G$ for any $m, n \in \mathbb{Z}$, so we can omit parentheses. We note the following easy fact:

Lemma 1.7.1. *Let G be a group and $m, n \in \mathbb{Z}$ such that* $\gcd(m, n) = 1$. *Then $mG \cap nG = mnG$.*

Proof. Clearly we have $mnG \subseteq mG \cap nG$. To prove the reverse inclusion, let $x \in mG \cap nG$, and write $x = my = nz$ with $y, z \in G$. Then $1 = km + \ln$ for some $k, l \in \mathbb{Z}$; hence, $y = kmy + lny = knz + lny \in nG$, which yields $x = my \in mnG$; thus, $mG \cap nG \subseteq mnG$. □

A subgroup A of a group G is called **pure** if

$$nG \cap A = nA$$

for all positive integers n. In other words, A is a pure subgroup of G if and only if for any given $y \in G$, the equation $nx = y$ is solvable in A whenever it is solvable in G. It is clear that a subgroup of a p-group G is pure exactly if it is p-pure. We have the following:

(a) A is pure in G if and only if A is p-pure in G for all primes p.
(b) Divisible subgroups are pure and p-divisible subgroups are p-pure.
(c) Direct summands of G are pure.
(d) If G/A is torsion-free, then A is pure in G. The converse holds whenever G is torsion-free.
(e) If $\{A_i\}_{i \in I}$ is an ascending chain of p-pure submodules of G, then $A = \bigcup_{i \in I} A_i$ is p-pure in G.
(f) A is p-pure in G if and only if $(G[p^n] + A)/A = (G/A)[p^n]$ for all $n < \omega$.

Lemma 1.7.2 (Prüfer [129]). *Let A and B be subgroups of G such that $A \subseteq B$. Then we have the following:*

1. *If A is pure in B and B is pure in G, then A is pure in G.*
2. *If B is pure in G, then B/A is pure in G/A.*
3. *If A is pure in G and B/A is pure in G/A, then B is pure in G.*

Proof. Let n be a positive integer. If $nB \cap A = nA$ and $nG \cap B = nB$, then $nG \cap A = nG \cap B \cap A = nB \cap A = nA$. Therefore, (1) holds.

To prove (2), assume that $b + A \in n(G/A) \cap B/A$. Then $b = ng + a$ for some $g \in G$, and $a \in A$; hence, $b - a = nb'$ for some $b' \in B$, and we obtain $b + A \in n(B/A)$.

Finally, suppose that A is pure in G, and B/A is pure in G/A. Let $x \in nG \cap B$, say $x = ng$ ($g \in G$). Then $x + A \in n(G/A) \cap B/A = n(B/A)$; thus, $x = nb + a$ for some $b \in B$ and $a \in A$. It follows that $a = ng - nb \in nG \cap A = nA$; hence, there is an $a' \in A$ such that $na' = ng - nb$, and we have $x = na' + nb \in nB$, as desired. □

Clearly, Lemma 1.7.2 remains valid if R-modules are considered and "pure" is replaced by "p-pure" with $p \in R$ a prime.

Proposition 1.7.3 (Szele). *Let A be a subgroup of a group G. Then A is contained in a pure subgroup B of G such that $|B| \le |A| \cdot \aleph_0$.*

Proof. Let $B_0 = A$ and define

$$X_0 = \{(n, b) \in \mathbb{N} \times B_0 : \text{ the equation } nx = b \text{ is solvable in } G\}.$$

For each $(n, b) \in X_0$, pick an element $g_{n,b} \in G$ satisfying $ng_{n,b} = b$, and let $B_1 = \langle B_0, g_{n,b} : (n, b) \in X_0\rangle$. Then we define X_1 to be the set of all $(n, b) \in \mathbb{N} \times B_1$ such that the equation $nx = b$ is solvable in G. For each $(n, b) \in X_1$, we let $h_{n,b} \in G$ such that $nh_{n,b} = b$ and define $B_2 = \langle B_1, h_{n,b} : (n, b) \in X_1\rangle$, et cetera. This yields a sequence

$$B_0 \subseteq B_1 \subseteq \cdots \subseteq B_i \subseteq \cdots \quad (i < \omega)$$

of subgroups of G. To show that $B = \bigcup_{i<\omega} B_i$ is pure in G, let $b \in B$, and suppose that the equation $nx = b$ is solvable in G. There is an $i < \omega$ such that $b \in B_i$; hence, $nx = b$ is solvable in B_{i+1}. Thus, B is a pure subgroup of G. Since $|B_i| \leq |A| \cdot \aleph_0$ for all $i < \omega$, we have $|B| \leq |A| \cdot \aleph_0$, and the proof is complete. □

Let G be a group and p a prime. A set X of nonzero elements of G is called **p-independent** if for all $x_1, \ldots, x_n \in X$, integers a_i with $a_i x_i \neq 0$ $(i = 1, \ldots, n)$, and $r \in \mathbb{N}$,

$$a_1 x_1 + \cdots + a_n x_n \in p^r G$$

implies that

$$p^r | a_i \quad \text{for all } i = 1, \ldots, n.$$

Notice that by Zorn's lemma, every group contains maximal p-independent subsets for all primes p.

Lemma 1.7.4 (Fuchs [25]). *Let G be a group, p a prime, and X a p-independent subset of G. Then we have the following:*
1. *X is independent.*
2. *X contains only elements of p-power order and infinite order.*
3. *$\langle X\rangle$ is p-pure in G.*

Proof. To prove (1), let $x_1, \ldots, x_n \in X$ and assume that $y = a_1 x_1 + \cdots + a_n x_n$ equals 0, where $a_1, \ldots, a_n \in \mathbb{Z}$. Then $y \in p^r G$ for all $r \in \mathbb{N}$. Therefore, if $a_i x_i \neq 0$ for some i, p-independence shows that $p^r | a_i$ for all $r \in \mathbb{N}$; thus, $a_i = 0$, which is impossible. Thus, X is independent.

For (2), let $x \in X$ be an element of finite order $m = p^s m'$, where $s < \omega$ and $p \nmid m'$. Then for every $r \in \mathbb{N}$, there are integers k and l such that $km' + lp^r = 1$; thus, $p^s x \in p^r G$. If $p^s x \neq 0$, then by p-independence, we have $p^r | p^s$ for all $r \in \mathbb{N}$, a contradiction.

To show that $B = \langle X\rangle$ is a p-pure subgroup of G, we let $y \in B \cap p^r G$ with $r \in \mathbb{N}$; say $y = a_1 x_1 + \cdots + a_n x_n$, where $a_i \in \mathbb{Z}$ and $x_i \in X$ with $a_i x_i \neq 0$ $(i = 1, \ldots, n)$. Then $p^r | a_i$ for all $i = 1, \ldots, n$; thus, $y \in p^r B$. Therefore, B is p-pure. □

Let G be a group, B a subgroup of G, and p a prime. Then B is called a ***p*-basic subgroup of** G if the following conditions hold:
(i) $B = \bigoplus_{i\in I}\langle b_i\rangle$ where each element b_i has p-power order or infinite order;
(ii) B is p-pure in G;
(iii) G/B is p-divisible.

In the case where G is a p-group, B is simply called a **basic subgroup of** G. The notion of a p-basic subgroup was introduced by Fuchs [25], extending Kulikov's [89] definition of a basic subgroup.

Example 1.7.5 (Kulikov [89]). Let p be a prime and G the torsion part of $\prod_{n=1}^{\infty} B_n$ with B_n a direct sum of copies of $\mathbb{Z}/p^n\mathbb{Z}$ $(n = 1, 2, \ldots)$. Then $B = \bigoplus_{n=1}^{\infty} B_n$ is a basic subgroup of G.

Proof. Let $r \in \mathbb{N}$ and pick $b \in p^rG \cap B$. Then $b = (p^r b_n)$ with $b_n \in B_n$ $(n \in \mathbb{N})$, and there is an $n_0 \in \mathbb{N}$ such that $p^r b_n = 0$ for all $n > n_0$. Thus, $b = p^r(b_1, \ldots, b_{n_0}, 0, \ldots) \in p^r B$, and it follows that B is pure in G.

To show that G/B is divisible, let $x = (x_n) \in G$. There is a $k \in \mathbb{N}$ such that $p^k x_n = 0$ for all $n \in \mathbb{N}$. Therefore, for all $i \in \mathbb{N}$, we have $x_{k+i} \in B_{k+i}[p^k]$; hence, $x_{k+i} = p y_{k+i}$ for some $y_{k+i} \in B_{k+i}$. Since $x - (0, \ldots, 0, x_{k+1}, x_{k+2}, \ldots) \in B$, we obtain $x + B = p(0, \ldots, 0, y_{k+1}, y_{k+2}, \ldots) + B$. Therefore, G/B is divisible. □

We prove the following important fact:

Theorem 1.7.6 (Fuchs [25]). *Let G be a group and p a prime. If $X = \{b_i : i \in I\}$ is a maximal p-independent subset of G, then $B = \langle X\rangle$ is a p-basic subgroup of G. Thus, every group contains p-basic subgroups for every prime p.*

Proof. By Lemma 1.7.4(1) the set X is independent, so we have $B = \bigoplus_{i\in I}\langle b_i\rangle$. Property (ii) follows from Lemma 1.7.4(3). To prove (iii), let $g \in G \setminus B$. By the maximality of X, there is an $r \in \mathbb{N}$ such that

$$a_0 g + a_1 b_{i_1} + \cdots + a_n b_{i_n} \in p^r G$$

with $a_k \in \mathbb{Z}$ and $i_k \in I$ satisfying $a_0 g \neq 0$, $a_k b_{i_k} \neq 0$ $(k = 1, \ldots, n)$ and $p^r \nmid a_j$ for some $j \in \{0, \ldots, n\}$. Since X is p-independent, we have $p^r \nmid a_0$. Therefore, we can write $a_0 = p^s a_0'$, where $0 \le s < r$ and $p \nmid a_0'$. Then $a_1 b_{i_1} + \cdots + a_n b_{i_n} \in p^s G$ implies that $a_k = p^s a_k'$ for some $a_k' \in \mathbb{Z}$ $(k = 1, \ldots, n)$; thus, $p^s(a_0' g + a_1' b_{i_1} + \cdots + a_n' b_{i_n}) = p^r g'$ for some $g' \in G$. Letting $\tilde{g} = a_0' g - p^{r-s} g' + a_1' b_{i_1} + \cdots + a_n' b_{i_n}$, we obtain

$$p^s \tilde{g} = 0.$$

Claim: If $z \in G \setminus B$ has order p^t $(t \in \mathbb{N})$, then $z + B \in p(G/B)$.

We prove the claim by induction on t. If $t = 1$, the paragraph above—with $g = z$—shows that $p \nmid a_0$; thus, $z + B \in p(G/B)$. Now let $t > 1$ and assume that the claim holds for all elements of order p^u with $1 \le u < t$. Considering the first paragraph of this proof again with $g = z$, we have $\tilde{g} + B \in p(G/B)$ by the induction hypothesis; thus, $z + B \in p(G/B)$ as desired.

It follows that for any $z \in G$ of finite order, we have $z+B \in p(G/B)$. Finally, suppose the order of z is infinite. If $z \notin B$, we consider $g = z$ above once more and obtain $\tilde{g}+B \in p(G/B)$ by applying the claim to $\tilde{g}$, since the order of $\tilde{g}$ is finite. Again, it follows that $z + B \in p(G/B)$. Therefore, G/B is p-divisible and the proof is complete. □

Exercises.

1. A subgroup A of G is pure if and only if $(G[n] + A)/A = (G/A)[n]$ for all $n \in \mathbb{N}$.
2. (Prüfer [129]) A subgroup A of G is pure if and only if each coset $x + A \in G/A$ contains an element having the same order as this coset.
3. If A is pure in G, then $(G/A)[n] \cong G[n]/A[n]$ for all $n \in \mathbb{N}$.
4. Let B be a p-basic subgroup of G. Then we have the following:
   (a) $G = B + p^n G$ for all $n < \omega$.
   (b) $B/p^n B \cong G/p^n G$ for all $n < \omega$.
   (c) $p^n G/p^n B \cong G/B$ for all $n < \omega$.

1.8 Bounded groups

A group G is called **bounded** if $nG = 0$ for some positive integer n. For example, consider a p-bounded group G for a given prime p. Then $pG = 0$ implies that G can be viewed as a vector space over the field $\mathbb{Z}/p\mathbb{Z}$; thus, a set of elements of G is independent if and only if it is linearly independent in the vector space G. We call an independent set X of the p-bounded group G with $G = \langle X\rangle$ a **basis of** G and obtain $G = \bigoplus_{x\in X} \langle x\rangle$. More generally, we have the following:

Theorem 1.8.1. *A bounded group is a direct sum of cyclic groups of prime power order.*

Proof. Suppose $nG = 0$ for some positive integer n. Then $G = \bigoplus_{p\in\mathbb{P}} t_p G$ by Theorem 1.3.2. We fix a prime p and let $G' = t_p G$. Since G' is a bounded p-group, we have $p^m G' = 0$ for some $m < \omega$. Letting B be a basic subgroup of G' (cf. Theorem 1.7.6), we obtain $G'/B = p^m(G'/B) = (p^m G'+B)/B = 0$; thus, $G' = B$ and the assertion follows. □

The following fact will be needed:

Lemma 1.8.2. *Let G be a group and p a prime. If $B = \bigoplus_{i\in I} \langle b_i\rangle$ is a p-basic subgroup of G, then the set $\{b_i + pG : i \in I\}$ is a basis of G/pG.*

Proof. First we show that the set $X = \{b_i : i \in I\}$ is p-independent. Let $r \in \mathbb{N}$ and assume that $y = a_1 b_{i_1} + \cdots + a_n b_{i_n} \in p^r G$, where $a_k \in \mathbb{Z}$ and $i_k \in I$ with $a_k b_{i_k} \neq 0$ $(k = 1, \ldots, n)$. Since B is p-pure, we can write $y = p^r(a'_1 b_{i_1} + \cdots + a'_n b_{i_n})$ for some $a'_k \in \mathbb{Z}$ $(k = 1, \ldots, n)$. The order of each element b_{i_k} is a p-power or infinite; thus, $a_k b_{i_k} = p^r a'_k b_{i_k}$ yields $p^r | a_k$. Hence, X is p-independent and, therefore, independent

(Lemma 1.7.4(1)). To prove independence of the set of cosets $b_i + pG$ ($i \in I$), assume that

$$\lambda_1 b_{i_1} + \cdots + \lambda_m b_{i_m} \in pG$$

with $0 \le \lambda_k < p$ and $i_k \in I$ ($k = 1, \ldots, m$). Then p-independence of X yields $\lambda_k = 0$ for all k.

It remains to prove that the cosets $b_i + pG$ ($i \in I$) generate G/B; so let $g \in G$. Since G/B is divisible, there is a $g' \in G$ such that $g + B = pg' + B$. But then we have $g + \mu_1 b_{i_1} + \cdots + \mu_s b_{i_s} \in pG$ for some $\mu_k \in \mathbb{Z}$ and $i_k \in I$ ($k = 1, \ldots, s$); thus, the assertion follows. Therefore, $\{b_i + pG : i \in I\}$ is a basis of G/pG. □

Exercises.

1. Let G be a torsion group. If t_pG is bounded for all primes p, then G is a direct sum of cyclic groups.
2. A pure finite cyclic subgroup of a group is a direct summand.
3. (Kulikov [88]) A bounded pure subgroup is a direct summand.

1.9 Finitely generated groups

In this section, we prove the famous structure theorem of finitely generated abelian groups. We start with a result on pure subgroups:

Lemma 1.9.1. *Let A be a pure subgroup of a group G such that G/A is a direct sum of cyclic groups. Then A is a direct summand of G. Specifically if $G/A = \bigoplus_{i\in I} \langle y_i + A\rangle$, then $G = A \oplus B$ with $B = \langle x_i : i \in I\rangle$ for some $\{x_i\}_{i\in I}$, where $x_i = y_i$ if $y_i + A$ has infinite order.*

Proof. We write $G/A = \bigoplus_{i\in I} \langle y_i + A\rangle$. Pick an $i \in I$ and let n_i denote the order of $y_i + A$ in G/A. If n_i is finite, then $n_i y_i = n_i a_i$ for some $a_i \in A$ since A is pure, and we let $x_i = y_i - a_i$. Clearly, $x_i + A = y_i + A$ and x_i has order n_i. In the case where n_i is infinite, we let $x_i = y_i$.

We define $B = \langle x_i : i \in I\rangle$ and have $G = A + B$. To show that this sum is direct, let $a \in A \cap B$ and write

$$a = k_{i_1} x_{i_1} + \cdots + k_{i_m} x_{i_m},$$

where $i_1, \ldots, i_m \in I$ and $k_{i_1}, \ldots, k_{i_m} \in \mathbb{Z}$. Then $0 + A = a + A = (k_{i_1} y_{i_1} + A) + \cdots + (k_{i_m} y_{i_m} + A)$. Letting $r \in \{1, \ldots, m\}$, we have $k_{i_r} y_{i_r} + A = 0$; thus, $k_{i_r} = 0$ in the case where n_{i_r} is infinite. If n_{i_r} is finite, we have $n_{i_r} | k_{i_r}$. In any case, we obtain $k_{i_r} x_{i_r} = 0$; hence, $a = 0$. Consequently, $G = A \oplus B$. □

As an application of Lemma 1.9.1, we have the following:

Theorem 1.9.2. *Let G be a finitely generated torsion-free group. Then G is a free group of finite rank.*

Proof. We prove the assertion by induction on the smallest number n of generators required to generate G. The case $n = 1$ being obvious, suppose that the claim holds for torsion-free groups generated by at most n elements. Assume that G requires $n+1$ generators, say $G = \langle x_1, \dots, x_n, x_{n+1}\rangle$. Then $A = \langle x_1, \dots, x_n\rangle$ is a free group of finite rank by the induction hypothesis, and G/A is cyclic. If $G/A \cong \mathbb{Z}$, then A is pure. Therefore, Lemma 1.9.1 shows that $G = A \oplus B$ with $B \cong \mathbb{Z}$. If $G/A \cong \mathbb{Z}/m\mathbb{Z}$ for some positive integer m, then $G \cong mG \subseteq A$, which implies that G is free of finite rank (Theorem 1.5.4). □

Now we are able to prove the main result of this section.

Theorem 1.9.3 (Fundamental theorem of finitely generated abelian groups). *A finitely generated group is a finite direct sum of cyclic groups, each of which is either infinite or of prime power order.*

Proof. The torsion-free group G/tG is finitely generated; hence, it is free and has finite rank by Theorem 1.9.2. Then Lemma 1.9.1 shows that $G = tG \oplus F$ with $F \cong \mathbb{Z}^n$ for some $n < \omega$. The torsion group $tG \cong G/F$ is finitely generated, so it is finite and, therefore, bounded. But then tG is a finite direct sum of cyclic groups of prime power order (Theorem 1.8.1) and the assertion follows. □

In Section 1.13 it will be shown that the decomposition of a finitely generated group as in Theorem 1.9.3 is unique up to the order in which the direct summands are written (see Corollary 1.13.1).

Exercises.

1. The group $\mathbb{Q}$ is not finitely generated.
2. A group is finite whenever it has only finitely many subgroups.
3. A finite group G is cyclic if and only if $|G[p]| \leq p$ for all primes p.
4. A surjective endomorphism of a finitely generated group is an automorphism.

1.10 Heights and isotypeness

Let G be an R-module and $p \in R$ a prime. Then for all ordinals α, submodules $p^\alpha G$ of G are defined inductively by

$$p^0 G = G,$$
$$p^{\alpha+1} G = p(p^\alpha G), \quad \text{and}$$
$$p^\alpha G = \bigcap_{\beta<\alpha} p^\beta G \quad \text{if } \alpha \text{ is a limit ordinal.}$$

The ***p*-height of an element** $x \in G$ is defined to be $|x|_p = \alpha$ whenever $x \in p^\alpha G \setminus p^{\alpha+1} G$ and $|x|_p = \infty$ if $x \in p^\infty G = \bigcap_\alpha p^\alpha G$. Occasionally, we write $|x|_p^G$ —instead of $|x|_p$—to emphasize the module in which the p-height is computed. The subscript p is dropped, and we may simply refer to the **height** of an element, if it is clear which prime p is used. Note that a homomorphism $f : G \to H$ does not decrease p-heights; that is, we have $|f(x)|_p^H \geq |x|_p^G$ for all $x \in G$. A key height-related characteristic of R-modules is their p-length. The **p-length** of G is the least ordinal λ such that $p^\lambda G = p^{\lambda+1}G$. Such a λ must exist by cardinality considerations. If G is a p-local group, we call λ the **length** of G. Notice that in this case, $p^\lambda G$ is the maximal divisible subgroup of G.

Much of what we will be looking at will involve sums of elements, so it is natural to ask how the height of a sum relates to the heights of the summands. We will use the **triangle inequality** quite frequently.

Lemma 1.10.1. *Let x and y be elements of an R-module G. Then for any prime $p \in R$,*

$$|x + y|_p \geq |x|_p \wedge |y|_p$$

and equality holds whenever $|x|_p \neq |y|_p$.

Proof. Letting $\alpha = |x|_p \wedge |y|_p$, we have $x, y \in p^\alpha G$. But then $x + y \in p^\alpha G$; hence, $|x + y|_p \geq \alpha$. It is clear that $|x + y| = \alpha$ whenever $|x|_p > \alpha$ or $|y|_p > \alpha$. □

Lemma 1.10.2. *Let G be an R-module, $x \in G$ and $p \in R$ a prime. Then $|rx|_p = |x|_p$ whenever $r \in R$ and $p \nmid r$.*

Proof. If $|x|_p = \infty$, the assertion is immediate. Therefore, assume that $|x|_p$ is an ordinal. There are $k, l \in R$ such that $kp + lr = 1$, so by Lemma 1.10.1, we have $|rx|_p \leq |lrx|_p = |x|_p \wedge |kpx|_p = |x|_p \leq |rx|_p$. □

In the following, we proceed with some elementary lemmas:

Lemma 1.10.3. *Let G be an R-module, $p \in R$ a prime, and $\beta < \alpha$ ordinals. Then we have the following:*
1. *$p^\alpha G \subseteq p^\beta G$.*
2. *If $x \in p^\alpha G$, then $x = py$ for some $y \in p^\beta G$.*

Proof. (1) is clear. To show (2), let $x \in p^\alpha G$. If $\alpha - 1$ exists, then $x \in p(p^{\alpha-1}G) \subseteq p(p^\beta G)$ by (1). If α is a limit ordinal, then $x \in p^{\beta+1}G$. In any case, the claim follows. □

Lemma 1.10.4. *Let G be an R-module, $p \in R$ a prime, and α an ordinal. Then*
1. *$p^\beta(p^\alpha G) = p^{\alpha+\beta}G$ for all ordinals β.*
2. *If $x \in p^\alpha G$, then $|x|_p^G = \alpha + |x|_p^{p^\alpha G}$.*

Proof. To show (1), we use transfinite induction on β. The case $\beta = 0$ being clear, let $\beta > 0$ and assume that the claim holds for all $\gamma < \beta$. If $\beta - 1$ exists, then $p^\beta(p^\alpha G) = p(p^{\beta-1}(p^\alpha G)) = p(p^{\alpha+\beta-1}G)$ by the induction hypothesis; thus, $p^\beta(p^\alpha G) = p^{\alpha+\beta}G$. If β is a limit ordinal, then $p^\beta(p^\alpha G) = \bigcap_{\gamma<\beta} p^\gamma(p^\alpha G) = \bigcap_{\gamma<\beta} p^{\alpha+\gamma}G$ by the induction hypothesis, and we obtain $p^\beta(p^\alpha G) = \bigcap_{\tau<\alpha+\beta} p^\tau G = p^{\alpha+\beta}G$.

To prove (2), let $x \in p^\alpha G$ and $\beta = |x|_p^{p^\alpha G}$. If $\beta = \infty$, then $|x|_p^G = \infty$, and if β is an ordinal, then $x \in p^\beta(p^\alpha G) \setminus p^{\beta+1}(p^\alpha G)$; thus, $|x|_p^G = \alpha + \beta$ by (1). □

The following basic results will be useful (see Crawley–Hales [14] and Fuchs [29]):

Lemma 1.10.5. *Let G be an R-module, $p \in R$ a prime, and α an ordinal. If A is a submodule of $p^\alpha G$, then $p^\alpha(G/A) = p^\alpha G/A$. In particular, we have $A = p^\alpha G$ whenever $p^\alpha(G/A) = 0$.*

Proof. We show that $p^\alpha(G/A) \subseteq p^\alpha G/A$ by transfinite induction on α. Let $\alpha > 0$ and assume that the inclusion holds for all $\beta < \alpha$. If $\alpha - 1$ exists, then $p^{\alpha-1}(G/A) \subseteq p^{\alpha-1}G/A$, and the induction step is immediate. If α is a limit ordinal and $x + A \in p^\alpha(G/A)$, then $x \in p^\beta G$ for all $\beta < \alpha$ by the induction hypothesis; thus, $x + A \in p^\alpha G/A$. □

Lemma 1.10.6. *Let G be an R-module and A a submodule of $G[p^n]$, where $p \in R$ is a prime and $n \in \mathbb{N}$. For every ordinal α we have the following:*

1. *If $x + A \in p^\alpha(G/A)$, then $p^n x \in p^\alpha G$.*
2. *If $p^\alpha G = 0$, then $p^\alpha(G/A) \subseteq G[p^n]/A$.*

Proof. The first claim is shown by transfinite induction on α. The case $\alpha = 0$ is clear, so let $\alpha > 0$ and assume that the assertion holds for all ordinals $< \alpha$. Let $x + A \in p^\alpha(G/A)$. If $\alpha - 1$ exists, then $x + A = py + A$ with $y + A \in p^{\alpha-1}(G/A)$. Therefore, by the induction hypothesis, we have $p^n y \in p^{\alpha-1}G$; thus, $p^n x = p^{n+1}y \in p^\alpha G$. If α is a limit ordinal, the induction step is immediate.

The second claim follows from (1). □

We will need the following useful fact on reduced p-groups:

Lemma 1.10.7. *Suppose that G is a reduced p-group such that its length λ is an infinite successor ordinal. Let β be the largest limit ordinal $< \lambda$. If $\alpha < \beta$, then $p^\alpha G/p^\beta G$ is infinite and $|p^\alpha G/p^\beta G| \geq |p^\beta G|$.*

Proof. We write $\lambda = \beta + n$ with $0 < n < \omega$ and let $\alpha < \beta$. If $p^\alpha G/p^\beta G$ is finite, then $p^m(p^\alpha G/p^\beta G) = 0$ for some $m < \omega$, and we obtain $p^{\alpha+m+n}G = p^{\beta+n}G = 0$, a contradiction since $\alpha + m + n < \lambda$. Thus, $p^\alpha G/p^\beta G$ is infinite.

Since $p^\beta G$ is bounded, we can write $p^\beta G = \bigoplus_{i\in I}\langle x_i\rangle$. Therefore, for each $i \in I$ there is a $y_i \in p^\alpha G$ such that $x_i = py_i$. Then $|p^\alpha G/p^\beta G| \geq |\bigoplus_{i\in I}\langle y_i\rangle/\langle x_i\rangle| \geq |I|$; thus, $|p^\alpha G/p^\beta G| \geq \max\{|I|, \aleph_0\} = |I| \cdot \aleph_0 \geq |p^\beta G|$. □

Let A be a submodule of an R-module G and $p \in R$ a prime. Then the p-heights of an element $x \in A$ computed in A and G do not necessarily coincide, even if A is p-pure in G. Notice that—in that case—we have $|x|_p^A = |x|_p^G$ whenever $|x|_p^G < \omega$ and $|x|_p^A \geq \omega$ if and only if $|x|_p^G \geq \omega$.

We call the submodule A of G p-**isotype** if

$$p^\alpha G \cap A = p^\alpha A$$

for all ordinals α. Clearly, A is p-isotype if and only if for every element $x \in A$ we have $|x|_p^A = |x|_p^G$, that is, the p-heights of x computed in A and G coincide. If A is p-isotype for all primes $p \in R$, then A is called **isotype**. The following items are easily verified:

(a) Direct summands of G are isotype.
(b) If $\{A_i\}_{i\in I}$ is an ascending chain of p-isotype submodules of G, then $A = \bigcup_{i\in I} A_i$ is p-isotype in G.

For an R-module G and a subset S of G, we define

$$S^0 = \{x \in G : rx \in S \text{ for some nonzero } r \in R\}.$$

The following fact will be useful:

Lemma 1.10.8. *If S is a submodule of an R-module G, then S^0 is an isotype submodule of G.*

Proof. Clearly, $A = S^0$ is a submodule of G. Letting $p \in R$ be a prime, we show that $p^\alpha G \cap A \subseteq p^\alpha A$ for all ordinals α. The case $\alpha = 0$ being trivial, let $\alpha > 0$ and assume that the assertion is true for all ordinals $\beta < \alpha$. Let $x \in p^\alpha G \cap A$.

If $\alpha - 1$ exists, then $x = py$ ($y \in p^{\alpha-1}G$) and $rx \in S$ for some nonzero $r \in R$. Then $(rp)y = rx \in S$; thus, $y \in A$, and we obtain $y \in p^{\alpha-1}G \cap A \subseteq p^{\alpha-1}A$ by induction. It follows that $x = py \in p^\alpha A$. If α is a limit ordinal, then for any $\beta < \alpha$, we have $x \in p^\beta G \cap A \subseteq p^\beta A$ by induction; thus, $x \in \bigcap_{\beta<\alpha} p^\beta A = p^\alpha A$. □

The concept of heights of elements is central to the classification of abelian groups and plays a major role in the invariants. In particular, one of the key characteristics of any element of a $\mathbb{Z}_p$-module is its p-Ulm sequence. Given an arbitrary R-module G and a prime $p \in R$, the **p-Ulm sequence of** $x \in G$ is the sequence

$$\|x\|_p = (|x|_p, |px|_p, |p^2x|_p, \dots)$$

of p-heights. Sometimes we write $\|x\|_p^G$ instead of $\|x\|_p$, or simply $\|x\|$ if p is understood, and the group is p-local. More generally, an **Ulm sequence** is a sequence $\overline{\alpha} = (\alpha_i)_{i<\omega}$, where each α_i is in Ord_∞ such that $\alpha_i < \alpha_{i+1}$ for all $i < \omega$. Recall that we adopt the convention that $\infty < \infty$. By $\overline{\alpha} \wedge \overline{\beta}$, we mean the Ulm sequence $\{\alpha_i \wedge \beta_i\}_{i<\omega}$ whenever $\overline{\beta} = (\beta_i)_{i<\omega}$. An **Ulm matrix** is an $\omega \times \omega$-matrix $M = [m_{p,i}]$ indexed over all primes and nonnegative integers such that each row $M_p = (m_{p,i})_{i<\omega}$ is an Ulm sequence. It will be convenient to define $M \wedge N = [m_{p,i} \wedge n_{p,i}]$, where $N = [n_{p,i}]$. For a group G, the **Ulm matrix of** $x \in G$ (written $\|x\|$ or $\|x\|_G$) is defined to be the Ulm matrix $[|p^ix|_p]$, where the rows are the p-Ulm sequences of x with $p \in \mathbb{P}$. If each entry of an Ulm matrix M equals ∞, we write $M = \overline{\infty}$.

Exercises.

1. If $x \in G$ and $|x|_p > 0$, then $|x|_p = \sup\{|y|_p + 1 : x = py\}$.
2. If G is a $\mathbb{Z}_p$-module with p-length λ and maximal divisible subgroup D, then $p^\infty G = p^\lambda G = D$.
3. A submodule A of G is p-isotype if and only if $(p^\alpha G)[p] \cap A = (p^\alpha A)[p]$ for all ordinals α.

4. Let A and B be submodules of G such that $A \subseteq B$.
   (a) If A is p-isotype in G, then A is p-isotype in B.
   (b) If A is p-isotype in B and B is p-isotype in G, then A is p-isotype in G.
5. If A is p-isotype in G, then $A/p^\alpha A$ is p-isotype in $G/p^\alpha A$ for all ordinals α.
6. (Irwin–Walker [67]) Let B be a subgroup of G. A subgroup H of G is called a **B-high subgroup** if $H \cap B = 0$ and H is maximal with respect to the property $H \cap B = 0$. Show that a $p^\alpha G$-high subgroup of a p-group G is isotype for all ordinals α.

1.11 Characteristics and examples of heights

A deep understanding of heights and their characteristics and possible quirks is necessary for what follows. Here we present examples, characteristics, and cautions. The reader who is already familiar with heights may skip this section.

Here are some simple examples of heights:

1. In any group, $|0|_p = \infty$ for all p.
2. If $n \in \mathbb{Z}$, $n \neq 0$, then $|n|_p$ = the largest k such that $p^k | n$.
3. If $x \in \mathbb{Q}$, then $|x|_p = \infty$ for all p.
4. If $x \in \mathbb{Z}/p\mathbb{Z}$ and $x \neq 0$, then $|x|_p = 0$ and $|x|_q = \infty$ for all $q \neq p$.
5. If $x \in \mathbb{Z}_p$, then $|x|_q = \infty$ for all $q \neq p$.
6. If $x \in D$ and D is divisible, then $|x|_p = \infty$ for all p.

The following characteristics of heights are straightforward to prove and will be used repeatedly. Many of them may seem redundant, but are all included here to give an indication of how they might be used. All heights here are p-heights for some fixed prime p:

(a) $|x| < |px|$ (recalling the convention that $\infty < \infty$).
(b) $|x| = |px|$ if and only if $|x| = \infty$.
(c) If $|x| \neq \infty$ and $|ax| > |x|$, then $p|a$.
(d) If $|x| = \infty$, then $x = py$ for some y such that $|y| = \infty$.
(e) If $|x + y| > |x| \wedge |y|$, then $|x| = |y|$.
(f) Suppose $G = A \oplus B$. If $x \in A$ and $y \in B$, then $|x + y|^G = |x|^A \wedge |y|^B$.

Note that the height of an element is highly dependent on the group. As a simple example, 1 has p-height 0 in $\mathbb{Z}$ for any p, but p-height ∞ in $\mathbb{Q}$, since $\mathbb{Q}$ provides $1/p, 1/p^2, \ldots$, and $\mathbb{Z}$ does not. Because of this, an isomorphism onto a subgroup may not preserve heights. As an example, consider $f(n) = pn$, an isomorphism from $\mathbb{Z}$ to a subgroup of $\mathbb{Z}$, for which $|f(n)|_p = |n|_p+1$. This means that any discussion of heights must be clear regarding which group is being used to compute the heights. Generally, however, we do not specify the group, but assume that the heights are relative to the largest group under discussion rather than to some subgroup. The concept of valuation, introduced later in this section, is a generalization of heights that avoids any ambiguities about the group under discussion.

The simple examples given so far may lead one to expect that $|px|_p = |x|_p+1$ as long as the heights are not ∞, in other words, that any $\|x\|_p$ has no gaps. This is certainly true for torsion-free groups, as we will see in Lemma 1.15.1. This is not true in general for mixed or torsion groups, and, in fact, the gaps in the Ulm sequences are central to the Ulm invariant used for the very first classification results. Here is a simple example of a gap. Consider $(1, p) \in G = \mathbb{Z}/p\mathbb{Z} \oplus \mathbb{Z}$. Then $|(1, p)|^G = |1|^{\mathbb{Z}/p\mathbb{Z}} \wedge |p|^{\mathbb{Z}} = 0 \wedge 1 = 0$, whereas $|p(1, p)|^G = |(0, p^2)|^G = |p^2|^{\mathbb{Z}} = 2$. Soon we will see that we may build examples with as many gaps of whatever size we wish. In fact, we will see that we can create groups with elements with heights of any ordinals we wish.

For any x and y in a group, the triangle inequality tells us that $|x + y| \geq |x| \wedge |y|$. Note the possibility of strict inequality. This can occur if $|x| = |y|$ and x and y combine in some way to get something of greater height. For an almost trivial example, let $y = (p^n - 1)x$. Then $|x| = |y|$ and $|x + y| = |p^n x| \geq |x| + n$. Clearly, if we can calculate the height of the sum from the heights of the summands, it will make our lives much easier. Consequently, we will be paying special attention to those cases in which $|x + y| = |x| \wedge |y|$.

Height ∞ will require special treatment. For example, sometimes we may wish to conclude that $p|a$ if $|ax|_p \geq |px|_p$. Yet, if $|x|_p = \infty$, $|px|_p = \infty$ as well, and yet $p \nmid 1$. Because of this, we will often break proofs into the ∞ case and the ordinal case.

Note the distinction between height ∞ and height α for an infinite ordinal α. In the first case, $x \in p^\beta G$ for any ordinal β, and in the second it is in $p^\alpha G$, but not $p^{\alpha+1}G$. In general, we will not use the term "infinite height," which might cause confusion.

Later, Theorem 2.7.3 will prove that we can find groups with any reasonable height structure whatsoever. This is a rich source of examples, counterexamples, and constructions. As corollaries, we will get existence proofs for some well-known groups. Even before we get to the formal construction, it is useful to look at these groups and keep them in mind as we explore groups and their heights. In particular, they are generally mixed groups, those that do not split into the direct sum of torsion and torsion-free subgroups.

First we define what we mean by a "reasonable height structure." For a prime p, we define a **(p-)valuation** on a group G to be a function v on G such that

1. $v(x)$ is an ordinal or ∞.
2. $v(x + y) \geq v(x) \wedge v(y)$ for all x and $y \in G$.
3. $v(px) > v(x)$, where we follow the convention that $\infty > \infty$.
4. $v(nx) = v(x)$ if n and p are relatively prime.

Note that height is a valuation. Furthermore, we will see that any valuation may become a height on some larger group. Specifically, if G is a $\mathbb{Z}_p$-module and v a p-valuation on G, then there is a $\mathbb{Z}_p$-module H such that G is a submodule of H, H is torsion over G, and for all $x \in G$, $|x|_p^H = v(x)$.

We will save the formal proof for the next chapter, but it is useful to introduce an informal construction here, as it provides a rich source of examples of mixed groups. We may think of the module H as generated by G and all sequences of the form

$$(\alpha_1, \dots, \alpha_n, x),$$

where $x \in G$, $x \neq 0$, $\alpha_1 < \dots < \alpha_n < v(x)$ are ordinals or ∞, and $\alpha_1 = \infty$ if $v(x) = \infty$. These sequences are subject to the relations

$$p(\alpha_1, \dots, \alpha_n, x) = (\alpha_2, \dots, \alpha_n, x)$$

and

$$p(\alpha, x) = x.$$

Using these relations, it may be shown that

$$|(\alpha_1, \dots, \alpha_n, x)|_p^H = \alpha_1$$

and

$$|x|_p^H = v(x),$$

as desired.

These sequences provide elements to prop up the elements of G to whatever height we wish. For example, to give some element x the height of ω, consider the sequences $(\alpha_1, \dots, \alpha_n, x)$, where $\alpha_1 < \dots < \alpha_n$ are nonnegative integers. For an example of an Ulm sequence with infinitely many gaps, we pick $y = 1 \in \mathbb{Z}_p$ and let $v(p^k y) = 2k$ for each $k < \omega$, and build an appropriate H. Note, for example, that $|y|_p^H = 0$, whereas $p^2(0, 1, py) = py$, giving py a height of at least 2.

Exercises.

1. Prove (f).
2. If v is a p-valuation on G, then $v(x) \geq |x|_p^G$ for all $x \in G$. [Hint: Section 1.10, Exercise 1.]
3. For every Ulm sequence $\overline{\alpha}$ and prime p, there is a group G and an $x \in G$ such that $\|x\|_p = \overline{\alpha}$.
4. For every prime p, there is a group containing elements x and y such that $\|x+y\|_p = (\alpha_i)_{i<\omega} = (0, 2, 4, \dots)$, whereas $|p^i x|_p > \alpha_i$ for infinitely many i and $|p^i y|_p > \alpha_i$ for infinitely many i.

1.12 Proper elements and height-preserving homomorphisms

Let A be a submodule of an R-module G and $p \in R$ a prime. Then an element $x \in G$ is called ***p*-proper** (or **proper** if p is understood) **with respect to** A if x has maximal

p-height in the coset $x + A$; that is, if $|x|_p \geq |x + a|_p$ for all $a \in A$. Notice that such an element always exists in cases where A is finite. The following observation will be useful:

Lemma 1.12.1. *Let A be a submodule of an R-module G, $p \in R$ a prime, $x \in G$ and α an ordinal. Then we have the following:*
1. *x is proper with respect to A if and only if $|x + a|_p = |x|_p \wedge |a|_p$ for all $a \in A$.*
2. *Assume that $x \notin A$ and $|x|_p = \alpha$. Then x is proper with respect to A if and only if $x \notin A + p^{\alpha+1}G$.*

Proof. To verify (1), suppose that x is proper with respect to A and let $a \in A$. Then $|a|_p \geq |x+a|_p \wedge |x|_p \geq |x+a|_p$; hence, $|x|_p \wedge |a|_p \geq |x+a|_p$. Therefore, we have $|x|_p \wedge |a|_p = |x+a|_p$. The converse is obvious. The second statement is clear. □

Lemma 1.12.2. *Let A be a submodule of an R-module G, $x \in G \setminus A$, and $p \in R$ a prime. Then the following conditions are equivalent:*
1. *The coset $x + A$ contains an element which is proper with respect to A;*
2. *There exists an element $y \in x + A$ such that $|y|_p^G = |x + A|_p^{G/A}$;*
3. *If α is an ordinal and $x + A \in p^\alpha(G/A)$, then $x \in p^\alpha G + A$.*

Proof. (1) implies (3). Suppose that $x + z$ is proper with respect to A, where $z \in A$. Let $\alpha > 0$ be an ordinal and assume that the claim is true for all $\beta < \alpha$. Suppose $x+A \in p^\alpha(G/A)$. If $\alpha-1$ exists, then there is a $y+A \in p^{\alpha-1}(G/A)$ such that $x+A = p(y+A)$. By induction, we have $y + A \in p^{\alpha-1}G + A$; hence, $x \in p^\alpha G + A$. If α is a limit ordinal, let $\beta < \alpha$. Then $x + A \in p^\beta(G/A)$; thus, $x \in p^\beta G + A$. Therefore, there exists a $y_\beta \in A$ such that $|x + y_\beta|_p \geq \beta$. But then $|x + z|_p \geq |x + y_\beta|_p \geq \beta$; therefore, $x \in p^\alpha G + A$, as desired.

(3) implies (2). Suppose that $|x + A|_p = \alpha$. Then from (3), there is a $y \in x + A$ such that $|y|_p \geq \alpha$. Therefore, $|x + A|_p = |y|_p$.

Finally, suppose that (2) holds. Then (1) is immediate since $|y|_p = |y+A|_p \geq |x+a|_p$ for all $a \in A$. □

Corollary 1.12.3. *Let A be a submodule of an R-module G, $x \in G \setminus A$, and $p \in R$ a prime. Then the following conditions are equivalent:*
1. *x is proper with respect to A;*
2. *$|x + A|_p^{G/A} = |x|_p^G$.*

Proof. If x is proper with respect to A, then by Lemma 1.12.2(2), there is a $y \in x+A$ such that $|y|_p = |x + A|_p$; hence, $|x|_p \geq |y|_p = |x + A|_p \geq |x|_p$. The converse is obvious. □

Suppose that S and T are submodules of R-modules G and H, respectively, and let $p \in R$ be a prime. Then we say that a homomorphism $f : S \to T$ **does not decrease p-heights** if $|f(x)|_p^H \geq |x|_p^G$ for all $x \in S$, and f is called **p-height-preserving** if $|f(x)|_p^H = |x|_p^G$ for all $x \in S$. If f is a p-height-preserving homomorphism for all primes $p \in R$, then we call f **height-preserving**. Notice that if $g : G \to H$ is an isomorphism, then the restriction $g \restriction S : S \to g(S)$ is a height-preserving isomorphism. Proper elements can be used to extend homomorphisms between submodules which preserve or do not decrease p-heights:

Lemma 1.12.4. *Let G and H be R-modules, $p \in R$ a prime, and $f : S \to T$ a homomorphism, where S and T are submodules of G and H, respectively. Let $a \in G \setminus S$ and $b \in H$ such that $pa \in S$, $pb \in T$ and $f(pa) = pb$. Then we have the following:*
1. *f extends to a homomorphism $g : \langle S, a\rangle \to \langle T, b\rangle$ by sending a onto b.*
2. *If f is an isomorphism and $b \notin T$, then the map g in (1) is an isomorphism.*
3. *If f is p-height-preserving, $|a|_p = |b|_p$, a is proper with respect to S, and b is proper with respect to T, then the map g in (1) is p-height-preserving.*
4. *If f does not decrease p-heights, $|a|_p \le |b|_p$ and a is proper with respect to S, then the map g in (1) does not decrease p-heights.*

Proof. Statements (1) and (2) were shown in Lemma 1.6.5 for $R = \mathbb{Z}$; the proof for arbitrary R carries over.

To verify (3), let $z = s + ra$, where $s \in S$ and $r \in R$. If $p|r$, then $z \in S$ and the p-heights $|f(z)|$ and $|z|$ coincide by assumption. If $p \nmid r$, then there are $k, l \in R$ such that $kp + lr = 1$; hence, $lz = l(ra + s) = a - kpa + ls \in a + S$. Therefore, we may write $lz = a + s'$ for some $s' \in S$. Since $p \nmid l$, we have $|z| = |lz| = |a + s'| = |a| \wedge |s'|$ since a is proper with respect to S. Likewise, $|g(z)| = |b| \wedge |f(s')| = |a| \wedge |s'|$ since b is proper with respect to T, and we conclude that $|g(z)| = |z|$.

A slight modification of the proof of (3) shows the last statement. □

Exercises.
1. If G and H are R-modules and $\phi : G \to H$ is a monomorphism, then $\phi(p^\alpha G) = p^\alpha \phi(G)$ for all primes $p \in R$ and ordinals α.
2. If G is an R-module and $G = A \oplus B$, then $p^\alpha G = p^\alpha A \oplus p^\alpha B$ for all primes $p \in R$ and ordinals α.
3. Let G be an R-module, $p \in R$ a prime, and α an ordinal. Then every coset $x + p^\alpha G \in G/p^\alpha G$ has an element p-proper with respect to $p^\alpha G$.

1.13 Ulm invariants

Ulm's theorem, which will be proved in the next section, was one of the first theorems to classify infinite groups by defining invariants that completely determine the groups in some class up to isomorphism. Here we define the invariant, which has subsequently been applied to broader classes of groups, as we shall see.

Let G be an R-module. Then for a given prime $p \in R$, the **Ulm invariants** of G are defined by

$$u_p(\alpha, G) = \dim(p^\alpha G)[p]/(p^{\alpha+1}G)[p]$$

if α is an ordinal and

$$u_p(\infty, G) = \dim(p^\infty G)[p]$$

with the groups considered as vector spaces over the field $R/(p)$. The subscript p may be dropped if it is clear which prime is used.

For instance, if G is a direct sum of cyclic p-groups and $n < \omega$, then $u(n, G)$ is the number of direct summands of order p^{n+1}, and we have $u(\alpha, G) = 0$ for all $\alpha \geq \omega$. Thus, two direct sums of cyclic p-groups are isomorphic exactly if they have the same Ulm invariants. Since the torsion-free rank of a group is an invariant (Corollary 1.4.3), we have the following:

Corollary 1.13.1. *The direct decomposition of a finitely generated group as in Theorem 1.9.3 is unique up to the order in which the cyclic direct summands of infinite and prime power order are written. Two finitely generated groups G and H are isomorphic if and only if $r_0(G) = r_0(H)$ and $u_p(n, G) = u_p(n, H)$ for all $n < \omega$ and primes p.*

The following hold for any group G and prime p:

(a) If $G = \bigoplus_{i \in I} G_i$, then $u_p(\alpha, G) = \sum_{i \in I} u_p(\alpha, G_i)$ for all $\alpha \in \mathrm{Ord}_\infty$.

(b) If α is an ordinal, then $r_p(p^\alpha G) \geq \sum_{\gamma \geq \alpha} u_p(\gamma, G)$.

Remark 1.13.2. If G is any group and p is a prime, then $t_p G = D \oplus A$ with D divisible and A reduced (Theorem 1.6.4); thus, (*a*) yields $u_p(\alpha, G) = u_p(\alpha, A)$ for all ordinals α.

Next we wish to establish a connection between Ulm invariants of a given group and any of its p-basic subgroups. The following lemma is needed (see [29, p. 168f]):

Lemma 1.13.3. *Suppose that $B = \bigoplus_{n<\omega} B_n$ is a p-basic subgroup of a group G, where B_0 is a direct sum of infinite cyclic groups and for $n \geq 1$, B_n is a direct sum of cyclic groups of order p^n. Let $n \geq 1$ and define $B_n^* = B_0 \oplus \bigoplus_{i=n+1}^{\infty} B_i$ and $G_n = B_n^* + p^n G$. Then we have the following:*

1. $G = B_1 \oplus \cdots \oplus B_n \oplus G_n$.
2. $G_n = B_{n+1} \oplus G_{n+1}$.
3. $G_n[p] = (p^n G)[p]$.

Proof. Let $n \geq 1$. If $x \in G$, then $x = b + p^n y$ for some $b \in B$ and $y \in G$ since G/B is p-divisible; thus, $G = (B_1 \oplus \cdots \oplus B_n) + G_n$. To show that this sum is direct, assume that $x \in (B_1 \oplus \cdots \oplus B_n) \cap G_n$. Then $x = b' + p^n z$ with $b' \in B_n^*$ and $z \in G$; therefore, $p^n z \in B$. Since B is p-pure in G, it follows that $p^n z = p^n b''$ for some $b'' \in B$. But then $p^n z \in B_n^*$, and we obtain $x \in (B_1 \oplus \cdots \oplus B_n) \cap B_n^* = 0$. Therefore, (1) holds.

Next we show (2). Letting $A = B_{n+1} + G_{n+1}$, we have $A \subseteq G_n$. Therefore, the modular law yields $G_n = (A + (B_1 \oplus \cdots \oplus B_n)) \cap G_n = A + ((B_1 \oplus \cdots \oplus B_n) \cap G_n) = A$. Since the sum $A = B_{n+1} + G_{n+1}$ is direct by (1), the assertion follows.

Finally, we verify (3). Notice that $p^nG \subseteq G_n$ by definition; thus, $(p^nG)[p] \subseteq G_n[p]$. To show the reverse inclusion, let $x \in G_n[p]$ and assume that $|x|_p^G = k < n$. We have $x = c+p^ng$ for some $c \in B_n^*$ and $g \in G$; hence, $|c|_p^G = k$. Now write $c = b_0+b_{n+1}+\cdots+b_{n+r}$, where $b_i \in B_i$ $(i = 0, n+1, \ldots, n+r)$. Since $|c|_p^{B_n^*} = |c|_p^G = k$, there is an index i such that $|b_i|_p^{B_i} = k$. Then $k < n$ shows that $pb_i \neq 0$; hence, $|pb_i|_p^{B_i} = k+1$. Consequently, we have $|pc|_p^G = |pc|_p^{B_n^*} = k+1$. On the other hand, $pc + p^{n+1}g = px = 0$ yields $pc \in p^{n+1}G$, a contradiction. Therefore, $|x|_p^G \geq n$ and (3) follows. □

Charles [13] proved the following result and its converse for p-groups:

Proposition 1.13.4. *Let G be a group, p a prime, and $B = \bigoplus_{n<\omega} B_n$, where B_0 is a direct sum of infinite cyclic groups and for $n \geq 1$, B_n is a direct sum of cyclic groups of order p^n. If B is a p-basic subgroup of G, then*

$$(p^nG)[p] = B_{n+1}[p] \oplus (p^{n+1}G)[p]$$

for all $n < \omega$.

Proof. The case $n = 0$ follows from Lemma 1.13.3(1), (3). For $n \geq 1$, use Lemma 1.13.3(2), (3). □

Corollary 1.13.5. *If B is a p-basic subgroup of a group G, then $u_p(n, G) = u_p(n, B)$ for all $n < \omega$.*

Proof. Let $n < \omega$ and write B as in Proposition 1.13.4. Then

$$u_p(n, G) = \dim(p^nG)[p]/(p^{n+1}G)[p] = \dim B_{n+1}[p] = r(B_{n+1}),$$

which coincides with the number of direct summands $\cong \mathbb{Z}/p^{n+1}\mathbb{Z}$ of B; thus, $u_p(n, G) = u_p(n, B)$. □

Lemma 1.13.6. *Let G be an R-module and p a prime. If β is a limit ordinal, then $u_p(\alpha, G) = u_p(\alpha, G/p^\beta G)$ whenever $\alpha < \beta$.*

Proof. Let $\alpha < \beta$. Let $A = (p^\alpha(G/p^\beta G))[p]/T$ with $T = (p^{\alpha+1}(G/p^\beta G))[p]$, and consider the homomorphism

$$\begin{aligned} \phi: \quad (p^\alpha G)[p] &\rightarrow \quad A \\ x &\mapsto (x + p^\beta G) + T. \end{aligned}$$

To show that ϕ is surjective, let $y = (x + p^\beta G) + T \in A$. Then $x \in p^\alpha G$ by Lemma 1.10.5 and $px \in p^\beta G$. Since $\alpha + 1 < \beta$, there is an $x' \in p^{\alpha+1}G$ such that $px = px'$. It follows that $x - x' \in (p^\alpha G)[p]$ and $\phi(x - x') = y$ since $x' + p^\beta G \in T$.

Next we verify that $\ker\phi = (p^{\alpha+1}G)[p]$. If $z \in \ker\phi$, then $z + p^\beta G \in p^{\alpha+1}G/p^\beta G$; thus, $z \in (p^{\alpha+1}G)[p]$ since $pz = 0$. Conversely, if $z \in (p^{\alpha+1}G)[p]$, then $\phi(z) \in (p^{\alpha+1}G/p^\beta G)[p]/T = 0$. Thus, $(p^\alpha G)[p]/(p^{\alpha+1}G)[p] \cong A$, as desired. □

Lemma 1.13.7. *Let G be a reduced p-group of length λ. Then we have the following:*
1. *If $u_p(\alpha, G) = 0$ for all ordinals $\alpha < \lambda$, then $G = 0$.*
2. *If $\alpha < \beta \leq \lambda$ with β a limit ordinal, then $u_p(\gamma, G) \neq 0$ for infinitely many $\alpha < \gamma < \beta$.*

Proof. To show (1), assume that $u_p(\alpha, G) = 0$ for all ordinals $\alpha < \lambda$. Then an easy induction shows that $G[p] = (p^\alpha G)[p]$ for all α; thus, $G[p] = (p^\lambda G)[p] = 0$ and the claim follows.

To prove (2), let $\beta \leq \lambda$ be a limit ordinal and $\alpha < \beta$. Assume that $u_p(\gamma, G) \neq 0$ only for finitely many γ between α and β; let γ' be the largest such ordinal, and set $\nu = \gamma' + 1$. Now let κ be any ordinal such that $\nu + \kappa < \beta$. Then $u(\kappa, p^\nu G/p^\beta G) = u(\nu + \kappa, G/p^\beta G) = u(\nu + \kappa, G) = 0$ by Lemma 1.13.6. Thus, all Ulm invariants of $p^\nu G/p^\beta G$ are 0. There is a κ' such that $\nu + \kappa' = \beta$. Therefore, by Lemma 1.10.5, we have $p^{\kappa'}(p^\nu G/p^\beta G) = 0$, and obtain $p^\nu G/p^\beta G = 0$ by (1). But then $p^\nu G = p^\beta G \subseteq p^{\nu+1} G$; thus, $p^\nu G = p^{\nu+1} G = p^\lambda G = 0$, a contradiction. □

Following Hill [47], we define **relative Ulm invariants** $u_{\alpha,p}(G, A)$ for a submodule A of G; for an ordinal α define

$$\phi : D_{\alpha,p}(A) = \frac{\{x \in A \cap p^\alpha G : px \in p^{\alpha+2} G\}}{A \cap p^{\alpha+1} G} \to \frac{(p^\alpha G)[p]}{(p^{\alpha+1} G)[p]}$$

by $\phi(x + A \cap p^{\alpha+1} G) = x - y + (p^{\alpha+1} G)[p]$, where y is any element of $p^{\alpha+1} G$ such that $px = py$. Then ϕ is an embedding and its image is denoted by $I_{\alpha,p}(A)$. The dimension of the cokernel of ϕ is

$$u_{\alpha,p}(G, A) = \dim(p^\alpha G)[p]/[(p^\alpha G)[p] \cap (A + p^{\alpha+1} G)].$$

The embedding

$$\psi : (A \cap p^\infty G)[p] \to (p^\infty G)[p]$$

gives rise to the definition $u_{\infty,p}(G, A) = \dim(p^\infty G)[p]/(A \cap p^\infty G)[p]$. Letting $I_{\infty,p}(A) = (A \cap p^\infty G)[p]$, we have the following:

Lemma 1.13.8. *Let A be a submodule of an R-module G, $p \in R$ a prime, and $\alpha \in \mathrm{Ord}_\infty$. Then*

$$u_p(\alpha, G) = \dim I_{\alpha,p}(A) + u_{\alpha,p}(G, A).$$

The following useful observation is easily verified:

Lemma 1.13.9. *Suppose that G and H are R-modules, $p \in R$ is a prime, and $\alpha \in \mathrm{Ord}_\infty$. If there is a p-height-preserving isomorphism between submodules S and T of G and H, respectively, then $I_{\alpha,p}(S) \cong I_{\alpha,p}(T)$.*

We mention the following basic but useful fact:

Lemma 1.13.10. *Let G be an R-module, $p \in R$ a prime, and β an ordinal. Then* $u_{\alpha,p}(G, p^\beta G) = 0$ *if* $\alpha \geq \beta$ *and* $u_{\alpha,p}(G, p^\beta G) = u_p(\alpha, G)$ *if* $\alpha < \beta$.

Proof. Let α be an ordinal. Clearly, $H = (p^\alpha G)[p] \cap (p^\beta G + p^{\alpha+1}G)$ equals $(p^\alpha G)[p]$ if $\alpha \geq \beta$, and $H = (p^{\alpha+1}G)[p]$ whenever $\alpha < \beta$. □

Exercises.
1. If A is a p-isotype submodule of G, then $u_p(\alpha, A) \leq u_p(\alpha, G)$ for all ordinals α.
2. If A and B are submodules of an R-module G with $A \subseteq B$, then $u_{\alpha,p}(G, A) \geq u_{\alpha,p}(G, B)$ for each ordinal α and prime $p \in R$.
3. Let B be a basic subgroup of a reduced p-group G. Show that $u_{n,p}(G, B) = 0$ for all $n < \omega$, and $u_{\alpha,p}(G, B) = u_p(\alpha, G)$ for all $\alpha \geq \omega$.
4. Let G be a p-group of length λ. Show that G contains an isotype subgroup of length $\sigma \leq \lambda$ if and only if either σ is a limit ordinal or $u_p(\sigma - 1, G) \neq 0$.

1.14 Ulm's theorem

Let G and H be R-modules, and suppose that I is a nonempty set of isomorphisms between submodules of G and H. Then we say that I **has the back-and-forth property** if for every $f \in I$ and $a \in G$ (or $b \in H$) there is a $g \in I$ extending f and $a \in domain(g)$ (or $b \in range(g)$). To prove Ulm's classification theorem, Kaplansky and Mackey [78] constructed a system I of height-preserving isomorphisms between subgroups of two countable reduced p-groups with identical Ulm invariants such that I has the back-and-forth property. In the following lemma, we isolate an important part of their proof in terms of relative Ulm invariants and include the nonreduced case as well:

Lemma 1.14.1. *Let G and H be R-modules, $p \in R$ a prime, and $f : S \to T$ a p-height-preserving isomorphism, where S and T are submodules of G and H, respectively. Let $a \in G \setminus S$ such that a is proper with respect to S and $pa \in S$. If $u_{\alpha,p}(G, S) = u_{\alpha,p}(H, T)$ for all $\alpha \in \mathrm{Ord}_\infty$, then there is an element $b \in H \setminus T$ which is proper with respect to T, such that f extends to a p-height-preserving isomorphism $g : \langle S, a\rangle \to \langle T, b\rangle$ by sending a onto b.*

Proof. First, suppose that the p-height of a is an ordinal, say $|a| = \alpha$. In the case where there exists an element $\tilde{a} \in a + S$ such that $|\tilde{a}| = \alpha$ and $|p\tilde{a}| > \alpha + 1$, we replace a by $\tilde{a}$.

Case I: $|pa| = \alpha + 1$. Then there is a $b \in p^\alpha H \setminus p^{\alpha+1}H$ such that $f(pa) = pb$. To show that b is proper with respect to T, assume that $|b + z| \geq \alpha + 1$ for some $z \in T$. Then $|z| = \alpha$. Letting $c = f^{-1}(z)$, we have $|a + c| \geq \alpha$; hence, $a + c$ is proper with respect to S. Then $|p(a+c)| = |p(b+z)| > \alpha+1$, contradicting our choice of a. Notice that $b \notin T$ since $|b| \neq \infty$. By Lemma 1.12.4, f can be extended to a p-height-preserving isomorphism $g : \langle S, a\rangle \to \langle T, b\rangle$ by sending a onto b.

Case II: $|pa| > \alpha + 1$. Then $f(pa) = pb'$ for some $b' \in p^{\alpha+1}H$. We write $pa = pa'$ with $a' \in p^{\alpha+1}G$ and have $a - a' \in (p^\alpha G)[p]$. Since a is proper, $a - a' \notin S + p^{\alpha+1}G$, and it follows that

$$u_{\alpha,p}(H,T) = u_{\alpha,p}(G,S) = \dim \frac{(p^\alpha G)[p]}{(p^\alpha G)[p] \cap (S + p^{\alpha+1}G)} \neq 0.$$

Therefore there is a $\tilde{b} \in (p^\alpha H)[p]$ with $\tilde{b} \notin T + p^{\alpha+1}H$. Letting $b = b' + \tilde{b}$, we obtain $b \notin T$ since $b' \in p^{\alpha+1}H$. Using the fact that $\tilde{b}$ is proper with respect to T, we obtain $|b| = \alpha = |\tilde{b}| \geq |\tilde{b} + z| = |b' + \tilde{b} + z| = |b + z|$ for all $z \in T$. Consequently, b is proper with respect to T. In addition, we have $pb = pb' + p\tilde{b} = f(pa)$, so again by Lemma 1.12.4, f extends to the desired map g.

Finally, suppose that $|a| = \infty$. Then $|f(pa)| = \infty$; hence, $f(pa) = pb'$ for some $b' \in p^\infty H$. In view of Lemma 1.12.4, it suffices to show that there is a $b \in H \setminus T$ such that $f(pa) = pb$ and $|b| = \infty$. If $b' \notin T$, we let $b = b'$ and are done. Therefore, suppose that $b' \in T$. Then $b' = f(y)$ for some $y \in S$; hence, $f(pa - py) = 0$. But then $p(a - y) = 0$. Thus, $a - y \in (p^\infty G)[p]$. Since $a - y \notin S$, we obtain

$$u_{\infty,p}(H,T) = u_{\infty,p}(G,S) = \dim \frac{(p^\infty G)[p]}{(p^\infty G \cap S)[p]} \neq 0.$$

Therefore, there is a $\tilde{b} \in (p^\infty H)[p] \setminus (p^\infty H \cap T)[p]$. Letting $b = b' + \tilde{b}$, we have $b \notin T$, $|b| = \infty$, and $pb = pb' + p\tilde{b} = f(pa)$, as desired. □

Now we are ready to prove the following Ulm's theorem.

Theorem 1.14.2 (Ulm [146]). *Two countable p-groups G and H are isomorphic if and only if $u_p(\alpha, G) = u_p(\alpha, H)$ for all $\alpha \in \mathrm{Ord}_\infty$.*

Proof. Suppose that G and H have the same Ulm invariants. Let I be the set of all height-preserving isomorphisms $f : S \to T$, where S and T are finite subgroups of G and H, respectively. We will show that I has the back-and-forth property. Clearly, I is nonempty. Therefore, let $f : S \to T$ be an element of I and assume that $a \in G \setminus S$. Since G is a p-group, we may assume that $pa \in S$. In addition, we can assume that a is proper with respect to S since S is finite. Notice that for any α, an ordinal or ∞, we have $I_\alpha(S) \cong I_\alpha(T)$ by Lemma 1.13.9. Therefore, since $I_\alpha(S)$ is finite, we obtain $u_\alpha(G,S) = u_\alpha(H,T)$ by Lemma 1.13.8. Then Lemma 1.14.1 shows that f extends to a map $g \in I$ with $a \in \mathit{domain}(g)$. By symmetry, the set I has the back-and-forth property. Since G and H are countable groups, this yields a step-by-step construction of an isomorphism $G \to H$. The converse of the statement is immediate. □

In view of Theorem 1.3.2, we obtain the following:

Corollary 1.14.3. *Two countable torsion groups are isomorphic if and only if they have the same Ulm invariants u_p ($p \in \mathbb{P}$).*

Exercises.

1. (Zippin [157]) Let G and H be countable p-groups having the same Ulm invariants. If α is an ordinal, then any isomorphism $p^\alpha G \to p^\alpha H$ extends to an isomorphism $G \to H$.
2. Give an example of p-groups G and H with the same Ulm invariants such that $G \not\cong H$. [Hint: Let G be the torsion part of $\prod_{n=1}^{\infty} \mathbb{Z}/p^n\mathbb{Z}$.]

1.15 Completely decomposable torsion-free groups

Now that we have a classification theorem for a class of torsion groups, we turn to torsion-free groups and Baer's classification theorem. We start with a basic observation on p-heights of elements in a torsion-free group G.

Lemma 1.15.1. *Let G be a torsion-free group, $x \in G$, and p a prime. Then we have the following:*

1. $|x|_p < \omega$ *or* $|x|_p = \infty$.
2. $|px|_p = |x|_p + 1$ *(we set* $\infty + 1 = \infty$*).*

Proof. Suppose that $x \in p^\omega G$. For every $n < \omega$, there is a $y_n \in G$ such that $x = p^{n+1} y_n$. Then $p(y_0 - p^n y_n) = 0$, which implies $y_0 = p^n y_n$ since G is torsion-free. It follows that $y_0 \in p^\omega G$; hence, $x = p y_0 \in p^{\omega+1} G$. Thus, $p^\omega G = p^{\omega+1} G$, and we have $|x|_p = \infty$.

To show (2), let $n < \omega$. If $px \in p^{n+1} G$, then $px = p(p^n y)$ for some $y \in G$; thus, $x = p^n y \in p^n G$. It follows that $|px|_p = |x|_p + 1$. □

It follows that the Ulm matrix of an element $x \in G$ is completely determined by the sequence of its p-heights

$$\chi_G(x) = (|x|_{p_i})_{i \in \mathbb{N}},$$

called the **characteristic of** x. More generally, by a **characteristic** we mean an ordered sequence $(h_1, \dots, h_i, \dots)$ of nonnegative integers and symbols ∞. We write $(h_1, \dots, h_i, \dots) \geq (k_1, \dots, k_i, \dots)$ whenever $h_i \geq k_i$ for all $i \in \mathbb{N}$. Two characteristics $(h_1, \dots, h_i, \dots)$ and $(k_1, \dots, k_i, \dots)$ are called **equivalent** if $\sum_{i \in \mathbb{N}} |h_i - k_i|$ is finite (we set $\infty - \infty = 0$). In other words, the sequences are equivalent if and only if $h_i \neq k_i$ only for finitely many i, and for such i, both h_i and k_i are finite. The resulting equivalence classes are called **types**. The **type of an element** $x \in G$, written $\mathbf{t}(x)$, is the type containing $\chi_G(x)$. The **typeset of** G is defined to be the set typeset$(G) = \{\mathbf{t}(x) : 0 \neq x \in G\}$. If every nonzero element of G has the same type $\mathbf{t}$, then we say that G **has type t** and write $\mathbf{t}(G) = \mathbf{t}$. In this case, G is called **homogeneous.** For example, torsion-free rank 1 groups (= the nontrivial subgroups of $\mathbb{Q}$; see Theorems 1.6.8, 1.6.9 and 1.6.7) are homogeneous.

For types $\mathbf{s}$ and $\mathbf{t}$, we write $\mathbf{s} \geq \mathbf{t}$ if $(h_1, \dots, h_i, \dots) \geq (k_1, \dots, k_i, \dots)$ for some elements $(h_1, \dots, h_i, \dots) \in \mathbf{s}$ and $(k_1, \dots, k_i, \dots) \in \mathbf{t}$. By $\mathbf{s} \wedge \mathbf{t}$, we mean the type of the characteristic $(h_1 \wedge k_1, \dots, h_i \wedge k_i, \dots)$. For G a torsion-free group, we have the following:
(a) If $x, y \in G$, then $mx = ny$ for some nonzero integers m, n implies $\mathbf{t}(x) = \mathbf{t}(y)$.
(b) If $x, y \in G$, then $\mathbf{t}(x + y) \geq \mathbf{t}(x) \wedge \mathbf{t}(y)$.
(c) If $G = A \oplus B$ and $x \in A$, $y \in B$, then $\mathbf{t}(x + y) = \mathbf{t}(x) \wedge \mathbf{t}(y)$.
(d) If $\alpha : G \to H$ is a homomorphism with H torsion-free, then $\mathbf{t}(x) \leq \mathbf{t}(\alpha(x))$ for all $x \in G$.

For a torsion-free group G and a type $\mathbf{t}$, we define

$$G(\mathbf{t}) = \{x \in G : \mathbf{t}(x) \geq \mathbf{t}\}$$

and

$$G^*(\mathbf{t}) = \langle x \in G : \mathbf{t}(x) > \mathbf{t} \rangle.$$

Then the **Baer invariants** of G are defined to be the torsion-free ranks of the factor groups

$$G_{\mathbf{t}} = G(\mathbf{t})/G^*(\mathbf{t})$$

for all types $\mathbf{t}$.

Lemma 1.15.2. *Let G, H be torsion-free rank 1 groups and assume that $g \in G$, $h \in H$ are nonzero elements such that $\chi_G(g) \leq \chi_H(h)$. Then the correspondence $g \mapsto h$ extends to a monomorphism $\alpha : G \to H$. The map α is an isomorphism if $\chi_G(g) = \chi_H(h)$.*

Proof. Given integers $m \neq 0$ and n, the equation $my = nh$ is solvable in H if the corresponding equation $mx = ng$ is solvable in G, and any such solutions are unique since the groups are torsion-free. Thus, $g \mapsto h$ extends to a homomorphism $\alpha : G \to H$. To show that α is injective, let $z \in \ker \alpha$. Then $mg = nz$ for some integers m and $n \neq 0$, and we have $m\alpha(g) = \alpha(mg) = \alpha(nz) = n\alpha(z) = 0$. But then $m = 0$ since $\alpha(g) = h \neq 0$. It follows that $z = 0$; thus, α is injective.

In the case $\chi_G(g) = \chi_H(h)$, in the first sentence of this proof, "if" can be replaced by "if and only if"; hence, α is bijective. □

Theorem 1.15.3 (Baer [6]). *Two torsion-free rank 1 groups G and H are isomorphic if and only if they have the same type. For every type $\mathbf{t}$, there is a subgroup A of $\mathbb{Q}$ which has type $\mathbf{t}$.*

Proof. Suppose that $\mathbf{t}(G) = \mathbf{t}(H)$ and let $g_1 \in G$, $h_1 \in H$ be nonzero elements. Then $\mathbf{t}(g_1) = \mathbf{t}(h_1)$; hence, $|g_1|_p = |h_1|_p$ for all but finitely many primes p, and $|g_1|_p \neq \infty$ exactly if $|h_1|_p \neq \infty$. Therefore, there are positive integers r and s such that $g = rg_1$ and $h = sh_1$ satisfy $\chi_G(g) = \chi_H(h)$. By Lemma 1.15.2, the correspondence $g \mapsto h$ extends to an isomorphism $G \to H$.

If $\mathbf{t} = [(h_1, \ldots, h_i, \ldots)]$ is a type, then let A be the subgroup of $\mathbb{Q}$ generated by the set $\{p_i^{-r_i} : 0 \le r_i \le h_i \ (i \in \mathbb{N})\}$. Then $\chi_A(1) = (h_1, \ldots, h_i, \ldots)$, and we have $\mathbf{t}(A) = \mathbf{t}$. □

A torsion-free group is called **completely decomposable** if it is a direct sum of rank 1 groups. Suppose G is completely decomposable, say $G = \bigoplus_{i \in I} G_i$, where each group G_i has rank 1. For any type $\mathbf{t}$, we define the sets $I_{\mathbf{t}} = \{i \in I : \mathbf{t}(G_i) \ge \mathbf{t}\}$ and $J_{\mathbf{t}} = \{i \in I : \mathbf{t}(G_i) > \mathbf{t}\}$. Note that $(x_i)_{i \in I} \in G(\mathbf{t})$ if and only if $x_i \in G_i(\mathbf{t})$ for all $i \in I$. Then we have $G(\mathbf{t}) = \bigoplus_{i \in I_{\mathbf{t}}} G_i$ and $G^*(\mathbf{t}) = \bigoplus_{i \in J_{\mathbf{t}}} G_i$ by (c); hence, the rank of $G_{\mathbf{t}}$ is equal to

$$|\{i \in I : \mathbf{t}(G_i) = \mathbf{t}\}|.$$

Since $G_{\mathbf{t}}$ does not depend on any particular decomposition of G, we obtain the following:

Theorem 1.15.4 (Baer [6]). *Let G be a completely decomposable group. If $G = \bigoplus_{i \in I} G_i = \bigoplus_{j \in J} H_j$ such that each group G_i and H_j has rank 1, then there is a bijection $\phi : I \to J$ such that $G_i \cong H_{\phi(i)}$ for all $i \in I$. Two completely decomposable groups are isomorphic if and only if they have the same Baer invariants.*

We state the following important fact, which will follow from a more general result (cf. Corollary 3.16.5):

Theorem 1.15.5 (Baer [6], Kulikov [90], Kaplansky [79]). *The class of completely decomposable groups is closed under direct summands.*

The following fact will be needed later:

Proposition 1.15.6. *Let G and H be torsion-free rank 1 groups. Then the following conditions are equivalent:*
1. *There is a monomorphism $G \to H$;*
2. *There is a nonzero homomorphism $G \to H$;*
3. $\mathbf{t}(G) \le \mathbf{t}(H)$.

Proof. Clearly (1) implies (2). Given any homomorphism $\alpha : G \to H$ and $x \in G$, we have $\chi_G(x) \le \chi_H(\alpha(x))$, thus (2) implies (3). Assuming that (3) is true, there are nonzero elements $g \in G$ and $h \in H$ such that $\chi_G(g) \le \chi_H(h)$. By Lemma 1.15.2, the correspondence $g \mapsto h$ extends to a monomorphism $G \to H$. Thus, (1) follows. □

Exercises.
1. For a subgroup A of a torsion-free group G, define A_* to be the intersection of all pure subgroups of G containing A, called the **purification** of A. Show the following:
   (a) A_* is a pure subgroup of G.
   (b) $A_* = A^0$ (see Section 1.10).
   (c) $A_*/A = t(G/A)$.
2. If G is torsion-free and $G = A \oplus B$, then $G(\mathbf{t}) = A(\mathbf{t}) \oplus B(\mathbf{t})$ and $G^*(\mathbf{t}) = A^*(\mathbf{t}) \oplus B^*(\mathbf{t})$ for all types $\mathbf{t}$.

3. Let G and H be torsion-free rank 1 groups. Then $G \cong H$ if and only if there is a homomorphism $G \to H$ whose cokernel is bounded.
4. Give an example of a completely decomposable group having a pure subgroup that is not completely decomposable.

1.16 Hom and Ext

Given two groups we may use them to form other useful groups, specifically the group of homomorphisms and the group of extensions. For homomorphisms $\alpha, \beta : A \to C$, we define their sum to be the map

$$\begin{array}{rccc} \alpha+\beta: & A & \to & C \\ & x & \mapsto & \alpha(x)+\beta(x) \end{array}$$

and obtain the **group of homomorphisms from** A **to** C, which is denoted by $\mathrm{Hom}(A, C)$. Notice that for $A = \mathbb{Z}$, we have an isomorphism

$$\mathrm{Hom}(\mathbb{Z}, C) \xrightarrow{\sim} C$$

given by $\phi \mapsto \phi(1)$. Likewise, we obtain an isomorphism

$$\mathrm{Hom}(\mathbb{Z}/n\mathbb{Z}, C) \xrightarrow{\sim} C[n].$$

The following items are easily verified:
(a) If C is torsion-free, then so is $\mathrm{Hom}(A, C)$.
(b) If A is torsion and C is torsion-free, then $\mathrm{Hom}(A, C) = 0$.
(c) If A is divisible, then $\mathrm{Hom}(A, C)$ is torsion-free.
(d) If A is torsion-free and divisible, then so is $\mathrm{Hom}(A, C)$.

It is clear that a given group G and $\alpha \in \mathrm{Hom}(A, B)$ give rise to homomorphisms

$$\begin{array}{rccc} \alpha^* : & \mathrm{Hom}(B, G) & \to & \mathrm{Hom}(A, G) \\ & \phi & \mapsto & \phi\alpha \end{array}$$

and

$$\begin{array}{rccc} \alpha_* : & \mathrm{Hom}(G, A) & \to & \mathrm{Hom}(G, B) \\ & \phi & \mapsto & \alpha\phi. \end{array}$$

Of particular note is when α is multiplication by a positive integer. The following simple—but useful—observation shows that we may look at such a multiplication either as a homomorphism or as the result of repeated additions:

Proposition 1.16.1. *Multiplication by a positive integer n on A or C induces multiplication by n on* $\mathrm{Hom}(A, C)$*; that is,* $(n1_A)^* = (n1_C)_* = n1_{\mathrm{Hom}(A,C)}$.

Proof. Consider the map $\alpha : A \to A$ given by $x \mapsto nx$. If $\phi \in \mathrm{Hom}(A, C)$, then $\phi\alpha(x) = \phi(nx) = n\phi(x)$; thus, $\alpha^* : \phi \mapsto n\phi$. Likewise, the map $\gamma : C \to C$ with $\gamma : y \mapsto ny$ induces $\gamma_* : \phi \mapsto n\phi$. □

Exactness of a short exact sequence induces exactness of Hom sequences, as shown in the following:

Theorem 1.16.2 (Cartan–Eilenberg [12]). *Given any group G, exactness of the sequence* $E : 0 \to A \xrightarrow{\alpha} B \xrightarrow{\beta} C \to 0$ *induces exactness of the sequences*

$$0 \to \mathrm{Hom}(C, G) \xrightarrow{\beta^*} \mathrm{Hom}(B, G) \xrightarrow{\alpha^*} \mathrm{Hom}(A, G)$$

and

$$0 \to \mathrm{Hom}(G, A) \xrightarrow{\alpha_*} \mathrm{Hom}(G, B) \xrightarrow{\beta_*} \mathrm{Hom}(G, C).$$

Proof. To show that β^* is injective, assume that $\phi \in \ker \beta^*$. Then $\phi\beta = 0$ implies that $\phi = 0$ since β is surjective; therefore, β^* is injective. For any $\phi \in \mathrm{Hom}(C, G)$, we have $\alpha^*\beta^*(\phi) = \phi\beta\alpha = 0$; thus, $\mathrm{im}\,\beta^* \subseteq \ker \alpha^*$. To prove the reverse inclusion, let $\phi \in \ker \alpha^*$. Then $\phi\alpha = 0$, so by Lemma 1.2.2 there is a $\nu \in \mathrm{Hom}(C, G)$ with $\nu\beta = \phi$; thus, $\phi \in \mathrm{im}\beta^*$. It follows that the first sequence is exact.

Letting $\phi \in \ker \alpha_*$, we obtain $\alpha\phi = 0$; thus, $\phi = 0$ since α is injective. Therefore, α_* is injective. If $\phi \in \mathrm{Hom}(G, A)$, then $\beta_*\alpha_*(\phi) = \beta\alpha\phi = 0$, which shows that $\mathrm{im}\,\alpha_* \subseteq \ker \beta_*$. For the reverse inclusion, let $\phi \in \ker \beta_*$. Then $\beta\phi = 0$, so by Lemma 1.2.1 there is a $\nu \in \mathrm{Hom}(G, A)$ such that $\alpha\nu = \phi$; thus, $\phi \in \mathrm{im}\alpha$. □

Now let A, A_i, C, C_i $(i \in I)$ be groups. If $\alpha : \bigoplus_i A_i \to C$ is a homomorphism and $\alpha_i = \alpha \restriction A_i$ $(i \in I)$, then $\alpha \mapsto (\alpha_i)_{i \in I}$ defines a map from $\mathrm{Hom}(\bigoplus_i A_i, C)$ to $\prod_i \mathrm{Hom}(A_i, C)$. A straightforward calculation shows that this map is an isomorphism. Likewise, given a homomorphism $\beta : A \to \prod_i C_i$ and projection maps $\pi_i : \prod_i C_i \to C_i$, the correspondence $\beta \mapsto (\pi_i\beta)_{i \in I}$ defines an isomorphism between $\mathrm{Hom}(A, \prod_i C_i)$ and $\prod_i \mathrm{Hom}(A, C_i)$. Thus, we have the following:

Theorem 1.16.3. *There are natural isomorphisms*

$$\mathrm{Hom}\left(\bigoplus_{i \in I} A_i, C\right) \cong \prod_{i \in I} \mathrm{Hom}(A_i, C)$$

and

$$\mathrm{Hom}\left(A, \prod_{i \in I} C_i\right) \cong \prod_{i \in I} \mathrm{Hom}(A, C_i).$$

Now we turn our attention to the group of extensions. We will define an equivalence relation and an addition that make it an abelian group. As we did with Hom, we will define morphisms between groups of extensions based on homomorphisms on the underlying groups. Finally, we will use these groups to extend the exact sequence of Theorem 1.16.2 above.

A short exact sequence of groups

$$E : 0 \to A \xrightarrow{\mu} B \xrightarrow{\nu} C \to 0$$

is called an **extension** of A by C. By a **morphism** between the extensions E and E' : $0 \to A' \xrightarrow{\mu'} B' \xrightarrow{\nu'} C' \to 0$, we mean a triple $\Gamma = (\alpha, \beta, \gamma)$ of homomorphisms such that the diagram

$$\begin{array}{ccccccccc} 0 & \to & A & \xrightarrow{\mu} & B & \xrightarrow{\nu} & C & \to & 0 \\ & & \downarrow \alpha & & \downarrow \beta & & \downarrow \gamma & & \\ 0 & \to & A' & \xrightarrow{\mu'} & B' & \xrightarrow{\nu'} & C' & \to & 0 \end{array}$$

is commutative, and we write $\Gamma : E \to E'$. If $\Gamma = (1_A, \beta, 1_C)$ for some $\beta : B \to B'$, then we call E and E' **equivalent**, written $E \equiv E'$. In this case, the map β is automatically an isomorphism by Lemma 1.2.3; thus, $\equiv$ is an equivalence relation. Then $\operatorname{Ext}(C, A)$ denotes the set of all equivalence classes of extensions of A by C. We say that the extension E **splits** if E is equivalent to the **split extension**

$$E_0 : 0 \to A \to A \oplus C \to C \to 0.$$

For instance, E splits whenever A is divisible or C is free (Theorems 1.6.3 and 1.5.3).

Given the extension E and a homomorphism $\alpha : A \to A'$, we obtain a **pushout diagram**, by which we mean the commutative diagram with exact rows

$$\begin{array}{cccccccccc} E: & 0 & \to & A & \xrightarrow{\mu} & B & \xrightarrow{\nu} & C & \to & 0 \\ & & & \downarrow \alpha & & \downarrow \beta & & \| & & \\ \alpha E: & 0 & \to & A' & \xrightarrow{\sigma} & X & \xrightarrow{\tau} & C & \to & 0, \end{array}$$

such that $X = (A' \oplus B)/L$ with $L = \{(-\alpha(a), \mu(a)) : a \in A\}$ and $\sigma : g \mapsto (g, 0) + L$, $\tau : (g, b) + L \mapsto \nu(b)$, $\beta : b \mapsto (0, b) + L$.

Note that any commutative diagram with exact rows

$$\begin{array}{cccccccccc} E: & 0 & \to & A & \xrightarrow{\mu} & B & \xrightarrow{\nu} & C & \to & 0 \\ & & & \downarrow \alpha & & \downarrow \beta' & & \downarrow \gamma' & & \\ E': & 0 & \to & A' & \xrightarrow{\mu'} & B' & \xrightarrow{\nu'} & C' & \to & 0 \end{array}$$

gives rise to the commutative diagram

$$\begin{array}{cccccccccc} E: & 0 & \to & A & \xrightarrow{\mu} & B & \xrightarrow{\nu} & C & \to & 0 \\ & & & \downarrow \alpha & & \downarrow \beta & & \| & & \\ \alpha E: & 0 & \to & A' & \xrightarrow{\sigma} & X & \xrightarrow{\tau} & C & \to & 0 \\ & & & \| & & \downarrow f & & \downarrow \gamma' & & \\ E': & 0 & \to & A' & \xrightarrow{\mu'} & B' & \xrightarrow{\nu'} & C' & \to & 0 \end{array}$$

with $f:(a',b)+L\mapsto\mu'(a')+\beta'(b)$, and we obtain $f\beta=\beta'$. Following Mac Lane's [111] terminology, we say that $\Gamma'=(\alpha,\beta',\gamma'):E\to E'$ can be **factored through** $\Gamma=(\alpha,\beta,1_C):E\to\alpha E$; that is, Γ' can be written as a composite

$$E\xrightarrow{\Gamma}\alpha E\xrightarrow{(1_{A'},f,\gamma')}E'.$$

In particular, any commutative diagram

$$\begin{array}{rccccccccc} E: & 0 & \to & A & \to & B & \to & C & \to & 0\\ & & & \downarrow\alpha & & \downarrow & & \| & & \\ \tilde E: & 0 & \to & A' & \to & \tilde B & \to & C & \to & 0\end{array}$$

yields $\alpha E\equiv\tilde E$. Thus, we have the following:

(i) $1_A E\equiv E$.

(ii) If $E\equiv E'$, then $\alpha E\equiv\alpha E'$ for any $\alpha\in\operatorname{Hom}(A,A')$.

(iii) $(\alpha'\alpha)E\equiv\alpha'(\alpha E)$ for any $\alpha\in\operatorname{Hom}(A,A')$ and $\alpha'\in\operatorname{Hom}(A',A'')$.

Then (ii) shows that for any $[E]\in\operatorname{Ext}(C,A)$ and $\alpha\in\operatorname{Hom}(A,A')$, there is a uniquely defined $[\alpha E]\in\operatorname{Ext}(C,A')$.

Dually, a homomorphism $\gamma:C'\to C$ gives rise to the commuting diagram with exact rows

$$\begin{array}{rccccccccc} E\gamma: & 0 & \to & A & \xrightarrow{\phi} & Y & \xrightarrow{\psi} & C' & \to & 0\\ & & & \| & & \downarrow\eta & & \downarrow\gamma & & \\ E: & 0 & \to & A & \xrightarrow{\mu} & B & \xrightarrow{\nu} & C & \to & 0,\end{array}$$

which is called a **pullback diagram**, where $Y=\{(b,c')\in B\oplus C':\nu(b)=\gamma(c')\}$ and $\phi:a\mapsto(\mu(a),0)$, $\psi:(b,c')\mapsto c'$, $\eta:(b,c')\mapsto b$. Then any morphism $\Gamma_1=(\alpha_1,\beta_1,\gamma):E^1\to E$ can be **factored through** $\bar\Gamma=(1_A,\eta,\gamma):E\gamma\to E$; that is, Γ_1 is a composite

$$E^1\xrightarrow{(\alpha_1,g,1_C)}E\gamma\xrightarrow{\bar\Gamma}E$$

for some map g. To see this, assume that

$$\begin{array}{rccccccccc} E^1: & 0 & \to & A_1 & \xrightarrow{\phi_1} & B_1 & \xrightarrow{\psi_1} & C' & \to & 0\\ & & & \downarrow\alpha_1 & & \downarrow\beta_1 & & \downarrow\gamma & & \\ E: & 0 & \to & A & \xrightarrow{\mu} & B & \xrightarrow{\nu} & C & \to & 0\end{array}$$

is a commutative diagram with exact rows. Then we obtain the commutative diagram

$$\begin{array}{rccccccccc} E^1: & 0 & \to & A_1 & \xrightarrow{\phi_1} & B_1 & \xrightarrow{\psi_1} & C' & \to & 0\\ & & & \downarrow\alpha_1 & & \downarrow g & & \| & & \\ E\gamma: & 0 & \to & A & \xrightarrow{\phi} & Y & \xrightarrow{\psi} & C' & \to & 0\\ & & & \| & & \downarrow\eta & & \downarrow\gamma & & \\ E: & 0 & \to & A & \xrightarrow{\mu} & B & \xrightarrow{\nu} & C & \to & 0\end{array}$$

by letting $g : b_1 \mapsto (\beta_1(b_1), \psi_1(b_1))$, and we have $\eta g = \beta_1$, proving the assertion. Therefore, any commutative diagram with exact rows

$$\begin{array}{ccccccccc} E'': & 0 & \to & A & \to & B'' & \to & C' & \to 0 \\ & & & \| & & \downarrow & & \downarrow \gamma & \\ E: & 0 & \to & A & \xrightarrow{\mu} & B & \xrightarrow{\nu} & C & \to 0 \end{array}$$

yields $E\gamma \equiv E''$. This shows the following:
(iv) $E1_C \equiv E$.
(v) If $E \equiv E'$, then $E\gamma \equiv E'\gamma$ for any $\gamma \in \operatorname{Hom}(C', C)$.
(vi) $E(\gamma\gamma') \equiv (E\gamma)\gamma'$ whenever $\gamma' \in \operatorname{Hom}(C'', C')$ and $\gamma \in \operatorname{Hom}(C', C)$.

Clearly, the diagram

$$\begin{array}{ccccccccc} E\gamma: & 0 & \to & A & \xrightarrow{\phi} & Y & \xrightarrow{\psi} & C' & \to 0 \\ & & & \downarrow \alpha & & \downarrow \beta\eta & & \downarrow \gamma & \\ \alpha E: & 0 & \to & A' & \xrightarrow{\sigma} & X & \xrightarrow{\tau} & C & \to 0 \end{array}$$

commutes, so by the observation above (with E^1 and E replaced by $E\gamma$ and αE, respectively), we obtain a commutative diagram

$$\begin{array}{ccccccccc} E\gamma: & 0 & \to & A & \xrightarrow{\phi} & Y & \xrightarrow{\psi} & C' & \to 0 \\ & & & \downarrow & & \downarrow & & \| & \\ (\alpha E)\gamma: & 0 & \to & A' & \to & Z & \to & C' & \to 0. \end{array}$$

On the other hand, $E\gamma$ and $\alpha : A \to A'$ generate a pushout diagram; thus, we obtain the equivalence

$$\alpha(E\gamma) \equiv (\alpha E)\gamma.$$

Proposition 1.16.4. *If*

$$\begin{array}{ccccccccc} E: & 0 & \to & A & \xrightarrow{\mu} & B & \xrightarrow{\nu} & C & \to 0 \\ & & & \downarrow \alpha & & \downarrow \beta & & \downarrow \gamma & \\ E': & 0 & \to & A' & \xrightarrow{\mu'} & B' & \xrightarrow{\nu'} & C' & \to 0 \end{array}$$

is a commutative diagram with exact rows, then $\alpha E \equiv E'\gamma$.

Proof. Factoring $\Gamma = (\alpha, \beta, \gamma) : E \to E'$ through $\tilde{\Gamma} : E \to \alpha E$, we write Γ as a composite $E \xrightarrow{\tilde{\Gamma}} \alpha E \xrightarrow{(1_{A'}, f, \gamma)} E'$ for some suitable map f. Since E' and γ generate a pullback diagram, we obtain $\alpha E \equiv E'\gamma$. □

Recall that the zero map on a group A is denoted by 0_A. We have the following basic fact:

Lemma 1.16.5. *Given an extension $E : 0 \to A \to B \to C \to 0$, we have $0_A E \equiv E_0$ and $E0_C \equiv E_0$, where E_0 is the split extension.*

Proof. The assertion follows from the commutativity of the diagram

$$\begin{array}{lcccccccc} E_0: & 0 & \to & A & \to & A\oplus C & \to & C & \to 0 \\ & & & \| & & \downarrow \alpha & & \downarrow 0_C & \\ E: & 0 & \to & A & \xrightarrow{\mu} & B & \xrightarrow{\nu} & C & \to 0 \\ & & & \downarrow 0_A & & \downarrow \beta & & \| & \\ E_0: & 0 & \to & A & \to & A\oplus C & \to & C & \to 0 \end{array}$$

with $\alpha : (a,c) \mapsto \mu(a)$ and $\beta : b \mapsto (0,\nu(b))$. □

Next we will show that the set $\mathrm{Ext}(C,A)$ can be turned into an abelian group using the **diagonal map**

$$\begin{array}{rccc} \Delta_G: & G & \to & G\oplus G \\ & x & \mapsto & (x,x) \end{array}$$

and the **codiagonal map**

$$\begin{array}{rccc} \nabla_G: & G\oplus G & \to & G \\ & (x,y) & \mapsto & x+y. \end{array}$$

Let

$$\phi_G : G\oplus(G\oplus G) \to (G\oplus G)\oplus G$$

denote the natural isomorphism and define the **direct sum of homomorphisms** $\sigma_i : G_i \to H_i$ $(i = 1,2)$ to be

$$\begin{array}{rccc} \sigma_1\oplus\sigma_2: & G_1\oplus G_2 & \to & H_1\oplus H_2 \\ & (x_1,x_2) & \mapsto & (\sigma_1(x_1),\sigma_2(x_2)). \end{array}$$

The following identities involving the maps above are easily verified:

(a) $(\Delta_G\oplus 1_G)\Delta_G = \phi_G(1_G\oplus\Delta_G)\Delta_G$.
(b) $\nabla_G(\nabla_G\oplus 1_G) = \nabla_G(1_G\oplus\nabla_G)\phi_G^{-1}$.
(c) If $\sigma_1,\sigma_2 \in \mathrm{Hom}(C,A)$, then $\sigma_1+\sigma_2 = \nabla_A(\sigma_1\oplus\sigma_2)\Delta_C$.
(d) If $\alpha \in \mathrm{Hom}(A,A')$, then $\alpha\nabla_A = \nabla_{A'}(\alpha\oplus\alpha)$.
(e) If $\gamma \in \mathrm{Hom}(C',C)$, then $\Delta_C\gamma = (\gamma\oplus\gamma)\Delta_{C'}$.

The **direct sum of extensions** $E_i : 0 \to A_i \xrightarrow{\mu_i} B_i \xrightarrow{\nu_i} C_i \to 0$ $(i = 1,2)$ is defined to be the induced extension

$$E_1\oplus E_2 : 0 \to A_1\oplus A_2 \xrightarrow{\mu_1\oplus\mu_2} B_1\oplus B_2 \xrightarrow{\nu_1\oplus\nu_2} C_1\oplus C_2 \to 0.$$

Letting

$$\begin{array}{rccc} \tau_G: & G_1\oplus G_2 & \to & G_2\oplus G_1 \\ & (x_1,x_2) & \mapsto & (x_2,x_1), \end{array}$$

we obtain a morphism $(\tau_A,\tau_B,\tau_C) : E_1\oplus E_2 \to E_2\oplus E_1$. We have the following:

(f) $\nabla_A \tau_A = \nabla_A$ and $\Delta_C = \tau_C \Delta_C$.
(g) If $\alpha_i \in \mathrm{Hom}(A_i, A_i')$ $(i = 1,2)$, then $(\alpha_1 \oplus \alpha_2)(E_1 \oplus E_2) \equiv \alpha_1 E_1 \oplus \alpha_2 E_2$.
(h) If $\gamma_i \in \mathrm{Hom}(C_i', C_i)$ $(i = 1,2)$, then $(E_1 \oplus E_2)(\gamma_1 \oplus \gamma_2) \equiv E_1\gamma_1 \oplus E_2\gamma_2$.

The sum of two extensions of A by C as defined in the next theorem is known as the **Baer sum.** Note the similarity to (c) above.

Theorem 1.16.6 (Mac Lane [111]). *Let A and C be groups. Then the set* $\mathrm{Ext}(C,A)$ *is an abelian group under the operation defined by* $[E_1] + [E_2] = [E_1 + E_2]$, *where*

$$E_1 + E_2 = \nabla_A(E_1 \oplus E_2)\Delta_C$$

with extensions E_1 and E_2 of A by C. The identity of $\mathrm{Ext}(C,A)$ *is the equivalence class of the split extension $E_0 : 0 \to A \to A \oplus C \to C \to 0$ and the inverse of $[E]$ is the equivalence class of $(-1_A)E$. If $\alpha, \alpha_i \in \mathrm{Hom}(A, A')$ and $\gamma, \gamma_i \in \mathrm{Hom}(C', C)$ $(i = 1,2)$, then*
1. $\alpha(E_1 + E_2) \equiv \alpha E_1 + \alpha E_2$, $(E_1 + E_2)\gamma \equiv E_1\gamma + E_2\gamma$,
2. $(\alpha_1 + \alpha_2)E \equiv \alpha_1 E + \alpha_2 E$, $E(\gamma_1 + \gamma_2) \equiv E\gamma_1 + E\gamma_2$.

Proof. Let E and E_i be extensions of A by C $(i = 1,2,3)$. We will start by proving (1) and (2). The equivalences in (1) are evident since

$$\alpha\nabla_A(E_1 \oplus E_2)\Delta_C = \nabla_{A'}(\alpha \oplus \alpha)(E_1 \oplus E_2)\Delta_C \equiv \nabla_{A'}(\alpha E_1 \oplus \alpha E_2)\Delta_C$$

and

$$\nabla_A(E_1 \oplus E_2)\Delta_C\gamma = \nabla_A(E_1 \oplus E_2)(\gamma \oplus \gamma)\Delta_{C'} \equiv \nabla_A(E_1\gamma \oplus E_2\gamma)\Delta_{C'}.$$

By Proposition 1.16.4, $(\Delta_A, \Delta_B, \Delta_C) : E \to E \oplus E$ yields $\Delta_A E \equiv (E \oplus E)\Delta_C$, and thus, we have $(\alpha_1+\alpha_2)E = \nabla_{A'}(\alpha_1 \oplus \alpha_2)\Delta_A E \equiv \nabla_{A'}(\alpha_1 \oplus \alpha_2)(E \oplus E)\Delta_C \equiv \nabla_{A'}(\alpha_1 E \oplus \alpha_2 E)\Delta_C = \alpha_1 E + \alpha_2 E$.

Likewise, $(\nabla_A, \nabla_B, \nabla_C) : E \oplus E \to E$ yields $E\nabla_C \equiv \nabla_A(E \oplus E)$; therefore, $E(\gamma_1 + \gamma_2) = E\nabla_C(\gamma_1 \oplus \gamma_2)\Delta_{C'} \equiv \nabla_A(E \oplus E)(\gamma_1 \oplus \gamma_2)\Delta_{C'} \equiv \nabla_A(E\gamma_1 \oplus E\gamma_2)\Delta_{C'} = E\gamma_1 + E\gamma_2$. This proves (2).

It remains to show that the Baer sum defines an abelian group structure on $\mathrm{Ext}(C,A)$. If $E_1' \equiv E_1$ and $E_2' \equiv E_2$, we obtain $E_1 \oplus E_2 \equiv E_1' \oplus E_2'$; thus, $(E_1 \oplus E_2)\Delta_C \equiv (E_1' \oplus E_2')\Delta_C$. Therefore, $E_1 + E_2 \equiv E_1' + E_2'$. Hence, our addition on $\mathrm{Ext}(C,A)$ is well-defined. To verify the associative law, we compute

$$\begin{aligned}(E_1 + E_2) + E_3 &= \nabla_A(\nabla_A(E_1 \oplus E_2)\Delta_C \oplus E_3)\Delta_C \\ &= \nabla_A(\nabla_A \oplus 1_A)[(E_1 \oplus E_2) \oplus E_3](\Delta_C \oplus 1_C)\Delta_C \\ &= \nabla_A(1_A \oplus \nabla_A)\phi_A^{-1}[(E_1 \oplus E_2) \oplus E_3]\phi_C(1_C \oplus \Delta_C)\Delta_C,\end{aligned}$$

which is equivalent to $\nabla_A(1_A \oplus \nabla_A)[E_1 \oplus (E_2 \oplus E_3)](1_C \oplus \Delta_C)\Delta_C$ by Proposition 1.16.4 since we have a morphism $(\phi_A, \phi_B, \phi_C) : E_1 \oplus (E_2 \oplus E_3) \to (E_1 \oplus E_2) \oplus E_3$. Thus, $(E_1 + E_2) + E_3 \equiv E_1 + (E_2 + E_3)$. To prove the commutative law, we compute

$$\nabla_A(E_1 \oplus E_2)\Delta_C = \nabla_A\tau_A(E_1 \oplus E_2)\Delta_C \equiv \nabla_A(E_2 \oplus E_1)\tau_C\Delta_C = \nabla_A(E_2 \oplus E_1)\Delta_C,$$

which shows that $E_1+E_2 \equiv E_2+E_1$. To prove that $[E_0]$ is the identity, we use Lemma 1.16.5 and the first equivalence in (2) to obtain $E + E_0 \equiv 1_A E + 0_A E \equiv (1_A + 0_A)E = 1_A E \equiv E$, as required. Finally, we have $E + (-1_A)E \equiv 1_A E + (-1_A)E \equiv (1_A - 1_A)E = 0_A E \equiv E_0$; thus, $[(-1_A)E]$ is the inverse of $[E]$. □

In the following, we often write $E \in \mathrm{Ext}(C,A)$ instead of $[E] \in \mathrm{Ext}(C,A)$. By Theorem 1.16.6(1), any homomorphisms $\alpha : A \to G$ and $\gamma : G \to C$ induce natural homomorphisms

$$\begin{array}{rccc} \alpha_* : & \mathrm{Ext}(C,A) & \to & \mathrm{Ext}(C,G) \\ & E & \mapsto & \alpha E \end{array}$$

and

$$\begin{array}{rccc} \gamma^* : & \mathrm{Ext}(C,A) & \to & \mathrm{Ext}(G,A) \\ & E & \mapsto & E\gamma . \end{array}$$

Then (iii) and (vi) yield

$$(\alpha'\alpha)_* = \alpha_* \alpha'_* \quad \text{for all } \alpha \in \mathrm{Hom}(A,A') \text{ and } \alpha' \in \mathrm{Hom}(A',A'').$$
$$(\gamma\gamma')^* = \gamma'^* \gamma^* \quad \text{for all } \gamma' \in \mathrm{Hom}(C'',C') \text{ and } \gamma \in \mathrm{Hom}(C',C).$$

Theorem 1.16.6(2) can be rephrased as follows:

Lemma 1.16.7. *Let $\alpha_i \in \mathrm{Hom}(A,A')$ and $\gamma_i \in \mathrm{Hom}(C',C)$ for $i = 1,2$. Then $(\alpha_1 + \alpha_2)_* = (\alpha_1)_* + (\alpha_2)_*$ and $(\gamma_1 + \gamma_2)^* = (\gamma_1)^* + (\gamma_2)^*$.*

Therefore, leaving E fixed, we obtain homomorphisms

$$\begin{array}{rccc} E^* : & \mathrm{Hom}(A,G) & \to & \mathrm{Ext}(C,G) \\ & \sigma & \mapsto & \sigma E \end{array}$$

and

$$\begin{array}{rccc} E_* : & \mathrm{Hom}(G,C) & \to & \mathrm{Ext}(G,A) \\ & \sigma & \mapsto & E\sigma , \end{array}$$

which can be used to extend the Hom sequences in Theorem 1.16.2. By the next result, the newly obtained sequences

$$\mathrm{Hom}(B,G) \to \mathrm{Hom}(A,G) \xrightarrow{E^*} \mathrm{Ext}(C,G)$$

and

$$\mathrm{Hom}(G,B) \to \mathrm{Hom}(G,C) \xrightarrow{E_*} \mathrm{Ext}(G,A)$$

are exact:

Lemma 1.16.8 (Mac Lane [111]). *Let $E : 0 \to A \xrightarrow{\alpha} B \xrightarrow{\beta} C \to 0$ be an exact sequence. Then we have the following:*

1. *Given $\sigma : A \to G$, the extension σE splits if and only if there is a map $\tau : B \to G$ such that the diagram*

$$\begin{array}{ccccccccc} 0 & \to & A & \xrightarrow{\alpha} & B & \xrightarrow{\beta} & C & \to & 0 \\ & & \sigma\downarrow & \swarrow\tau & & & & & \\ & & G & & & & & & \end{array}$$

is commutative. In particular, αE splits.

2. *Given $\sigma : G \to C$, the extension $E\sigma$ splits if and only if there is a map $\tau : G \to B$ such that the diagram*

$$\begin{array}{ccccccccc} & & & & & & G & & \\ & & & & & \tau\swarrow & \downarrow\sigma & & \\ 0 & \to & A & \xrightarrow{\alpha} & B & \xrightarrow{\beta} & C & \to & 0 \end{array}$$

is commutative. In particular, $E\beta$ splits.

Proof. (1): If there is such a map $\tau : B \to G$, then we have a commutative diagram

$$\begin{array}{cccccccccc} E: & 0 & \to & A & \xrightarrow{\alpha} & B & \xrightarrow{\beta} & C & \to & 0 \\ & & & \downarrow\sigma & & \downarrow\nu & & \| & & \\ E_0: & 0 & \to & G & \xrightarrow{\mu} & G\oplus C & \xrightarrow{\phi} & C & \to & 0 \end{array}$$

with $\nu : b \mapsto (\tau(b), \beta(b))$, $\mu : g \mapsto (g, 0)$ and $\phi : (g, c) \mapsto c$; thus, $\sigma E \equiv E_0$. Conversely, assume that $\sigma E \equiv E_0$. Then we have a composite $E \overset{(\sigma,\nu',1_C)}{\longrightarrow} \sigma E \overset{(1_G,\nu'',1_C)}{\longrightarrow} E_0$ and define $\tau = \pi\nu''\nu' : B \to G$ with $\pi : G \oplus C \to G$ the projection map. We consider the resulting diagram and obtain $\tau\alpha = \pi\mu\sigma = \sigma$, as required.

(2): Assuming that there is a map $\sigma : G \to B$, making the diagram in (2) commutative, we have

$$\begin{array}{cccccccccc} E_0: & 0 & \to & A & \xrightarrow{\mu} & A\oplus G & \xrightarrow{\phi} & G & \to & 0 \\ & & & \| & & \downarrow\nu & & \downarrow\sigma & & \\ E: & 0 & \to & A & \xrightarrow{\alpha} & B & \xrightarrow{\beta} & C & \to & 0 \end{array}$$

with $\nu : (a, g) \mapsto \alpha(a) + \tau(g)$, $\mu : a \mapsto (a, 0)$ and $\phi : (a, g) \mapsto g$; hence, $E\sigma \equiv E_0$. Conversely, suppose that $E\sigma \equiv E_0$. Then there is a composite $E_0 \overset{(1_A,\nu',1_C)}{\longrightarrow} E\sigma \overset{(1_A,\nu'',\sigma)}{\longrightarrow} E$. We define $\tau = \nu''\nu'\iota$ with $\iota : G \to A \oplus G$ the inclusion map and obtain $\beta\tau = \sigma\phi\iota = \sigma$. □

Corollary 1.16.9. *Let $E : 0 \to A \xrightarrow{\alpha} B \xrightarrow{\beta} C \to 0$ be an exact sequence. Then the following statements are equivalent:*

1. *E splits;*
2. *There is a map $\tau : B \to A$ such that $\tau\alpha = 1_A$;*
3. *There is a map $\tau : C \to B$ such that $\beta\tau = 1_C$.*

Now we are ready to prove the following important result:

Theorem 1.16.10 (Cartan–Eilenberg [12]). *Given any group G, exactness of the sequence $E : 0 \to A \xrightarrow{\alpha} B \xrightarrow{\beta} C \to 0$ induces exactness of the* Hom-Ext *sequences*

$$0 \to \mathrm{Hom}(C,G) \to \mathrm{Hom}(B,G) \to \mathrm{Hom}(A,G) \xrightarrow{E^*} \mathrm{Ext}(C,G)$$
$$\xrightarrow{\beta^*} \mathrm{Ext}(B,G) \xrightarrow{\alpha^*} \mathrm{Ext}(A,G) \to 0$$

and

$$0 \to \mathrm{Hom}(G,A) \to \mathrm{Hom}(G,B) \to \mathrm{Hom}(G,C) \xrightarrow{E_*} \mathrm{Ext}(G,A)$$
$$\xrightarrow{\alpha_*} \mathrm{Ext}(G,B) \xrightarrow{\beta_*} \mathrm{Ext}(G,C) \to 0.$$

Proof. By Theorem 1.16.2 and Lemma 1.16.8, it suffices to prove exactness at the groups of extensions. To show that $\mathrm{im}\, E^* \subseteq \ker \beta^*$, let $\sigma \in \mathrm{Hom}(A,G)$. Then $\beta^* E^*(\sigma) = \sigma E\beta = 0$ since $E\beta$ splits by Lemma 1.16.8(2). Now let $E_1 : 0 \to G \xrightarrow{\phi} H \xrightarrow{\psi} C \to 0 \in \ker \beta^*$. Then $E_1\beta$ splits, so by Lemma 1.16.8(2), there is a $\tau : B \to H$ such that $\psi\tau = \beta$; thus, $\psi\tau\alpha = \beta\alpha = 0$. By Lemma 1.2.1 there is a map $\rho : A \to G$ with $\phi\rho = \tau\alpha$; hence, the diagram

$$\begin{array}{ccccccccc}
E: & 0 & \to & A & \xrightarrow{\alpha} & B & \xrightarrow{\beta} & C & \to 0 \\
 & & & \downarrow \rho & & \downarrow \tau & & \| & \\
E_1: & 0 & \to & G & \xrightarrow{\phi} & H & \xrightarrow{\psi} & C & \to 0
\end{array}$$

is commutative. But then $E_1 = \rho E$, and we obtain $\ker \beta^* \subseteq \mathrm{im} E^*$.

Next we prove exactness at $\mathrm{Ext}(B,G)$. Letting $E_1 \in \mathrm{Ext}(C,G)$, we have $\beta^*(E_1) \in \ker \alpha^*$ since $\alpha^*\beta^*(E_1) = E_1\beta\alpha = 0$. Now assume that $E_2 : 0 \to G \xrightarrow{\phi} H \xrightarrow{\psi} B \to 0 \in \ker \alpha^*$. Since $E_2\alpha$ splits, Lemma 1.16.8(2) shows that there is a $\tau : A \to H$ such that $\psi\tau = \alpha$, and τ is injective since α is. Then $\beta\psi\tau = \beta\alpha = 0$; thus, $h + \tau(A) \mapsto \beta\psi(h)$ defines a map $\sigma : H/\tau(A) \to C$, and we obtain a commutative diagram

$$\begin{array}{cccccccccc}
 & & & 0 & & 0 & & 0 & & \\
 & & & \downarrow & & \downarrow & & \downarrow & & \\
 & 0 & \to & 0 & \to & A & = & A & \to & 0 \\
 & & & \downarrow & & \downarrow \tau & & \downarrow \alpha & & \\
E_2: & 0 & \to & G & \xrightarrow{\phi} & H & \xrightarrow{\psi} & B & \to & 0 \\
 & & & \| & & \downarrow \pi & & \downarrow \beta & & \\
E_3: & 0 & \to & G & \xrightarrow{\pi\phi} & H/\tau(A) & \xrightarrow{\sigma} & C & \to & 0. \\
 & & & \downarrow & & \downarrow & & \downarrow & & \\
 & & & 0 & & 0 & & 0 & &
\end{array}$$

By Lemma 1.2.4, the row E_3 is exact, so we have $E_2 = E_3\beta$, and therefore, $E_2 \in \mathrm{im}\beta^*$.

To show that α^* is surjective, let $E' \in \text{Ext}(A, G)$. By Theorem 1.5.1, there is an epimorphism $\phi : F \to B$ with F free, and we set $K = \ker\phi$. Now let $L = \phi^{-1}(\alpha(A))$. Then $l \mapsto \alpha^{-1}\phi(l)$ defines a surjective homomorphism $\phi' : L \to A$ with $\ker\phi' = K$, so we obtain a commutative diagram

$$\begin{array}{lccccccc} E_1: & 0 & \to & K & \to & L & \stackrel{\phi'}{\longrightarrow} & A & \to 0 \\ & & & \| & & \downarrow & & \downarrow \alpha & \\ E_2: & 0 & \to & K & \to & F & \stackrel{\phi}{\longrightarrow} & B & \to 0 \end{array}$$

with exact rows. Thus, $E_1 = E_2\alpha$. By the first part of this proof, the sequence $\text{Hom}(K, G) \stackrel{E_1^*}{\longrightarrow} \text{Ext}(A, G) \to \text{Ext}(L, G)$ is exact. As a subgroup of a free group, L is free; hence, $\text{Ext}(L, G) = 0$, and so we have $E' = E_1^*(\eta)$ for some $\eta \in \text{Hom}(K, G)$. Then $\alpha^*(\eta E_2) = \eta E_2\alpha = \eta E_1 = E'$; thus, α^* is surjective. This proves exactness of the first sequence.

Next we show exactness at $\text{Ext}(G, A)$. The extension αE splits by Lemma 1.16.8(1). Therefore, for any $\rho \in \text{Hom}(G, C)$, we have $\alpha_* E_*(\rho) = \alpha E\rho = 0$. Therefore, $\text{im}\, E_* \subseteq \ker\alpha_*$. Now let $E_1 : 0 \to A \stackrel{\phi}{\longrightarrow} H \stackrel{\psi}{\longrightarrow} G \to 0 \in \ker\alpha_*$. Then αE_1 splits, so by Lemma 1.16.8(1), there is a $\tau : H \to B$ such that $\tau\phi = \alpha$. This implies that $\beta\tau\phi = \beta\alpha = 0$; thus, Lemma 1.2.2 yields a map $\nu : G \to C$ such that $\nu\psi = \beta\tau$. Hence, we obtain a commutative diagram

$$\begin{array}{lccccccccc} E_1: & 0 & \to & A & \stackrel{\phi}{\longrightarrow} & H & \stackrel{\psi}{\longrightarrow} & G & \to & 0 \\ & & & \| & & \downarrow \tau & & \downarrow \nu & & \\ E: & 0 & \to & A & \stackrel{\alpha}{\longrightarrow} & B & \stackrel{\beta}{\longrightarrow} & C & \to & 0. \end{array}$$

But then $E_1 = E\nu$; thus, $E_1 \in \text{im}E_*$ as desired.

It is clear that $\text{im}\,\alpha_* \subseteq \ker\beta_*$ since for any $E_1 \in \text{Ext}(G, A)$, we have $\beta_*\alpha_*(E_1) = \beta\alpha E_1 = 0$. Suppose $E_2 : 0 \to B \stackrel{\phi}{\longrightarrow} H \stackrel{\psi}{\longrightarrow} G \to 0 \in \ker\beta_*$. Then βE_2 splits, so Lemma 1.16.8(1) shows that $\beta = \tau\phi$ for some $\tau : H \to C$ which is surjective. Since $\tau\phi\alpha = \beta\alpha = 0$, we obtain a map $\mu = \phi\alpha : A \to \ker\tau$. Letting $\nu : \ker\tau \to H$ denote the inclusion map, we obtain a commutative diagram

$$\begin{array}{lccccccccc} & & & 0 & & 0 & & 0 & & \\ & & & \downarrow & & \downarrow & & \downarrow & & \\ E': & 0 & \to & A & \stackrel{\mu}{\longrightarrow} & \ker\tau & \stackrel{\psi\nu}{\longrightarrow} & G & \to & 0 \\ & & & \downarrow \alpha & & \downarrow \nu & & \| & & \\ E_2: & 0 & \to & B & \stackrel{\phi}{\longrightarrow} & H & \stackrel{\psi}{\longrightarrow} & G & \to & 0 \\ & & & \downarrow \beta & & \downarrow \tau & & \downarrow & & \\ & 0 & \to & C & = & C & \to & 0 & \to & 0 \\ & & & \downarrow & & \downarrow & & \downarrow & & \\ & & & 0 & & 0 & & 0 & & \end{array}$$

with exact second row. By Lemma 1.2.4, the row E' is exact, and we have $\alpha E' = E_2$. Thus, $E_2 \in \operatorname{im}\alpha_*$.

To show that β_* is surjective, let $E_1 : 0 \to C \xrightarrow{\phi} D \xrightarrow{\psi} G \to 0 \in \operatorname{Ext}(G, C)$. Since $\operatorname{Ext}(D, A) \xrightarrow{\phi^*} \operatorname{Ext}(C, A) \to 0$ is exact, there is an $E' \in \operatorname{Ext}(D, A)$ such that $E'\phi = E$; hence, there is a commutative diagram

$$\begin{array}{ccccccccc}
E: & 0 & \to & A & \xrightarrow{\alpha} & B & \xrightarrow{\beta} & C & \to 0 \\
 & & & \| & & \downarrow \beta' & & \downarrow \phi & \\
E': & 0 & \to & A & \xrightarrow{\sigma} & M & \xrightarrow{\tau} & D & \to 0 \\
 & & & & & \downarrow \psi\tau & & \downarrow \psi & \\
 & & & & & G & = & G &
\end{array}$$

with β' injective (Lemma 1.2.3(2)). To show that the middle column yields an extension $\tilde{E} \in \operatorname{Ext}(G, B)$, we need to show that $\operatorname{im}\beta' = \ker\psi\tau$. Clearly, $\operatorname{im}\beta' \subseteq \ker\psi\tau$ since $\psi\tau\beta' = \psi\phi\beta = 0$. Letting $m \in \ker\psi\tau$, we have $\tau(m) \in \ker\psi = \operatorname{im}\phi$; thus, $\tau(m) = \phi\beta(b)$ for some $b \in B$. Then $\tau\beta'(b) = \tau(m)$ shows that $\beta'(b) - m \in \ker\tau = \operatorname{im}\sigma$. Therefore, there is an $a \in A$ with $\sigma(a) = \beta'(b) - m$; hence, $m = \beta'(b) - \beta'\alpha(a) \in \operatorname{im}\beta'$, as desired. It follows that $E_1 = \beta\tilde{E}$; thus, β_* is surjective. This completes the proof. □

In terms of direct sums and direct products, we obtain the following:

Theorem 1.16.11. *There are natural isomorphisms:*

$$\operatorname{Ext}\Big(\bigoplus_{i\in I} C_i, A\Big) \cong \prod_{i\in I} \operatorname{Ext}(C_i, A)$$

and

$$\operatorname{Ext}\Big(C, \prod_{j\in J} A_j\Big) \cong \prod_{j\in J} \operatorname{Ext}(C, A_j).$$

Proof. Since every group is an epimorphic image of a free group, we obtain extensions $0 \to B_i \to F_i \to C_i \to 0$ with F_i free. Recalling that extensions by free groups split, we obtain the exact sequence $\operatorname{Hom}(F_i, A) \to \operatorname{Hom}(B_i, A) \to \operatorname{Ext}(C_i, A) \to \operatorname{Ext}(F_i, A) = 0$. This yields a diagram

$$\begin{array}{ccccccc}
\operatorname{Hom}(\bigoplus F_i, A) & \xrightarrow{f_1} & \operatorname{Hom}(\bigoplus B_i, A) & \xrightarrow{f_2} & \operatorname{Ext}(\bigoplus C_i, A) & \to & 0 \\
\phi_1 \downarrow & & \phi_2 \downarrow & & \phi_3 \downarrow & & \\
\prod \operatorname{Hom}(F_i, A) & \xrightarrow{g_1} & \prod \operatorname{Hom}(B_i, A) & \xrightarrow{g_2} & \prod \operatorname{Ext}(C_i, A) & \to & 0
\end{array}$$

with exact rows and isomorphisms ϕ_1 and ϕ_2 (Theorem 1.16.3), making the first square commute. We define the map ϕ_3 by $\phi_3(f_2(x)) = g_2\phi_2(x)$ $(x \in \operatorname{Hom}(\bigoplus B_i, A))$. It is straightforward to verify that ϕ_3 is well-defined. Since ϕ_3 is a homomorphism, it is an isomorphism by Lemma 1.2.3. The proof of the second assertion is similar. □

We have the following facts:

Proposition 1.16.12. *Let A and C be groups. Then we have the following:*
1. *Multiplication by a positive integer n on A or C induces multiplication by n on* $\mathrm{Ext}(C,A)$*; that is,* $(n1_A)_* = (n1_C)^* = n1_{\mathrm{Ext}(C,A)}$.
2. *If α is an automorphism of A, then α_* is an automorphism of* $\mathrm{Ext}(C,A)$.
3. *If γ is an automorphism of C, then γ^* is an automorphism of* $\mathrm{Ext}(C,A)$.

Proof. To prove (1), let $E \in \mathrm{Ext}(C,A)$. Since $(1_A)_* : E \mapsto 1_A E = E$, we have $(n1_A)_* = n(1_A)_*$ by Lemma 1.16.7; thus, $(n1_A)_* : E \mapsto nE$. Likewise, $(n1_C)^*$ maps E onto nE since $E = E1_C$.

(2) is true since $\alpha\beta = \beta\alpha = 1_A$ implies $\alpha_*\beta_* = \beta_*\alpha_* = 1_{\mathrm{Ext}(C,A)}$. Likewise, (3) holds since η^* is the inverse of γ^* whenever η is the inverse of γ. □

As an application of the above proposition we obtain the following:

Proposition 1.16.13. *Let A, C be groups and n a positive integer. Then the following hold:*
1. *If $nA = 0$ or $nC = 0$, then* $n\mathrm{Ext}(C,A) = 0$.
2. $\mathrm{Ext}(\mathbb{Z}/n\mathbb{Z}, A) \cong A/nA$.
3. *If $nA = A$, then* $n\mathrm{Ext}(C,A) = \mathrm{Ext}(C,A)$*. Thus,* $\mathrm{Ext}(C,A)$ *is divisible whenever A is.*
4. *If $nA = A$ and $A[n] = 0$, then* $n\mathrm{Ext}(C,A) = \mathrm{Ext}(C,A)$ *and* $\mathrm{Ext}(C,A)[n] = 0$*. Hence,* $\mathrm{Ext}(C,A)$ *is divisible torsion-free whenever A is.*
5. *If $C[n] = 0$, then* $n\mathrm{Ext}(C,A) = \mathrm{Ext}(C,A)$*. Thus,* $\mathrm{Ext}(C,A)$ *is divisible if C is torsion-free.*
6. *If $nC = 0$ and $C[n] = 0$, then* $n\mathrm{Ext}(C,A) = \mathrm{Ext}(C,A)$ *and* $\mathrm{Ext}(C,A)[n] = 0$*. Therefore,* $\mathrm{Ext}(C,A)$ *is divisible torsion-free if C is.*

Proof. (1): Let $E \in \mathrm{Ext}(C,A)$. Then $(0_A)_*$ maps E onto E_0 (Lemma 1.16.5) and Proposition 1.16.12 shows that $(n1_A)_*$ maps E onto nE. Thus, $nE = 0$ provided that $n1_A = 0_A$. Likewise, we have $nE = 0$ if $n1_C = 0_C$.

(2): The extension $0 \to \mathbb{Z} \xrightarrow{\sigma} \mathbb{Z} \to \mathbb{Z}/n\mathbb{Z} \to 0$ with $\sigma = n1_A$ induces the exact sequence $\mathrm{Hom}(\mathbb{Z},A) \xrightarrow{\sigma^*} \mathrm{Hom}(\mathbb{Z},A) \to \mathrm{Ext}(\mathbb{Z}/n\mathbb{Z},A) \to \mathrm{Ext}(\mathbb{Z},A) = 0$. Since $\sigma^* : \phi \mapsto n\phi$ (Proposition 1.16.1) and $\mathrm{Hom}(\mathbb{Z},A) \cong A$, we obtain $\mathrm{Ext}(\mathbb{Z}/n\mathbb{Z},A) \cong \mathrm{Hom}(\mathbb{Z},A)/n\mathrm{Hom}(\mathbb{Z},A) \cong A/nA$.

(3): Since the sequence $A \xrightarrow{\beta} A \to 0$ with $\beta = n1_A$ is exact, the induced sequence $\mathrm{Ext}(C,A) \xrightarrow{\beta_*} \mathrm{Ext}(C,A) \to 0$ is exact, and we have $\beta_* = n1_{\mathrm{Ext}(C,A)}$.

(4) and (6) follow from Proposition 1.16.12(2), (3).

(5): Exactness of $0 \to C \xrightarrow{\alpha} C$ with $\alpha = n1_A$ induces exactness of $\mathrm{Ext}(C,A) \xrightarrow{\alpha^*} \mathrm{Ext}(C,A) \to 0$, where $\alpha^* = n1_{\mathrm{Ext}(C,A)}$. □

Exercises.
1. If $A = nA$ for some $n \in \mathbb{N}$, then $\mathrm{Hom}(A,C)[n] = 0$ for any group C.
2. If A is any group, then $\mathrm{Hom}(A,\mathbb{Q}) \cong \prod_{r_0(A)} \mathbb{Q}$.

3. Show that $\mathrm{Hom}(A, C) \cong \mathrm{Hom}(A/tA, C)$ whenever C is torsion-free.
4. If A is torsion, then $\mathrm{Hom}(A, C) \cong \prod_{p\in\mathbb{P}} \mathrm{Hom}(t_pA, t_pC)$.
5. If C is any group and $n \in \mathbb{N}$, then we have $\mathrm{Ext}(C, \mathbb{Z}/n\mathbb{Z}) \cong \mathrm{Ext}(C[n], \mathbb{Z}/n\mathbb{Z})$.
6. Given a group A, we have $\mathrm{Ext}(C, A) = 0$ for all groups C if and only if A is divisible.
7. Given a group C, we have $\mathrm{Ext}(C, A) = 0$ for all groups A if and only if C is free.
8. Let A and C be torsion-free rank 1 groups. Then $A \cong C$ if and only if $\mathrm{Hom}(A, C) \neq 0$ and $\mathrm{Hom}(C, A) \neq 0$.

1.17 Tensor products and p-localizations

Many of the problems we solve are best addressed by first localizing to simplify the problem by focusing on a particular prime, and then extending the local result to the more general, global result. Here we construct the tensor product, which will facilitate localization. Let A, C and G be groups. A map $b : A \times C \to G$ is called **bilinear** if it satisfies

$$b(a_1 + a_2, c) = b(a_1, c) + b(a_2, c) \quad \text{and} \quad b(a, c_1 + c_2) = b(a, c_1) + b(a, c_2)$$

for all $a, a_1, a_2 \in A$ and $c, c_1, c_2 \in C$.

Now let F be the free group with free generators (a, c), where $a \in A$, $c \in C$. Letting Y be the subgroup generated by all elements of the form

$$(a_1 + a_2, c) - (a_1, c) - (a_2, c) \text{ and } (a, c_1 + c_2) - (a, c_1) - (a, c_2)$$

with $a, a_1, a_2 \in A$ and $c, c_1, c_2 \in C$, we define the **tensor product of** A **and** C to be the group

$$A \otimes C = F/Y.$$

The coset $(a, c) + Y$ is written $a \otimes c$. Then every element of $A \otimes C$ can be expressed as a (not necessarily unique) finite sum of the form $\sum_{i=1}^{m} a_i \otimes c_i$ ($a_i \in A, c_i \in C$), and we have
(a) $(a_1 + a_2) \otimes c = a_1 \otimes c + a_2 \otimes c$
(b) $a \otimes (c_1 + c_2) = a \otimes c_1 + a \otimes c_2$
(c) $na \otimes c = n(a \otimes c) = a \otimes nc$

for all $a, a_1, a_2 \in A$, $c, c_1, c_2 \in C$ and integers n. The following properties are evident for any $a \in A$, $c \in C$ and $m, n \in \mathbb{N}$:
(d) $a \otimes 0 = 0 = 0 \otimes c$.
(e) $(-a) \otimes c = -(a \otimes c) = a \otimes (-c)$.
(f) If $ma = 0$ and $nc = 0$, then $\gcd(m, n)(a \otimes c) = 0$.
(g) If $ma = 0$ and $m|c$, then $a \otimes c = 0$.
(h) If $m|a$ and $n|c$, then $mn|a \otimes c$.

The **tensor map** is defined to be

$$\begin{array}{rccc} t: & A \times C & \to & A \otimes C \\ & (a,c) & \mapsto & a \otimes c \end{array}$$

and is clearly a bilinear map. We have the following:

Theorem 1.17.1. *Let A and C be groups and let $t : A \times C \to A \otimes C$ be the tensor map. If $f : A \times C \to G$ is a bilinear map, then there exists a unique homomorphism $\phi : A \otimes C \to G$ such that the diagram*

$$\begin{array}{ccc} A \times C & \xrightarrow{t} & A \otimes C \\ f \searrow & & \swarrow \phi \\ & G & \end{array}$$

is commutative.

Proof. Letting F be the free group on the set $A \times C$, the correspondence $(a,c) \mapsto f(a,c)$ uniquely defines a homomorphism $\varphi : F \to G$. Let Y be the subgroup of F generated by all elements of the form $(a_1 + a_2, c) - (a_1, c) - (a_2, c)$ and $(a, c_1 + c_2) - (a, c_1) - (a, c_2)$ with $a, a_1, a_2 \in A$ and $c, c_1, c_2 \in C$. Then φ induces the homomorphism

$$\begin{array}{rccc} \phi: & F/Y = A \otimes C & \to & G \\ & x + Y & \mapsto & \varphi(x) \end{array}$$

since $Y \subseteq \ker \varphi$. Letting $(a,c) \in A \times C$, we have $\phi t(a,c) = \phi(a \otimes c) = \phi((a,c) + Y) = f(a,c)$; thus, $\phi t = f$. To prove the uniqueness claim, assume that $\psi : A \otimes C \to G$ is a homomorphism with $\psi t = f$. Then $\psi(a \otimes c) = \psi t(a,c) = f(a,c)$; hence, $\psi = \phi$. □

The next result contains some basic properties of tensor products:

Proposition 1.17.2. *Let A and C be groups. Then we have the following:*
1. $A \otimes C \cong C \otimes A$.
2. $\mathbb{Z} \otimes C \cong C$.
3. *If A or C is a p-group (torsion), then $A \otimes C$ is a p-group (torsion).*
4. *If A is a p-group and C is p-divisible, then $A \otimes C = 0$.*
5. *If A or C is p-divisible, then $A \otimes C$ is p-divisible.*

Proof. To prove (1), let F be the free group on the set $A \times C$. Then $(a,c) \mapsto c \otimes a$ defines an epimorphism $\mu : F \to C \otimes A$ with $F/\ker \mu = A \otimes C$.

To show (2), consider the map $\sigma : C \to \mathbb{Z} \otimes C$ given by $c \mapsto 1 \otimes c$. Then σ is surjective since elements of $\mathbb{Z} \otimes C$ have the form $\sum_{i=1}^{n} (r_i \otimes c_i) = \sum_{i=1}^{n} (1 \otimes r_i c_i) = 1 \otimes (\sum_{i=1}^{n} r_i c_i)$ with $r_i \in \mathbb{Z}$ and $c_i \in C$ $(i = 1, \dots, n)$. The map $f : \mathbb{Z} \times C \to C$ given by $(m,c) \mapsto mc$ is bilinear, so by Theorem 1.17.1 there is a homomorphism ϕ making the diagram

$$\begin{array}{ccc} \mathbb{Z} \times C & \xrightarrow{t} & \mathbb{Z} \otimes C \\ f \searrow & & \swarrow \phi \\ & C & \end{array}$$

commute. Then $\phi : 1 \otimes c \mapsto f(1, c) = c$ shows that ϕ and σ are inverse to each other; thus, σ is an isomorphism.

For (3), use (c) and (d), whereas (4) and (5) follow from (g) and (h), respectively. □

Tensor products commute with direct sums:

Theorem 1.17.3. *If $A = \bigoplus_{i \in I} A_i$ and $C = \bigoplus_{j \in J} C_j$, then $A \otimes C \cong \bigoplus_{i,j}(A_i \otimes C_j)$.*

Proof. We claim that by Proposition 1.17.2(1), it suffices to prove that $A \otimes C \cong \bigoplus_{i \in I}(A_i \otimes C)$. Assume that we know $A \otimes C \cong \bigoplus_i (A_i \otimes C)$. Since $A_i \otimes C \cong C \otimes A_i$, we have $A \otimes C \cong \bigoplus_i (A_i \otimes C) \cong \bigoplus_i (C \otimes A_i) \cong \bigoplus_i (\bigoplus_j (C_j \otimes A_i))$. The last isomorphism holds because for fixed i, $C \otimes A_i \cong \bigoplus_j (C_j \otimes A_i)$ by our assumption. Therefore, $A \otimes C \cong \bigoplus_{i,j}(A_i \otimes C_j)$ follows, proving the claim.

For $i \in I$ let $\sigma_i : A_i \to A$ be the inclusion and $\pi_i : A \to A_i$ the projection. Since

$$\begin{array}{rccc} f: & A \times C & \to & B = \bigoplus_{i \in I}(A_i \otimes C) \\ & (a, c) & \mapsto & \sum(\pi_i(a) \otimes c) \end{array}$$

is a bilinear map, there is a homomorphism $\phi : A \otimes C \to B$ such that $f = \phi t$, where $t : A \times C \to A \otimes C$ is the tensor map. Clearly, ϕ is surjective since the elements $\pi_i(a) \otimes c$ generate B.

For each $i \in I$, there is a homomorphism $\eta_i : A_i \otimes C \to A \otimes C$, making the diagram

$$\begin{array}{ccc} A_i \times C & \xrightarrow{t_i} & A_i \otimes C \\ f_i \searrow & & \swarrow \eta_i \\ & A \otimes C & \end{array}$$

commute, where t_i is the tensor map and $f_i : (a_i, c) \mapsto \sigma_i(a_i) \otimes c$; thus, $\eta_i : a_i \otimes c \mapsto \sigma_i(a_i) \otimes c$. This yields a homomorphism

$$\begin{array}{rccc} \eta : & B & \to & A \otimes C \\ & \sum x_i & \mapsto & \sum \eta_i(x_i) \end{array}$$

with $x_i \in A_i \otimes C$. But then $\eta f(a, c) = \eta(\sum(\pi_i(a) \otimes c)) = \sum \eta_i(\pi_i(a) \otimes c) = \sum(\sigma_i \pi_i(a) \otimes c) = a \otimes c$; hence, $\eta\phi(a \otimes c) = \eta\phi t(a, c) = \eta f(a, c) = a \otimes c$. This shows that ϕ is injective; thus, ϕ is an isomorphism. □

Given homomorphisms $\alpha : A \to A'$ and $\gamma : C \to C'$, we obtain a bilinear map $f : A \times C \to A' \otimes C'$ defined by $(a, c) \mapsto \alpha(a) \otimes \gamma(c)$. By Theorem 1.17.1, we obtain a homomorphism $\phi : A \otimes C \to A' \otimes C'$ such that the diagram

$$\begin{array}{ccc} A \times C & \xrightarrow{t} & A \otimes C \\ f \searrow & & \swarrow \phi \\ & A' \otimes C' & \end{array}$$

is commutative. Writing $\alpha \otimes \gamma$ instead of ϕ, we have

$$\alpha \otimes \gamma : a \otimes c \mapsto \alpha(a) \otimes \gamma(a).$$

Therefore, given a group G, the map $\alpha : A \to A'$ induces the homomorphism

$$\alpha_* = \alpha \otimes 1_G : A \otimes G \to A' \otimes G.$$

Tensoring with G preserves right-exactness of short exact sequences:

Theorem 1.17.4 (Cartan–Eilenberg [12]). *Given any group G, exactness of the sequence $0 \to A \xrightarrow{\alpha} B \xrightarrow{\beta} C \to 0$ induces exactness of the sequence*

$$A \otimes G \xrightarrow{\alpha_*} B \otimes G \xrightarrow{\beta_*} C \otimes G \to 0.$$

Proof. Let $a \in A$ and $g \in G$. Then $\beta_*\alpha_*(a \otimes g) = \beta_*(\alpha(a) \otimes g) = \beta\alpha(a) \otimes g = 0$; thus, $\operatorname{im}\alpha_* \subseteq \ker\beta_*$. Therefore, β_* induces the homomorphism

$$\eta : K = (B \otimes G)/\operatorname{im}\alpha_* \to C \otimes G$$

with $\eta : (b \otimes g) + \operatorname{im}\alpha_* \mapsto \beta_*(b \otimes g)$. To prove that $\ker\beta_* \subseteq \operatorname{im}\alpha_*$, we show that η is injective. For every $c \in C$, there is a $b \in B$ such that $\beta(b) = c$ and if $\beta(b) = \beta(b')$ for some $b' \in B$, then $(b \otimes g) - (b' \otimes g) = (b - b') \otimes g = \alpha(a) \otimes g$ for some $a \in A$. Thus, the map

$$\begin{array}{rccc} f: & C \times G & \to & K \\ & (c, g) & \mapsto & (b \otimes g) + \operatorname{im}\alpha_* \end{array}$$

is well-defined. Then f is a bilinear map, so there is a homomorphism $\phi : C \otimes G \to K$ with $\phi(c \otimes g) = (b \otimes g) + \operatorname{im}\alpha_*$ (Theorem 1.17.1). Then $\phi\eta = 1_K$; thus, η is injective.

It is clear that β_* is surjective since $C \otimes G$ is generated by elements of the form $\beta(b) \otimes g$. □

Of particular interest is tensoring of G with $\mathbb{Z}_p$ (= the p-localization of $\mathbb{Z}$ at the prime p), resulting in

$$G_p = G \otimes \mathbb{Z}_p,$$

which is called the ***p*-localization of** G. An examination of these groups is in order since they form the basis for many of the classification results:

Lemma 1.17.5. *Let G be a group and p a prime. Then we have the following:*
1. *Every element of G_p can be written as $x \otimes \frac{1}{n}$ for some $x \in G$ and $n \in \mathbb{Z}$ with $p \nmid n$.*
2. *G_p is a p-local group, that is, a $\mathbb{Z}_p$-module by defining $r(x \otimes z) = x \otimes rz$ for $r \in \mathbb{Z}_p$. Thus, every element of G_p has a unique q-divisor for any prime $q \neq p$.*

3. *Multiplication by a prime $\neq p$ in G_p is an automorphism. Hence, G_p has no q-torsion for any prime $q \neq p$.*

Proof. An arbitrary element of G_p is of the form $x_1 \otimes \frac{m_1}{n_1} + \cdots + x_k \otimes \frac{m_k}{n_k}$, where $x_i \in G$ and $m_i, n_i \in \mathbb{Z}$ with $p \nmid n_i$ for $i = 1, \ldots, k$. This reduces to $[\sum_{i=1}^k m_i(\prod_{j\neq i} n_j)x_i] \otimes \frac{1}{n}$ with $n = \prod_{j=1}^k n_j$.

To prove (2), we fix $r \in \mathbb{Z}_p$ and consider the bilinear map $f : G \times \mathbb{Z}_p \to G_p$ given by $(x, z) \mapsto x \otimes rz$. By Theorem 1.17.1, we obtain a homomorphism $\phi : G_p \to G_p$ which maps $x \otimes z$ onto $x \otimes rz$. Thus, for any given $x \in G$ and $r, z \in \mathbb{Z}_p$, the product $r(x \otimes z) = x \otimes rz$ is well-defined. This yields a $\mathbb{Z}_p$-module structure of G_p.

(3) is a key characteristic of p-local groups, and we include a proof for the sake of completeness. Let q be a prime $\neq p$. The map $\sigma : G_p \to G_p$ given by $y \mapsto qy$ is clearly surjective, and since σ is a $\mathbb{Z}_p$-module homomorphism, $K = \ker \sigma$ is a submodule of G_p. If $y \in K$, then $y' = \frac{1}{q}y \in K$, and we have $y = qy' = 0$; hence, $K = 0$. Therefore, σ is an automorphism. □

This lemma leads us to multiple ways of characterizing p-local groups:

Theorem 1.17.6. *Let A be a group. Then the following are equivalent:*
1. *A is a $\mathbb{Z}_p$-module;*
2. *$A \cong A \otimes \mathbb{Z}_p$;*
3. *$A \cong C \otimes \mathbb{Z}_p$ for some group C;*
4. *For every $q \neq p$, A is uniquely q-divisible.*

Proof. Suppose A is a $\mathbb{Z}_p$-module. Define $f : A \times \mathbb{Z}_p \to A$ by $f(a, r) = ra$ for all $a \in A$ and $r \in \mathbb{Z}_p$. Then Theorem 1.17.1 defines a homomorphism $\phi : A \otimes \mathbb{Z}_p \to A$, which is clearly surjective. Let $a \otimes \frac{1}{n} \in \ker \phi$ for some $a \in A$ and n not divisible by p. Then $0 = f(a, \frac{1}{n}) = \frac{1}{n}a$, and hence, $a = 0$ and $a \otimes \frac{1}{n} = 0$. Thus, ϕ is an isomorphism, proving (1) implies (2). Clearly, (2) implies (3), and (3) implies (4) by Lemma 1.17.5(2). Now suppose (4). Let $a \in A$ and $m/n \in \mathbb{Z}_p$. Since n is relatively prime to p, there is a unique $b \in A$ such that $a = nb$. Define $\frac{m}{n}a = mb$. It is straightforward to verify that this defines a $\mathbb{Z}_p$-module, proving (4) implies (1). □

Note that (4) implies an important property of p-local groups: They have no q-torsion for any prime $q \neq p$.

As we will see later, heights and torsion elements are key characteristics for classifying groups. By converting G to G_p, we remove all torsion—except for p-torsion—and make all q-heights infinite for all primes $q \neq p$. We will soon see that localization preserves p-heights, exact sequences, and Ulm invariants.

The natural map

$$\begin{array}{rccl} v_p : & G & \to & G_p = G \otimes \mathbb{Z}_p \\ & x & \mapsto & x \otimes 1 \end{array}$$

is called the **localization map**.

Now we take a closer look at the mapping. First we note that for any $x \otimes \frac{1}{n} \in G_p$, we have $n(x \otimes \frac{1}{n}) \in v_p(G)$. On the other hand, suppose $p^k(x \otimes \frac{1}{n}) = y \otimes 1 \in v_p(G)$. Then since n and p^k are relatively prime, for some integers a and b, $x \otimes \frac{1}{n} = a(x \otimes 1) + b(y \otimes 1) \in v_p(G)$. This gives us the following characteristics of $v_p(G)$:

(i) If $p^k z \in v_p(G)$ for some $z \in G_p$, then $z \in v_p(G)$.

(j) Every element of G_p has a multiple in $v_p(G)$.

The following result will be needed:

Lemma 1.17.7. *Let $f : G \to H$ be a homomorphism and p a prime such that* $\ker f$ *is p-divisible and* $\operatorname{coker} f$ *has no p-torsion. Then we have the following:*
1. *If α is an ordinal, then $p^\alpha H/f(p^\alpha G)$ has no p-torsion.*
2. *f is p-height-preserving.*

Proof. The first assertion is proved using transfinite induction. For $\alpha = 0$, there is nothing to do, so let $\alpha > 0$ and assume that the claim holds for all ordinals $< \alpha$. Let $y \in p^\alpha H$ such that $p^n y \in f(p^\alpha G)$ for some positive integer n. We need to show that $y \in f(p^\alpha G)$.

If $\alpha - 1$ exists, we have $y = pz$ for some $z \in p^{\alpha-1}H$. Then $p^{n+1}z = p^n y \in f(p^\alpha G) \subseteq f(p^{\alpha-1}G)$; hence, $z \in f(p^{\alpha-1}G)$ by the induction hypothesis. It follows that $z = f(x)$ for some $x \in p^{\alpha-1}G$; hence, $y = pz = pf(x) = f(px) \in f(p^\alpha G)$, as desired. If α is a limit ordinal, then $p^n y \in f(p^\alpha G) = f(\bigcap_{\beta<\alpha} p^\beta G)$. Then for any $\beta < \alpha$, we have $y \in p^\beta H$ and $p^n y \in f(p^\beta G)$. Therefore, by the induction hypothesis, there is an $x_\beta \in p^\beta G$ such that $y = f(x_\beta)$. Letting $\gamma < \alpha$, we have $f(x_\gamma - x_\beta) = 0$; hence, $x_\gamma - x_\beta \in \ker f = p^\infty \ker f \subseteq p^\infty G$ since $\ker f$ is p-divisible. Consequently, $|x_\gamma|_p \geq |x_\gamma - x_\beta|_p \wedge |x_\beta|_p = |x_\beta|_p \geq \beta$, and we obtain $x_\gamma \in \bigcap_{\beta<\alpha} p^\beta G = p^\alpha G$. Therefore, $y = f(x_\gamma) \in f(p^\alpha G)$ and the induction is complete.

To prove (2), let $x \in G$ and suppose that $y = f(x)$ has p-height α. We need to show that $|x|_p \geq \alpha$. Assuming that $|x|_p = \beta$, an ordinal $< \alpha$, we obtain $y \in p^{\beta+1}H$ and $py = f(px) \in f(p^{\beta+1}G)$. Since $p^{\beta+1}H/f(p^{\beta+1}G)$ has no p-torsion by (1), there is a $z \in p^{\beta+1}G$ such that $y = f(z)$. Then $f(x - z) = 0$, so $x - z \in \ker f$, which is p-divisible by assumption. It follows that $|x|_p \geq |x - z|_p \wedge |z|_p = |z|_p \geq \beta + 1$, a contradiction. Therefore, f is p-height-preserving. □

Next, we prove some basic properties of the map $v_p : G \to G_p$ (cf. [28, p. 101]).

Lemma 1.17.8. *Let $v_p : G \to G_p$ be the localization map. Then we have the following:*
1. $\operatorname{coker} v_p$ *is torsion with no p-torsion, hence p-divisible.*
2. $t(G_p) \subseteq v_p(G)$.
3. $\ker v_p$ *is torsion with no p-torsion, hence p-divisible. In particular,* $\ker v_p = \bigoplus_{q \in \mathbb{P}\setminus\{p\}} t_q G$.
4. *v_p maps $t_p G$ isomorphically onto $t(G_p)$.*
5. *v_p is p-height-preserving.*

Proof. By (i) and (j), $\operatorname{coker} \nu_p$ is torsion with no p-torsion; hence, $\operatorname{coker} \nu_p = \bigoplus_{q \in \mathbb{P}\setminus\{p\}} t_q(\operatorname{coker} \nu_p)$ is p-divisible, proving (1).

To verify (2), let $x \in t(G_p)$. Then, by Lemma 1.17.5(3), $p^n x = 0 \in \nu_p(G)$ for some positive integer n. Therefore, by (1), we have $x \in \nu_p(G)$, as claimed.

We let $a \in \ker \nu_p$ and consider the natural map $\mu_p : \langle a \rangle \to \langle a \rangle_p$. By (1), $\operatorname{coker} \mu_p \cong \langle a \rangle_p$ is torsion with no p-torsion. Now if $\langle a \rangle \cong \mathbb{Z}$, then $\langle a \rangle_p \cong \mathbb{Z}_p$, which is impossible. Thus, $\langle a \rangle$ is finite; hence, $\langle a \rangle \cong \mathbb{Z}/p^r\mathbb{Z} \times \mathbb{Z}/n\mathbb{Z}$ for some nonnegative integer r and positive integer n such that $p \nmid n$. Since $\langle a \rangle_p \cong (\mathbb{Z}/p^r\mathbb{Z} \times \mathbb{Z}/n\mathbb{Z}) \otimes \mathbb{Z}_p \cong (\mathbb{Z}/p^r\mathbb{Z} \otimes \mathbb{Z}_p) \times (\mathbb{Z}/n\mathbb{Z} \otimes \mathbb{Z}_p) \cong (\mathbb{Z}/p^r\mathbb{Z}) \times (\mathbb{Z}/n\mathbb{Z} \otimes \mathbb{Z}_p)$ has no p-torsion, we obtain $r = 0$, and therefore, $na = 0$. Thus, $\ker \nu_p$ is torsion with no p-torsion; hence, $\ker \nu_p \subseteq \bigoplus_{q \in \mathbb{P}\setminus\{p\}} t_q G$. Now let $x \in t_q G$ for some prime $q \neq p$. Then $q^m x = 0$ for some positive integer m, and we have $q^m(x \otimes 1) = q^m x \otimes 1 = 0$; hence, $x \otimes 1 = 0$ since G_p is a p-local group. Consequently, $x \in \ker \nu_p$. Thus, $\ker \nu_p = \bigoplus_{q \in \mathbb{P}\setminus\{p\}} t_q G$ and the proof of (3) is complete.

To prove (4), consider the homomorphism $\phi = \nu_p \restriction t_p G : t_p G \to t(G_p)$. By (3) we have $\ker \phi = \ker \nu_p \cap t_p G = 0$; hence, ϕ is injective. Letting $z \in t(G_p)$, we have $z = x \otimes 1$ for some $x \in G$ by (2). Since $p^n x \otimes 1 = p^n z = 0$ for some positive integer n, $p^n x$ is an element of $\ker \nu_p$, which is p-divisible by (3). Therefore, there is a $y \in \ker \nu_p$ such that $p^n x = p^n y$. It follows that $x - y \in t_p G$, and $\phi(x - y) = (x - y) \otimes 1 = x \otimes 1 - y \otimes 1 = z$. Therefore, ϕ is surjective.

(5) follows from Lemma 1.17.7(2) since $\ker \nu_p$ is p-divisible by (3) and $\operatorname{coker} \nu_p$ has no p-torsion by (1). □

Dieudonné [17] proved that tensoring with a torsion-free group preserves exactness of short exact sequences. We will need the following special case, which shows that for any subgroup A of a group G, A_p can be viewed as a subgroup of G_p, and $(G/A)_p$ may be identified with G_p/A_p:

Theorem 1.17.9. *Let* $0 \to A \xrightarrow{\alpha} B \xrightarrow{\beta} C \to 0$ *be an exact sequence. Then p-localization by a prime p preserves exactness; that is, the induced sequence*

$$0 \to A_p \xrightarrow{\alpha_*} B_p \xrightarrow{\beta_*} C_p \to 0$$

is exact.

Proof. By Theorem 1.17.4, it suffices to prove that α_* is injective. Let $x \in \ker \alpha_*$ and write $x = a \otimes \frac{1}{n}$ for some $a \in A$ and positive integer n with $p \nmid n$. Then $\alpha(a) \otimes 1 = n(\alpha(a) \otimes \frac{1}{n}) = n\alpha_*(x) = 0$; hence, $\alpha(a) \in \ker \mu_p$, where $\mu_p : B \to B_p$ denotes the localization map. By Lemma 1.17.8(3), there is a positive integer m such that $p \nmid m$ and $m\alpha(a) = 0$. This yields $ma = 0$ since $\alpha(ma) = 0$; hence, $x = ma \otimes \frac{1}{mn} = 0$. Therefore, α_* is injective. □

Finally, we determine how p-localization of a group G affects (relative) Ulm invariants. Recall that the relative Ulm invariants for a subgroup A of G are defined by

$$u_{\alpha,p}(G, A) = \dim(p^\alpha G)[p]/[(p^\alpha G)[p] \cap (A + p^{\alpha+1} G)] \quad (\alpha \text{ an ordinal})$$

and

$$u_{\infty,p}(G,A) = \dim(p^\infty G)[p]/(A \cap p^\infty G)[p].$$

First, we show the following:

Lemma 1.17.10. *Let G be a group. Then $p^\alpha(G_p) = (p^\alpha G)_p$ for all primes p and ordinals α.*

Proof. Letting $z \in p^\alpha(G_p)$, we write $z = x \otimes \frac{1}{n}$ for some $x \in G$ and positive integer n with $p \nmid n$. Then $nz = x \otimes 1 \in p^\alpha(G_p)$, so by Lemma 1.17.8(5), we have $x \in p^\alpha G$. This yields $z = x \otimes \frac{1}{n} \in (p^\alpha G)_p$. Conversely, let $z \in (p^\alpha G)_p$. Then there is an $x \in p^\alpha G$ such that $z = x \otimes \frac{1}{n}$ for some positive integer n with $p \nmid n$. Since the map $v_p : G \to G_p$ does not decrease p-heights, we have $x \otimes 1 \in p^\alpha(G_p)$; thus, $z = \frac{1}{n}(x \otimes 1) \in p^\alpha(G_p)$. □

The following result on relative Ulm invariants is taken from Fuchs [28, Lemma 16] and will be needed in Section 3.7.

Lemma 1.17.11. *Let A be a subgroup of G, p a prime, and $\alpha \in \mathrm{Ord}_\infty$. Then $u_{\alpha,p}(G,A) = u_{\alpha,p}(G_p, A_p)$.*

Proof. Let α be an ordinal. By Lemma 1.17.8(4), $v_p : G \to G_p$ maps $G[p]$ isomorphically onto $G_p[p]$. Thus, v_p maps $(A + p^{\alpha+1}G)[p]$ onto $(A + p^{\alpha+1}G)_p[p]$ which coincides with $(A_p + p^{\alpha+1}(G_p))[p]$ by Lemma 1.17.10. To show that v_p maps $(p^\alpha G)[p]$ onto $(p^\alpha(G_p))[p]$, let $y \in (p^\alpha(G_p))[p]$. Then by Lemma 1.17.8(4), there is an $x \in G[p]$ such that $v_p(x) = y$, and we have $x \in p^\alpha G$ by Lemma 1.17.8(5). Consequently, the underlying groups of the invariants $u_{\alpha,p}(G,A)$ and $u_{\alpha,p}(G_p, A_p)$ are isomorphic; hence, the invariants coincide.

Finally, $u_{\infty,p}(G,A) = u_{\infty,p}(G_p, A_p)$ since v_p maps $A[p]$ onto $A_p[p]$ and $(p^\infty G)[p]$ onto $(p^\infty(G_p))[p]$. □

Exercises.

1. For $m, n \in \mathbb{N}$, describe $\mathbb{Z}/m\mathbb{Z} \otimes \mathbb{Z}/n\mathbb{Z}$.
2. There is a natural isomorphism $\mathrm{Hom}(A \otimes B, C) \xrightarrow{\sim} \mathrm{Hom}(A, \mathrm{Hom}(B, C))$.
3. Let A' and C' be subgroups of A and C, respectively. Then $a' \otimes b' \mapsto a' \otimes b'$ induces a homomorphism $\phi : A' \otimes C' \to A \otimes C$, but ϕ need not be injective.
4. If A is a p-isotype subgroup of G, then A_p is a p-isotype subgroup of G_p.
5. If G is a torsion group, then $G_p \cong t_p G$.
6. The localization map $v_p : G \to G_p$ has the following universal property: If H is a $\mathbb{Z}_p$-module and $\phi : G \to H$ is any homomorphism, then there is a unique homomorphism $\psi : G_p \to H$ such that $\phi = \psi v_p$.

2 Simply presented groups

We introduce a key characteristic of subgroups, called nice, and a major class of groups, called simply presented. We develop several essential simply presented groups, including mixed groups. The simply presented torsion groups have many and varied equivalent characterizations. The class also has some useful characteristics, such as being closed under direct summands. Simply presented torsion groups are classified using the Ulm invariants.

2.1 *p*-Nice submodules

As the proof of Ulm's Theorem 1.14.2 shows, proper elements are useful to extend p-height-preserving isomorphisms between finite subgroups. Repeated application of this technique results in an isomorphism between two countable p-groups if they have the same Ulm invariants. This gives rise to the following definition of submodules A, sharing the handy property, that every coset of A possesses an element that is proper with respect to A. Let G be an R-module and $p \in R$ a prime. Then a submodule A of G is called **p-nice (in G)** if every coset of A contains an element that is p-proper with respect to A. That is, if for every $g \in G$, there is an $x \in g + A$ such that $|x|_p \geq |g + a|_p$ for all $a \in A$. In the case where G is a torsion $\mathbb{Z}_p$-module (that is, a p-group), A is simply called **nice (in G)**. The definition of niceness was introduced by Hill [47]. We note the following basic properties, which are easily verified:

(a) If A is p-nice in G and B is a submodule of G with B/A finite, then B is p-nice in G.
(b) Suppose that A_i are submodules of G_i ($i \in I$). Then $\bigoplus_{i \in I} A_i$ is p-nice in $\bigoplus_{i \in I} G_i$ if and only if each A_i is p-nice in G_i.

In terms of p-groups, most results of this section can be found in Fuchs [29]. We start with an application of Lemma 1.12.2:

Proposition 2.1.1. *Let A be a submodule of an R-module G, and let $p \in R$ be a prime. Then the following are equivalent:*

1. *A is p-nice in G;*
2. *For each $x \in G \setminus A$, there exists an $a \in A$ such that $|x + A|_p^{G/A} = |x + a|_p^G$;*
3. *$p^\alpha(G/A) = (p^\alpha G + A)/A$ for all ordinals α.*

We use the above proposition to prove the following results:

Lemma 2.1.2. *Let G be an R-module and $p \in R$ a prime. If β is an ordinal, then $p^\beta G$ is p-nice in G.*

Proof. If $\alpha \leq \beta$, then $p^\alpha(G/p^\beta G) = p^\alpha G/p^\beta G$ by Lemma 1.10.5. If $\alpha \geq \beta$, $p^\alpha(G/p^\beta G) = 0$ follows. Thus, $p^\alpha(G/p^\beta G) = (p^\alpha G + p^\beta G)/p^\beta G$ for all ordinals α. □

https://doi.org/10.1515/9783110427684-002

Proposition 2.1.3. *Assume that A is a submodule of an R-module G, $p \in R$ a prime, and β an ordinal. Let $\overline{G} = G/p^\beta G$ and $\overline{A} = (A + p^\beta G)/p^\beta G$. Then we have the following:*
1. *If A is p-nice in G, then $\overline{A}$ is p-nice in $\overline{G}$.*
2. *If A is p-isotype in G, then $\overline{A}$ is p-isotype in $\overline{G}$.*

Proof. Suppose that A is p-nice in G, and let $\alpha \le \beta$. Using Lemma 1.10.5, we have

$$\begin{aligned} p^\alpha(\overline{G}/\overline{A})/[(p^\alpha\overline{G}+\overline{A})/\overline{A})] &= p^\alpha[(\overline{G}/\overline{A})/(p^\alpha\overline{G}+\overline{A})/\overline{A})] \\ &= p^\alpha[(\overline{G}/\overline{A})/(((p^\alpha G + A)/p^\beta G)/\overline{A})] \\ &\cong p^\alpha[G/(p^\alpha G + A)] \\ &\cong p^\alpha[(G/A)/((p^\alpha G + A)/A)] \\ &= p^\alpha(G/A)/[(p^\alpha G + A)/A] = 0. \end{aligned}$$

If $\alpha > \beta$, then $\alpha = \beta + \gamma$ for some ordinal γ, and we obtain $p^\alpha(\overline{G}/\overline{A}) = p^\gamma p^\beta(\overline{G}/\overline{A}) = p^\gamma((p^\beta\overline{G}+\overline{A})/\overline{A}) = 0$. Therefore, $\overline{A}$ is p-nice in $\overline{G}$.

Now assume that A is p-isotype in G and let $a \in A \setminus p^\beta G$. Since $p^\beta G$ is p-nice (Lemma 2.1.2), there is a $g \in p^\beta G$ such that $|a + p^\beta G|_p^{\overline{G}} = |a + g|_p^G$ by Proposition 2.1.1. Then $|a + g|_p^G = |a|_p^G \wedge |g|_p^G = |a|_p^G = |a|_p^A \le |a + p^\beta G|_p^{\overline{A}} \le |a + p^\beta G|_p^{\overline{G}}$; thus, $\overline{A}$ is p-isotype in $\overline{G}$. □

The following facts will be useful:

Lemma 2.1.4. *Suppose that A and B are submodules of an R-module G such that $A \subseteq B$, and let $p \in R$ be a prime. Then we have the following:*
1. *If B is p-nice in G, then B/A is p-nice in G/A.*
2. *If A is p-nice in G and B/A is p-nice in G/A, then B is p-nice in G.*

Proof. To prove (1), we show by transfinite induction that

$$p^\alpha((G/A)/(B/A)) \subseteq (p^\alpha(G/A) + B/A)/(B/A)$$

for all ordinals α. The case $\alpha = 0$ being trivial, let $\alpha > 0$, and assume that the inclusion holds for all ordinals $\beta < \alpha$. Suppose that $x + A + B/A \in p^\alpha((G/A)/(B/A))$. If $\alpha - 1$ exists, then there is a $y + A + B/A \in p^{\alpha-1}((G/A)/(B/A))$ such that

$$py + A + B/A = x + A + B/A. \tag{$*$}$$

By the induction hypothesis, we have $y + A \in p^{\alpha-1}(G/A) + B/A$; consequently, $x + A \in p^\alpha(G/A) + B/A$. Finally, assume that α is a limit ordinal, and let $\beta < \alpha$. Then there is a $y + A + B/A \in p^\beta((G/A)/(B/A))$ satisfying ($*$). By the induction hypothesis, we obtain $y + A \in p^\beta(G/A) + B/A$; thus, there is a $b \in B$ such that $x + b + A \in p^\beta(G/A)$. But then $x + B \in p^\beta(G/B)$, which shows that $x + B \in p^\alpha(G/B)$. Since B is p-nice in G, there is a $b' \in B$ such that $x + b' \in p^\alpha G$. Thus, $x + b' + A \in p^\alpha(G/A)$, and again, it follows that $x + A \in p^\alpha(G/A) + B/A$.

To show the second assertion, let $x \in G\setminus B$, and use Proposition 2.1.1(2) repeatedly. Then $|x+A+B/A|_p^{(G/A)/(B/A)} = |x+b+A|_p^{G/A}$ for some $b \in B$ and $|x+b+A|_p^{G/A} = |x+b+a|_p^G$ for some $a \in A$. This yields $|x+b+a|_p^G = |x+B|_p^{G/B}$; therefore, B is p-nice in G. □

Niceness is not transitive (see, for example, Fuchs [29] Chapter 11, Section 2, Exercise 7), but we have the following:

Lemma 2.1.5. *Let G be an R-module, $p \in R$ a prime, and suppose that A, B and C are submodules of G such that $A \subseteq B \subseteq C$. If every coset $b + A$ ($b \in B$) has an element of maximal p-height and every coset $c + B$ ($c \in C$) has an element of maximal p-height, then every coset $z + A$ ($z \in C$) has an element of maximal p-height. Here all p-heights are computed in the module G.*

Proof. Let $z \in C$. Then the coset $z + A$ has an element of maximal p-height if $z \in B$, so let us assume that $z \notin B$. By assumption, there is a $b \in B$ such that $|z+b| \geq |z+x|$ for all $x \in B$. If $b \in A$, the assertion follows. If $b \notin A$, there is an $a \in A$ such that $|b+a| \geq |b+y|$ for all $y \in A$. Let $c \in A$. Then $|z-a| = |b+a-(z+b)| \geq |b+a| \wedge |z+b| \geq |b-c| \wedge |z+c| = |z+b-(z+c)| \wedge |z+c| \geq |z+b| \wedge |z+c| = |z+c|$. Consequently, $z-a$ is an element of maximal p-height in the coset $z + A$. □

Lemma 2.1.6. *Let G be an R-module, $p \in R$ a prime, and β an ordinal. If A is a submodule of $p^\beta G$, then A is p-nice in $p^\beta G$ if and only if A is p-nice in G.*

Proof. Suppose A is p-nice in $p^\beta G$. Let $x \in G$. If $x \notin p^\beta G$, then for all $a \in A$ $|x+a|^G = |x|^G \wedge |a|^G = |x|^G$, proving x proper with respect to A. So suppose $x \in p^\beta G$. Since A is p-nice in $p^\beta G$, there is some $a \in A$ such that $|x+a|^{p^\beta G} \geq |x+a'|^{p^\beta G}$ for all $a' \in A$. Since $x+a$ and $x+a' \in p^\beta G$, we have $|x+a|^G = \beta + |x+a|^{p^\beta G} \geq \beta + |x+a'|^{p^\beta G} = |x+a'|^G$ by Lemma 1.10.4(2). Now suppose A is p-nice in G. Let $x \in p^\beta G$. Then there is $a \in A$ with $|x+a|^G \geq |x+a'|^G$ for all $a' \in A$. Then by Lemma 1.10.4(2), $\beta + |x+a|^{p^\beta G} \geq \beta + |x+a'|^{p^\beta G}$; so by left cancellation of ordinals, $|x+a|^{p^\beta G} \geq |x+a'|^{p^\beta G}$ for all $a' \in A$. □

Exercises.

1. If A is a nice subgroup of a reduced p-group G, then $A = G$ whenever G/A is divisible.
2. A pure subgroup of a p-group needs not be nice.
3. (Hill [47]) Let A be a submodule of an R-module G, $p \in R$ a prime, and β an ordinal.
   (a) If A is p-nice in G, then $A \cap p^\beta G$ is p-nice in $p^\beta G$.
   (b) Let $\overline{A}$ and $\overline{G}$ be as in Proposition 2.1.3. If $\overline{A}$ is p-nice in $\overline{G}$ and $A \cap p^\beta G$ is p-nice in $p^\beta G$, then A is p-nice in G.

2.2 Families of subgroups and the third axiom of countability

In [47], Hill introduced the following axioms of countability for p-groups:

Axiom 1. A group is countable.

Axiom 2. A group is a direct sum of countable groups.

Axiom 3. A group G has a system $\mathcal{C}$ of nice subgroups satisfying the following conditions:

(H1) $0 \in \mathcal{C}$;
(H2) if $N_i \in \mathcal{C}$ $(i \in I)$, then $\sum_{i \in I} N_i \in \mathcal{C}$;
(H3) if $N \in \mathcal{C}$ and S is any countable subset of G, then there is an $M \in \mathcal{C}$ such that $\langle S, N \rangle \subseteq M$ and $|M/N| \leq \aleph_0$.

Clearly we have the following:

$$\text{Axiom 1} \Rightarrow \text{Axiom 2} \Rightarrow \text{Axiom 3}.$$

Axiom 3 is also called **the third axiom of countability** or **Hill's condition.** Given any R-module G, a system $\mathcal{C}$ of submodules of G satisfying (H1)–(H3) is called an $H(\aleph_0)$**-family** in G (see Fuchs–Hill [30], in which $H(\kappa)$-families were introduced for arbitrary infinite cardinals κ). In the case where G is a p-group, then an $H(\aleph_0)$-family of nice subgroups is simply called **a nice system.** For instance, a divisible p-group G has a nice system since every subgroup of G is nice. In [38], Griffith replaced (H2) by the weaker condition

(G2) $\mathcal{C}$ is closed under ascending unions.

A system $\mathcal{C}$ of subgroups of a group G is called a $G(\aleph_0)$**-family** in G if it satisfies (H1), (G2), and (H3). Obviously, every $H(\aleph_0)$-family in G is a $G(\aleph_0)$-family. A well-ordered ascending chain of subgroups $N_0 \subseteq N_1 \subseteq \cdots \subseteq N_\alpha \subseteq \cdots$ $(\alpha < \lambda)$ is called **smooth** if for every limit ordinal $\alpha < \lambda$ we have $N_\alpha = \bigcup_{\beta<\alpha} N_\beta$. A smooth chain of subgroups

$$0 = N_0 \subseteq N_1 \subseteq \cdots \subseteq N_\alpha \subseteq \cdots \subseteq N_\lambda = G$$

is called an $F(\aleph_0)$**-family** in G if $|N_{\alpha+1}/N_\alpha| \leq \aleph_0$ for all ordinals $\alpha < \lambda$. If in addition each quotient $N_{\alpha+1}/N_\alpha$ is cyclic of infinite or prime order, this chain is said to be a **composition series for** G. Following Fuchs [27], we call a composition series $\{N_\alpha\}_{\alpha\leq\lambda}$ for a p-group G **nice** if each subgroup N_α is nice in G. The following fact can be found in [27]:

Lemma 2.2.1. *Let G be a p-group and α an ordinal. If $p^\alpha G$ and $G/p^\alpha G$ have a nice composition series, then so does G.*

Proof. The assertion follows from Lemmas 2.1.4(2) and 2.1.6. □

The next two observations follow from remarks in Megibben–Ullery [116] and Fuchs–Hill [30], respectively:

Lemma 2.2.2. *Every $G(\aleph_0)$-family contains a $G(\aleph_0)$-family of pure subgroups.*

Proof. Suppose that $\mathcal{C}$ is a $G(\aleph_0)$-family in G, and let $\mathcal{C}' = \{A \in \mathcal{C} : A \text{ is pure in } G\}$. Clearly, $\mathcal{C}'$ satisfies (H1) and (G2). To prove that (H3) holds for $\mathcal{C}'$, let S be a countable subset of G and $N \in \mathcal{C}'$. Then there is a $B_1 \in \mathcal{C}$ containing $\langle S, N\rangle$ such that $|B_1/N| \leq \aleph_0$. By Proposition 1.7.3, B_1/N is contained in some countable pure subgroup C_1/N of G/N, and by Lemma 1.7.2, C_1 is pure in G. Since C_1 is a countable extension of N, there is a $B_2 \in \mathcal{C}$ containing C_1 such that $|B_2/N| \leq \aleph_0$. Again, B_2/N is contained in some countable pure subgroup C_2/N of G/N. Then C_2 is pure in G, et cetera. This yields an ascending sequence

$$B_1 \subseteq C_1 \subseteq B_2 \subseteq C_2 \subseteq \cdots$$

of subgroups of G, and we define $M = \bigcup_{i=1}^{\infty} B_i$. Then $M \in \mathcal{C}$ and $|M/N| \leq \aleph_0$. As the union of an ascending sequence of pure subgroups, M is pure in G; hence, $M \in \mathcal{C}'$. Thus, $\mathcal{C}'$ is a $G(\aleph_0)$-family in G. □

Lemma 2.2.3. *Every $G(\aleph_0)$-family contains an $F(\aleph_0)$-family.*

Proof. Let G be a group and suppose that $\mathcal{C}$ is a $G(\aleph_0)$-family in G. We may assume that G is infinite and write $G = \{g_\beta : \beta < \lambda\}$ for some limit ordinal λ. Starting with $N_0 = 0$, it is easy to construct inductively a well-ordered ascending chain $\{N_\alpha\}_{\alpha \leq \lambda}$ of subgroups of G such that (i) $N_\alpha \in \mathcal{C}$, (ii) $\{g_\beta : \beta < \alpha\} \subseteq N_\alpha$, (iii) $|N_\alpha/N_{\alpha-1}| \leq \aleph_0$ if $\alpha - 1$ exists, and (iv) $N_\alpha = \bigcup_{\beta<\alpha} N_\beta$ if α is a limit ordinal. Then $N_\lambda = G$; thus, $\{N_\alpha\}_{\alpha\leq\lambda}$ is an $F(\aleph_0)$-family in G. □

Assume that G is a p-group. Since finite extensions of nice subgroups of G are nice, we may turn an $F(\aleph_0)$-family of nice subgroups for G into a nice composition series by adding intermediate subgroups as necessary, and we have the following:

Corollary 2.2.4. *If a p-group G has a nice system, then G has a nice composition series.*

The following fact will be useful:

Lemma 2.2.5. *Let G be a group, and let A be a countable subgroup of G. If $\mathcal{C} = \{N_i/A : i \in I\}$ is an $H(\aleph_0)$-family in G/A, then $\mathcal{C}' = \{N_i : i \in I\} \cup \{0\}$ is an $H(\aleph_0)$-family in G.*

Proof. It is clear that $\mathcal{C}'$ satisfies (H1) and (H2). Let $N \in \mathcal{C}'$ and S a countable subset of G. If $N = 0$, there is an $M/A \in \mathcal{C}$ containing $\langle S, A\rangle/A$ such that $|(M/A)/(A/A)| \leq \aleph_0$; thus, $M \in \mathcal{C}'$, $\langle S, N\rangle \subseteq M$ and $|M/N| = |M| = |M/A| \cdot |A| \leq \aleph_0$. In the case where $N = N_i$ for some $i \in I$, we have $N/A \in \mathcal{C}$, so there is an $M/A \in \mathcal{C}$ containing $\langle S, A\rangle/A$ and N/A such that $|(M/A)/(N/A)| \leq \aleph_0$. Again, we have $M \in \mathcal{C}'$, $\langle S, N\rangle \subseteq M$ and $|M/N| \leq \aleph_0$. Thus, $\mathcal{C}'$ satisfies (H3). □

In view of Lemma 2.1.4(2) we obtain the following:

Corollary 2.2.6. *Let G be a p-group, and assume that A is a countable nice subgroup of G. If G/A has a nice system, then so does G.*

Let $\mathcal{B}$ be a nonempty collection of subsets of a given group G. We say that a subgroup H of G is **$\mathcal{B}$-torsion** if $H/\langle X \cap H\rangle$ is torsion for some $X \in \mathcal{B}$. Then we have the following:

Proposition 2.2.7. *Let G be a group and $\mathcal{B}$ any nonempty collection of subsets of G. Then the following conditions are equivalent:*
1. *G is $\mathcal{B}$-torsion;*
2. *G has an $H(\aleph_0)$-family of $\mathcal{B}$-torsion subgroups;*
3. *G has a $G(\aleph_0)$-family of $\mathcal{B}$-torsion subgroups;*
4. *G has an $F(\aleph_0)$-family of $\mathcal{B}$-torsion subgroups.*

Proof. By Lemma 2.2.3, we need only to verify that (1) implies (2). Therefore, suppose that $G/\langle X\rangle$ is torsion for some $X \in \mathcal{B}$. Let $\mathcal{C}$ be the set of all subgroups A of G such that $A/\langle X \cap A\rangle$ is torsion. We show that $\mathcal{C}$ is an $H(\aleph_0)$-family in G. Clearly, $0 \in \mathcal{C}$. To prove (H2), let $\{N_i : i \in I\}$ be a collection of groups in $\mathcal{C}$. If $x \in N = \sum_{i\in I} N_i$, then $x = \sum_{i\in J} x_i$ for a finite subset J of I and elements $x_i \in N_i$; hence, there is a positive integer n such that $nx_i \in \langle X \cap N_i\rangle$ for all $i \in J$. Then

$$nx \in \sum_{i\in J}\langle X \cap N_i\rangle \subseteq \langle X \cap N\rangle;$$

hence, $N/\langle X \cap N\rangle$ is torsion, and we have $N \in \mathcal{C}$. Therefore, (H2) is true. To prove (H3), let $S = \{s_i : i < \omega\}$ be a countable subset of G and $H \in \mathcal{C}$. For every $i < \omega$, there is a positive integer n_i such that $n_i s_i \in \langle X_i\rangle$ for some finite subset X_i of X. Then $H_i = \langle s_i, X_i\rangle$ is in $\mathcal{C}$ since $H_i/\langle X \cap H_i\rangle$ is torsion. It follows that the group $M = H + \sum_{i<\omega} H_i$ is a countable extension of H containing S. By (H2), we have $M \in \mathcal{C}$; thus, (H3) is satisfied. Consequently, $\mathcal{C}$ is an $H(\aleph_0)$-family in G. □

For an application of the above result, see Corollary 3.1.1. The next result follows from the proof of [47, Theorems 5.2 and 5.3]:

Theorem 2.2.8 (Hill [47]). *Let G be a group. Then we have the following:*
1. *If $G = \bigoplus_{i\in I} G_i$, where each group G_i ($i \in I$) has an $H(\aleph_0)$-family $\mathcal{C}_i$, then $\mathcal{C} = \{\bigoplus_{i\in I} N_i : N_i \in \mathcal{C}_i$ for all $i \in I\}$ is an $H(\aleph_0)$-family in G.*
2. *If G has an $H(\aleph_0)$-family $\mathcal{C}$ and $G = A \oplus B$ for some subgroups A and B of G, then $\mathcal{C}' = \{N' : N'$ is a subgroup of A with $N = N' \oplus (N \cap B)$ for some $N \in \mathcal{C}\}$ is an $H(\aleph_0)$-family in A.*

Proof. First, we assume that $G = \bigoplus_{i\in I} G_i$ as in (1) and will show that the set $\mathcal{C} = \{\bigoplus_{i\in I} N_i : N_i \in \mathcal{C}_i$ for all $i \in I\}$ is an $H(\aleph_0)$-family in G. Clearly, $0 \in \mathcal{C}$, and $\mathcal{C}$ is closed under arbitrary group unions. To verify (H3), let S be a countable subset of G and $N \in \mathcal{C}$, say $N = \bigoplus_{i\in I} N_i$, where $N_i \in \mathcal{C}_i$ for all $i \in I$. Letting $\pi_i : G \to G_i$ be the projection map and $S_i = \pi_i(\langle S\rangle)$, we obtain a countable subset J of I such that $S \subseteq \sum_{i\in J} S_i$. Then for every $i \in J$ there is an $M_i \in \mathcal{C}_i$ such that $\langle S_i, N_i\rangle \subseteq M_i$ and $|M_i/N_i| \le \aleph_0$. For $i \in I \setminus J$, we set $M_i = N_i$. Then $M = \bigoplus_{i\in I} M_i$ satisfies $M \in \mathcal{C}$, $\langle S, N\rangle \subseteq \sum_{i\in I}\langle S_i, N_i\rangle \subseteq M$ and $|M/N| = |\bigoplus_{i\in J} M_i/N_i| \le \aleph_0$; hence, (H3) holds.

To prove the second assertion, suppose that $G = A \oplus B$ has an $H(\aleph_0)$-family $\mathcal{C}$. We will show that

$$\mathcal{C}' = \{N' : N' \text{ is a subgroup of } A \text{ with } N = N' \oplus (N \cap B) \text{ for some } N \in \mathcal{C}\}$$

is an $H(\aleph_0)$-family in A. It is clear that $0 \in \mathcal{C}'$. Letting $N_i' \in \mathcal{C}'$ $(i \in I)$, we have $N_i = N_i' \oplus (N_i \cap B)$ for some $N_i \in \mathcal{C}$ $(i \in I)$. Then $\sum_{i\in I} N_i \subseteq (\sum_{i\in I} N_i') \oplus \sum_{i\in I}(N_i \cap B)$. Thus,

$$\sum_{i\in I} N_i' \oplus \left(\left(\sum_{i\in I} N_i\right) \cap B\right) = \sum_{i\in I} N_i' \oplus \sum_{i\in I}(N_i \cap B) = \sum_{i\in I} N_i \in \mathcal{C},$$

and therefore, $\sum_{i\in I} N_i' \in \mathcal{C}'$. It follows that (H2) holds.

To show that $\mathcal{C}'$ satisfies (H3), let S be a countable subset of A and $N' \in \mathcal{C}'$. Then $N = N' \oplus (N \cap B)$ for some $N \in \mathcal{C}$, so there exists an $N_1 \in \mathcal{C}$ such that $\langle S, N\rangle \subseteq N_1$ and $|N_1/N| \le \aleph_0$. This yields subgroups A_1 of A and B_1 of B such that

$$N_1 \subseteq A_1 \oplus B_1 \text{ and } |(A_1 \oplus B_1)/N| \le \aleph_0.$$

Since A_1 is a countable extension of $N' \subseteq N$, there is an $N_2 \in \mathcal{C}$ such that $\langle A_1, N\rangle \subseteq N_2$ and $|N_2/N| \le \aleph_0$. Then there are subgroups A_2 of A and B_2 of B such that $N_2 \subseteq A_2 \oplus B_2$ and $|(A_2 \oplus B_2)/N| \le \aleph_0$, et cetera. This yields sequences of groups $N_i \in \mathcal{C}$, $A_i \subseteq A$ and $B_i \subseteq B$ such that

$$|N_i/N| \le \aleph_0, \quad N_i \subseteq A_i \oplus B_i \quad \text{and} \quad \langle A_i, N\rangle \subseteq N_{i+1} \quad (i = 1, 2, \dots).$$

Let $M = \sum_{i=1}^{\infty} N_i$. Then $M \in \mathcal{C}$ and $M = (M \cap A) \oplus (M \cap B)$; hence, $M \cap A \in \mathcal{C}'$. Finally, $\langle S, N'\rangle \subseteq M \cap A$ and $|(M \cap A)/N'| = |M/(N' \oplus (M \cap B))| \le |M/N| \le \aleph_0$. Therefore, $\mathcal{C}'$ is an $H(\aleph_0)$-family in A. □

The class of p-groups with nice systems possesses the following important property:

Corollary 2.2.9 (Hill [47]). *The class of p-groups with nice systems is closed under arbitrary direct sums and direct summands.*

Proof. By Section 2.1(b), the assertion follows from Theorem 2.2.8. □

For additional applications of Theorem 2.2.8, see Corollaries 3.8.20, 3.9.3 and 3.12.14.

Exercises.

1. If a p-group G has a nice system and A is a countable subgroup of G, then G/A has a nice system.
2. Let G be a p-group and $n \in \mathbb{N}$. Show the following:
   (a) p^nG has a nice system if and only if G has one.
   (b) p^nG has a nice composition series if and only if G has one.

3. (Hill [47]) Let G be a p-group and β any ordinal. If $G/p^\beta G$ has a nice system and if $p^{\beta+1}G = 0$, then G has a nice system. [Hint: Section 2.1, Exercise 3(b).]
4. Show that every torsion-free group has a $G(\aleph_0)$-family of pure subgroups.

2.3 Classification of p-groups with nice systems

Suppose that A and B are p-nice submodules of R-modules G and H, respectively, such that there is a p-height-preserving isomorphism $\phi : A \to B$. Under which conditions does ϕ extend to an isomorphism between G and H? The next extension result was proved by Hill [47] and Walker [148] assuming that G/A and H/B are reduced p-groups. It was then extended by Warfield [154] to include the nonreduced case. As Warfield mentioned in [154], the arguments carry over without significant change.

As in the proof of Ulm's theorem, we first need a lemma to show that we can extend an isomorphism of submodules to a single element. Building on the earlier Lemma 1.14.1, we choose the image of that element to satisfy certain conditions and maintain the Ulm invariants.

Here we assume that G/A and H/B are arbitrary (not necessarily reduced) p-groups, slightly extending [148, Lemma 2.6]. It will be convenient to write $u_\alpha(G,A) = \dim(p^\alpha G)[p]/A(\alpha)$ if α is an ordinal, and $u_\infty(G,A) = \dim(p^\infty G)[p]/A(\infty)$, where

$$A(\alpha) = (p^\alpha G)[p] \cap (A + p^{\alpha+1}G)$$

and

$$A(\infty) = (p^\infty G)[p] \cap A.$$

Lemma 2.3.1 (Walker [148]). *Let G and H be R-modules, where $R = \mathbb{Z}$ or $R = \mathbb{Z}_p$ for some prime p. Suppose that $\phi : A \to B$ is a p-height-preserving isomorphism, where A and B are p-nice submodules of G and H, respectively. Assume that G/A and H/B are p-groups, and that for each $\alpha \in \mathrm{Ord}_\infty$, we have $u_\alpha(G,A) = u_\alpha(H,B)$ and an isomorphism $f_\alpha : (p^\alpha G)[p]/A(\alpha) \to (p^\alpha H)[p]/B(\alpha)$. Let $x \in G$. Then ϕ extends to a p-height-preserving isomorphism $\phi' : A_1 = \langle A, x\rangle \to B_1 = \langle B, y\rangle$ for some $y \in H$ such that*
1. *A_1/A and B_1/B are finite;*
2. *f_α maps $A_1(\alpha)/A(\alpha)$ onto $B_1(\alpha)/B(\alpha)$ for all α;*
3. *$u_\alpha(G,A_1) = u_\alpha(H,B_1)$ for all α.*

Proof. We can assume that $x \in G \setminus A$, $px \in A$, and that x is proper with respect to A since A is p-nice. First, we suppose that the p-height of x is an ordinal, say $|x| = \beta$. If there is an $x' \in x + A$ that is proper with respect to A such that $|px'| > \beta + 1$, we replace x by x'.

Case I: $|px| > \beta + 1$. Then there is a $g \in p^{\beta+1}G$ such that $px = pg$. It follows that $x - g \in (p^\beta G)[p]$ and $|x - g| = \beta$; hence, $x - g \notin A + p^{\beta+1}G$ since x is proper with respect to A. Therefore, there is a $z \in (p^\beta H)[p] \setminus B(\beta)$ such that

$$f_\beta(x - g + A(\beta)) = z + B(\beta),$$

and we have $|z| = \beta$ since $z \notin B + p^{\beta+1}H$. Now write $\phi(px) = pw$ for some $w \in p^{\beta+1}H$, and let $y = z + w$. Then $|y| = \beta$ and $y \notin B + p^{\beta+1}H$, so y is proper with respect to B. By Lemma 1.12.4, ϕ extends to a p-height-preserving isomorphism $\phi' : A_1 = \langle A, x\rangle \to B_1 = \langle B, y\rangle$ by sending x onto y. We have $x - g \in A_1(\beta) \setminus A(\beta)$ and $z \in B_1(\beta) \setminus B(\beta)$; hence, f_β maps $0 \neq x - g + A(\beta) \in A_1(\beta)/A(\beta)$ onto $z + B(\beta) \in B_1(\beta)/B(\beta)$. Notice that the natural map $A_1/A \to A_1(\beta)/A(\beta)$ is surjective; hence, $\dim[A_1(\beta)/A(\beta)] = 1$. Likewise, we have $\dim[B_1(\beta)/B(\beta)] = 1$. Therefore, f_β maps $A_1(\beta)/A(\beta)$ onto $B_1(\beta)/B(\beta)$.

Case II: $|px| = \beta+1$. Then $\phi(px) = py$ for some $y \in H$ such that $|y| = \beta$. If $|y+b| \geq \beta+1$ for some $b \in B$, then $|b| = \beta$, and $b = \phi(a)$ for some $a \in A$ such that $|a| = \beta$. But then $x+a$ is proper with respect to A and $|p(x+a)| > \beta+1$, contradicting our choice of x. Therefore, y is proper with respect to B, and we have $y \notin B$ since $|y| \neq \infty$. By Lemma 1.12.4, ϕ extends to a p-height-preserving isomorphism $\phi' : A_1 = \langle A, x\rangle \to B_1 = \langle B, y\rangle$ by sending x onto y. To show that $A_1(\beta) = A(\beta)$, let $a + rx + g \in A_1(\beta)$, where $a \in A$, $r \in R$ and $g \in p^{\beta+1}G$. If $p|r$, then $a + rx + g \in A(\beta)$, and if $p \nmid r$, then there are $s, t \in R$ such that $sp + tr = 1$. Thus, $|a + rx| = |ta + trx| = |ta - spx + x| \leq |x| = \beta$. Letting $a' = ta - spx$, we have $a' + x = t(a + rx)$, and so $|a' + x| = \beta$ since $|a + rx + g| \geq \beta$. Then $a' + x$ is proper with respect to A, and $|p(a' + x)| = |pg| > \beta + 1$, again contradicting our choice of x. Therefore, we have $A_1(\beta) = A(\beta)$. Likewise, we obtain $B_1(\beta) = B(\beta)$.

In either case, we constructed $\phi' : A_1 \to B_1$—as required—such that f_β maps $A_1(\beta)/A(\beta)$ onto $B_1(\beta)/B(\beta)$. Now assume that $\alpha \neq \beta$ is an ordinal. Notice that A_1 and B_1 are p-nice since finite extensions of p-nice submodules are p-nice. Then if $\alpha < \beta$, we have $x \in p^{\alpha+1}G$; hence, $A_1(\alpha) = A(\alpha)$ and likewise, $B_1(\alpha) = B(\alpha)$. If $\alpha > \beta$, let $a + rx + g \in A_1(\alpha)$, where $a \in A$, $r \in R$ and $g \in p^{\alpha+1}G$. If $p|r$, then $a + rx + g \in A(\alpha)$. If $p \nmid r$, then there are $s, t \in R$ such that $sp + tr = 1$. Hence, $|a + rx| = |ta + trx| = |ta - spx + x| \leq |x| = \beta < \alpha$, therefore, $|a + rx + g| < \alpha$, a contradiction. Thus, it follows that $A_1(\alpha) = A(\alpha)$. Likewise, we have $B_1(\alpha) = B(\alpha)$. Now consider the case $\alpha = \infty$, and let $a \in A$ and $r \in R$ such that $a + rx \in A_1(\infty)$. If $p|r$, then $a + rx \in A(\infty)$, and if $p \nmid r$, then again, there are $s, t \in R$ such that $sp + tr = 1$, and we have $|a + rx| = |ta + trx| = |ta - spx + x| \leq |x| = \beta < \infty$, a contradiction. Therefore, $A_1(\infty) = A(\infty)$, and by symmetry, $B_1(\infty) = B(\infty)$. Thus, (2) is true whenever $|x| \neq \infty$.

Finally, suppose that $|x| = \infty$. First we assume that $A_1(\infty) = A(\infty)$. Let $\phi' : A_1 = \langle A, x\rangle \to B_1 = \langle B, y\rangle$ be any p-height-preserving isomorphism extending ϕ by sending x onto some $y \in H$ (see Lemma 1.14.1). To show that $B_1(\infty) = B(\infty)$, let $b + ry \in B_1(\infty)$ with $b \in B$ and $r \in R$. Then $\phi^{-1}(b) + rx \in A_1(\infty) = A(\infty)$; thus $p|r$, and we obtain $B_1(\infty) = B(\infty)$.

Now assume that $A_1(\infty) \neq A(\infty)$. We wish to find an element $x' \in x + A$ such that $x' \in (p^\infty G)[p]$. There exists a $z \in A_1(\infty) \setminus A(\infty)$, say $z = rx + a \in (p^\infty G)[p]$ with $r \in R$ $(p \nmid r)$ and $a \in A$. Then $1 = sr + tp$ for some $s, t \in R$; thus, $sz = srx + sa = (1 - tp)x + sa =$

$x + a'$ with $a' = sa - tpx \in A$. Consequently, $x' = x + a'$ is an element of $(p^\infty G)[p]$, as desired. Let $y \in (p^\infty H)[p]$ such that $f_\infty(x' + A(\infty)) = y + B(\infty)$. By Lemma 1.12.4, ϕ extends to a p-height-preserving isomorphism $\phi' : A_1 = \langle A, x'\rangle \to B_1 = \langle B, y\rangle$ by sending x' onto y. We have $A_1(\infty) = \langle x', A(\infty)\rangle$, and $B_1(\infty) = \langle y, B(\infty)\rangle$; thus, f_∞ maps $A_1(\infty)/A(\infty)$ onto $B_1(\infty)/B(\infty)$, as required.

Since $|x| = \infty = |y|$, it is clear that, in any case, we have $A_1(\alpha) = A(\alpha)$ and $B_1(\alpha) = B(\alpha)$ for each ordinal α. Therefore, (2) is true if $|x| = \infty$. (3) follows from (2), so the proof is complete. □

Now we can prove the following result:

Theorem 2.3.2 (Hill [47], Walker [148], Warfield [154]). *Let G and H be R-modules, where $R = \mathbb{Z}$ or $R = \mathbb{Z}_p$ for some prime p. Suppose that A and B are p-nice submodules of G and H, respectively, such that G/A and H/B are p-groups with nice systems and $u_\alpha(G, A) = u_\alpha(H, B)$ for every $\alpha \in \mathrm{Ord}_\infty$. Then any p-height-preserving isomorphism $\phi : A \to B$ can be extended to an isomorphism $G \to H$.*

Proof. Let $\mathcal{C}$ and $\mathcal{D}$ be nice systems for G/A and H/B, respectively. For any α, an ordinal or ∞, let $f_\alpha : (p^\alpha G)[p]/A(\alpha) \to (p^\alpha H)[p]/B(\alpha)$ be an isomorphism, and define $\mathcal{F}$ to be the system of all p-height-preserving isomorphisms $C \to D$ extending ϕ satisfying the following properties:

1. $C/A \in \mathcal{C}$ and $D/B \in \mathcal{D}$;
2. f_α maps $C(\alpha)/A(\alpha)$ onto $D(\alpha)/B(\alpha)$ for all α.

The system $\mathcal{F}$ is partially ordered in an obvious manner and is nonempty since $\phi \in \mathcal{F}$. Thus, $\mathcal{F}$ possesses a maximal element $\phi_0 : C \to D$ by Zorn's lemma. We will show that $C = G$ and $D = H$. Assume that there is an $x \in G \setminus C$. By (2), we have $u_\alpha(G, C) = u_\alpha(H, D)$ for every α, an ordinal or ∞; hence, f_α induces an isomorphism $g_\alpha : (p^\alpha G)[p]/C(\alpha) \to (p^\alpha H)[p]/D(\alpha)$. By Lemma 2.1.4(2), the modules C and D are p-nice. Then Lemma 2.3.1 yields a p-height-preserving isomorphism $\phi_1 : A_1 = \langle C, x\rangle \to B_1$ such that A_1/C and B_1/D are finite and g_α maps $A_1(\alpha)/C(\alpha)$ onto $B_1(\alpha)/D(\alpha)$ for all α. Consequently, each map f_α maps $A_1(\alpha)/A(\alpha)$ onto $B_1(\alpha)/B(\alpha)$. Since $C/A \in \mathcal{C}$ and $\langle A, x\rangle/A$ is finite, there is a $C_1/A \in \mathcal{C}$ such that $A_1/A \subseteq C_1/A$ and $|C_1/A| \le \aleph_0$. We write $C_1 = \langle A_1, x_{11}, x_{12}, \dots\rangle$. Using Lemma 2.3.1 again, we extend ϕ_1 to a p-height-preserving isomorphism

$$\phi_1' : A_1' = \langle A_1, x_{11}\rangle \to B_2 = \langle B_1, y\rangle$$

such that f_α maps $A_1'(\alpha)/A(\alpha)$ onto $B_2(\alpha)/B(\alpha)$ for each α. Since $D/B \in \mathcal{D}$ and $\langle B, y\rangle/B$ is finite, there is a $D_2/B \in \mathcal{D}$ such that $B_2/B \subseteq D_2/B$ and $|D_2/B_2| \le \aleph_0$, and we write $D_2 = \langle B_2, y_{21}, y_{22}, \dots\rangle$. Again, we apply Lemma 2.3.1 and extend ϕ_1' to a p-height-preserving isomorphism

$$\phi_2 : A_2 = \langle A_1', z\rangle \to B_2' = \langle B_2, y_{21}\rangle$$

so that f_α maps $A_2(\alpha)/A(\alpha)$ onto $B_2'(\alpha)/B(\alpha)$ for each α. Then there is a $C_2/A \in \mathcal{C}$ such that $A_2/A \subseteq C_2/A$ and $|C_2/A_2| \le \aleph_0$, and we write $C_2 = \langle A_2, x_{21}, x_{22}, \dots\rangle$. The map ϕ_2 extends to $\phi_2' : A_2' = \langle A_2, x_{12}, x_{21}\rangle \to B_3$ such that f_α maps $A_2'(\alpha)/A(\alpha)$ onto $B_3(\alpha)/B(\alpha)$ for each α. There is a $D_3/B \in \mathcal{D}$ such that $B_3/B \subseteq D_3/B$ and $|D_3/B_3| \le \aleph_0$, so we write $D_3 = \langle B_3, y_{31}, y_{32}, \dots\rangle$ and extend ϕ_2' to $\phi_3 : A_3 \to B_3' = \langle B_3, y_{22}, y_{31}\rangle$. Continuing in this way, with each A_n' adding new elements from each C_i with $1 \le i \le n$, and likewise for B_n'; this yields sequences

$$C \subseteq A_1 \subseteq A_1' \subseteq A_2 \subseteq A_2' \subseteq \cdots$$

and

$$D \subseteq B_1 \subseteq B_1' \subseteq B_2 \subseteq B_2' \subseteq \cdots,$$

and a p-height-preserving isomorphism

$$\tilde{\phi} : \tilde{C} = \bigcup_{i=1}^{\infty} A_i = \bigcup_{i=1}^{\infty} C_i \to \tilde{D} = \bigcup_{i=1}^{\infty} B_i = \bigcup_{i=1}^{\infty} D_i$$

extending ϕ. Since $\mathcal{C}$ and $\mathcal{D}$ are nice systems, we have $\tilde{C}/A \in \mathcal{C}$ and $\tilde{D}/B \in \mathcal{D}$. Moreover, f_α induces an isomorphism $\tilde{C}(\alpha)/A(\alpha) \to \tilde{D}(\alpha)/B(\alpha)$ for all α, and we conclude that $\tilde{\phi} \in \mathcal{F}$. Since $x \in \tilde{C}$, this contradicts the maximality of ϕ; hence, $C = G$. By symmetry, we have $D = H$, as desired. □

As a corollary, we obtain Hill's generalization of Ulm's theorem, the classification of p-groups having nice systems in terms of Ulm invariants:

Theorem 2.3.3 (Hill [47]). *Let G and H be p-groups having nice systems. Then G and H are isomorphic if and only if* $u(\alpha, G) = u(\alpha, H)$ *for each* $\alpha \in \mathrm{Ord}_\infty$.

Another application of Theorem 2.3.2 is the following extension theorem, which will be needed in Section 2.9:

Theorem 2.3.4 (Warfield [154]). *Let G and H be* $\mathbb{Z}_p$*-modules and A a p-nice submodule of G such that G/A is a p-group with a nice system. If* $\phi : A \to H$ *is a homomorphism which does not decrease p-heights, then* ϕ *extends to a homomorphism* $G \to H$.

Proof. If $a \in A$, then $|\phi(a)|_p^H \ge |a|_p^G$ yields $|a|_p^G \wedge |\phi(a) + h|_p^H = |a|_p^G \wedge |h|_p^H$ for all $h \in H$; thus, the map

$$\begin{array}{rccc} \varphi : & A \oplus H & \to & A \oplus H \\ & (a,h) & \mapsto & (a, \phi(a) + h) \end{array}$$

is a p-height-preserving isomorphism. By Theorem 2.3.2, φ extends to an automorphism φ' of $G \oplus H$. Let $\sigma : G \to G \oplus H$ be the natural inclusion and $\mu : G \oplus H \to H$ the natural projection. Then $\mu\varphi'\sigma : G \to H$ extends ϕ, as desired. □

Exercises.

1. Find a p-group which does not have a nice system. [Hint: Section 1.14, Exercise 2.]
2. (Fuchs [27]) Let G and H be reduced p-groups, and assume that $\phi : A \to B$ is a height-preserving isomorphism, where A is a nice subgroup of G and B is any subgroup of H. If G/A has a nice composition series such that $u_\alpha(G, A) \le u_\alpha(H, B)$ for all ordinals α, then ϕ extends to a height-preserving isomorphism $G \to H'$ for some subgroup H' of H.
3. (Fuchs [27]) Let G be a reduced p-group with nice composition series, and let H be any p-group. Then G is isomorphic to an isotype subgroup of H if and only if $u(\alpha, G) \le u(\alpha, H)$ for all ordinals α. [Hint: Exercise 2.]
4. (Fuchs [27]) Let G and H be p-groups and $\phi : A \to H$ a homomorphism, where A is a nice subgroup of G. If G/A has a nice composition series and if ϕ does not decrease heights, then ϕ extends to a homomorphism $G \to H$ which does not decrease heights either. [Hint: Exercise 2.]

2.4 Totally projective p-groups and generalized Prüfer groups

This section presents another way of looking at the groups we just classified, using a homological approach introduced by Nunke [124]. A p-group G is called **totally projective** if it satisfies

$$p^\alpha \mathrm{Ext}(G/p^\alpha G, H) = 0$$

for all ordinals α and groups H. A torsion group is said to be **totally projective** if each of its p-torsion parts is totally projective. The following properties are immediate:

(a) Divisible p-groups are totally projective.
(b) Cyclic p-groups are totally projective.
(c) The class of totally projective p-groups is closed under arbitrary direct sums and direct summands.
(d) If G is a totally projective p-group, then so is $G/p^\sigma G$ for all ordinals σ.

Following Nunke [124], we will construct a reduced p-group H_α for every ordinal α such that

(i) H_α has length α;
(ii) $p^\alpha H_{\alpha+1}$ is cyclic of order p and $H_{\alpha+1}/p^\alpha H_{\alpha+1} \cong H_\alpha$;
(iii) $H_\alpha = \bigoplus_{\beta<\alpha} H_\beta$ if α is a limit ordinal.

The group H_α is called a **generalized Prüfer group (of length α)**. It will be shown in Proposition 2.4.3 that H_α is totally projective.

Starting with $H_0 = 0$, let $H_n = \mathbb{Z}/p^n\mathbb{Z}$ for all positive integers n. Arguing inductively, we define H_α as in (iii) whenever α is a limit ordinal. Now let α be an ordinal $\ge \omega$, and assume that the groups H_β ($\beta \le \alpha$) are already constructed. To construct $H_{\alpha+1}$, we consider the following two cases:

Case I: α is a limit ordinal. Then $H_\alpha = \bigoplus_{\beta<\alpha} H_\beta \cong \bigoplus_{\beta<\alpha} H_{\beta+1}/p^\beta H_{\beta+1}$. Letting $C = \mathbb{Z}/p\mathbb{Z}$, we define $f : B = \bigoplus_{\beta<\alpha} p^\beta H_{\beta+1} \to C$ to be the sum of the isomorphisms $f_\beta : p^\beta H_{\beta+1} \to C$ (cf. (ii)). Then define $H_{\alpha+1}$ via the pushout diagram

$$\begin{array}{ccccccccc} 0 & \to & B & \to & \bigoplus_{\beta<\alpha} H_{\beta+1} & \to & H_\alpha & \to & 0 \\ & & f\downarrow & & \downarrow & & \| & & \\ 0 & \to & C & \xrightarrow{\sigma} & H_{\alpha+1} & \xrightarrow{\tau} & H_\alpha & \to & 0. \end{array}$$

For every $\beta < \alpha$ and $c \in C$, there is an $x \in p^\beta H_{\beta+1}$ such that $f(x) = c$. Since the diagram commutes, $\sigma(c) \in p^\beta H_{\alpha+1}$. Thus, for any $\beta < \alpha$, $\sigma(C) \subseteq p^\beta H_{\alpha+1}$. Therefore, we have $\sigma(C) \subseteq p^\alpha H_{\alpha+1}$. On the other hand, exactness of the second row yields $p^\alpha H_{\alpha+1} \subseteq \sigma(C)$ since $\tau(p^\alpha H_{\alpha+1}) \subseteq p^\alpha H_\alpha = 0$. Thus, $p^\alpha H_{\alpha+1} = \sigma(C) \cong \mathbb{Z}/p\mathbb{Z}$, so $H_{\alpha+1}$ has length $\alpha+1$ and condition (ii) holds.

Case II: α is a successor ordinal. Letting $D = \mathbb{Z}/p^2\mathbb{Z}$, we have an epimorphism $g : D \to p^{\alpha-1}H_\alpha$. Then the induced map $g_* : \operatorname{Ext}(H_{\alpha-1}, D) \to \operatorname{Ext}(H_{\alpha-1}, p^{\alpha-1}H_\alpha)$ is an epimorphism (see Theorem 1.16.10). Therefore, there exists a group $H_{\alpha+1}$ such that the diagram

$$\begin{array}{ccccccccc} 0 & \to & D & \xrightarrow{\phi} & H_{\alpha+1} & \xrightarrow{\psi} & H_{\alpha-1} & \to & 0 \\ & & g\downarrow & & h\downarrow & & \| & & \\ 0 & \to & p^{\alpha-1}H_\alpha & \to & H_\alpha & \to & H_{\alpha-1} & \to & 0 \end{array}$$

is commutative with both rows exact. It is clear that $K = \ker h$ contains $\phi(pD)$, and if $k \in K$, then $k \in \ker\psi = \operatorname{im}\phi$, and therefore, $k = \phi(d)$ for some $d \in D$. Thus, $d \in \ker g = pD$. Consequently, K coincides with $\phi(pD)$.

Since $p^{\alpha-1}H_{\alpha-1} = 0$, we have $p^{\alpha-1}H_{\alpha+1} \subseteq \ker\psi = \operatorname{im}\phi$. Thus, $p^{\alpha+1}H_{\alpha+1} = 0$. We claim $p^{\alpha-1}H_{\alpha+1} = \phi(D)$. Suppose—by way of contradiction—that $p^{\alpha-1}H_{\alpha+1} \neq \phi(D)$. Since the only subgroups of D are 0, pD, and D, we then have $p^{\alpha-1}H_{\alpha+1} \subseteq p\phi(D)$. The map h is an epimorphism (Lemma 1.2.3(3)), and finite subgroups are nice. Therefore, it follows—by Proposition 2.1.1(3)—that $p^{\alpha-1}H_\alpha \cong p^{\alpha-1}(H_{\alpha+1}/K) = (p^{\alpha-1}H_{\alpha+1} + K)/K = 0$, which is impossible. Therefore, we have $p^{\alpha-1}H_{\alpha+1} = \phi(D)$. Consequently, we obtain $p^\alpha H_{\alpha+1} \cong \mathbb{Z}/p\mathbb{Z}$, and $H_\alpha \cong H_{\alpha+1}/K = H_{\alpha+1}/\phi(pD) = H_{\alpha+1}/p^\alpha H_{\alpha+1}$, as required. This completes the construction of the groups H_α. The group $H_{\omega+1}$ is called the **Prüfer group** (see Exercise 2).

The following observations were made by Nunke [124]:

Lemma 2.4.1. *If α is an ordinal and $n \in \mathbb{N}$, then we have the following:*
1. *$p^\alpha H_{\alpha+n}$ is cyclic of order p^n.*
2. *$H_{\alpha+n}/p^\alpha H_{\alpha+n} \cong H_\alpha$.*

Proof. Since $H_{\alpha+n+1}/p^{\alpha+n}H_{\alpha+n+1} \cong H_{\alpha+n}$, we have $p^\alpha H_{\alpha+n+1}/p^{\alpha+n}H_{\alpha+n+1} \cong p^\alpha H_{\alpha+n}$, so the first claim follows by induction on n. To prove (2), use

$$\frac{H_{\alpha+n+1}}{p^\alpha H_{\alpha+n+1}} \cong \frac{H_{\alpha+n+1}/p^{\alpha+n}H_{\alpha+n+1}}{p^\alpha H_{\alpha+n+1}/p^{\alpha+n}H_{\alpha+n+1}} \cong \frac{H_{\alpha+n}}{p^\alpha H_{\alpha+n}}.$$

Then the assertion follows by induction on n as well. □

The next result will be needed:

Lemma 2.4.2 (Nunke [124]). *Let G be a p-group and σ an ordinal such that $p^{\sigma+1}G = 0$. If $G/p^\sigma G$ is totally projective, then so is G.*

Proof. Assume that H is a group. Then we have an exact sequence $A \xrightarrow{\phi} B \xrightarrow{\psi} C \to D \to 0$ with $A = \mathrm{Hom}(p^\sigma G, H)$, $B = \mathrm{Ext}(G/p^\sigma G, H)$, $C = \mathrm{Ext}(G, H)$, and $D = \mathrm{Ext}(p^\sigma G, H)$ by Theorem 1.16.10. Then $\phi(A) \subseteq B[p]$ since $pA = 0$, so we obtain the exact sequence

$$0 \to B[p]/\phi(A) \to B/\phi(A) \to pB \to 0.$$

Since $p^\sigma B = 0$, we have $p^\sigma(B/\phi(A)) \subseteq B[p]/\phi(A)$ by Lemma 1.10.6(2); therefore, $p^{\sigma+1}\psi(B) \cong p^{\sigma+1}(B/\phi(A)) = 0$. Notice that $pC \subseteq \psi(B)$ since $D \cong C/\psi(B)$ and $pD = 0$. If $\sigma < \omega$, then $p^{\sigma+1}C \subseteq \psi(p^\sigma B) = 0$, and if $\sigma \geq \omega$, we obtain $p^{\sigma+1}C = p^{1+(\sigma+1)}C = p^{\sigma+1}(pC) \subseteq p^{\sigma+1}\psi(B) = 0$ using Lemma 1.10.4(1). If $\beta \leq \sigma$, then

$$p^\beta \mathrm{Ext}(G/p^\beta G, H) \cong p^\beta \mathrm{Ext}((G/p^\sigma G)/(p^\beta(G/p^\sigma G)), H) = 0$$

by Lemma 1.10.5. For any $\beta > \sigma$, we have $p^\beta G = 0$. Therefore, $p^\beta \mathrm{Ext}(G/p^\beta G, H) \cong p^\beta \mathrm{Ext}(G, H) = p^\beta C = 0$, so we conclude that G is totally projective. □

Proposition 2.4.3 (Nunke [124]). *Direct summands of direct sums of generalized Prüfer groups are totally projective.*

Proof. In view of (c), it suffices to prove that each generalized Prüfer group H_α is totally projective. Let $\alpha > 0$, and assume that the assertion is true for all $\beta < \alpha$. The induction step is immediate if α is a limit ordinal. Therefore, suppose that $\alpha - 1$ exists. Since $H_{\alpha-1} \cong H_\alpha/p^{\alpha-1}H_\alpha$ is totally projective by the induction hypothesis, Lemma 2.4.2 shows that H_α is totally projective. □

Proposition 2.4.4 (Hill [47]). *Direct summands of direct sums of generalized Prüfer groups have nice systems.*

Proof. By Corollary 2.2.9, it suffices to show that each group H_α has a nice system. Using transfinite induction, we let $\alpha > 0$, and assume that H_β has a nice system for all $\beta < \alpha$. If α is a limit ordinal, then $H_\alpha = \bigoplus_{\beta<\alpha} H_\beta$ has a nice system by Theorem 2.2.8(1). If $\alpha - 1$ exists, then $H_{\alpha-1} \cong H_\alpha/p^{\alpha-1}H_\alpha$ has a nice system; thus, H_α has a nice system by Corollary 2.2.6. □

Exercises.

1. Show that the generalized Prüfer groups are uniquely determined up to isomorphism by properties (ii–iii).
2. Show that the Prüfer group $H_{\omega+1}$ can be written as $\langle x_0, x_1, \ldots, x_n, \ldots : px_0 = 0$ and $p^n x_n = x_0$ for all $n = 1, 2, \ldots \rangle$.
3. The generalized Prüfer group H_α has cardinality $|\alpha|$ for every $\alpha \geq \omega$.
4. (a) Show that $u_p(\alpha, H_{\omega+1}) = 1$ if $\alpha \leq \omega$, and $u_p(\alpha, H_{\omega+1}) = 0$ otherwise.
   (b) Show that $u_p(\beta, H_\alpha) \neq 0$ whenever $\beta < \alpha$.
   (c) Compute the Ulm invariants of $H_{\omega+n}$ for $1 < n < \omega$.
5. (Nunke [123]) Let p be a prime. For every ordinal α, there is a short exact sequence $0 \to Z_\alpha \to G_\alpha \to H_\alpha \to 0$ such that $p^\alpha G_\alpha = Z_\alpha \cong \mathbb{Z}$.

2.5 Defining relations and simple presentations

Let G be a group, X a set of generators of G, and let F_X be the free group on the set X. Then the correspondence $x \mapsto x$ $(x \in X)$ gives rise to an epimorphism $\phi : F_X \to G$ with kernel H, so we can identify G with F_X/H. The subgroup H of F_X consists of all elements $r_1x_1 + \cdots + r_kx_k$ of F_X $(r_i \in \mathbb{Z}, x_i \in X)$ such that

$$r_1x_1 + \cdots + r_kx_k = 0 \quad \text{holds in } G.$$

Equations of this form are called **defining relations.** Given a subset S of H with $\langle S \rangle = H$, we let Λ be the corresponding set of defining relations; that is, $s = 0$ is a defining relation in Λ if and only if $s \in S$. It is clear that the group G is completely determined by *any* given set X of generators and any such corresponding set Λ of defining relations. Therefore, we let $\langle X; \Lambda \rangle = F_X/H$ and call the equation

$$G = \langle X; \Lambda \rangle$$

a **presentation of** G.

Notice that for every group G, there exists a presentation $G = \langle X; \Lambda \rangle$ with all defining relations in Λ involving at most *three* generators, as follows. Starting with the free group F on the underlying set X of G, we have $G \cong F/H$, where H is the subgroup of F generated by all elements $x + y - z$ such that the equation $x + y = z$ holds in G. On the other hand, any direct sum of cyclic groups can be defined in terms of generators and relations involving at most *one* generator. Of particular interest in abelian group theory is the class of groups in between. Following Fuchs [29], we call a group G **simply presented** if it can be defined in terms of a set X of generators and a set Λ of defining relations involving at most *two* generators; that is, each relation in Λ has the form

$$nx = 0 \quad \text{or} \quad mx = ny,$$

where $x, y \in X$, and $m, n \in \mathbb{Z}$. In this case, we call $G = \langle X; \Lambda \rangle$ a **simple presentation.** The following is immediate:

Lemma 2.5.1. *Let X be a set and $S \subseteq F_X$, such that each element of S has the form nx or $mx - ny$ with $x, y \in X$ and $m, n \in \mathbb{Z}$. Then $G = F_X/\langle S\rangle$ is a simply presented group with simple presentation $G = \langle X; \Lambda\rangle$, where $\Lambda = \{s = 0 : s \in S\}$.*

If the relation $nx = 0$ occurs in Λ for $n = p_k p_{k-1} \cdots p_1$, where each p_i is a prime, we may replace it by the relations

$$p_1 x = x_1,\ p_2 x_1 = x_2, \ldots,\ p_k x_{k-1} = 0$$

and add the elements $x_1, \ldots, x_{k-1}$ to X. In the case where the relation $nx = 0$ does not occur in Λ with such n, but $x \in X$ has order n in G, we add the relations above to Λ and elements $x_1, \ldots, x_{k-1}$ to X. Likewise, we can replace each relation of the form $mx = ny$ by relations, each involving two generators with 1 and a prime number as coefficients, adding more generators to X if necessary. Thus, we may assume that every defining relation in Λ is of the form

$$px = 0 \quad \text{or} \quad px = y,$$

where $x, y \in X$ and p is a prime (various primes p may occur) such that for any $x \in X$, we have $px \in X$ whenever $px \neq 0$ and $x \in t_p G$. Of course, it is possible that after such modifications, a generator in X becomes 0 in G and different generators in X become identical in G. This situation gives rise to the following definition: A simple presentation $G = \langle X; \Lambda\rangle$ is called **faithful** if the following conditions hold:

(i) Every defining relation in Λ is of the form $px = 0$ or $px = y$, where $x, y \in X$ and p is a prime (different primes p may occur);
(ii) If $x \in X$, then $x \neq 0$ in G;
(iii) If x and y are distinct elements of X, then $x \neq y$ in G.

Every simply presented group possesses a faithful simple presentation:

Lemma 2.5.2. *Let $G = \langle X, \Lambda\rangle$ be a simply presented group. Then there is a simple presentation of G which is faithful.*

Proof. We may assume that each defining relation in Λ is of the form $px = 0$ or $px = y$, involving elements $x, y \in X$ and primes p. Then $G = F_X/H$, where H is the subgroup of the free group F_X generated by all elements corresponding to the defining relations in Λ. Now delete those $x \in X$ which become 0 in G (and replace x in a defining relation by 0), and give the same name to elements in X which become identical in G (and adjust the corresponding defining relations). Let X', Λ', and H' denote the resulting modifications of X, Λ, and H, respectively. Then the correspondence $x \mapsto x$ ($x \in X'$) induces a natural epimorphism $F_{X'} \to G$, whose kernel contains H', yielding an epimorphism

$$\phi : G' = F_{X'}/H' \to G.$$

On the other hand, the kernel of the natural map $F_X \to G'$ contains H, yielding an epimorphism $\psi : G \to G'$. Since $\phi\psi = 1_G$, we have $G \cong G' = \langle X'; \Lambda' \rangle$, as desired. □

From now on, all simple presentations $G = \langle X; \Lambda \rangle$ are assumed to be faithful and satisfy the condition: for any $x \in X$, we have $px \in X$ whenever $px \neq 0$ and $x \in t_pG$, unless stated otherwise. Adding these elements to X automatically adds the appropriate elements to Λ. We exhibit some basic examples of simply presented groups:

1. Cyclic groups of order p^n are simply presented: $\mathbb{Z}/p^n\mathbb{Z} = \langle x_1, \ldots, x_n; px_1 = 0, px_2 = x_1, \ldots, px_n = x_{n-1} \rangle$.
2. Quasicyclic p-groups are simply presented: $\mathbb{Z}(p^\infty) = \langle x_1, \ldots, x_n, \ldots; px_1 = 0, px_2 = x_1, \ldots, px_n = x_{n-1}, \ldots \rangle$.
3. The Prüfer group $H_{\omega+1}$ is simply presented (see Exercise 2 in Section 2.4).

The following fact is immediate:

Lemma 2.5.3. *Direct sums of simply presented groups are simply presented.*

The building blocks of simply presented groups are those of torsion-free rank ≤ 1:

Theorem 2.5.4 (Warfield [152]). *Every simply presented group is a direct sum of simply presented groups of torsion-free rank ≤ 1.*

Proof. Let $G = \langle X; \Lambda \rangle$ be a simply presented group. Again, we write $G = F_X/H$, where H is the subgroup of the free group F_X generated by all elements corresponding to the defining relations in Λ. We define a relation on X by $x \sim y$ if $mx = ny \neq 0$ holds in G for some integers m and n. Then $\sim$ is an equivalence relation with resulting equivalence classes X_i $(i \in I)$, and we can write $F_X = \bigoplus_{i \in I} F_{X_i}$. Letting $H_i = H \cap F_{X_i}$ $(i \in I)$, we obtain $G \cong \bigoplus_{i \in I} F_{X_i}/H_i$. Moreover, every group F_{X_i}/H_i is simply presented and has torsion-free rank ≤ 1. □

Since every group is an epimorphic image of a free group, the factor group of a simply presented group is not necessarily simply presented. However, we have the following:

Proposition 2.5.5. *Let $G = \langle X; \Lambda \rangle$ be a simply presented group and A a subgroup of G generated by elements of the form px or $px - y$ with $x, y \in X$ and p a prime (various primes p may occur). Then G/A is simply presented.*

Proof. The correspondence $x \mapsto x$ with $x \in X$ defines a natural epimorphism $F_X \to G$ with kernel H which is generated by all elements corresponding to the defining relations in Λ. Thus, we have an isomorphism

$$\phi : G \to F_X/H$$

given by $\sum \lambda_i x_i \mapsto \sum \lambda_i x_i + H$ $(\lambda_i \in \mathbb{Z}, x_i \in X)$. Then ϕ maps A onto the factor group $(L + H)/H$, where L is the subgroup of F_X generated by elements corresponding to generators of A, and hence of the form px or $px - y$ with $x, y \in X$. Therefore, $G/A \cong (F_X/H)/((L+H)/H) \cong F_X/(L+H)$ is a simply presented group. □

Exercises.

1. (Fomin [23]) Prove that the group $\langle z_1, \ldots, z_n, \ldots : pz_1 = \cdots = p^n z_n = \cdots \rangle$ does not split.
2. Find a faithful simple presentation $\mathbb{Z} = \langle X; \Lambda \rangle$ with nonempty Λ.
3. Find simple presentations of $H_{\omega+n}$ for $1 < n < \omega$.

2.6 Simply presented torsion-free groups

The torsion-free case is easy to handle. We start with the following:

Lemma 2.6.1. *Torsion-free groups of rank 1 are simply presented.*

Proof. Suppose that G is a torsion-free group of rank 1, that is, a nontrivial subgroup of $\mathbb{Q}$. Let F be the free group on the underlying set X of G and H the subgroup of F generated by all elements $mx - ny$ (where $x, y \in X$ and $m, n \in \mathbb{Z}$) such that $mx = ny$ holds in G. We claim that H is the kernel of the natural epimorphism $\phi : F \to G$. Clearly, H is a subset of $\ker \phi$. Now assume that $0 \neq z \in \ker \phi$. Then

$$z = \lambda_1 x_1 + \cdots + \lambda_n x_n$$

for some nonzero integers $\lambda_1, \ldots, \lambda_n$ and $x_1, \ldots, x_n \in X$. There is a positive integer m such that $m\lambda_i \phi(x_i) \in \mathbb{Z}$ for all $i = 1, \ldots, n$, so there exist integers r_i such that $\frac{r_i}{m} = \lambda_i \phi(x_i)$ for $i = 1, \ldots, n$. It follows that $\frac{r_1}{m} + \cdots + \frac{r_n}{m} = \phi(\lambda_1 x_1 + \cdots + \lambda_n x_n) = 0$. Therefore, since $\frac{r_1}{m}, \ldots, \frac{r_n}{m} \in X$, we obtain $z = (\lambda_1 x_1 - \frac{r_1}{m}) + \cdots + (\lambda_n x_n - \frac{r_n}{m}) \in H$. □

Since completely decomposable groups are defined to be direct sums of torsion-free rank 1 groups, we obtain the following:

Corollary 2.6.2. *Completely decomposable groups are simply presented.*

Using Theorems 2.5.4 and 1.15.4, we conclude with a classification result:

Corollary 2.6.3. *A torsion-free group is simply presented if and only if it is completely decomposable. Two torsion-free simply presented groups are isomorphic if and only if they have the same Baer invariants.*

Exercises.

1. Find faithful simple presentations of $\mathbb{Q}$ and $\mathbb{Z}_p$.
2. If G is a torsion-free simply presented group such that $G \oplus G \cong H \oplus H$ for some group H, then $G \cong H$.

2.7 Richman–Walker groups and variations

We will now construct important examples of simply presented groups. Of particular interest are Walker's [149] p-groups P_α and Warfield's [152] mixed groups H_X.

First, we consider the p-local case (p a fixed prime), and let G be a **free $\mathbb{Z}_p$-module on a set** Y; that is, $G = \bigoplus_{y\in Y}\langle y\rangle$ with $\langle y\rangle \cong \mathbb{Z}_p$ $(y \in Y)$. Letting T be a given subset of G such that each element of T has the form px or $px - y$ with $x, y \in Y$, we define A to be the submodule of G generated by T. Then G/A is called a **simply presented $\mathbb{Z}_p$-module**. As expected, we have the following:

Lemma 2.7.1. *A simply presented $\mathbb{Z}_p$-module is a simply presented group.*

Proof. Let B be a simply presented $\mathbb{Z}_p$-module. Then $B = G/A$, where $G = \bigoplus_{y\in Y}\langle y\rangle$ with $\langle y\rangle \cong \mathbb{Z}_p$ for all $y \in Y$, and A is the submodule of G generated by a set T of elements of the form px or $px - y$ with $x, y \in Y$. For $y \in Y$ let X_y be the underlying set of the module $\langle y\rangle$ and define $X = \dot{\bigcup}_{y\in Y} X_y$. By Lemma 2.6.1, its proof and Lemma 2.5.3, the group G has a simple presentation $G = \langle X; \Lambda\rangle$. Let $t \in T$ and $q \neq p$ a prime. If $t = px$ $(x \in Y)$, then $\frac{1}{q}t = px'$ with $x' = \frac{1}{q}x \in X$, and if $t = px - y$ $(x, y \in Y)$, then $\frac{1}{q}t = px' - y'$ with $x', y' \in X$. Since the subgroup A of G is generated by such elements $\frac{1}{q}t$, Proposition 2.5.5 shows that $B = G/A$ is a simply presented group. □

We start with a slightly generalized version of the Richman–Walker construction. Recall from Section 1.11 that we define a **p-valuation** on a group G to be a function v on G such that

1. $v(x)$ is an ordinal or ∞.
2. $v(x + y) \geq v(x) \wedge v(y)$ for all x and $y \in G$.
3. $v(px) > v(x)$, where we follow the convention that $\infty > \infty$.
4. $v(nx) = v(x)$ if n and p are relatively prime.

Theorem 2.7.2 (Richman–Walker [134]). *Let G be a $\mathbb{Z}_p$-module, $S \subseteq G$, and v a p-valuation on G. Then there is a $\mathbb{Z}_p$-module H with submodule $G' \cong G$, where H is torsion over G', H/G' is simply presented, and for every $x \in S$ and its image $x' \in G'$, $|x'|_p^H = v(x)$.*

Proof. Let B be the free $\mathbb{Z}_p$-module generated by all finite sequences of the form $(\alpha_1, \ldots, \alpha_n, x)$, where $x \in S$, $x \neq 0$, $\alpha_1 < \cdots < \alpha_n < v(x)$ are ordinals or ∞, and $\alpha_1 = \infty$ if $v(x) = \infty$. Let A be the submodule of $B \oplus G$ generated by the set of all elements of the form $p(\alpha_1, \ldots, \alpha_n, x) - (\alpha_2, \ldots, \alpha_n, x)$ or $p(\alpha, x) - x$. Let $H = (B \oplus G)/A$. Then $G \cong G' = (A + G)/A$ via $g \mapsto g + A$ since $A \cap G = 0$ and H/G' is torsion. We have $H/G' \cong (B \oplus G)/(A + G) \cong B/(B \cap (A + G))$, and $B \cap (A + G)$ is generated by all elements of the form $p(\alpha_1, \ldots, \alpha_n, x) - (\alpha_2, \ldots, \alpha_n, x)$. Thus H/G' is simply presented by Lemma 2.7.1.

We next determine the height of an arbitrary element of H. We will show that $|(\alpha_1, \ldots, \alpha_n, x) + A| \geq \alpha_1$, by induction on α_1. If $\beta < \alpha_1$, then $(\alpha_1, \ldots, \alpha_n, x) + A = p(\beta, \alpha_1, \ldots,$

$\alpha_n, x)+A$, so by the induction hypothesis, $|(\alpha_1,\ldots,\alpha_n,x)+A| > |(\beta,\alpha_1,\ldots,\alpha_n,x)+A| \geq \beta$. Since β was arbitrary, it follows that $|(\alpha_1,\ldots,\alpha_n,x)+A| \geq \alpha_1$, completing the induction.

Every element of $B \oplus G$ may be written in the form

$$a = a_1(\alpha_{11},\ldots,\alpha_{1n_1},x_1) + \cdots + a_k(\alpha_{k1},\ldots,\alpha_{kn_k},x_k) + x_{k+1},$$

where $x_1,\ldots,x_{k+1} \in G$, the $(\alpha_{i1},\ldots,\alpha_{in_i},x_i)$ are distinct, and $a_1,\ldots,a_k \in \mathbb{Z}_p \setminus \{0\}$. Since p-heights are unaffected by multiplication by integers relatively prime to p, we may assume $a_1,\ldots,a_k \in \mathbb{Z}$. If $0 < a_i < p$ for all $i = 1,\ldots,k$, we say a is in canonical form. We define $h(a) = \min\{\alpha_{11},\ldots,\alpha_{k1},v(x_{k+1})\}$. We prove by induction on m, the maximum number of ordinals in the terms, that there is an $a' \in a + A$ in the canonical form and $h(a') \geq h(a)$. If $m = 0$, we have an element of G which is immediately in the form. Suppose it holds for $m-1$, and write $a_i = a_i' + pb_i$, where $0 \leq a_i' < p$ for $i = 1,\ldots,n$. Let $a' = a_1'(\alpha_{11},\ldots,\alpha_{1n_1},x_1)+\cdots+a_k'(\alpha_{k1},\ldots,\alpha_{kn_k},x_k)+b_1(\alpha_{12},\ldots,\alpha_{1n_1},x_1)+\cdots+b_k(\alpha_{k2},\ldots,\alpha_{kn_k},x_k)+x_{k+1}$. Then $a+A = a'+A$, and each term of a' is either already in the canonical form, or can be put in the canonical form by the induction hypothesis. Gather terms and repeat as necessary to put a' in canonical form. We also note that $h(a') \geq h(a)$, completing the induction. The same argument shows that $h(a') \geq h(b)$ for all $b \in a+A$.

Now we prove that the canonical form is unique. Suppose, on the contrary,

$$(a_1 - a_1')(\alpha_{11},\ldots,\alpha_{1n_1},x_1) + \cdots + (a_k - a_k')(\alpha_{k1},\ldots,\alpha_{kn_k},x_k) + x_{k+1} \in A,$$

where $0 \leq a_i < p$ and $0 \leq a_i' < p$ for $i = 1,\ldots,k$. But this has no nonzero coefficients divisible by p, whereas every element of A does.

Suppose a is in canonical form. From the claim $|a + A| \geq h(a)$. We will prove by induction on $h(a)$ that $|a+A| = h(a)$. Suppose $a+A = pb+A$ for some $b \in B\oplus G$. The case of $h(a) = \infty$ follows immediately from the claim, so we may assume $h(a) \neq \infty$. Write $b+A = b_1(\beta_{11},\ldots,\beta_{1m_1},y_1)+\cdots+b_k(\beta_{k1},\ldots,\beta_{km_k},y_k)+y_{k+1}+A$, where $0 < b_i < p$ for $i = 1,\ldots,k$. Then $pb+A = a'+A$, where $a' = b_1(\beta_{12},\ldots,\beta_{1m_1},y_1)+\cdots+b_k(\beta_{k2},\ldots,\beta_{km_k},y_k)+py_{k+1}+A$. Note that $h(a') > h(b)$. Let $a'' \in a' + A$ be in canonical form. Then, $h(a'') \geq h(a')$. Now $a''+A = a'+A = pb+A = a+A$. By the uniqueness of the canonical form, we have $h(a) = h(a'') \geq h(a') > h(b)$. By the induction hypothesis, $|b+A| = h(b) < h(a)$. This completes the induction. As a special case, we find that for all $x \in S$, $|x'| = v(x)$ with $x' = x + A$, the image of x in G', as desired. □

If we take $S = G$ we get the usual form of the theorem:

Theorem 2.7.3 (Richman–Walker [134]). *Let G be a $\mathbb{Z}_p$-module and v a p-valuation on G. Then there is a $\mathbb{Z}_p$-module H with submodule $G' \cong G$, where H is torsion over G', H/G' is simply presented, and for every $x \in G$ and its image $x' \in G'$, $|x'|_p^H = v(x)$.*

The group H in Theorem 2.7.3 is called a **Richman–Walker group**. Recall that this was introduced in Section 1.11, without proof, as an example of a group with any desired height structure.

Theorem 2.7.4 (Walker [148]). *Let p be a prime. If α is an ordinal, then there exists a simply presented p-group P_α such that $p^\alpha P_\alpha$ is cyclic of order p.*

Proof. Given α, let $G = \langle x\rangle$ be generated by some x of order p, and define a valuation on G by $v(ax) = \alpha$ for all a not divisible by p. Let $S = \{x\}$. Construct H as in Theorem 2.7.2, and call it P_α. Then the construction shows that $|ax| = \alpha$ for all a relatively prime to p and $|y| < \alpha$ for all $y \in P_\alpha \setminus \langle x\rangle$. Thus, $p^\alpha P_\alpha = \langle x\rangle$, cyclic of order p. By Proposition 2.5.5, the group P_α is simply presented. □

Let us take a closer look at this important group P_α, called a **Walker group**. Since x is an element of height α, we may replace x by α in the representation, obtaining the simple presentation $P_\alpha = \langle X; \Lambda\rangle$, where X is the set of all finite sequences of ordinals $(\alpha_1, \dots, \alpha_n)$ with $\alpha = \alpha_n > \alpha_{n-1} > \dots > \alpha_1$, and Λ is the set of all relations of the forms

$$p(\alpha_1, \dots, \alpha_n) = (\alpha_2, \dots, \alpha_n) \quad (n > 1)$$

and

$$p(\alpha) = 0.$$

Note that $u_p(\beta, P_\alpha) = 1$ if $\beta = \alpha$ and 0 otherwise.

Corollary 2.7.5 (Warfield [152, 154]). *Let p be a prime and $\overline{\alpha}$ an Ulm sequence. Then there is a simply presented $\mathbb{Z}_p$-module H_x of torsion-free rank 1 containing an element of infinite order with Ulm sequence $\overline{\alpha}$.*

Proof. Let $G = \langle x\rangle$ for some x of infinite order, and write $\overline{\alpha} = (\alpha_k)$. Define a valuation on $\langle x\rangle$ by $v(p^k x) = \alpha_k$, and construct H_x as in Theorem 2.7.3. By construction, H_x is simply presented. Since $p^n(\alpha_1, \dots, \alpha_n, x) \in \langle x\rangle$, $H_x/\langle x\rangle$ is torsion, and so, since x has infinite order, H_x has torsion-free rank 1. Also, $|p^k x| = v(p^k x) = \alpha_k$, so the Ulm sequence of x is $\overline{\alpha}$. □

We may extend this result to general groups:

Theorem 2.7.6. *Let G be a group and v_p a p-valuation on G for each prime p. Then there is a group H such that G is a subgroup of H, H/G is torsion, H/G is simply presented, and for all $x \in G$ and p prime, $|x|_p^H = v_p(x)$.*

Proof. For each prime p, let $B(p)$ be the free group generated by all finite sequences $(\alpha_{p1}, \dots, \alpha_{pn}, x)$ with $0 \neq x \in G$ and $\alpha_{p1} < \dots < \alpha_{pn} < v_p(x)$ ordinals or ∞ such that $\alpha_{p1} = \infty$ if $v_p(x) = \infty$. Let $B = \bigoplus_{p\in\mathbb{P}} B(p)$ and A the subgroup of $B \oplus G$ generated by all elements of the form $p(\alpha_{p1}, \dots, \alpha_{pn}, x) - (\alpha_{p2}, \dots, \alpha_{pn}, x)$ or $p(\alpha_{pn}, x) - x$. Let $H = (B \oplus G)/A$, and proceed as in the proof of Theorem 2.7.2. Then H/G' is a simply presented torsion group with $G' = (A + G)/A \cong G$ and $|x'|_p^H = v_p(x)$ for every $x \in G$, its image x' in G' and $p \in \mathbb{P}$. □

Corollary 2.7.7 (Warfield [152, 154]). *Let M be an Ulm matrix. Then there is a simply presented group H_x of torsion-free rank 1 containing an element of infinite order with Ulm matrix M.*

Proof. Let $G = \langle x \rangle$ for some x of infinite order, and write $M = [m_{p,k}]$. Define $v_p(ap^k x) = m_{p,k}$ if $a \in \mathbb{Z}$ and $p \nmid a$. Construct H_x as in the theorem. The rest follows as before. □

We will find the groups P_α and H_x useful.

Exercises.
1. (Walker [148]) Show that $P_\alpha / p^\alpha P_\alpha \cong \bigoplus_{\beta<\alpha} P_\beta$ for all $\alpha > 0$.
2. (Walker [148]) The Walker groups P_α are totally projective.
3. (Walker [149]) If $\alpha = \beta + \delta$, then $p^\beta P_\alpha \cong P_\delta$.

2.8 Simply presented torsion groups

By Lemma 2.6.1, the group $\mathbb{Z}_p$ (= the p-localization of $\mathbb{Z}$ at the prime p) is simply presented. Since a torsion group G is a direct sum of its p-torsion parts $t_p G \cong G \otimes \mathbb{Z}_p$, we may examine torsion groups locally, one prime at a time. The following result is needed; it shows that simple presentations are maintained under such localization:

Lemma 2.8.1. *Let G be a simply presented group and p a prime. Then $G_p = G \otimes \mathbb{Z}_p$ is a simply presented group.*

Proof. We can write $G = F/H$, where F is a free group on a set X and H is a subgroup of F generated by elements of the form nx or $mx - ny$ (where $x, y \in X$ and $m, n \in \mathbb{Z}$). Then G_p can be identified with F_p/H_p by Theorem 1.17.9. Letting Z be the free group on the underlying set X' of F_p, we have a natural epimorphism $Z \to F_p$, whose kernel is denoted by K. This yields an isomorphism

$$f : Z/K \to F_p,$$

and we can write $f^{-1}(H_p) = (K + L)/K$ for some subgroup L of Z which is generated by elements of the form nx or $mx - ny$ (where $x, y \in X'$ and $m, n \in \mathbb{Z}$). But then $K + L$ is generated by elements of the same form since K is, following from the fact that F_p is simply presented (Corollary 2.6.2). Consequently, the group $G_p \cong (Z/K)/((K + L)/K) \cong Z/(K + L)$ is simply presented. □

By Lemmas 2.8.1 and 2.5.3, a torsion group is simply presented if and only if each of its p-torsion parts is simply presented. Thus, for the torsion case, it suffices to study simply presented p-groups; they were introduced by Crawley and Hales [14], who called them *T-groups*.

Now let G be a simply presented p-group with (faithful) presentation $G = \langle X; \Lambda \rangle$. Then a partial order $\leq$ is defined in X by

$$y \leq x \quad \text{if } p^n x = y \text{ for some nonnegative integer } n.$$

This partial order satisfies the minimum condition since G is a p-group, and the set M of all minimal elements consists exactly of those $x \in X$ satisfying the relation $px = 0$ in G. It is clear that in case G is a reduced p-group, $\leq$ also satisfies the maximum condition with those elements of height 0. For each $z \in X$, we define

$$X_z = \{x : x \in X \text{ and } x \geq z\}.$$

Then $\{X_z : z \in M\}$ is a partition of X, so the next result is immediate.

Lemma 2.8.2 (Crawley–Hales [15]). *Let $G = \langle X; \Lambda\rangle$ be a simply presented p-group. Then $G = \bigoplus_{z \in M} \langle X_z\rangle$.*

Proposition 2.8.3 (Crawley–Hales [15]). *Let $G = \langle X; \Lambda\rangle$ be a simply presented p-group, and let Y be a subset of X. Then the subgroup $\langle Y\rangle$ of G and the factor group $G/\langle Y\rangle$ are simply presented. Specifically, $\langle Y\rangle \cong \langle Y', \Lambda_1\rangle$, where $Y' = \langle Y\rangle \cap X$, and Λ_1 consists of those elements of Λ involving only elements of Y', and $G/\langle Y\rangle \cong \langle X_2, \Lambda_2\rangle$, where $X_2 = X \setminus \langle Y\rangle$, and Λ_2 is the set of relations in Λ involving only elements of X_2 and $px = 0$ for all $x \in X_2$ with $px \in \langle Y\rangle$.*

Proof. We write $G = F_X/H$, where H is the subgroup of F_X generated by all elements corresponding to the defining relations in Λ. Let $X_1 = \langle Y\rangle \cap X$ and H_1 the subgroup of F_{X_1} generated by all elements of the form px or $px - y$ $(x, y \in X_1)$ which appear as generators in H. We will show that $\langle Y\rangle \cong G_1 = F_{X_1}/H_1$. Clearly, the correspondence $x \mapsto x$ $(x \in X_1)$ gives rise to an epimorphism $F_{X_1} \to \langle Y\rangle \cong (F_Y + H)/H$ whose kernel contains H_1, yielding an epimorphism

$$\phi : G_1 \to \langle Y\rangle.$$

Now define $Z = \{z : z \in X \setminus X_1 \text{ and } pz = w_z \in X_1\}$ and $X_z = \{x \in X : x \geq z\}$ for each $z \in Z$. Letting H_z be any group isomorphic to $\langle X_z\rangle$, we obtain an isomorphism $\psi_z : \langle X_z\rangle \to H_z$. Then we let

$$A = G_1 \oplus \bigoplus_{z \in Z} H_z$$

and N the subgroup of A generated by all elements of the form

$$(-w_z, 0, \dots, 0, p\psi_z(z), 0, \dots)$$

[which we write simply as $p\psi_z(z) - w_z$] with $z \in Z$. To show that the sum $G_1 + N$ is direct, let $a \in G_1 \cap N$, say $a = \sum r_i(p\psi_{z_i}(z_i) - w_{z_i})$ with $r_i \in \mathbb{Z}$ and $z_i \in Z$. Then $a - \sum r_i w_{z_i} = \sum r_i p\psi_{z_i}(z_i)$ is an element of $G_1 \cap \bigoplus_{z \in Z} H_z = 0$; thus, $a = \sum r_i w_{z_i}$ and $\sum r_i p\psi_{z_i}(z_i) = 0$. Then for $i = 1, \dots, n$, we obtain $r_i p\psi_{z_i}(z_i) = 0$; thus, $r_i w_{z_i} = r_i pz_i = 0$, which yields $a = 0$.

Now there is an epimorphism $f : F_X \to A/N$ such that $x \mapsto (x + H_1, 0, \dots) + N$ if $x \in X_1$, $x \mapsto (0, \dots, 0, \psi_z(x), 0, \dots) + N$ if $x \in X_z$ $(z \in Z)$, and $x \mapsto 0$ whenever

$x \in X \setminus (X_1 \cup \bigcup_{z \in Z} X_z)$. Since we may verify by taking cases that $H \subseteq \ker f$, we obtain a map $G = F_X/H \to A/N$ sending $\langle Y \rangle$ onto $(G_1 + N)/N \cong G_1$. This yields a map

$$\psi : \langle Y \rangle \to G_1$$

such that $\phi\psi = 1_{\langle Y \rangle}$; thus, $\langle Y \rangle \cong G_1$.

To prove the second assertion, let $X_2 = X \setminus \langle Y \rangle$, and define H_2 to be the subgroup of F_{X_2} generated by all elements of the form px or $px - y$ $(x, y \in X_2)$ which appear as generators in H, and by all elements px $(x \in X_2)$ whenever $px \in \langle Y \rangle$. Then $x \mapsto x$ $(x \in X_2)$ gives rise to an epimorphism $F_{X_2} \to G/\langle Y \rangle$ whose kernel contains H_2, which yields a map

$$\phi_2 : G_2 = F_{X_2}/H_2 \to G/\langle Y \rangle.$$

Finally, $x \mapsto x$ $(x \in X_2)$ and $x \mapsto 0$ $(x \in X \cap \langle Y \rangle)$ induce a map $F_X \to G_2$ whose kernel contains H, which yields a map $F_X/H \to G_2$ whose kernel contains $\langle Y \rangle$. Therefore, there is a map $\psi_2 : G/\langle Y \rangle \to G_2$ such that $\phi_2\psi_2 = 1_{G/\langle Y \rangle}$; thus, $G/\langle Y \rangle \cong G_2$. □

We now introduce a concept that generalizes the canonical form we saw in the Richman–Walker groups. Let p be a prime. Following Walker [149], we call a subset X of a group G a ***p*-basis of** G if

1. $px \in X$ whenever $x \in X$ such that $px \neq 0$;
2. for every $a \in G$, there are unique integers r_x with $0 \leq r_x < p$ $(x \in X)$ such that $a = \sum_{x \in X} r_x x$.

The notion of a p-basis was introduced by Crawley and Hales [14], who called it a *T-basis*. For example, $X = \{a, pa, p^2a, \dots, p^{n-1}a\}$ is a p-basis of a cyclic group $\langle a \rangle$ of order p^n, and $X = \{1, p, p^2, \dots\}$ is a p-basis of $\mathbb{Z}$. Notice that a p-basis X does not contain 0, and $X = \emptyset$ is a p-basis of $G = 0$. Clearly, any basis of the $\mathbb{Z}/p\mathbb{Z}$-vector space G/pG is a p-basis of G/pG. We have the following:

Proposition 2.8.4 (Crawley–Hales [16]). *Let G be a p-group and X a subset of G. Then G is simply presented with simple presentation $G = \langle X; \Lambda \rangle$ if and only if X is a p-basis of G.*

Proof. Suppose that $G = \langle X; \Lambda \rangle$ is simply presented. By construction of $\langle X; \Lambda \rangle$, condition (1) holds. To prove (2), let $a \in G$. Then we can write $a = \sum_{i=1}^n r_i x_i$ since X generates G. In fact, since $p^n x \in X$ for all $x \in X$ and $n \geq 0$ with $p^n x \neq 0$, we may assume $0 \leq r_i < p$. To prove uniqueness, assume that $a = \sum_{i=1}^n r_i x_i = \sum_{i=1}^n s_i x_i$ with $x_i \in X$ and $0 \leq r_i, s_i < p$. The claim being trivial for $n = 1$, let $n > 1$, and suppose the assertion is true for $n - 1$. We may assume that x_1 is not p-divisible by any of the other elements x_i. By Lemma 1.12.4, the zero map $\langle px_1, x_2, \dots, x_n \rangle \to \mathbb{Z}(p^\infty)$ extends to a homomorphism $\phi : \langle x_1, x_2, \dots, x_n \rangle \to \mathbb{Z}(p^\infty)$ by sending x_1 onto an element of order p, obtaining $\phi(a) = r_1\phi(x_1) = s_1\phi(x_1)$. Then $r_1 = s_1$, and we have $r_i = s_i$ $(i \neq 1)$ by the induction hypothesis, as desired. Therefore, X is a p-basis of G.

Conversely, assume that G has a p-basis $X = \{x_i : i \in I\}$. Let F_X be the free group on X and H the subgroup of F_X generated by all elements of the form $p^n x$ whenever $p^n x = 0$ in G $(x \in X)$ and $p^m x - y$ whenever $p^m x = y$ in G $(x, y \in X)$. Then the epimorphism $\phi : F_X \to G$ induces the epimorphism

$$f: \begin{array}{rcl} G' = F_X/H & \to & G \\ \sum \lambda_i x_i + H & \mapsto & \sum \lambda_i x_i. \end{array}$$

Let $a \in \ker f$. The group G' is simply presented. Therefore, X is a p-basis of G' by the first part of this proof, and we can write $a = \sum \lambda_i x_i + H$ with $0 \leq \lambda_i < p$. Since $\sum \lambda_i x_i = 0$ in G and X is a p-basis of G, we have $\lambda_i = 0$ by uniqueness of λ_i. Thus, $\ker f = 0$, and we obtain $G \cong G'$. □

Proposition 2.8.5. *Let $G = \langle X; \Lambda \rangle$ be a simply presented p-group. Then we have the following:*
1. *If $a = r_1x_1 + \cdots + r_nx_n$, where $x_1, \ldots, x_n$ are distinct elements of X and $0 < r_1, \ldots, r_n < p$ (see Proposition 2.8.4), then for every ordinal α, we have $a \in p^\alpha G$ if and only if $x_i \in p^\alpha G$ for all $i = 1, \ldots, n$.*
2. *If $x \in X$ and $|x|_p^G > 0$, then there exists a $y \in X$ such that $x = py$.*
3. *The groups $p^\alpha G$ and $G/p^\alpha G$ are simply presented for all ordinals α. In particular, the divisible part D of G and G/D are simply presented.*
4. *If Y is a subset of X, then $A = \langle Y \rangle$ is a nice subgroup of G.*
5. *If G is a reduced group of length λ, then G is a direct sum of simply presented groups G_i $(i \in I)$, such that if λ is a limit ordinal, each group G_i has length $< \lambda$, and if λ is a successor ordinal, each subgroup $p^{\lambda-1}G_i$ is cyclic of order p.*

Proof. We prove that $a \in p^\alpha G$ implies that $x_i \in p^\alpha G$ for all $i = 1, \ldots, n$ by transfinite induction on α. Suppose the assertion holds for all ordinals $< \alpha$. It suffices to assume that α is a successor ordinal. Therefore, we write $a = pb$ for some $b = t_1y_1 + \cdots + t_ky_k \in p^{\alpha-1}G$ for $0 < t_i < p$ and distinct elements $y_i \in X$ for $i = 1, \ldots, k$. By the induction hypothesis, we have $y_1, \ldots, y_k \in p^{\alpha-1}G$. Then $pb = t_1(py_1) + \cdots + t_k(py_k)$, and we rewrite this sum as $pb = u_1z_1 + \cdots + u_lz_l$, using distinct elements $z_i \in X \cap p^\alpha G$, and $0 < u_i < p$ $(i = 1, \ldots, l)$. By uniqueness of the representation of a, we have $\{x_1, \ldots, x_n\} = \{z_1, \ldots, z_l\} \subseteq p^\alpha G$.

To show (2), let $x \in X$ and assume that $x = pg$ for some $g \in G$. Since X is a p-basis of G (Proposition 2.8.4), we can write $g = \sum_{i \in I} r_i x_i$ with $0 < r_i < p$ and $x_i \in X$ $(i \in I)$. Then $px_i \in X$ for all $i \in J = \{i \in I : px_i \neq 0\}$, and the equation $x = \sum_{i \in J} r_i p x_i$ reduces to $x = px_j$ for some $j \in J$.

The third assertion follows from (1) and Proposition 2.8.3.

To prove (4), let $g \in G$. By Proposition 2.8.4, we have $g = r_1x_1 + \cdots + r_mx_m + s_1y_1 + \cdots + s_ny_n$ for elements $x_i \in X \setminus Y$, $y_j \in Y$, and $0 \leq r_i, s_j < p$ $(i = 1, \ldots, m, j = 1, \ldots, n)$. To show that $b = r_1x_1 + \cdots + r_mx_m$ is an element of maximal p-height in the coset $g + A$, let $a \in A$, and write $a = t_1y_1' + \cdots + t_ly_l'$ for some $y_i' \in Y$ and $0 < t_i < p$ $(i = 1, \ldots, l)$. Then by (1), we have

$$|b| = |r_1x_1| \wedge \cdots \wedge |r_mx_m| \geq |r_1x_1| \wedge \cdots \wedge |r_mx_m| \wedge |t_1y_1'| \wedge \cdots \wedge |t_ly_l'| = |b + a|.$$

Therefore, A is nice in G.

To show (5), write $G = \bigoplus_{z \in M} \langle X_z \rangle$ as in Lemma 2.8.2, and let α be any ordinal. If $p^\alpha \langle X_z \rangle$ has nonzero elements, then $z \in p^\alpha \langle X_z \rangle$ by (1). Letting $\beta = |z|_p$, we obtain $p^\beta \langle X_z \rangle = \langle z \rangle$; thus, the length of $\langle X_z \rangle$ is $\beta + 1$, and the first assertion follows. If $\lambda - 1$ exists, let $M' = \{z \in M : |z|_p = \lambda - 1\}$, pick $z' \in M'$, and define $G_1 = \langle X_{z'} \rangle \oplus \bigoplus_{z \in M \setminus M'} \langle X_z \rangle$. Then $G = G_1 \oplus \bigoplus_{z \in M' \setminus \{z'\}} \langle X_z \rangle$. □

Lemma 2.8.6 (Crawley–Hales [16]). *Let $G = \langle X; \Lambda \rangle$ be a reduced simply presented p-group, and let $X_0 = \{x \in X : |x|_p^G = 0\}$. Then we have the following:*
1. *$G = \langle X_0 \rangle$.*
2. *The set $\{x + pG : x \in X_0\}$ is a basis of G/pG.*
3. *If $Y \subseteq X_0$ and $a \in \langle Y \rangle$ with $|a|_p^G > 0$, then $|a|_p^{\langle Y \rangle} > 0$.*

Proof. To prove (1), let $g \in G$, and write $g = \sum_{i \in I} r_ix_i$ with $r_i \in \mathbb{Z}$, $x_i \in X$ $(i \in I)$. We fix $i \in I$, and let $y_1 = x_i$. If $|y_1|_p \neq 0$, there is a $y_2 \in X$ such that $y_1 = py_2$ by Proposition 2.8.5(2), and in case $|y_2|_p \neq 0$, there is a $y_3 \in X$ with $y_2 = py_3$, et cetera. Since any decreasing sequence of ordinals becomes stationary, we must have $|y_n|_p = 0$ for some $n \in \mathbb{N}$, i. e., $y_n \in X_0$. Therefore, $g \in \langle X_0 \rangle$.

To show (2), it suffices to verify that the subset $\{x + pG : x \in X_0\}$ of G/pG is independent. Assume that $\sum_{i \in I} \lambda_ix_i \in pG$ with $0 \leq \lambda_i < p$ and $x_i \in X_0$ $(i \in I)$. Then for each $i \in I$, we have $\lambda_i = 0$, since otherwise $x_i \in pG$, by Proposition 2.8.5(1), is a contradiction.

To prove the last assertion, let $Y \subseteq X_0$ and $a \in \langle Y \rangle$ with $|a|_p^G > 0$. Then $a = \sum_{i \in I} \mu_iy_i$ with $0 \neq \mu_i \in \mathbb{Z}$ and $y_i \in Y$. For each $i \in I$, we write $\mu_i = s_ip + r_i$ with $0 \leq r_i < p$. Then $\sum_{i \in I} r_iy_i \in pG$. Therefore, if $r_i \neq 0$, then $y_i \in pG$ by Proposition 2.8.5(1), which is impossible. Thus, $r_i = 0$ for all $i \in I$, and we obtain $a \in p\langle Y \rangle$. □

The next result follows from Proposition 2.8.5(4), taking $\mathcal{C}$ to be all $\langle Y \rangle$ where $Y \subseteq X$ and $G = \langle X; \Lambda \rangle$:

Proposition 2.8.7 (Crawley–Hales [14], Hill [47]). *Simply presented p-groups have nice systems.*

Consequently, Hill's Theorem 2.3.3 yields the following classification of simply presented torsion groups:

Corollary 2.8.8 (Crawley–Hales [14]). *Two simply presented torsion groups are isomorphic if and only if they have the same Ulm invariants.*

The following useful fact was extracted from [27, p. 99]:

Lemma 2.8.9. *Let p be a prime, σ a limit ordinal, and B a direct sum of cyclic p-groups, say* $B \cong \bigoplus_{i\in I} \mathbb{Z}/p^{n_i}\mathbb{Z}$ $(n_i \in \mathbb{N})$. *Let* $A = \langle X; \Lambda\rangle$ *be a reduced simply presented p-group of length σ, and let M be the set of all minimal elements in X. For every* $\gamma < \sigma$, *let* $M_\gamma = \{x \in M : |x|_p \geq \gamma\}$, *and assume that* $|M_\gamma| \geq |I|$. *Then there exists a simply presented p-group C such that* $C/p^\sigma C \cong A$ *and* $p^\sigma C \cong B$.

Proof. Let y_{i0} be a generator of $\mathbb{Z}/p^{n_i}\mathbb{Z}$ $(i \in I)$, and write $B = \langle Y, \Lambda_Y\rangle$, where $Y = \{y_{i0}, y_{i1}, \ldots, y_{i(n_i-1)} : i \in I\}$, and Λ_Y consists of relations $py_{ij} = y_{i(j+1)}$ with $j = 0, \ldots, n_i - 2$ and $py_{i(n_i-1)} = 0$ $(i \in I)$. As usual, we assume the presentation of A is faithful as well. For (1), we define a simply presented p-group C with presentation $C = \langle X'; \Lambda'\rangle$, where $X' = X \cup Y$. We wish to define Λ' with the following properties:

1. If $px = y$ in is Λ, then it is in Λ';
2. $\Lambda_Y \subseteq \Lambda'$;
3. For every $i \in I$ and $\gamma < \sigma$ there is at least one $x \in M_\gamma$ such that $px = y_{i0} \in \Lambda'$;
4. For every $x \in M$ there is a unique y_{i0} such that $px = y_{i0}$ is in Λ'.

First, for each $\gamma < \sigma$, we use induction on γ to construct injective functions $\phi_\gamma : Y_0 = \{y_{i0} : i \in I\} \to M_\gamma$, with the property that if $\phi_\alpha(y_{i0}) \in M_\gamma$ for some $\alpha < \gamma$, then $\phi_\gamma(y_{i0}) = \phi_\alpha(y_{i0})$. Let $\gamma < \sigma$, and assume that suitable maps $\phi_\alpha : Y_0 \to M_\alpha$ are already constructed for all $\alpha < \gamma$. We define $\phi_\gamma(y_{i0})$ for each $y_{i0} \in Y_0$. If $\phi_\alpha(y_{i0}) \in M_\gamma$ for some $\alpha < \gamma$, let $\phi_\gamma(y_{i0}) = \phi_\alpha(y_{i0})$. If not (that is, if $|\phi_\alpha(y_{i0})| < \gamma$ for all $\alpha < \gamma$), let $\phi_\gamma(y_{i0}) = x$ for some $x \in M_\gamma$ in such a way that ϕ_γ is injective. This can be done since $|M_\gamma| \geq |I|$. This completes the induction.

Now construct Λ' to include the following:

(a) all relations of the form $px = y$, where $y \in Y_0$ and $\phi_\gamma(y) = x$ for some $\gamma < \sigma$;
(b) one relation $px = y$ for each $x \in M$ that does not appear in any of the relations in (a), for some $y \in Y_0$;
(c) the relations defined by properties (1) and (2) above.

We confirm that Λ' satisfies properties (1–4). By construction, we have (1) and (2) immediately; (3) because ϕ_γ is a function, and the existence of the y_{i0} in (4) by construction. It remains to prove y_{i0} is unique. If $x \notin \bigcup_{\alpha<\sigma} \phi_\alpha(Y_0)$, then by (b), x is assigned by construction to a unique $y \in Y_0$. Now suppose for some $\alpha \leq \beta < \sigma$, we have $x = \phi_\alpha(y) = \phi_\beta(y')$. If $\alpha = \beta$, then $y = y'$ since ϕ_α is injective. If $\alpha < \beta$, we consider two cases. If $|x|_p \geq \beta$, then $x \in M_\beta$, so $\phi_\beta(y) = \phi_\alpha(y) = \phi_\beta(y')$, and again, we get $y = y'$. Suppose $|x|_p < \beta$. Then $x \notin M_\beta$, whereas $\phi_\beta(y') \in M_\beta$, a contradiction.

We claim that the subset X of X' has no elements in C of p-height $\geq \sigma$. To prove this, write $A = F_X/H$ and $C = F_{X'}/K$. Then the correspondences $x \mapsto x$ $(x \in X)$ and $y \mapsto 0$ $(y \in Y)$ give rise to an epimorphism $F_{X'} \to F_X/H$ whose kernel contains K, inducing a natural epimorphism $F_{X'}/K \to F_X/H$. This shows that $|x+K|_p \leq |x+H|_p < \sigma$ whenever $x \in X$, as claimed.

This proves $|x|_p^C \le |x|_p^A$ for all $x \in X$. We prove by induction on $|x|_p^C$ that $|x|_p^C = |x|_p^A$ for all $x \in X$. Suppose $|x|_p^A > |x|_p^C = \alpha$ for some $x \in X$. Then in A, $x = py$ for some $y \in X$ with $|y|_p^A \ge \alpha$ by Proposition 2.8.5(2) applied to $p^\alpha A$. Then $x = py$ in C as well, so $|y|_p^C < |x|_p^C = \alpha$. From the induction hypothesis, we get $|y|_p^C = |y|_p^A \ge \alpha$, a contradiction.

Now consider any y_{i0}. For every $\gamma < \sigma$ there is an $x \in X$ such that $y_{i0} = px$ and $|x|_p^C = |x|_p^A \ge \gamma$. Hence, $|y_{i0}|_p^C \ge \sigma$. Every element of Y is a multiple of some y_{i0}, so $|y|_p^C \ge \sigma$ for all $y \in Y$.

Since all of the relations are in the proper form, to prove this presentation is faithful, it suffices to prove that the elements of X' are distinct and nonzero in C. If $x \in X$, then $|x + K|_p \le |x + H|_p < \sigma$ shows that $x \notin K$; thus, $x \ne 0$ in C. Likewise, we see that the elements of X are distinct in C. Next suppose some $y \in Y$ is 0 in C. Then $y \in K$, which is generated by expressions of the following forms: $px - z$ where $x, z \in X$ and $px \ne 0$ in A; $px = y_{i0}$ for some $x \in X$ with $px = 0$ in A; any expression that reduces to 0 in B. Suppose there is a nonzero term of the form $px - y_{i0}$. Then there must be another term involving px to allow the sum to reduce to y_{i0}. But the only possibility is $px - y_{j0}$ with $j \ne i$, but this is impossible based on our construction of Λ'. Thus, the only terms involve relations in Y which say $y = 0$ in B. Likewise, the elements of Y are distinct in C since $y_{i0} - y_{j0} \in K$ implies $px = y_{i0}$ and $px' = y_{j0}$ are in Λ' for some $x, x' \in X$, which forces $x = x'$, which violates (4) unless $y_{i0} = y_{j0}$. Clearly, the elements of X and Y are distinct in C since their heights are different.

Next we show that $p^\sigma C \cong B$. By Proposition 2.8.5(1), we have $p^\sigma C = \langle X''\rangle$ with $X'' = \{x \in X' : |x|_p^C \ge \sigma\}$. Then $X'' = \langle X''\rangle \cap X'$, so Proposition 2.8.3 yields $p^\sigma C \cong \langle X'', \Lambda_1\rangle$, where Λ_1 consists of those elements of Λ' involving only elements of X''. Assume Λ' has a relation $px = y$ $(x, y \in X'')$; that is, $x, y \in X' = X \cup Y$ with $|x|_p^C, |y|_p^C \ge \sigma$. Then $x \in X$ is impossible since X has no elements in C of p-height $\ge \sigma$. Thus, $px = y$ is one of the relations $py_{i0} = y_{i1}, \dots, py_{i(n_i-2)} = y_{i(n_i-1)}$ for some $i \in I$. It follows that $p^\sigma C \cong \langle X'', \Lambda_1\rangle = \langle Y, \Lambda_Y\rangle = B$.

Likewise, Proposition 2.8.3 shows that $C/p^\sigma C \cong \langle X_2, \Lambda_2\rangle$, where $X_2 = X' \setminus p^\sigma C = \{x \in X' : |x|_p^C < \sigma\}$, and Λ_2 consists of all relations in Λ' involving only elements of X_2 and $px = 0$ for all $x \in X_2$ with $px \in p^\sigma C$. Then $\Lambda_2 = \Lambda$, and we have $C/p^\sigma C \cong \langle X_2, \Lambda_2\rangle \cong \langle X, \Lambda\rangle = A$, as desired. □

Lemma 2.8.10. *Let G be a p-group and σ a limit ordinal. If $G/p^\sigma G$ is simply presented such that $p^\sigma G$ is cyclic, then G is simply presented.*

Proof. Write $G/p^\sigma G = \langle X, \Lambda\rangle$, and $p^\sigma G = \langle z\rangle$. Then $|z|_p^{p^\sigma G} = 0$, so $|z|_p^G = \sigma$ by Lemma 1.10.4(2). Let $\gamma < \sigma$. Since $|z|_p^{p^\gamma G} > 0$, there is a $y \in p^\gamma G$ such that $z = py$. Since $y \notin p^\sigma G$, $\bar{y} = y + p^\sigma G$ is a nonzero element of $G/p^\sigma G$, so $\bar{y} = a_1x_1 + \cdots + a_nx_n$ for distinct elements $x_1, \dots, x_n$ of X and $0 < a_i < p$ for all $1 \le i \le n$. Since $\bar{y} \in p^\gamma G/p^\sigma G = p^\gamma(G/p^\sigma G)$ (Lemma 1.10.5), we have $x_i \in p^\gamma(G/p^\sigma G)$ for all i. Thus, $\bar{y} \in \langle X_\gamma\rangle$, where $X_\gamma = \{x \in X : |x|_p^{G/p^\sigma G} \ge \gamma\}$. Let M_γ be the set of all minimal elements x in X such that $|x|_p^{G/p^\sigma G} \ge \gamma$. Then $|M_\gamma| \ge 1$ since $X_\gamma \ne \emptyset$. Therefore, by Lemma 2.8.9, applied to

$p^\sigma G$ and $G/p^\sigma G$, there is a simply presented p-group C, such that $C/p^\sigma C \cong G/p^\sigma G$ and $p^\sigma C \cong p^\sigma G$. Since simply presented p-groups have nice systems (Proposition 2.8.7) and $u_\alpha(C, p^\sigma C) = u_\alpha(G, p^\sigma G)$ for all ordinals α (Lemmas 1.13.6 and 1.13.10), we obtain $C \cong G$ by Theorem 2.3.2. □

As an application of Lemma 2.8.10, we have the following:

Proposition 2.8.11. *Generalized Prüfer groups are simply presented.*

Proof. The assertion is shown by transfinite induction on α. For $n < \omega$, the claim is true since $H_n \cong \mathbb{Z}/p^n\mathbb{Z}$. Suppose $\alpha \geq \omega$, and assume that the claim holds for all ordinals $< \alpha$. If α is a limit ordinal, then $H_\alpha = \bigoplus_{\beta<\alpha} H_\beta$ is simply presented by induction and Lemma 2.5.3. If α is a successor ordinal, then $\alpha = \beta + n$ with β a limit ordinal and $0 < n < \omega$. Since by Lemma 2.4.1, $p^\beta H_{\beta+n} \cong \mathbb{Z}/p^n\mathbb{Z}$ and $H_{\beta+n}/p^\beta H_{\beta+n} \cong H_\beta$ is simply presented by induction, Lemma 2.8.10 shows that $H_{\beta+n}$ is simply presented, as desired. □

Proposition 2.8.12. *Let G be an infinite reduced simply presented p-group. Then $|G| = \sum_\alpha u_p(\alpha, G)$.*

Proof. We prove the claim by induction on the length λ of G. If $\lambda < \omega$, then G is bounded. Therefore, it is a direct sum of cyclic groups and the assertion is immediate. Suppose $\lambda \geq \omega$, and assume the claim holds for all ordinals $< \lambda$. If λ is a limit ordinal, then G is a direct sum of simply presented p-groups of lengths $< \lambda$ (Proposition 2.8.5(5)). Thus, $|G| = \sum_\alpha u_p(\alpha, G)$ by the induction hypothesis and Section 1.13(a). If $\lambda - 1$ exists, we let β be the largest limit ordinal $< \lambda$. Then $G/p^\beta G$ is infinite and $|G/p^\beta G| \geq |p^\beta G| \geq \sum_{\alpha\geq\beta} u_p(\alpha, G)$ by Lemma 1.10.7 and Section 1.13(b). By the induction hypothesis and Lemma 1.13.6, we have $|G/p^\beta G| = \sum_{\alpha<\beta} u_p(\alpha, G/p^\beta G) = \sum_{\alpha<\beta} u_p(\alpha, G)$. Thus,

$$|G| = |G/p^\beta G| \cdot |p^\beta G| = |G/p^\beta G| = \sum_{\alpha<\beta} u_p(\alpha, G),$$

which equals $\sum_{\alpha<\beta} u_p(\alpha, G) + \sum_{\alpha\geq\beta} u_p(\alpha, G)$ since $G/p^\beta G$ is infinite. □

Let f be a function from the ordinals to the cardinals that vanishes beyond some ordinal. We say that f **has length** λ, provided that λ is an ordinal such that

$$\lambda = \sup\{\alpha + 1 : \alpha < \lambda \text{ and } f(\alpha) \neq 0\},$$

and we call f **λ-admissible** if, in addition, we have

$$\sum_{\sigma\leq\beta<\sigma+\omega} f(\beta) \geq \sum_{\beta\geq\sigma+\omega} f(\beta) \quad \text{for all ordinals } \sigma \text{ with } \sigma + \omega < \lambda.$$

For example, any p-group G gives rise to its **Ulm function** u_G defined by $u_G(\alpha) = u_p(\alpha, G)$ for all ordinals α. If G has length λ, then u_G has length λ:

Lemma 2.8.13. *If G is a p-group of length λ, then λ is the length of u_G.*

Proof. We can assume that G is reduced (see Remark 1.13.2). If λ is a limit ordinal, the claim follows from Lemma 1.13.7(2). If $\lambda - 1$ exists, we have $p^{\lambda-1}G \neq 0$. Then $(p^{\lambda-1}G)[p] \neq 0$; thus, $u_p(\lambda - 1, G) \neq 0$. □

For countable p-groups, we can say a bit more:

Lemma 2.8.14. *Let G be a countable p-group of length λ. Then u_G is a λ-admissible function.*

Proof. Again, we may assume that G is reduced. Suppose that $\sigma + \omega < \lambda$. Since $\sigma + \omega$ is a limit ordinal, we have $\sum_{\sigma \leq \beta < \sigma+\omega} u_G(\beta) = \aleph_0$ by Lemma 1.13.7(2). □

Crawley and Hales [14] were able to prove that a function f is λ-admissible if and only if there is a reduced simply presented p-group G with $u_G = f$. To establish this result, a set-theoretic lemma is needed, which can be found in Fuchs [27]:

Lemma 2.8.15. *Let f be a λ-admissible function, where λ is a limit ordinal. Then there are λ_i-admissible functions f_i with $\lambda_i < \lambda$ $(i \in I)$ such that $f = \sum_{i\in I} f_i$.*

Proof. For $\sigma < \lambda$, we consider $\mathfrak{m}_\sigma = \sum_{\beta\geq\sigma} f(\beta)$ and set $\mathfrak{m} = \min\{\mathfrak{m}_\sigma : \sigma < \lambda\}$. Let γ be the smallest ordinal such that $\mathfrak{m}_\gamma = \mathfrak{m}$. Such a γ must exist since decreasing sequences of ordinals become stationary. We define $I = \{\nu : \gamma \leq \nu < \lambda\}$. Let $\mu < \lambda$ be a limit ordinal or $\mu = 0$, and consider the cardinal

$$\mathfrak{n}_\mu = \min\left\{\sum_{\mu+k\leq\beta<\mu+\omega} f(\beta) : k < \omega\right\}.$$

If $\mu+\omega = \lambda$, then $\mathfrak{n}_\mu$ is infinite since λ is the length of f, and in the case where $\mu+\omega < \lambda$, $\mathfrak{n}_\mu$ is infinite since f is a λ-admissible function. Assuming that $\mathfrak{n}_\mu < |I|$, we can write I as the union of $> \mathfrak{n}_\mu$ pairwise disjoint intervals $[\alpha, \alpha + \omega)$, each containing an ordinal β_α with $f(\beta_\alpha) \neq 0$; thus, $\sum_{\beta\geq\gamma} f(\beta) > \mathfrak{n}_\mu$. But this is impossible since $\mathfrak{n}_\mu \geq \sum_{\beta\geq\mu+\omega} f(\beta) \geq \mathfrak{m} = \sum_{\beta\geq\gamma} f(\beta)$; thus, $\mathfrak{n}_\mu \geq |I|$.

Now we concentrate on the interval $[\mu, \mu + \omega)$. Let ϕ_μ be the restriction of f to this interval. We claim that there are functions $\phi_{\mu\nu}$ $(\nu \in I)$ such that ϕ_μ can be written as the sum

$$\phi_\mu = \sum_{\nu\in I} \phi_{\mu\nu},$$

and such that

$$\sum_{\mu+k\leq\beta<\mu+\omega} \phi_{\mu\nu}(\beta) \geq \mathfrak{n}_\mu \quad \text{for all } k < \omega, \nu \in I.$$

First note that the interval consists of countably many elements $\mu + i$ $(i < \omega)$, and we may associate a value with each $i < \omega$: Let $\kappa_i = f(\mu + i)$ for each $i < \omega$. Let $\delta = |I|$ and $I' = \{\alpha : \alpha < \delta\}$. Now for each $\alpha \in I'$, we define a function η_α on the interval.

In the case where κ_i is finite, we define $\eta_0(\mu+i) = \kappa_i$, and $\eta_\alpha(\mu+i) = 0$ for $0 < \alpha < \delta$. If κ_i is infinite, then we define $\eta_\alpha(\mu+i) = \kappa_i$ if $\alpha < \kappa_i$ and $\eta_\alpha(\mu+i) = 0$ whenever $\kappa_i \le \alpha < \delta$. Then for each $i < \omega$, if κ_i is finite, $\sum_{\alpha\in I'} \eta_\alpha(\mu+i) = \eta_0(\mu+i) = \kappa_i$, and if κ_i is infinite, $\sum_{\alpha\in I'} \eta_\alpha(\mu+i) = \sum \kappa_i$, where the sum is taken over all $\alpha \in I'$ with $\alpha < \kappa_i$, which sums to κ_i. Thus, $\sum_{\alpha\in I'} \eta_\alpha(\mu+i) = f(\mu+i)$ for all i.

For elements $\alpha \in I'$ and $k < \omega$, we define $J_k = \{i : k \le i < \omega, \kappa_i \ge \alpha\}$. Since $\sum_{i\in J_k} f(\mu+i)$ differs from $\sum_{k\le i<\omega} f(\mu+i)$ in only countably many terms, all less than $\alpha < \delta = |I| \le \mathfrak{n}_\mu$, we have $\sum_{k\le i<\omega} \eta_\alpha(\mu+i) \ge \sum_{i\in J_k} \eta_\alpha(\mu+i) = \sum_{i\in J_k} f(\mu+i) = \sum_{k\le i<\omega} f(\mu+i) \ge \mathfrak{n}_\mu$.

Now let $I \to I'$ be any bijection given by $\nu \mapsto \nu'$, and define $\phi_{\mu\nu} = \eta_{\nu'}$ for $\nu \in I$. Then the functions $\phi_{\mu\nu}$ $(\mu \in I)$ satisfy the desired properties.

We extend ϕ_μ to the entire range of f by defining $\phi_\mu(\beta) = 0$ if $\beta < \mu$ or $\beta \ge \mu+\omega$, and set

$$\psi_\nu = \sum_\mu \phi_{\mu\nu} \quad \text{for each } \nu \in I.$$

Notice that $f = \sum_\mu \phi_\mu = \sum_\nu \psi_\nu$. For every $\nu \in I$, we define a function f_ν of length $\lambda_\nu < \lambda$ by

$$f_\nu(\beta) = \begin{cases} \psi_\nu(\beta) & \text{if } \beta < \nu, \\ \sum_{\alpha\le\nu} \psi_\alpha(\beta) & \text{if } \beta = \nu, \\ 0 & \text{if } \beta > \nu. \end{cases}$$

It is clear that $\sum_\nu f_\nu(\beta) = \sum_{\nu>\beta} \psi_\nu(\beta) + \sum_{\alpha\le\beta} \psi_\alpha(\beta) = f(\beta)$ for all β.

To prove that each f_ν is λ_ν-admissible, let σ be an ordinal with $\sigma+\omega < \nu+1$, and write $\sigma = \rho + k$ for some ρ (a limit ordinal or 0) and $k < \omega$. Then

$$\sum_{\sigma\le\beta<\sigma+\omega} \psi_\nu(\beta) = \sum_\mu \sum_{\sigma\le\beta<\sigma+\omega} \phi_{\mu\nu}(\beta) \ge \sum_{\rho+k\le\beta<\rho+\omega} \phi_{\rho\nu}(\beta) \ge \mathfrak{n}_\rho \ge \sum_{\beta\ge\sigma+\omega} f(\beta).$$

Thus, we obtain

$$\sum_{\sigma\le\beta<\sigma+\omega} f_\nu(\beta) = \sum_{\sigma\le\beta<\sigma+\omega} \psi_\nu(\beta) \ge \sum_{\beta\ge\sigma+\omega} f(\beta) = \sum_{\beta\ge\sigma+\omega} \sum_\nu \psi_\nu(\beta) \ge \sum_{\beta\ge\sigma+\omega} f_\nu(\beta)$$

and the proof is complete. □

Based on the proof in [27, Theorem 83.6], we are now able to show the following:

Theorem 2.8.16 (Crawley–Hales [14]). *Let f be a λ-admissible function. Then there exists a reduced simply presented p-group G such that $u_G = f$.*

Proof. We will prove the result by transfinite induction on λ. If $\lambda < \omega$, then we let $G = \bigoplus_{i<\lambda} \bigoplus_{f(i)} \mathbb{Z}/p^{i+1}\mathbb{Z}$ and have $u_G(i) = f(i)$ for all $i < \lambda$.

Now suppose $\lambda \ge \omega$. If λ is a limit ordinal, then by Lemma 2.8.15, we can write $f = \sum_{i\in I} f_i$, where each f_i is a λ_i-admissible function with $\lambda_i < \lambda$. By the induction hypothesis, there are reduced simply presented p-groups G_i with $u_{G_i} = f_i$ $(i \in I)$. Thus, $G = \bigoplus_{i\in I} G_i$ satisfies $u_G = f$. Now assume that $\lambda = \sigma + n$, where σ is a limit ordinal and $0 < n < \omega$. Define a function g by $g(\alpha) = f(\alpha)$ if $\alpha < \sigma$, and $g(\alpha) = 0$ if $\alpha \geq \sigma$. Then g is a σ-admissible function. Therefore, by induction there is a reduced simply presented p-group A of length σ, such that $u_A = g$. Now let $B = \bigoplus_{i<n} \bigoplus_{f(\sigma+i)} \mathbb{Z}/p^{i+1}\mathbb{Z}$. Note that if I is the summation index for B, then $|I| = \sum_{i<n} f(\sigma + i)$. Since A is simply presented, A may be represented as some $\langle X; \Lambda\rangle$. Let $M = \{x \in X : px = 0\}$, and for any $\gamma < \sigma$, $M_\gamma = \{x \in M : |x| \geq \gamma\} = \bigcup_{\alpha\geq\gamma}\{x \in M : |x| = \alpha\}$. Suppose $x_1, \ldots, x_n \in M$, where $|x_i| = \alpha$ for each $1 \leq i \leq n$ and $a_1x_1 + \cdots + a_nx_n \in p^{\alpha+1}A$, where $a_1, \ldots, a_n$ are all relatively prime to p. Then by Proposition 2.8.5, $a_ix_i \in p^{\alpha+1}A$ for all $1 \leq i \leq n$. In other words, their images are independent in $(p^\alpha A)[p]/(p^{\alpha+1}A)[p]$, and so $|\{x \in M_\gamma : |x| = \alpha\}| \geq u_A(\alpha) = g(\alpha)$. Taking sums, $|M_\gamma| \geq \sum_{\beta\geq\gamma} g(\beta) = \sum_{\gamma\leq\beta\leq\sigma} f(\beta)$. Since $\gamma + \omega \leq \sigma < \lambda$, we have

$$\sum_{\gamma\leq\beta\leq\sigma} f(\beta) \geq \sum_{\gamma\leq\beta<\gamma+\omega} f(\beta) \geq \sum_{\beta\geq\gamma+\omega} f(\beta) \geq \sum_{\beta\geq\sigma} f(\beta).$$

Thus, $|M_\gamma| \geq \sum_{\beta\geq\sigma} f(\beta) \geq \sum_{i<n} f(\sigma + i) = |I|$. Then Lemma 2.8.9 yields a reduced simply presented p-group G such that $G/p^\sigma G \cong A$ and $p^\sigma G \cong B$. If $\alpha < \sigma$, then $u_p(\alpha, G) = u_p(\alpha, A) = f(\alpha)$ by Lemma 1.13.6. Finally, for every $i < n$, we have $u_p(\sigma+i, G) = u_p(i, p^\sigma G) = u_p(i, B) = f(\sigma + i)$. Thus, $u_G = f$ and the proof is complete. □

The converse of Theorem 2.8.16 will follow from a stronger result (Theorem 2.8.19). First, we prove the useful fact that countable p-groups are simply presented:

Theorem 2.8.17 (Crawley–Hales [14]). *Countable p-groups are simply presented.*

Proof. Let G be a countable p-group of length λ. Then $G = D \oplus A$ with D divisible and A reduced. By Lemma 2.8.14, u_A is λ-admissible. Therefore, by Theorem 2.8.16, there is a reduced simply presented p-group H with $u_A = u_H$. Since A and H have nice systems (see Proposition 2.8.7), we obtain $A \cong H$ by Theorem 2.3.3. By the structure theorem for divisible groups (Theorem 1.6.7) D is simply presented. Thus, G is isomorphic to the simply presented p-group $D \oplus H$. □

Lemma 2.8.18 (Crawley–Hales [16]). *Let G be a reduced simply presented p-group. If $G = A \oplus A'$, then A contains a subset Z such that $A = \langle Z\rangle$ and $\{z + pA : z \in Z\}$ is a basis of A/pA.*

Proof. Assuming that $G = \langle X; \Lambda\rangle$, we let $X_0 = \{x \in X : |x|_p^G = 0\}$ and have $G = \langle X_0\rangle$ by Lemma 2.8.6(1). Now write $X_0 = \{x_\sigma : \sigma < \tau\}$. Then $x_0 = a_0 + a_0'$ with $a_0 \in A$, $a_0' \in A'$, and we can write $a_0 = \sum_{i\in I_0} \lambda_{0i}x_{0i}$ and $a_0' = \sum_{i\in I_0} \mu_{0i}x_{0i}$ for elements $x_{0i} \in X_0$ and $\lambda_{0i}, \mu_{0i} \in \mathbb{Z}$ $(i \in I_0)$. For each $i \in I_0$, we write $x_{0i} = a_1 + a_1'$ with $a_1 \in A$, $a_1' \in A'$, and obtain $a_1 = \sum_{i\in I_1} \lambda_{1i}x_{1i}$ and $a_1' = \sum_{i\in I_1} \mu_{1i}x_{1i}$, where $x_{1i} \in X_0$ and $\lambda_{1i}, \mu_{1i} \in \mathbb{Z}$ $(i \in I_1)$, et cetera. We let $X_0' = \bigcup_{k=0}^{\infty}\{x_{ki} : i \in I_k\}$, $R_0 = 0$, and $R_1 = \langle X_0'\rangle$. Proceeding inductively, we successively construct $\tilde{X}_\alpha$, starting with x_α, as was done for X_0' and x_0. Then we let $X_\alpha' = \tilde{X}_\alpha \setminus \bigcup_{\beta<\alpha} X_\beta'$ and $R_\alpha = \langle\bigcup_{\beta<\alpha} X_\beta'\rangle$. In this way, we obtain pairwise disjoint countable subsets X'_σ of X_0 with $x_\sigma \in \bigcup_{\beta\le\sigma} X'_\beta$ and subgroups $R_\sigma = \langle \bigcup_{\beta<\sigma} X'_\beta \rangle$ of G such that

$$(R_\sigma \cap A) \oplus (R_\sigma \cap A') = R_\sigma$$

for all $\sigma < \tau$. Clearly, we have $A = \bigcup_\sigma (R_\sigma \cap A)$.

Next we will construct an ascending chain $Z_0 \subseteq Z_1 \subseteq \cdots \subseteq Z_\sigma \subseteq \cdots (\sigma < \tau)$ of subsets of A such that—for each $\sigma < \tau$—we have $\langle Z_\sigma \rangle = R_\sigma \cap A$ and $\{z + pA : z \in Z_\sigma\}$ is a set of independent elements in A/pA. The case $\sigma = 0$ being trivial, let $\sigma > 0$, and assume that sets Z_β are already constructed for all $\beta < \sigma$. First, suppose $\sigma - 1$ exists. Then

$$(R_\sigma \cap A)/(R_{\sigma-1} \cap A) \oplus (R_\sigma \cap A')/(R_{\sigma-1} \cap A') \cong R_\sigma/R_{\sigma-1} = (\langle X'_{\sigma-1} \rangle + R_{\sigma-1})/R_{\sigma-1};$$

thus, $R_\sigma/R_{\sigma-1}$ is a countable p-group. Furthermore, by Proposition 2.8.3, $R_{\sigma-1}$ and R_σ are simply presented, and the generating set of the former is a subset of that of the latter, and so $R_{\sigma-1}$ is nice in R_σ by Proposition 2.8.5(4). Then $R_\sigma/R_{\sigma-1}$ is reduced by Proposition 2.1.1. By Theorem 2.8.17, $C = (R_\sigma \cap A)/(R_{\sigma-1} \cap A)$ is a reduced simply presented p-group. This group is generated by a set $Y_0 = \{x + (R_{\sigma-1} \cap A) : x \in W_\sigma\}$ for some countable subset W_σ of $R_\sigma \cap A$ such that every element $y = r_1y_1 + \cdots + r_ny_n$ with $0 < r_i < p$ and $y_i \in Y_0$ $(i = 1, \dots, n)$ satisfies $|y|_p^C = 0$ (see Lemma 2.8.6(1) and Proposition 2.8.5(1)). Letting $Z_\sigma = Z_{\sigma-1} \cup W_\sigma$, we obtain $\langle Z_\sigma \rangle = R_\sigma \cap A$.

To show that the set of cosets $z + pA$ with $z \in Z_\sigma$ is independent, let $a = \sum_{z\in Z_\sigma} \lambda_z z$ with $0 \le \lambda_z < p$, and suppose that $a \in pA$. Then $|a|_p^G > 0$; thus, $|a|_p^{R_\sigma} > 0$ by Lemma 2.8.6(3). Therefore, $|a|_p^{R_\sigma\cap A} > 0$. On the other hand, $a + (R_{\sigma-1} \cap A) = \sum_{z\in W_\sigma} \lambda_z z + (R_{\sigma-1} \cap A)$. Assuming $\lambda_z \ne 0$ for some $z \in W_\sigma$, we obtain $|a + (R_{\sigma-1} \cap A)|_p^C = 0$; hence, $|a|_p^{R_\sigma\cap A} = 0$, which is impossible. Thus, $\lambda_z = 0$ for all $z \in W_\sigma$, and we have $\sum_{z\in Z_{\sigma-1}} \lambda_z z + pA = 0$, which yields $\lambda_z z + pA = 0$ for all $z \in Z_{\sigma-1}$ by the induction hypothesis, proving independence.

If σ is a limit ordinal, we let $Z_\sigma = \bigcup_{\beta<\sigma} Z_\beta$, obtaining $\langle Z_\sigma \rangle = \bigcup_{\beta<\sigma} (R_\beta \cap A) = R_\sigma \cap A$, as required. This completes the induction.

Since $A = \bigcup_\sigma (R_\sigma \cap A)$ it follows that $Z = \bigcup_\sigma Z_\sigma$ generates A and that $\{z + pA : z \in Z\}$ is a basis of A/pA. □

Theorem 2.8.19 (Crawley–Hales [16]). *Let G be a direct summand of a reduced simply presented p-group. If G has length λ, then u_G is a λ-admissible function.*

Proof. Suppose $H = G \oplus G'$ is a reduced simply presented p-group. To show that u_G is λ-admissible, assume that $\sigma + \omega < \lambda$. Then $p^\sigma H = p^\sigma G \oplus p^\sigma G'$ is simply presented (Proposition 2.8.5(3)). Therefore, by Lemma 2.8.18, the group $A = p^\sigma G$ is generated by some subset Z such that $\{z + pA : z \in Z\}$ is a basis of A/pA. Let B be a p-basic subgroup of A, which exists by Theorem 1.7.6. Then we can write $B = \bigoplus_{i\in I} \langle b_i \rangle$. Therefore, by

Lemma 1.8.2, the set $\{b_i + pA : i \in I\}$ is a basis of A/pA; thus, $|I| = |Z|$. Notice that $|Z| = |A|$ since A is infinite. Then by Corollary 1.13.5, we have

$$\sum_{n<\omega} u_p(n, A) = \sum_{n<\omega} u_p(n, B) = |I| = |A| \geq \sum_{\alpha\geq\omega} u_p(\alpha, A);$$

thus, since $u_p(\sigma + n, G) = u_p(n, A)$, $\sum_{\sigma\leq\beta<\sigma+\omega} u_p(\beta, G) \geq \sum_{\beta\geq\sigma+\omega} u_p(\beta, G)$. □

Now we can prove the following important fact:

Theorem 2.8.20 (Crawley–Hales [14]). *Direct summands of simply presented p-groups are simply presented.*

Proof. Let $K = G \oplus G'$ be a simply presented p-group. Writing $G = D \oplus R$ and $G' = D' \oplus R'$ with D, D' divisible and R, R' reduced, we have $K = (D \oplus D') \oplus (R \oplus R')$ and $R \oplus R' \cong K/(D \oplus D')$ is simply presented by Proposition 2.8.5(3). If λ is the length of R, then u_R is a λ-admissible function (Theorem 2.8.19). By Theorem 2.8.16, there is a reduced simply presented p-group H such that $u_R = u_H$. Since H and R have nice systems (see Proposition 2.8.7 and Corollary 2.2.9) they are isomorphic by Theorem 2.3.3; thus, R is simply presented. But then G is simply presented. □

Corollary 2.8.21. *The class of simply presented torsion groups is closed under direct summands.*

Exercises.

1. Let G be a simply presented torsion group. If $G \oplus G \cong H \oplus H$, then $G \cong H$.
2. (Crawley–Hales [15]) For a given $n \in \mathbb{N}$, let G_i be simply presented p-groups ($i \in I$) such that $p^{\alpha_i} G_i = \langle x_i \rangle \cong \mathbb{Z}/p^n\mathbb{Z}$ for each $i \in I$. Then $(\bigoplus_{i\in I} G_i)/\langle x_j - x_k : j, k \in I\rangle$ is simply presented.
3. Let G be a p-group and α an ordinal. If $p^\alpha G$ and $G/p^\alpha G$ are simply presented, then G is simply presented.
4. Let G be a simply presented group and H a torsion group. If G is isomorphic to a direct summand of H, and if H is isomorphic to a direct summand of G, then $G \cong H$.
5. There is a simply presented p-group $G = \langle X; \Lambda\rangle$ such that $G[p] \neq \langle Y\rangle$, where $Y = \{x \in X : px = 0\}$.

2.9 Balanced-projective p-groups

Following Fuchs [27], we call a subgroup of a p-group **balanced** if it is both nice and isotype. Then a short exact sequence

$$0 \to A \to B \to C \to 0$$

of p-groups is said to be **balanced-exact** if A is a balanced subgroup of B. We have the following:

Proposition 2.9.1 (Fuchs [27]). *Let* $E : 0 \to A \to B \xrightarrow{\nu} C \to 0$ *be an exact sequence of p-groups. Then the following statements are equivalent:*

1. *E is balanced-exact;*
2. *The sequence* $0 \to p^\alpha A \to p^\alpha B \to p^\alpha C \to 0$ *is exact for every ordinal* α*;*
3. *The sequence* $0 \to A/p^\alpha A \to B/p^\alpha B \to C/p^\alpha C \to 0$ *is exact for every ordinal* α*;*
4. $\nu((p^\alpha B)[p]) = (p^\alpha C)[p]$ *for all ordinals* α.

Proof. Evidently, (1) and (2) are equivalent. For every ordinal α, we have a commutative diagram, with all columns exact:

$$\begin{array}{ccccccccc} & & 0 & & 0 & & 0 & & \\ & & \downarrow & & \downarrow & & \downarrow & & \\ 0 & \to & p^\alpha A & \to & p^\alpha B & \to & p^\alpha C & \to & 0 \\ & & \downarrow & & \downarrow & & \downarrow & & \\ 0 & \to & A & \to & B & \to & C & \to & 0 \\ & & \downarrow & & \downarrow & & \downarrow & & \\ 0 & \to & A/p^\alpha A & \to & B/p^\alpha B & \to & C/p^\alpha C & \to & 0. \\ & & \downarrow & & \downarrow & & \downarrow & & \\ & & 0 & & 0 & & 0 & & \end{array}$$

Therefore, by Lemma 1.2.4 (the 3×3-lemma), all three rows are exact whenever the first two or the last two rows are exact; thus, (2) $\Leftrightarrow$ (3). To prove that (1) implies (4), let α be an ordinal, and notice that ν maps $(p^\alpha B)[p]$ into $(p^\alpha C)[p]$. Letting $c \in (p^\alpha C)[p]$, we have $c = \nu(b)$ for some $b \in p^\alpha B$ by (2); hence, $pb \in p^{\alpha+1}B \cap A = p^{\alpha+1}A$. Therefore, there is an $a \in p^\alpha A$ such that $pa = pb$. Then $b - a \in (p^\alpha B)[p]$ and $\nu(b-a) = c$; thus, ν maps $(p^\alpha B)[p]$ onto $(p^\alpha C)[p]$.

To show that (4) implies (1), we first prove that $A \cap p^\alpha B = p^\alpha A$ for all ordinals α by induction. Let $\alpha > 0$ and assume that equality holds for all $\beta < \alpha$. If α is a limit ordinal, the claim is immediate. Therefore, suppose that $\alpha - 1$ exists, and let $x \in A \cap p^\alpha B$. Then $x = py$ for some $y \in p^{\alpha-1}B$; thus, $\nu(y) \in p^{\alpha-1}C$ and $p\nu(y) = \nu(x) = 0$. It follows that $\nu(y) \in (p^{\alpha-1}C)[p]$. Therefore, by (4), there is a $z \in (p^{\alpha-1}B)[p]$ such that $\nu(z) = \nu(y)$. But then $y - z \in p^{\alpha-1}B \cap \ker \nu = p^{\alpha-1}B \cap A$, which equals $p^{\alpha-1}A$ by the induction hypothesis. Therefore, we have $x = py = p(y-z) \in p^\alpha A$; thus, $A \cap p^\alpha B \subseteq p^\alpha A$ and the assertion follows. It remains to prove that for all ordinals α, we have $\nu(p^\alpha B) = p^\alpha C$. Since C is a p-group, it suffices to show that

$$(p^\alpha C)[p^n] \subseteq \nu(p^\alpha B)$$

for all ordinals α and positive integers n. By assumption, the claim is true for $n = 1$. Therefore, let $n > 1$, and assume that each inclusion holds for $n-1$. Let α be an ordinal and $c \in (p^\alpha C)[p^n]$. Then there is a $b \in B$ such that $\nu(b) = c$. Since $pc \in (p^{\alpha+1}C)[p^{n-1}]$, there is an $x \in p^{\alpha+1}B$ such that $\nu(x) = pc$ by the induction hypothesis. Letting $b' \in p^\alpha B$ such that $pb' = x$, we have $\nu(b - b') = c - \nu(b') \in (p^\alpha C)[p]$. Therefore, by (4), there is a $b'' \in p^\alpha B$ so that $\nu(b'') = c - \nu(b')$. But then $c = \nu(b' + b'') \in \nu(p^\alpha B)$, as desired. Therefore, E is balanced-exact. □

A p-group G is called **balanced-projective** if it has the projective property relative to all balanced-exact sequences; that is, if for every balanced-exact sequence $0 \to A \to B \xrightarrow{\nu} C \to 0$ and homomorphism $\gamma : G \to C$, there is a homomorphism $\tau : G \to B$ such that the diagram

$$\begin{array}{ccccccccc} & & & & & & G & & \\ & & & & & \tau\swarrow & \downarrow\gamma & & \\ 0 & \to & A & \to & B & \xrightarrow{\nu} & C & \to & 0 \end{array}$$

is commutative. The projectivity property yields the following useful fact:

Lemma 2.9.2. *The class of balanced-projective p-groups is closed under arbitrary direct sums and direct summands.*

Proof. Let $G = H \oplus H'$ be a balanced-projective p-group, $0 \to A \to B \xrightarrow{\nu} C \to 0$ a balanced-exact sequence, and $\phi : H \to C$ a homomorphism. Letting $\sigma : H \to G$ be the inclusion and $\pi : G \to H$ the projection map, we obtain $\tau : G \to B$ such that $\nu\tau = \phi\pi$, thus $\nu\tau\sigma = \phi\pi\sigma = \phi$, as required. The proof of the other assertion is similar. □

Groups with nice composition series are balanced-projective:

Proposition 2.9.3 (Fuchs [27]). *Let G be a p-group. If G has a nice composition series, then G is balanced-projective.*

Proof. Suppose that $\{N_\alpha\}_{\alpha \le \lambda}$ is a nice composition series for G. Letting $0 \to A \to B \xrightarrow{\nu} C \to 0$ be a balanced-exact sequence and $\gamma : G \to C$ a homomorphism, we construct step by step a homomorphism $\tau_\alpha : N_\alpha \to B$ which does not decrease p-heights and extends τ_β for all $\beta < \alpha$ such that $\nu\tau_\alpha = \gamma \restriction N_\alpha$. Clearly, this will yield a homomorphism $\tau : G \to B$ with $\nu\tau = \gamma$. Let $0 < \alpha \le \lambda$, and suppose that a suitable map $\tau_\beta : N_\beta \to B$ has already been constructed for all $\beta < \alpha$. If α is a limit ordinal, the induction step is immediate, so assume that $\alpha - 1$ exists. Since $N_{\alpha-1}$ is nice, there is an $x_\alpha \in G$ such that $N_\alpha = \langle N_{\alpha-1}, x_\alpha\rangle$ and x_α is proper with respect to $N_{\alpha-1}$. If $\sigma = |x_\alpha|$, then by Proposition 2.9.1, there is an element $b \in p^\sigma B$ such that $\nu(b) = \gamma(x_\alpha)$; hence, by induction, $\nu(pb) = \nu\tau_{\alpha-1}(px_\alpha)$. Then

$$pb - \tau_{\alpha-1}(px_\alpha) \in A \cap p^{\sigma+1}B = p^{\sigma+1}A;$$

thus, $pb - \tau_{\alpha-1}(px_\alpha) = pa$ for some $a \in p^\sigma A$. By Lemma 1.12.4, the correspondence $x_\alpha \mapsto b - a$ gives rise to a homomorphism $\tau_\alpha : N_\alpha \to B$ which extends $\tau_{\alpha-1}$ and does not decrease p-heights. This completes the induction. □

Lemma 2.9.4. *Let G be a reduced p-group of length λ. Then there is a balanced-exact sequence $0 \to A \to B \to G \to 0$, where B is a direct sum of generalized Prüfer groups of lengths $\le \lambda$.*

Proof. For every $0 \neq x \in (p^\alpha G)[p^n]$ with $\alpha + n \leq \lambda$, we define $H(x) = H_{\alpha+n}$, and let y be a generator of the cyclic group $p^\alpha H_{\alpha+n} \cong \mathbb{Z}/p^n\mathbb{Z}$ (see Lemma 2.4.1(1)). Then $y \mapsto x$ gives rise to a homomorphism $p^\alpha H_{\alpha+n} \to G$ which does not decrease p-heights. Therefore, by Theorem 2.3.4, it can be extended to a homomorphism $\phi_x : H(x) \to G$. Define $\phi : B = \bigoplus H(x) \to G$ to be the sum of the homomorphisms ϕ_x. Then ϕ is surjective, and $\phi((p^\alpha B)[p]) = (p^\alpha G)[p]$; thus, $A = \ker \phi$ is balanced in B by Proposition 2.9.1. □

Consequently, we have the following:

Proposition 2.9.5 (Fuchs [27]). *Every balanced-projective p-group G is a direct summand of a direct sum of copies of* $\mathbb{Z}(p^\infty)$ *and generalized Prüfer groups.*

Proof. By Lemma 2.9.2, we may assume that G is reduced. Consider the exact sequence as in Lemma 2.9.4, where $\nu : B \to G$. Then there is a $\pi : G \to B$ such that $\nu\pi = 1_G$. Hence, by Lemma 1.16.8(2), the sequence splits. □

We are now able to prove the converse of Proposition 2.4.3 (see [27, Theorem 82.3]):

Theorem 2.9.6 (Nunke [124]). *Let G be a totally projective p-group. Then G is a direct summand of a direct sum of copies of* $\mathbb{Z}(p^\infty)$ *and generalized Prüfer groups.*

Proof. Again, we can assume that G is reduced. The claim is proved by transfinite induction on the length λ of G. Let $\lambda > 0$, and assume that the assertion holds for all ordinals $< \lambda$. If $\lambda - 1$ exists, then by the induction hypothesis and Proposition 2.4.4, the totally projective p-group $G/p^{\lambda-1}G$ has a nice system and, therefore, a nice composition series (Corollary 2.2.4). Since the p-bounded group $p^{\lambda-1}G$ has a nice composition series, G has a nice composition series as well (Lemma 2.2.1). Thus, G is a direct summand of a direct sum of generalized Prüfer groups (Propositions 2.9.3, 2.9.5).

Now assume that λ is a limit ordinal, and let $E : 0 \to A \to B \to G \to 0$ be a balanced-exact sequence as in Lemma 2.9.4. In view of that lemma, it suffices to prove that E splits. Let $\alpha < \lambda$. By Proposition 2.9.1, the induced sequence $E_\alpha : 0 \to A/p^\alpha A \to B/p^\alpha B \to G/p^\alpha G \to 0$ is exact. Letting $\pi : A \to A/p^\alpha A$ and $\gamma : G \to G/p^\alpha G$ be the natural maps, we have a commutative diagram

$$
\begin{array}{rcccccccc}
E: & 0 & \to & A & \to & B & \xrightarrow{\beta} & G & \to & 0 \\
 & & & \downarrow \pi & & \downarrow \phi & & \| & & \\
(E_\alpha)\gamma: & 0 & \to & A/p^\alpha A & \to & Y & \to & G & \to & 0 \\
 & & & \| & & \downarrow & & \downarrow \gamma & & \\
E_\alpha: & 0 & \to & A/p^\alpha A & \to & B/p^\alpha B & \to & G/p^\alpha G & \to & 0,
\end{array}
$$

where the lower part forms a pullback diagram, and $\phi : b \mapsto (b + p^\alpha B, \beta(b))$. By the induction hypothesis, $G/p^\alpha G$ is a direct summand of a direct sum of generalized Prüfer groups. Therefore, as before, we conclude that it has a nice composition series. Thus, it is balanced-projective (Proposition 2.9.3). But then E_α splits by Corollary 1.16.9(3) ⇒ (1) since it is balanced-exact (Proposition 2.1.3). Therefore, $(E_\alpha)\gamma$ splits by Corollary 1.16.9(1) ⇔ (2). Let $\psi : p^\alpha A \to A$ be the inclusion map. Then exactness of

$$\mathrm{Ext}(G, p^\alpha A) \xrightarrow{\psi_*} \mathrm{Ext}(G, A) \xrightarrow{\pi_*} \mathrm{Ext}(G, A/p^\alpha A)$$

(Theorem 1.16.10), and $\pi E \equiv (E_\alpha)\gamma = 0$, yield $E \in \operatorname{im} \psi_*$. Every group is an epimorphic image of a free group. Thus, we have an exact sequence $0 \to D \to F \to G \to 0$ for some free group F, and obtain the commutative diagram with exact first row (by Theorem 1.16.10 again)

$$\begin{array}{ccccc} \mathrm{Hom}(D, p^\alpha A) & \xrightarrow{\nu} & \mathrm{Ext}(G, p^\alpha A) & \to & \mathrm{Ext}(F, p^\alpha A) = 0 \\ \downarrow \rho & & \downarrow \psi_* & & \\ \mathrm{Hom}(D, A) & \xrightarrow{\mu} & \mathrm{Ext}(G, A). & & \end{array}$$

Since ν is surjective, there is an $f \in \mathrm{Hom}(D, p^\alpha A)$ such that $\psi_* \nu(f) = E$. Subgroups of free groups are free (Theorem 1.5.4), so we can write $D \cong \bigoplus_{i \in I} \mathbb{Z}$. Then ρ maps $\mathrm{Hom}(D, p^\alpha A) \cong \prod_{i \in I} \mathrm{Hom}(\mathbb{Z}, p^\alpha A) \cong \prod_{i \in I} p^\alpha A$ into $p^\alpha \mathrm{Hom}(D, A)$. Thus, $E = \mu\rho(f) \in p^\alpha \mathrm{Ext}(G, A)$. Since $\alpha < \lambda$ was arbitrary, it follows that $E \in p^\lambda \mathrm{Ext}(G, A) = p^\lambda \mathrm{Ext}(G/p^\lambda G, A) = 0$ since G is totally projective. □

Exercises.

1. (Fuchs [27]) Let A and B be subgroups of a p-group G such that $A \subseteq B$.
 (a) If A is balanced in G, then A is balanced in B.
 (b) If B is balanced in G, then B/A is balanced in G/A.
 (c) If A is balanced in B and B/A is balanced in G/A, then B is balanced in G.
 (d) If A is balanced in B and B is balanced in G, then A is balanced in G.
2. Let A be a balanced subgroup of G. If A and G/A have nice composition series, then G has a nice composition series.
3. (Fuchs [27]) Suppose that

$$\begin{array}{ccccccccc} & & 0 & \to & H & \to & G & & \\ & & & & \downarrow \phi & & \downarrow \psi & & \\ 0 & \to & A & \to & B & \xrightarrow{\nu} & C & \to & 0 \end{array}$$

 is a commutative diagram of p-groups with exact rows such that A is balanced in B. If ϕ does not decrease heights (that is, if $|\phi(h)|_p^B \geq |h|_p^G$ for all $h \in H$), $x \in G$ is proper with respect to H, and $px \in H$, then ϕ extends to a homomorphism $\phi' : \langle H, x \rangle \to B$ which does not decrease heights such that $\nu\phi'(x) = \psi(x)$.

2.10 The class of simply presented p-groups

The next theorem contains a variety of characterizations of simply presented p-groups, a class for which we have a classification result. Most of these we already established. The results are due to Crawley and Hales [14], Fuchs [27], Griffith [38], Hill [47], Nunke [124], Parker and Walker [125] and Walker [149].

Theorem 2.10.1. *The following conditions are equivalent for a p-group G:*
1. *G is simply presented;*
2. *G has a p-basis;*
3. *G is totally projective;*
4. *G has a nice system;*
5. *G has a $G(\aleph_0)$-family of nice subgroups;*
6. *G has a nice composition series;*
7. *G is balanced-projective;*
8. *G is a direct summand of a direct sum of copies of $\mathbb{Z}(p^\infty)$ and generalized Prüfer groups;*
9. *G is a direct summand of a direct sum of copies of $\mathbb{Z}(p^\infty)$ and Walker groups;*
10. *G belongs to the smallest class $\mathcal{C}$ of p-groups such that* (i) *$\mathbb{Z}/p\mathbb{Z} \in \mathcal{C}$ and $\mathbb{Z}(p^\infty) \in \mathcal{C}$,* (ii) *$\mathcal{C}$ is closed under arbitrary direct sums and direct summands, and* (iii) *given any ordinal α and group A, $A \in \mathcal{C}$ if and only if $p^\alpha A \in \mathcal{C}$ and $A/p^\alpha A \in \mathcal{C}$.*

Proof. We have (4) ⇔ (5) ⇔ (6) ⇔ (7) ⇔ (8) by the results cited earlier. Specifically, (4) ⇒ (5) by examination, (5) ⇒ (6) by Lemma 2.2.3, (6) ⇒ (7) by Proposition 2.9.3, (7) ⇒ (8) by Proposition 2.9.5, and (8) ⇒ (4) by Proposition 2.4.4. Then the equivalences (1) ⇔ (2) ⇔ (3) ⇔ (4) ⇔ (8) follow from Propositions 2.4.3, 2.8.4, 2.8.7, 2.8.11 and Theorems 2.8.20, 2.9.6.

Clearly, (9) ⇒ (1) by Theorem 2.8.20. To prove the converse, assume that G is a simply presented p-group, and let D be the set of all $\alpha \in \mathrm{Ord}_\infty$ such that $u(\alpha, G) \neq 0$. If $\alpha \in D$ is an ordinal, let P_α be the Walker group as in Theorem 2.7.4, and let $P_\alpha = \mathbb{Z}(p^\infty)$ whenever $\alpha = \infty$. Define

$$H = \bigoplus_{\mathfrak{m}} \bigoplus_{\alpha \in D} P_\alpha,$$

where $\mathfrak{m} = |G| \cdot \aleph_0$. Then H and $G \oplus H$ have the same Ulm invariants; thus, $H \cong G \oplus H$ (Corollary 2.8.8). It follows that (1) ⇒ (9).

It remains to show that (3) ⇔ (10). Therefore, assume that $\mathcal{C}$ is any class of p-groups satisfying conditions (i–iii) in (10), and let $\mathcal{D}$ be the class of all totally projective p-groups, equivalently any of (1) through (8). The generalized Prüfer groups H_α belong to $\mathcal{C}$ as an easy induction on α shows. Therefore, by (3) ⇒ (8), we have $\mathcal{D} \subseteq \mathcal{C}$. Since $\mathcal{D}$ satisfies conditions (i–iii) in (10) (by (1), (8), (6), Lemma 2.2.1 and Proposition 2.8.5(3)), we conclude that $G \in \mathcal{D}$ if and only if G satisfies (10). □

The following uniqueness theorem can be found in Fuchs [27], formulated in terms of (reduced) totally projective p-groups:

Theorem 2.10.2. *Any class $\mathcal{C}$ of p-groups that contains all simply presented p-groups, is closed under direct sums, and in which the Ulm invariants distinguish between nonisomorphic groups coincides with the class of simply presented p-groups.*

Proof. Let $G \in \mathcal{C}$, and consider the simply presented p-group H, as defined in the proof of (1) ⇒ (9) of Theorem 2.10.1. Since the Ulm invariants of H and $G \oplus H$ are identical, we have $H \cong G \oplus H$ by definition of the class $\mathcal{C}$. Then Theorem 2.8.20 shows that G is simply presented. □

Exercises.

1. (Warfield [152]) Let G be a torsion group and A a subgroup of finite index. Then G is simply presented if and only if A is simply presented.
2. (Warfield [152]) Let G be a torsion group and A a subgroup of countable index. If A is simply presented, then so is G.
3. A balanced subgroup of a simply presented p-group is a direct summand.
4. Kaplansky [78] defined a p-group G to be **transitive**, if for any $x, y \in G$ with $\|x\|_p = \|y\|_p$ (see Section 1.10) there is an automorphism α of G satisfying $\alpha(x) = y$. Show that p-groups with nice systems are transitive.
5. (Wallace [150]) Suppose G is a reduced p-group containing a subgroup A having a nice system. If G/A is countable, then G has a nice system.
6. (Walker [149]) Any short exact sequence of p-groups, relative to which the Walker groups P_α have the projective property, is balanced-exact.

2.11 Going further

After Ulm's theorem was extended to the class of totally projective p-groups, more general classes of groups were investigated to further generalize this classification result. For instance, Warfield [151, 153] introduced and investigated p-local and global balanced-projective groups. For the p-local case, the definition of a balanced-exact sequence was extended to arbitrary R-modules, where R is a discrete valuation ring with p a generator of the maximal ideal of R (for example, $R = \mathbb{Z}_p$). Then a **balanced-projective module** is defined to be an R-module having the projective property relative to all balanced-exact sequences of R-modules. Introducing a class of p-groups including the totally projective p-groups, Warfield [151, 153] defined an S-**group** to be a p-group which is isomorphic to the torsion part of a balanced-projective $\mathbb{Z}_p$-module and established a classification theorem in terms of Ulm invariants and additional numerical invariants.

For the global case, height sequences are needed. A **height sequence** is a sequence $s = (s_p)_{p\in\mathbb{P}}$ with $s_p \in \mathrm{Ord}_\infty$ for all primes p. Given a group G, any height sequence s gives rise to the subgroup

$$G(s) = \{x \in G : |x|_p^G \geq s_p \text{ for all } p \in \mathbb{P}\}.$$

Then an exact sequence of groups $0 \to A \to B \to C \to 0$ is called **balanced-exact** if the induced sequence $0 \to A(s) \to B(s) \to C(s) \to 0$ is exact for all height sequences s.

Warfield [154] introduced the class of **balanced-projective groups** (= the groups having the projective property relative to all balanced-exact sequences), which includes all totally projective p-groups and torsion-free completely decomposable groups, and established characterizations and a classification theorem. A variety of additional characterizations of balanced projective groups was proved by Lane and Megibben [93].

The theory of simply presented p-groups, introduced by Crawley and Hales [14], was extended by Warfield [151, 152] to modules over a discrete valuation ring and to arbitrary abelian groups (see Chapter 3). Since a simply presented group is a direct sum of simply presented groups of torsion-free rank ≤ 1 (Theorem 2.5.4), rank 1 groups (= groups of torsion-free rank 1) are of particular interest. Megibben [114] classified rank 1 groups whose torsion parts are direct sums of countable groups, and Wallace [150] proved a classification theorem for rank 1 groups with totally projective p-torsion parts. Warfield [155] classified rank 1 modules over a discrete valuation ring possessing an $H(\aleph_0)$-family of p-nice submodules. The theory of simply presented p-groups was extended to valuated p-groups by Hunter, Richman, and Walker (see [64] and [63]).

In [46], Hill established a new characterization of totally projective p-groups. Hill and Ullery [55] introduced and investigated **almost totally projective groups** (= p-groups possessing a weak Axiom 3 system of nice subgroups; see Section 3.12, Exercise 3). Hill [42] investigated IT-**groups** (= the isotype subgroups of totally projective p-groups); see also the paper [48] by Hill and Megibben. In [45], Hill classified A-**groups**, which are certain isotype subgroups of totally projective p-groups.

Generalizations of completely decomposable groups include **Butler groups**, which are defined to be pure subgroups of completely decomposable groups of finite rank. Their study was initiated by Butler [11]; for more information, see Fuchs' book [29]. The class of all Butler groups contains **almost completely decomposable groups** (= finite rank torsion-free groups containing a completely decomposable subgroup of finite index), which were defined by Lady [91] and have been investigated extensively by Blagoveshchenskaya, Mader, Mutzbauer, and others (see the book by Mader [112]).

Problems.

1. Which p-groups (besides the totally projective p-groups) can be classified by Ulm invariants?
2. (Fuchs [27]) Characterize the p-groups that can be embedded in totally projective p-groups.
3. Is there an Axiom 3 characterization of isotype subgroups of totally projective p-groups?

3 Warfield groups

We explore the classifications of mixed groups, specifically Warfield groups, which are direct summands of simply presented groups. They may also be thought of in terms of a decomposition basis. Any Warfield group G has a nice decomposition basis X such that $G/\langle X\rangle$ is torsion and simply presented. With the aid of the Warfield invariants, we prove both local and global classifications for this class of groups. We find many other ways of characterizing Warfield groups. The Warfield groups form the unique class that includes the simply presented groups, is closed under direct sums and direct summands, and is classified by these invariants along with the Ulm invariants.

3.1 Groups with decomposition bases

Here we study a large class of groups that includes the torsion and torsion-free groups examined earlier. A **Warfield group** is a direct summand of a simply presented group. We will see that one important characteristic of Warfield groups is that each has a decomposition basis. Hence, we start our discussion by introducing the concept of a decomposition basis, which helps us characterize the torsion-free aspects of a mixed group.

Let G be an R-module and I a (possibly empty) set. Then a direct sum $\bigoplus_{i\in I} A_i \subseteq G$ is called a **valuated coproduct in** G if for all primes $p \in R$ and ordinals α, we have

$$\left(\bigoplus_{i\in I} A_i\right) \cap p^\alpha G = \bigoplus_{i\in I}(A_i \cap p^\alpha G).$$

If in addition $A_i = \langle x_i\rangle$, where x_i is a nonzero element of infinite order ($i \in I$), then $X = \{x_i : i \in I\}$ is called a **decomposition set in** G. Then X is called a **decomposition basis for** G if in addition $G/\langle X\rangle$ is torsion.

As an application of Proposition 2.2.7, we obtain the following characterization of groups with decomposition bases:

Corollary 3.1.1. *Let G be a group and suppose that $\mathcal{B}$ is the set of all decomposition sets in G. Then the following conditions are equivalent:*
1. *G has a decomposition basis;*
2. *G has an $H(\aleph_0)$-family of $\mathcal{B}$-torsion subgroups;*
3. *G has a $G(\aleph_0)$-family of $\mathcal{B}$-torsion subgroups;*
4. *G has an $F(\aleph_0)$-family of $\mathcal{B}$-torsion subgroups.*

Torsion-free groups with decomposition basis are easy to describe. The following result can be found in [113]:

Theorem 3.1.2. *Let G be a torsion-free group. Then G has a decomposition basis if and only if G is completely decomposable.*

https://doi.org/10.1515/9783110427684-003

Proof. Clearly, a completely decomposable group has a decomposition basis. To prove the converse, assume that G has a decomposition basis X. Let $x \in X$, and recall that $\langle x\rangle^0 = \{g \in G : ng \in \langle x\rangle$ for some nonzero integer $n\}$ (see Section 1.10). First, we show that this subgroup of G has rank 1. Letting $g, h \in \langle x\rangle^0$, we have $mg, nh \in \langle x\rangle$ for some nonzero $m, n \in \mathbb{Z}$; thus, $mg = kx$ and $nh = lx$ for some nonzero $k, l \in \mathbb{Z}$. It follows that $lmg = knh$. To show that the sum $\sum_{z\in X}\langle z\rangle^0$ is direct, let

$$y \in \langle x\rangle^0 \cap \sum_{z\in X\setminus\{x\}} \langle z\rangle^0.$$

This yields elements $z_1, \dots, z_t \in X \setminus \{x\}$ such that $y \in \sum_{i=1}^t \langle z_i\rangle^0$. But then there is a positive integer s such that $sy \in \langle x\rangle \cap (\bigoplus_{i=1}^t \langle z_i\rangle) = 0$. It follows that $y = 0$ since G is torsion-free; thus, the sum is direct.

It remains to show that $G = \bigoplus_{z\in X}\langle z\rangle^0$, so let $g \in G$. Since $G/\langle X\rangle$ is torsion, there is a positive integer r such that $rg = \sum_{i=1}^u k_i x_i$ for $x_1, \dots, x_u \in X$ with nonzero integers $k_1, \dots, k_u$. Now write $r = p_1^{r_1} \cdots p_v^{r_v}$ for primes p_j and positive integers r_j ($j = 1, \dots, v$). Then for every $i \in \{1, \dots, u\}$ and $j \in \{1, \dots, v\}$, we have

$$|k_i x_i|_{p_j} \geq |rg|_{p_j} = |p_j^{r_j} g|_{p_j} \geq r_j.$$

Letting $A_i = \langle x_i\rangle^0$, we obtain $k_i x_i \in p_j^{r_j} G \cap A_i = p_j^{r_j} A_i$ (see Lemma 1.10.8), so it follows that

$$k_i x_i \in \bigcap_{j=1}^{v} p_j^{r_j} A_i.$$

By Lemma 1.7.1, we obtain $k_i x_i \in rA_i$, and we can write $k_i x_i = ry_i$ with $y_i \in A_i$. This yields $r(g - \sum_{i=1}^u y_i) = 0$, which implies $g = \sum_{i=1}^u y_i \in \bigoplus_{z\in X}\langle z\rangle^0$. This completes the proof. □

Let G be an R-module, A a submodule of G, $p \in R$ a prime, α an ordinal, and $\overline{A} = \{x \in A \cap p^\alpha G : px \in p^{\alpha+2} G\}/(A \cap p^{\alpha+1} G)$. Recall that $I_{\alpha,p}(A)$ denotes the image of the map

$$\phi : \overline{A} \to (p^\alpha G)[p]/(p^{\alpha+1} G)[p],$$

which is defined by $\phi(x + A \cap p^{\alpha+1} G) = x - y + (p^{\alpha+1} G)[p]$, where $x \in A \cap p^\alpha G$ and y is any element in $p^{\alpha+1} G$ such that $px = py$. The following observation will be useful:

Lemma 3.1.3. *Let G be an R-module, A a submodule of G, $p \in R$ a prime, and α an ordinal. Then we have the following:*
1. *If B is a submodule of A, then $I_{\alpha,p}(B) \subseteq I_{\alpha,p}(A)$.*
2. *If $A = \bigoplus_{i\in I} A_i$ is a valuated coproduct in G, then $I_{\alpha,p}(A) = \bigoplus_{i\in I} I_{\alpha,p}(A_i)$.*

Proof. The first assertion is clear. For (2), we need to show that $\phi(\overline{A}) = \bigoplus_{i\in I} \phi_i(\overline{A_i})$, where each map ϕ_i is defined accordingly. Clearly, $\sum \phi_i(\overline{A_i}) \subseteq \phi(\overline{A})$. Conversely, if $x \in$

$A \cap p^\alpha G$ such that $px \in p^{\alpha+2}G$, then $x = \sum x_i$ ($x_i \in A_i$), $x \in \bigoplus(A_i \cap p^\alpha G)$, and $px \in \bigoplus(A_i \cap p^{\alpha+2}G)$. Then for each $i \in I$, there exists a $y_i \in p^{\alpha+1}G$ such that $px_i = py_i$; hence, $y = \sum y_i \in p^{\alpha+1}G$, and $px = py$. It follows that $\phi(x + A \cap p^{\alpha+1}G) = x - y + (p^{\alpha+1}G)[p] = \sum \phi_i(x_i + A_i \cap p^{\alpha+1}G) \in \sum \phi_i(\overline{A_i})$. Therefore, we have $\phi(\overline{A}) = \sum_{i\in I} \phi_i(\overline{A_i})$. To show that this sum is direct, let $i \in I$ and $z \in \phi_i(\overline{A_i}) \cap \sum_{j\neq i} \phi_j(\overline{A_j})$, say, $z = x_i - y_i + (p^{\alpha+1}G)[p] = \sum_{j\neq i}(x_j - y_j) + (p^{\alpha+1}G)[p]$. Then $x_i - \sum_{j\neq i} x_j \in \bigoplus(A_i \cap p^{\alpha+1}G)$, and we have $x_i \in p^{\alpha+1}G$. Therefore, $x_i - y_i \in (p^{\alpha+1}G)[p]$; hence, $z = 0$. □

Let G be an R-module, $x \in G$, $p \in R$ a prime, and α an ordinal. Then we say that the **Ulm sequence $\|x\|_p$ has a gap at** α if there is an $n < \omega$ such that $|p^n x|_p = \alpha$ and $|p^{n+1}x|_p > \alpha + 1$.

Lemma 3.1.4. *Let G be an R-module, $p \in R$ a prime, $x \in G$, and α an ordinal. Then $\|x\|_p$ has a gap at α exactly if $I_{\alpha,p}(\langle x\rangle) \neq 0$. If G has a decomposition basis X, then the dimension of $I_{\alpha,p}(\langle X\rangle)$ is the number of elements $x \in X$ such that $\|x\|_p$ has a gap at α.*

Proof. Suppose that $p^n x \in p^\alpha G \setminus p^{\alpha+1}G$ and $p^{n+1}x \in p^{\alpha+2}G$ for some $n < \omega$. Then there exists a $y \in p^{\alpha+1}G$ such that $p^{n+1}x = py$ and $p^n x - y \notin p^{\alpha+1}G$; hence, $I_{\alpha,p}(\langle x\rangle) \neq 0$. Conversely, suppose that $I_{\alpha,p}(\langle x\rangle) \neq 0$. Then there is a nonzero $m \in R$ such that $pmx \in p^{\alpha+2}G$ and $mx - y \in (p^\alpha G)[p] \setminus (p^{\alpha+1}G)[p]$ for some $y \in p^{\alpha+1}G$. It follows that $mx \in p^\alpha G \setminus p^{\alpha+1}G$; thus, $\|x\|_p$ has a gap at α.

If G has a decomposition basis X, then we have $I_{\alpha,p}(\langle X\rangle) = \bigoplus_{x\in X} I_\alpha(\langle x\rangle)$ by Lemma 3.1.3. Since $I_{\alpha,p}(\langle x\rangle)$ is either 0 or 1, the second assertion follows. □

If X and Y are decomposition bases for an R-module G, then Y is called a **subordinate of** X if, for every $y \in Y$, there is an $x \in X$ such that $y = mx$ for some nonzero $m \in R$. Following Warfield [152], we call X **a lower decomposition basis for** G if, for every prime $p \in R$ and ordinal α, $I_{\alpha,p}(\langle X\rangle)$ is finite dimensional or $u_p(\alpha, G) = u_{\alpha,p}(G, \langle X\rangle)$. To prove that every decomposition basis has a subordinate that is a lower decomposition basis, the following fact will be useful (see the proof of [152, Lemma 11]):

Lemma 3.1.5. *Let G be an R-module with decomposition basis X, $p \in R$ a prime, and α an ordinal. Suppose X_α is a subset of X such that $|X_\alpha| = \dim I_{\alpha,p}(\langle X\rangle)$, and $X' = \{m_x x : x \in X\}$ is a subordinate of X such that $\|x\|_p$ has a gap at α for all $x \in X_\alpha$ and $\|y\|_p$ has no gap at α for all $y \in X'_\alpha = \{m_x x : x \in X_\alpha\}$. Then the following hold:*
1. $\dim I_{\alpha,p}(\langle X'\rangle) \le \dim[I_{\alpha,p}(\langle X\rangle)/I_{\alpha,p}(\langle X'\rangle)]$.
2. *If the dimension of $I_{\alpha,p}(\langle X\rangle)$ is infinite, then $u_p(\alpha, G) = u_{\alpha,p}(G, \langle X'\rangle)$.*

Proof. Letting $Y = X \setminus X_\alpha$ and $Y' = \{m_x x : x \in X \setminus X_\alpha\}$, we have

$$\frac{I_{\alpha,p}(\langle X\rangle)}{I_{\alpha,p}(\langle X'\rangle)} = \frac{I_{\alpha,p}(\langle X_\alpha\rangle) \oplus I_{\alpha,p}(\langle Y\rangle)}{I_{\alpha,p}(\langle X'_\alpha\rangle) \oplus I_{\alpha,p}(\langle Y'\rangle)} \cong I_{\alpha,p}(\langle X_\alpha\rangle) \oplus \frac{I_{\alpha,p}(\langle Y\rangle)}{I_{\alpha,p}(\langle Y'\rangle)}.$$

Hence, $\dim I_{\alpha,p}(\langle X\rangle)/I_{\alpha,p}(\langle X'\rangle) \ge \dim I_{\alpha,p}(\langle X_\alpha\rangle) = \dim I_{\alpha,p}(\langle X\rangle) \ge \dim I_{\alpha,p}(\langle X'\rangle)$, and the first assertion follows.

Notice that (1) yields

$$\dim I_{\alpha,p}(\langle X'\rangle) \le \dim I_{\alpha,p}(\langle X\rangle)/I_{\alpha,p}(\langle X'\rangle) \le \dim[(p^\alpha G)[p]/(p^{\alpha+1}G)[p]]/I_{\alpha,p}(\langle X'\rangle),$$

which coincides with $u_{\alpha,p}(G,\langle X'\rangle)$. If the dimension of $I_{\alpha,p}(\langle X\rangle)$ is infinite, then so is $u_p(\alpha,G)$. But then the equation

$$u_p(\alpha,G) = \dim I_{\alpha,p}(\langle X'\rangle) + u_{\alpha,p}(G,\langle X'\rangle)$$

shows that $u_p(\alpha,G) = u_{\alpha,p}(G,\langle X'\rangle)$, as desired. □

Lemma 3.1.6 (Warfield [152]). *Suppose G is a group with decomposition basis X. Then X has a subordinate X' that is a lower decomposition basis for G. Moreover, every subordinate of X' is a lower decomposition basis for G.*

Proof. For finite X, the assertion is trivial, so assume that X is infinite. We write $X = \dot\bigcup_{\lambda<\kappa} X^*_\lambda$, where each set X^*_λ is countable, and let $A = \{(\alpha,p) : I_{\alpha,p}(\langle X\rangle)$ is infinite dimensional$\}$. Using transfinite induction, we will construct countable sets X_λ ($\lambda<\kappa$) such that $X = \dot\bigcup_{\lambda<\kappa} X_\lambda$ and for all $\lambda<\kappa$ and $(\alpha,p)\in A$, the dimension of $I_{\alpha,p}(\langle X_\lambda\rangle)$ is either 0 or infinite, with the exception of at most one λ for each $(\alpha,p)\in A$.

In this proof, we will say that an element $x\in X$ **has a gap at** $(\alpha,p)\in A$ if $\|x\|_p$ has a gap at α. First, let $\lambda=0$, and define $A_1 = \{(\alpha,p)\in A : I_{\alpha,p}(\langle X^*_0\rangle)\ne 0\}$. Since this set is countable, we can write $A_1 = \{a_{11}, a_{12}, a_{13}, \dots\}$. If $a_{11} = (\alpha,p)$ and the dimension of $I_{\alpha,p}(\langle X^*_0\rangle)$ is infinite, let $X^{11}_0 = X^*_0$. Otherwise, we add countably infinitely many elements $x\in\bigcup_{\nu>0} X^*_\nu$ to X^*_0 such that x has a gap at a_{11}, obtaining the improved set X^{11}_0. Then let $A_2 = \{(\alpha,p)\in A : I_{\alpha,p}(\langle X^{11}_0\rangle)\ne 0\}$, and write $A_2 = \{a_{21}, a_{22}, a_{23}, \dots\}$. If $a_{12} = (\beta,q)$ and $I_{\beta,q}(\langle X^{11}_0\rangle)$ is not infinite dimensional, we add countably infinitely many elements $x\in\bigcup_{\sigma>0} X^*_\sigma$ to X^{11}_0 such that x has a gap at a_{12} if this is possible. If only finitely many such elements are available, we add all of those to the set. The resulting set is denoted by X^{12}_0. Now define $A_3 = \{(\alpha,p)\in A : I_{\alpha,p}(\langle X^{12}_0\rangle)\ne 0\}$, and write $A_3 = \{a_{31}, a_{32}, a_{33}, \dots\}$. For $a_{21}\in A_2$, we construct the set X^{21}_0, define A_4, and pick a_{13}, et cetera. This yields a sequence of countable sets

$$X^*_0 \subseteq X^{11}_0 \subseteq X^{12}_0 \subseteq X^{21}_0 \subseteq X^{13}_0 \subseteq X^{22}_0 \subseteq X^{31}_0 \subseteq \cdots,$$

and we let $X_0 = \bigcup_{i,j\in\mathbb{N}} X^{ij}_0$.

Now let $0<\lambda<\kappa$, and assume that we have already constructed countable subsets X_α of X ($\alpha<\lambda$) having the required properties, such that $\dot\bigcup_{\alpha<\lambda} X^*_\alpha \subseteq \dot\bigcup_{\alpha<\lambda} X_\alpha$. To construct X_λ, let $Z = X\setminus\bigcup_{\alpha<\lambda} X_\alpha$ and write $Z = \dot\bigcup_{\beta\ge\lambda}(Z\cap X^*_\beta)$. Then enlarge $Z\cap X^*_\lambda$ accordingly by adding elements from $Z\cap X^*_\beta$ for $\beta>\lambda$, obtaining the desired countable set X_λ. To show that $\dot\bigcup_{\alpha\le\lambda} X^*_\alpha \subseteq \dot\bigcup_{\alpha\le\lambda} X_\alpha$, let $x\in X^*_\lambda$. If $x\notin Z$, then $x\in\bigcup_{\alpha<\lambda} X_\alpha$. Thus, $X^*_\lambda \subseteq \bigcup_{\alpha<\lambda} X_\alpha \cup (Z\cap X^*_\lambda) \subseteq \bigcup_{\alpha<\lambda} X_\alpha \cup X_\lambda$, and the claim follows by the induction hypothesis. Hence, $X = \dot\bigcup_{\lambda<\kappa} X_\lambda$.

Now fix $\lambda < \kappa$, and let $h(\lambda) = \{(\alpha, p) : I_{\alpha,p}(\langle X_\lambda \rangle)$ is infinite dimensional$\}$. We will construct a set X'_λ so that $X' = \bigcup_{\lambda<\kappa} X'_\lambda$ is the desired subordinate of X. If $h(\lambda)$ is empty, let $X'_\lambda = X_\lambda$. Otherwise, we can write $h(\lambda) = \{a_i : i \in I\}$, where $I = \mathbb{N}$ or $I = \{1, \dots, n\}$ for some positive integer n. Let $Y = X_\lambda$. Then there is an infinite subset Y_{11} of Y such that x has a gap at a_1 for all $x \in Y_{11}$. Let $x_{11} \in Y_{11}$. If the set Y_{11} has an infinite number of elements x having a gap at a_2, we let Y'_{12} be an infinite subset of Y_{11} such that each element of Y'_{12} has a gap at a_2 and $Y_{12} = Y_{11} \setminus Y'_{12}$ is an infinite set containing the element x_{11}. If Y_{11} has only a finite number of elements x having a gap at a_2, let $Y_{12} = Y_{11}$. Note that in either case $Y \setminus Y_{12}$ is infinite. Let $x_{12} \in Y_{12}$ such that $x_{12} \neq x_{11}$. Then for a_3, we construct Y_{13} accordingly, and pick $x_{13} \in Y_{13} \setminus \{x_{11}, x_{12}\}$, et cetera. We define $Y_1 = \bigcap_{i\in\mathbb{N}} Y_{1i}$. This set is infinite since it contains $\{x_{11}, x_{12}, x_{13}, \dots\}$, and all elements $x \in Y_1$ have a gap at a_1. Furthermore, note that if infinitely many elements of Y have a gap at a_i for $i > 1$, the same is true of $Y \setminus Y_1$. Using the same technique as before, we can find an infinite subset Y_2 of $Y \setminus Y_1$ such that each of its elements has a gap at a_2. Continuing in this manner, we obtain $Y = \dot{\bigcup}_{i\in I} Y_i \mathbin{\dot\cup} \tilde{Y}$, where each set Y_i is infinite and each of its elements has a gap at a_i. Now replace every element x of Y_i by a suitable nonzero multiple x' having no gap at a_i, and let $Y'_i = \{x' : x \in Y_i\}$. Letting $X'_\lambda = \bigcup_{i\in I} Y'_i \cup \tilde{Y}$, we obtain the subordinate $X' = \bigcup_{\lambda<\kappa} X'_\lambda$ of X.

To verify that X' is a lower decomposition basis for G, let (α, p) be any element of A, and write $\Lambda = \{\lambda : (\alpha, p) \in h(\lambda)\}$. Define $X(\alpha, p)$ to be the union of all (infinite) subsets Y_i of X_λ ($\lambda \in \Lambda$) belonging to (α, p). Since $I_{\alpha,p}(\langle X_\lambda \rangle)$ cannot have positive finite dimension for infinitely many $\lambda < \kappa$, the dimension of $I_{\alpha,p}(\langle X \rangle)$ coincides with the cardinality of $X(\alpha, p)$, and we obtain $u_p(\alpha, G) = u_{\alpha,p}(G, \langle X' \rangle)$ by Lemma 3.1.5. By construction, every subordinate of X' is a lower decomposition basis for G as well. This completes the proof. □

Exercises.

1. (Lane–Megibben [93]) Let $N = A \oplus B$ be a valuated coproduct in a $\mathbb{Z}_p$-module G. If N is p-nice in G, then A is p-nice in G.
2. (Lane–Megibben [93]) A decomposition basis X for an R-module G is called a K-**basis of** G if $|p^n x|_p^G = |x|_p^G + n$ for all $x \in X$, primes $p \in R$, and $n < \omega$. Show: If X is a K-basis of a $\mathbb{Z}_p$-module G, then $\langle X \rangle$ is p-nice in G.

3.2 The local classification theorem

We now classify certain local mixed groups. We start by defining the Stanton invariant, based on a construction of Hill and Megibben. Let G be a $\mathbb{Z}_p$-module and $\overline{\alpha} = (\alpha_i)_{i<\omega}$, $\overline{\beta} = (\beta_i)_{i<\omega}$ Ulm sequences. For $k < \omega$ let $p^k\overline{\alpha} = (\alpha_{i+k})_{i<\omega}$. We write $\overline{\alpha} \geq \overline{\beta}$ whenever $\alpha_i \geq \beta_i$ for all $i < \omega$. The following two submodules of G will be needed (see Hill–Megibben [49]): $G(\overline{\alpha}) = \{x \in G : \|x\|_p \geq \overline{\alpha}\}$, and $G(\overline{\alpha}^*) = \langle G(\overline{\alpha}^*)^\diamond \rangle$ where

$$G(\overline{\alpha}^*)^\diamond = \{x \in G(\overline{\alpha}) : |p^i x|_p \neq \alpha_i \text{ for infinitely many } i \text{ or } x \text{ is torsion}\}.$$

The following items are easily verified:

(a) If $x \in G(\overline{\alpha}^*)$, then $x = z_1 + \cdots + z_n$ for some $z_1, \ldots, z_n \in G(\overline{\alpha}^*)^\diamond$.

(b) If $\alpha_i = \infty$ for some i, then $G(\overline{\alpha}^*) = t(G(\overline{\alpha}))$. Otherwise, we have $G(\overline{\alpha}^*) = \langle x \in G(\overline{\alpha}) : |p^i x|_p \neq \alpha_i$ for infinitely many $i\rangle$.

(c) If $x \in G(\overline{\alpha}^*)^\diamond$, then $px \in G(p\overline{\alpha}^*)^\diamond$.

(d) If $x \in G(\overline{\alpha}^*)$, then $px \in G(p\overline{\alpha}^*)$.

(e) $G(\overline{\alpha})/G(\overline{\alpha}^*)$ is a vector space over $\mathbb{Z}/p\mathbb{Z}$ (if $\alpha_i \neq \infty$ for all i) or over $\mathbb{Q}$ (if $\alpha_i = \infty$ for some i).

Note that it is possible to have $z \in G(\overline{\alpha}^*)$, though $z \notin G(\overline{\alpha}^*)^\diamond$. In other words, $G(\overline{\alpha}^*)^\diamond$ is not necessarily a group. For example, consider $G = H_x \oplus H_y$, where $\|x\|_p = (0, 3, 4, 7, 8, \ldots)$ and $\|y\|_p = (1, 2, 5, 6, 9, \ldots)$ (see Corollary 2.7.5). Then if $\overline{\alpha} = (0, 2, 4, \ldots)$, $x, y \in G(\overline{\alpha}^*)^\diamond$; hence, $x + y \in G(\overline{\alpha}^*)$, though $\|x + y\|_p = \overline{\alpha}$.

It is clear that a direct sum $A = \bigoplus_{i \in I} A_i$ in a $\mathbb{Z}_p$-module G is a valuated coproduct in G if and only if $A \cap G(\overline{\alpha}) = \bigoplus_{i \in I}(A_i \cap G(\overline{\alpha}))$ for all Ulm sequences $\overline{\alpha}$. Following Hill and Megibben [49], we call the valuated coproduct $A = \bigoplus_{i \in I} A_i$ in G a **$*$-valuated coproduct in** G if

$$A \cap G(\overline{\alpha}^*) = \bigoplus_{i \in I}(A_i \cap G(\overline{\alpha}^*))$$

for all Ulm sequences $\overline{\alpha}$.

Lemma 3.2.1. *Let G be a $\mathbb{Z}_p$-module with decomposition basis X. Then the following hold:*

1. *If $y \in G(\overline{\alpha}^*)$ for some Ulm sequence $\overline{\alpha}$, and $y = a_1 x_1 + \cdots + a_n x_n$ with $x_1, \ldots, x_n \in X$ and $a_1, \ldots, a_n \in \mathbb{Z}_p$, then $a_i x_i \in G(\overline{\alpha}^*)^\diamond$ for all $i = 1, \ldots, n$.*
2. *$\langle X\rangle = \bigoplus_{x \in X}\langle x\rangle$ is a $*$-valuated coproduct in G.*

Proof. For (1), we first note that the result is immediate if $\alpha_i = \infty$ for some i by (b), so we may assume $\alpha_i \neq \infty$ for all i. Suppose $\{x_1, \ldots, x_n\} \subseteq X$, and $y = a_1 x_1 + \cdots + a_n x_n \in G(\overline{\alpha}^*)$ for nonzero $a_1, \ldots, a_n \in \mathbb{Z}_p$. By (a), there are $z_1, \ldots, z_m \in G(\overline{\alpha}^*)^\diamond$ such that $a_1 x_1 + \cdots + a_n x_n = z_1 + \cdots + z_m$. Then for all $1 \le j \le m$, $|p^k z_j| \ge \alpha_k$ for all k, and there are infinitely many k such that $|p^k z_j| > \alpha_k$. Since $\{x_1, \ldots, x_n\}$ is a decomposition set, $|p^k a_i x_i| \ge \alpha_k$ for all i and k. Furthermore, for each $1 \le i \le n$, either $|p^k a_i x_i| > \alpha_k$ for infinitely many k, or there is some $r < \omega$ such that $\|p^r a_i x_i\|_p = p^r \overline{\alpha}$. Since X is a decomposition basis, there is an $s < \omega$ such that $p^s z_1, \ldots, p^s z_m \in \langle X\rangle$. Choose such an s so that for each $1 \le i \le n$ either $\|p^s a_i x_i\| = p^s \overline{\alpha}$, or $|p^{k+s} a_i x_i| > \alpha_{k+s}$ for infinitely many k. Then, replacing $p^s z_j$ with z_j for each $1 \le j \le m$, $p^s a_i$ with a_i for each $1 \le i \le n$ and $p^s \overline{\alpha}$ with $\overline{\alpha}$, we may assume $z_j \in \langle X\rangle$ for all $1 \le j \le m$, and for all $1 \le i \le n$ either $|p^k a_i x_i| > \alpha_k$ for infinitely many k or $|p^k a_i x_i| = \alpha_k$ for all k. Let $Y = \{y_1, \ldots, y_l\} \subseteq X$, where $l \ge n$ and $y_i = x_i$ for all $1 \le i \le n$, and $z_j \in \langle Y\rangle$ for $j = 1, \ldots, m$.

For each $1 \le j \le m$, write $z_j = a_{1j}y_1 + \cdots + a_{lj}y_l$. Then $a_1x_1 + \cdots + a_nx_n = z_1 + \cdots + z_m = \sum_{j=1}^m a_{1j}y_1 + \cdots + \sum_{j=1}^m a_{lj}y_l$. By independence, $a_i = \sum_{j=1}^m a_{ij}$ for all $1 \le i \le n$.

Suppose for some t, $1 \le t \le n$, $|p^k a_t x_t| = \alpha_k$ for all k. Let $1 \le j \le m$. Choose k such that $|p^k z_j| > \alpha_k$. Then $\alpha_k < |p^k a_{1j}y_1 + \cdots + p^k a_{lj}y_l| = \min_i\{|p^k a_{ij}y_i|\}$, so in particular, $|p^k a_{tj}y_t| > \alpha_k = |p^k a_t x_t|$. Since $x_t = y_t$, $|a_{tj}|_p > |a_t|_p$ computed in $\mathbb{Z}_p$. Since j was arbitrary, this holds for all j, contradicting $a_t = \sum_j a_{tj}$. This completes the proof of (1).

To prove (2), let $y \in \langle X\rangle \cap G(\overline{\alpha}^*)$, say $y = a_1x_1 + \cdots + a_nx_n$ for some $x_1, \ldots, x_n \in X$. Then by (1), $a_ix_i \in G(\overline{\alpha}^*)^\diamond \subseteq G(\overline{\alpha}^*)$ for all $1 \le i \le n$. Therefore, we have $y \in \bigoplus_{x\in X}(\langle x\rangle \cap G(\overline{\alpha}^*))$. □

Note that as a special case, if $x \in X$ then $x \in G(\overline{\alpha}^*)$ implies $x \in G(\overline{\alpha}^*)^\diamond$.

Lemma 3.2.2 (Stanton [139], Hunter–Richman–Walker [65]). *Let G be a $\mathbb{Z}_p$-module and $\overline{\alpha}$ an Ulm sequence. Then the map $f : G(\overline{\alpha})/G(\overline{\alpha}^*) \to G(p\overline{\alpha})/G(p\overline{\alpha}^*)$ defined by $x + G(\overline{\alpha}^*) \mapsto px + G(p\overline{\alpha}^*)$ is an isomorphism.*

Proof. Let $x \in G(\overline{\alpha})$. If $x \in G(\overline{\alpha}^*)$, then $px \in G(p\overline{\alpha}^*)$; hence, f is well-defined. Clearly, f is a homomorphism. To prove that f is injective, suppose that $px \in G(p\overline{\alpha}^*)$. Then $px = z_1 + \cdots + z_n$, where $z_1, \ldots, z_n \in G(p\overline{\alpha}^*)^\diamond$. Let $j \in \{1, \ldots, n\}$ and $z = z_j$. Since $z \in p^{\alpha_1}G \subseteq p(p^{\alpha_0}G)$, there is a $z' \in p^{\alpha_0}G$ such that $pz' = z$. Then $p^{i+1}z' = p^iz \in p^{\alpha_{i+1}}G$ for all $i < \omega$; hence, $z' \in G(\overline{\alpha})$. If z is torsion, then so is z', and if $|p^iz| \ne \alpha_{i+1}$ for infinitely many i, then $|p^{i+1}z'| = |p^iz| \ne \alpha_{i+1}$ for infinitely many i. Therefore, $z' \in G(\overline{\alpha}^*)^\diamond$. Letting $y = z_1' + \cdots + z_n'$, we have $y \in G(\overline{\alpha}^*)$ and $p(x - y) = 0$; hence, $x - y \in G(\overline{\alpha}^*)$. But then $x = (x - y) + y \in G(\overline{\alpha}^*)$. Therefore, f is injective.

To prove that f is surjective, let $y \in G(p\overline{\alpha})$. Then there is an $x \in p^{\alpha_0}G$ such that $y = px$, and it follows that $p^{i+1}x = p^iy \in p^{\alpha_{i+1}}G$ for all $i < \omega$; hence, $x \in G(\overline{\alpha})$. □

Two Ulm sequences $\overline{\alpha}$ and $\overline{\beta}$ are called **equivalent**, and we write $\overline{\alpha} \sim \overline{\beta}$ if $p^k\overline{\alpha} = p^l\overline{\beta}$ for some $k, l < \omega$. Let e be an equivalence class of Ulm sequences and $\overline{\alpha} \in e$. Then Lemma 3.2.2 shows that the cardinal number

$$s(e,G) = \lim_{i\to\infty} r(G(p^i\overline{\alpha})/G(p^i\overline{\alpha}^*))$$

is well-defined and coincides with the rank of $G(\overline{\alpha})/G(\overline{\alpha}^*)$. Notice that by (e), we automatically have $r = r_p$ whenever $(\infty, \infty, \ldots) \notin e$, and $r = r_0$ otherwise. The cardinal number $s(e, G)$ is called the ***e*-th Stanton invariant of** G.

Now suppose that the $\mathbb{Z}_p$-module G has a decomposition basis X. Then for an equivalence class e of Ulm sequences, define

$$w_X(e,G) = \left|\{x \in X : \|x\|_p \in e\}\right|.$$

This is called the ***e*-th Warfield invariant of** G. For a $\mathbb{Z}_p$-module with decomposition basis, the two invariants coincide:

Theorem 3.2.3 (Stanton [139]). *If G is a $\mathbb{Z}_p$-module with decomposition basis X, then $s(e, G) = w_X(e, G)$ for all e.*

Proof. Let e be an equivalence class of Ulm sequences, and

$$X_e = \{x \in X : \|x\|_p \in e\}.$$

First, we suppose that $(\infty, \infty, \dots) \in e$. Let $\overline{\alpha} = (\infty, \infty, \dots)$. For every $x \in X_e$, there is an $m < \omega$ such that $|p^m x| = \infty$. By replacing $p^m x$ with x, we may assume $\|x\|_p = \overline{\alpha}$ for all $x \in X_e$. If $x \in X_e$, then $x \in G(\overline{\alpha}) \setminus G(\overline{\alpha}^*)^\diamond$ and, therefore, $x \notin G(\overline{\alpha}^*)$ by Lemma 3.2.1(1). Now $\bigoplus_{x \in X_e} \langle x \rangle$ is a $*$-valuated coproduct in G (Lemma 3.2.1(2)); thus, $\sum_{x \in X_e} \langle x + G(\overline{\alpha}^*) \rangle = \bigoplus_{x \in X_e} \langle x + G(\overline{\alpha}^*) \rangle$. Then

$$s(e, G) = r(G(\overline{\alpha})/G(\overline{\alpha}^*)) \geq |X_e| = w_X(e, G).$$

To verify that $s(e, G) \leq w_X(e, G)$, we assume that $s(e, G) > w_X(e, G)$. Then there is an independent set S in $G(\overline{\alpha})/G(\overline{\alpha}^*)$ properly containing $\overline{X}_e = \{x + G(\overline{\alpha}^*) : x \in X_e\}$. Therefore, there exists an element $\overline{y} = y + G(\overline{\alpha}^*) \in S \setminus \overline{X}_e$. Notice that $\overline{y}$ has infinite order since $G(\overline{\alpha})/G(\overline{\alpha}^*)$ is torsion-free for this $\overline{\alpha}$. Since X is a decomposition basis for G, there is a positive integer a such that $ay = a_1 x_1 + \cdots + a_r x_r$ for some $x_1, \dots, x_r \in X$ and integers $a_1, \dots, a_r$. We have $ay \in G(\overline{\alpha})$; hence, $\|a_i x_i\|_p \geq \overline{\alpha} = (\infty, \infty, \dots)$. Therefore, $x_i \in X_e$ for all $i \in \{1, \dots, r\}$. But then

$$ay + G(\overline{\alpha}^*) = a_1 x_1 + \cdots + a_r x_r + G(\overline{\alpha}^*)$$

with $ay \notin G(\overline{\alpha}^*)$ contradicts independence of the set S. Consequently, we obtain $s(e, G) \leq w_X(e, G)$.

Now assume that $\overline{\alpha} \not\to (\infty, \infty, \dots)$. For every $x \in X_e$, there is an integer $m \geq 0$ such that $\|p^m x\|_p = p^n \overline{\alpha}$ for some $n \geq 0$. By replacing $p^m x$ with x, we can assume that $X_e = \bigcup_{i<\omega} X_{p^i \overline{\alpha}}$, where $X_{\overline{\beta}} = \{x \in X_e : \|x\|_p = \overline{\beta}\}$. Let $Y_0 = X_{\overline{\alpha}}$, $Y_1 = X_{p\overline{\alpha}} \cup \{px : x \in Y_0\}$, $Y_2 = X_{p^2 \overline{\alpha}} \cup \{px : x \in Y_1\}$, et cetera. Then $z \in Y_i$ if and only if $\|z\|_p = p^i \overline{\alpha}$ and $z = p^j x$ for some $j < \omega$ and $x \in X_e$. Note that the map $x \mapsto px$ maps each Y_i injectively into Y_{i+1}. Hence,

$$|Y_0| \leq |Y_1| \leq \cdots \leq |Y_i| \leq \cdots \quad (i < \omega),$$

and we have $|X_e| = \lim_{i \to \infty} |Y_i|$. Notice that if $x \in Y_i$, then $x \in G(p^i \overline{\alpha}) \setminus G(p^i \overline{\alpha}^*)$ by Lemma 3.2.1(1), and since $\bigoplus_{x \in Y_i} \langle x \rangle$ is a $*$-valuated coproduct in G (Lemma 3.2.1(2)), we obtain $\sum_{x \in Y_i} \langle x + G(p^i \overline{\alpha}^*) \rangle = \bigoplus_{x \in Y_i} \langle x + G(p^i \overline{\alpha}^*) \rangle$. We have $s(e, G) = r(G(p^i \overline{\alpha})/G(p^i \overline{\alpha}^*)) \geq |Y_i|$ for all $i < \omega$. Hence, $s(e, G) \geq |X_e| = w_X(e, G)$.

To prove that $s(e, G) \leq w_X(e, G)$, assume that $s(e, G) > w_X(e, G)$. For $i < \omega$ let

$$\phi_i : G(\overline{\alpha})/G(\overline{\alpha}^*) \to G(p^i \overline{\alpha})/G(p^i \overline{\alpha}^*)$$

denote the isomorphism defined by $x + G(\overline{\alpha}^*) \mapsto p^i x + G(p^i\overline{\alpha}^*)$ (see Lemma 3.2.2), and define $\overline{Y_i} = \{x + G(p^i\overline{\alpha}^*) : x \in Y_i\}$. Then the set $S' = \bigcup_{i<\omega} \phi_i^{-1}(\overline{Y_i})$ is independent in $G(\overline{\alpha})/G(\overline{\alpha}^*)$ since

$$\overline{Y_0} = \phi_0^{-1}(\overline{Y_0}) \subseteq \phi_1^{-1}(\overline{Y_1}) \subseteq \phi_2^{-1}(\overline{Y_2}) \subseteq \cdots,$$

and we obtain $|S'| = \lim_{i\to\infty} |\phi_i^{-1}(\overline{Y_i})| = \lim_{i\to\infty} |Y_i| = |X_e|$. It follows that there exists an independent set S in $G(\overline{\alpha})/G(\overline{\alpha}^*)$ properly containing S'; hence, there is an element $\overline{y} = y + G(\overline{\alpha}^*) \in S \setminus S'$. Consequently, the subset $\{\phi_i(\overline{y})\} \cup \overline{Y_i}$ of $\phi_i(S)$ is independent for each $i < \omega$.

Since X is a decomposition basis for G, there is a positive integer $a = p^k a'$ ($k < \omega$, and a' an integer relatively prime to p) such that

$$ay = a_1x_1 + \cdots + a_rx_r + b_1y_1 + \cdots + b_sy_s,$$

where $a_1, \ldots, a_r, b_1, \ldots, b_s \in \mathbb{Z}$, $x_1, \ldots, x_r \in X_e$, and $y_1, \ldots, y_s \in X$ with $\|a_1x_1\|_p = \cdots = \|a_rx_r\|_p = p^k\overline{\alpha}$, and $b_1y_1, \ldots, b_sy_s \in G(p^k\overline{\alpha}^*)$. Then

$$ay + G(p^k\overline{\alpha}^*) = a_1x_1 + \cdots + a_rx_r + G(p^k\overline{\alpha}^*)$$

with $ay \notin G(p^k\overline{\alpha}^*)$. Let $1 \le i \le r$, and write $a_i = a_i' p^{l_i}$ for some $l_i \le k$ and $\gcd(a_i', p) = 1$. Then $a_ix_i = a_i' p^{l_i} x_i$, where $p^{l_i}x_i \in Y_k$. Thus, each a_ix_i is a multiple of an element of Y_k, contradicting independence of the set $\{\phi_k(\overline{y})\} \cup \overline{Y_k}$. Therefore, $s(e, G) \le w_X(e, G)$, and the proof is complete. □

This result allows us to drop the subscript X.

Lemma 3.2.4. *Suppose G and H are $\mathbb{Z}_p$-modules with decomposition bases X and Y, respectively. Then there is a height-preserving isomorphism $\langle X'\rangle \to \langle Y'\rangle$ for some subordinates X' of X and Y' of Y if and only if $w(e, G) = w(e, H)$ for all e.*

Proof. Suppose that G and H have the same Warfield invariants. Then there exists a bijection $f : X \to Y$ such that $\|x\|_p \sim \|f(x)\|_p$ for all $x \in X$. Thus, for each $x \in X$ there are positive integers $m(x)$, $n(x)$ such that $\|p^{m(x)}x\|_p = \|p^{n(x)}f(x)\|_p$. Then the sets $X' = \{p^{m(x)}x : x \in X\}$ and $Y' = \{p^{n(x)}f(x) : x \in X\}$ are the desired subordinates. The converse is immediate. □

The global version of the following result (see Lemma 3.4.2) can be found in Hunter–Richman [61]:

Lemma 3.2.5 (Hunter–Richman [61]). *Let G be a $\mathbb{Z}_p$-module with decomposition basis X. If Y is a finite subset of X, then $\langle Y\rangle$ is p-nice in G.*

Proof. First, we will prove that for $y \in X$, $\langle y\rangle$ is p-nice. Let $g \in G \setminus \langle y\rangle$. Then for some $k > 0$, $p^kg = ay + x$ for some $a \in \mathbb{Z}$ and $x \in \langle X \setminus \{y\}\rangle$. Next we define $g' = g - (a/p^k)y$ if

$p^k|a$, and $g' = g$ otherwise. This gives us $p^k g' = a'y + x$, where either $a' = 0$ or $a' = a$ is not divisible by p^k.

If g' has maximal height in $g + \langle y \rangle$, we are done. Suppose, to the contrary, that the coset $g + \langle y \rangle$ has no element proper with respect to $\langle y \rangle$. Then $g' + \langle y \rangle = g + \langle y \rangle$ takes on infinitely many heights; thus, $|g'| < |g' + my| < |g' + ny|$ for some $m, n \in \mathbb{Z}_p$. This can happen only if $|g'| = |my| = |p^i y|$ for some i.

Therefore, we obtain $|g'+my| < |p^k g'+p^k my| = |(a'+p^k m)y+x| = |(a'+p^k m)y| \wedge |x| \leq |(a' + p^k m)y| \leq |p^{k+i} y|$. The last inequality comes from noticing that $|p^k my| = |p^{k+i} y|$, and a' is either 0 or not divisible by p^k. Also, $|g' + my| = |(g' + my) - (g' + ny)| = |(m - n)y| = |p^j y|$ for some $j < i + k$. This gives only finitely many choices for j; thus, only finitely many choices for $|g' + my|$. Consequently, $g + \langle y \rangle = g' + \langle y \rangle$ takes on only finitely many distinct heights, and so $\langle y \rangle$ is p-nice.

We will now prove by induction on n that any n elements from a decomposition basis generate a p-nice submodule. Let $Y \subseteq X$ and $y \in Y$. Then $\langle y \rangle$ is p-nice in G. By the induction hypothesis, $\langle Y \rangle / \langle y \rangle$ is p-nice in $G/\langle y \rangle$ since the image of Y under the standard mapping has fewer elements. Then by Lemma 2.1.4(2), $\langle Y \rangle$ is p-nice in G. □

Lemma 3.2.6. *Let G be a $\mathbb{Z}_p$-module with decomposition basis X, and suppose that X' is a subordinate of X. If $x \in \langle X \rangle$, then the coset $x + \langle X' \rangle$ has an element of maximal p-height.*

Proof. There is a finite subset Y of X such that $x \in \langle Y \rangle$. We write $X' = \{p^{m(a)} a : a \in X\}$, and $Y' = \{p^{m(a)} a : a \in Y\}$. By Lemma 3.2.5, there is a $y \in \langle Y' \rangle$ such that $x + y$ has maximal p-height in the coset $x + \langle Y' \rangle$. Let $z \in \langle X' \rangle$ and write $z = z_1 + z_2$, where $z_1 \in \langle Y' \rangle$ and $z_2 \in \langle X' \setminus Y' \rangle$. Since $\langle X' \setminus Y' \rangle$ is a subset of $\langle X \setminus Y \rangle$, we have $|x + y| \geq |x + z_1| \geq |x + z_1| \wedge |z_2| = |x + z_1 + z_2| = |x + z|$. Therefore, $x + y$ has maximal p-height in the coset $x + \langle X' \rangle$. □

A decomposition set X in a $\mathbb{Z}_p$-module G is called a **nice decomposition set** if $\langle X \rangle$ is p-nice in G. If in addition, X is a decomposition basis for G, we call X a **nice decomposition basis**. The global version of the first part of the next lemma (see Lemma 3.4.3) can be found in Fuchs [28]:

Lemma 3.2.7. *Suppose G is a $\mathbb{Z}_p$-module with nice decomposition basis X, and let X' be a subordinate of X. Then we have the following:*
1. *X' is a nice decomposition basis for G.*
2. *If $G/\langle X \rangle$ is simply presented, then so is $G/\langle X' \rangle$.*

Proof. By Lemma 3.2.6, every coset $x + \langle X' \rangle$ with $x \in \langle X \rangle$ has an element of maximal p-height. Then the first assertion follows from Lemma 2.1.5. If $G/\langle X \rangle$ is a simply presented p-group, it has a nice composition series $\{N_\alpha / \langle X \rangle : \alpha \leq \lambda\}$ by Theorem 2.10.1. Since $\langle X \rangle$ is p-nice in G, each submodule N_α is p-nice in G as well by Lemma 2.1.4(2). Then by Lemma 2.1.4(1), $N_\alpha / \langle X' \rangle$ is p-nice in $G/\langle X' \rangle$; hence, we obtain a nice composition series for $G/\langle X' \rangle$, and the assertion follows from Theorem 2.10.1. □

A **Warfield module** is defined to be a direct summand of a simply presented $\mathbb{Z}_p$-module. We will see later that Warfield modules are exactly those $\mathbb{Z}_p$-modules G with a nice decomposition basis X such that $G/\langle X\rangle$ is simply presented. We will use this form in the following local classification theorem for Warfield modules:

Theorem 3.2.8 (Warfield [154]). *Let G and H be $\mathbb{Z}_p$-modules with nice decomposition bases X and Y, respectively, such that $G/\langle X\rangle$ and $H/\langle Y\rangle$ are simply presented. Then $G \cong H$ if and only if for every $\alpha \in \mathrm{Ord}_\infty$ and equivalence class e of Ulm sequences, we have $u(\alpha, G) = u(\alpha, H)$ and $w(e, G) = w(e, H)$.*

Proof. By Theorem 3.2.3, only sufficiency needs to be proved. Therefore, suppose G and H have the same Ulm and Warfield invariants. By Lemmas 3.1.6 and 3.2.4, the decomposition bases X and Y have subordinates X' and Y', respectively, that are lower decomposition bases such that there is a height-preserving isomorphism $f : \langle X'\rangle \to \langle Y'\rangle$. Let α be an ordinal. Then $u(\alpha, G) = \dim I_\alpha(\langle X'\rangle) + u_\alpha(G, \langle X'\rangle)$ yields

$$\dim I_\alpha(\langle X'\rangle) + u_\alpha(G, \langle X'\rangle) = \dim I_\alpha(\langle Y'\rangle) + u_\alpha(H, \langle Y'\rangle).$$

If $x \in X'$, then $\|x\|_p$ has a gap at α exactly if $\|f(x)\|_p$ has a gap at α as well; hence, $I_\alpha(\langle X'\rangle) \cong I_\alpha(\langle Y'\rangle)$. Then

$$u_\alpha(G, \langle X'\rangle) = u_\alpha(H, \langle Y'\rangle)$$

in the case where $I_\alpha(\langle X'\rangle)$ has finite dimension. If the dimension of $I_\alpha(\langle X'\rangle)$ is infinite this equality holds as well since X' and Y' are lower decomposition bases. By Lemma 3.2.7, X' and Y' are nice decomposition bases and the factor groups $G/\langle X'\rangle$ and $H/\langle Y'\rangle$ are simply presented. Notice that $u_\infty(G, \langle X'\rangle) = u(\infty, G) = u(\infty, H) = u_\infty(H, \langle Y'\rangle)$ since $\langle X'\rangle$ and $\langle Y'\rangle$ are torsion-free. Then, by Theorem 2.3.2, f can be extended to an isomorphism $G \to H$. □

Exercises.

1. (Warfield [155]) Let G and H be $\mathbb{Z}_p$-modules of torsion-free rank 1 that have $H(\aleph_0)$-families of p-nice submodules. Then $G \cong H$ if and only if $u(\alpha, G) = u(\alpha, H)$ for all $\alpha \in \mathrm{Ord}_\infty$ and $\|x\|_p^G \sim \|y\|_p^H$ for some (hence any) elements $x \in G$, $y \in H$ of infinite order.
2. (Warfield [155]) Let G be a simply presented $\mathbb{Z}_p$-module. Show the following:
   (a) G is a direct sum of simply presented $\mathbb{Z}_p$-modules of torsion-free rank ≤ 1.
   (b) G has an $H(\aleph_0)$-family of p-nice submodules.
3. (Warfield [155]) Let G be a $\mathbb{Z}_p$-module of torsion-free rank 1. Then G has an $H(\aleph_0)$-family of p-nice submodules if and only if G is a direct summand of a simply presented $\mathbb{Z}_p$-module of torsion-free rank 1. [Hint: Exercises 1 and 2(b).]

3.3 The class of Warfield modules

In this section, we state various characterizations of Warfield modules (Theorems 3.3.1–3.3.4). The global versions of these theorems and others will be proved independently in Section 3.7 and following sections (summarized in Theorem 3.16.1).

A short exact sequence $0 \to A \to B \to C \to 0$ of $\mathbb{Z}_p$-modules is called **sequentially pure** if for each Ulm sequence $\overline{\alpha}$, the sequence

$$0 \to A(\overline{\alpha}) \to B(\overline{\alpha}) \to C(\overline{\alpha}) \to 0$$

is exact. Then we say that a $\mathbb{Z}_p$-module G is **sequentially-pure-projective** if G has the projective property relative to all sequentially pure exact sequences; that is, if for every sequentially pure sequence $0 \to A \to B \xrightarrow{\nu} C \to 0$ and homomorphism $\gamma : G \to C$ there exists a homomorphism $\tau : G \to B$ such that the diagram

$$\begin{array}{ccccccccc} & & & & & & G & & \\ & & & & & \tau \swarrow & \downarrow \gamma & & \\ 0 & \to & A & \to & B & \xrightarrow{\nu} & C & \to & 0 \end{array}$$

is commutative.

Theorem 3.3.1 (Warfield [155]). *Let G be a $\mathbb{Z}_p$-module. Then the following conditions are equivalent:*

1. *G is a Warfield module;*
2. *G has a nice decomposition basis X such that $G/\langle X\rangle$ is simply presented;*
3. *G is sequentially-pure-projective.*

This theorem, which will be proved more generally in Theorem 3.14.2, allows us to use (2) as an alternative characterization of Warfield modules. Warfield modules are classified in Theorem 3.2.8 using (2), whereas (1) tells us that the class of Warfield modules is closed under direct sums and direct summands.

An R-module G is called **of simple isomorphism type** if every nonzero submodule of G is isomorphic to G. For instance, if $R = \mathbb{Z}_p$, the R-modules of simple isomorphism type are exactly the cyclic modules whose order is p or infinite. Then a smooth chain of submodules

$$0 = N_0 \subseteq N_1 \subseteq \cdots \subseteq N_\alpha \subseteq \cdots \subseteq N_\lambda = G$$

is called a **composition series for** G if $N_{\alpha+1}/N_\alpha$ is of simple isomorphism type for all ordinals $\alpha < \lambda$. It is clear that for $R = \mathbb{Z}$, this definition coincides with the definition of "composition series" for groups given in Section 2.2.

A p-nice submodule A of a $\mathbb{Z}_p$-module G is called **quasi-sequentially nice** if for each $x \in G$, there is an $n < \omega$ and an $a \in A$ such that the Ulm sequences of $p^n x + a$ (in G) and $p^n x + A$ (in G/A) are identical.

Theorem 3.3.2 (Moore [120]). *Let G be a $\mathbb{Z}_p$-module. Then the following conditions are equivalent:*

1. *G has a nice decomposition basis X such that $G/\langle X\rangle$ is simply presented;*
2. *G has a composition series of quasi-sequentially nice submodules.*

Let G be a $\mathbb{Z}_p$-module. Following Hill and Megibben [49], we call an element $x \in G$ **primitive** if $x \notin G(\overline{\alpha}^*)$ whenever $\overline{\alpha}$ is the Ulm sequence of x. Then a p-nice submodule A of G is called **knice** if for every finite subset S of G, there is a finite (possibly empty) collection of primitive elements $y_1, \ldots, y_m \in G$ and an $r < \omega$ such that $A \oplus \langle y_1\rangle \oplus \cdots \oplus \langle y_m\rangle$ is a $*$-valuated coproduct in G containing $p^r\langle S\rangle$. A $\mathbb{Z}_p$-module is called an **Axiom 3 module** if it has an $H(\aleph_0)$-family of knice submodules.

Theorem 3.3.3 (Hill–Megibben [49]). *Let G be a $\mathbb{Z}_p$-module. Then the following conditions are equivalent:*

1. *G has a nice decomposition basis X such that $G/\langle X\rangle$ is simply presented;*
2. *G is an Axiom 3 module;*
3. *G has an $H(\aleph_0)$-family of p-nice submodules and possesses a decomposition basis.*

It should be pointed out that the proof of [49, Theorem 4.1] implicitly contains additional characterizations of Warfield modules in terms of composition series (see [102, Theorem 3.4]):

Theorem 3.3.4. *Let G be a $\mathbb{Z}_p$-module. Then the following conditions are equivalent:*

1. *G has a composition series of quasi-sequentially nice submodules;*
2. *G has a composition series $\{N_\alpha\}_{\alpha\le\lambda}$ of p-nice submodules such that if $N_{\alpha+1}/N_\alpha$ is infinite, there is an $x_\alpha \in G$ such that $N_{\alpha+1} = N_\alpha \oplus \langle x_\alpha\rangle$ is a valuated coproduct in G;*
3. *G has a composition series $\{N_\alpha\}_{\alpha\le\lambda}$ of knice submodules such that if $N_{\alpha+1}/N_\alpha$ is infinite, there is a primitive element $x_\alpha \in G$ such that $N_{\alpha+1} = N_\alpha \oplus \langle x_\alpha\rangle$ is a $*$-valuated coproduct in G;*
4. *G has a composition series of knice submodules.*

To establish the uniqueness theorem for Warfield modules, the following fact will be useful (see Stanton [140] or Jacoby–Loth [71]):

Theorem 3.3.5. *If G is a $\mathbb{Z}_p$-module, there are simply presented $\mathbb{Z}_p$-modules H and T (where T is torsion) such that $G \oplus T$ and H have the same Ulm and Warfield invariants. If G is torsion-free, we may take H to be completely decomposable and $T = 0$.*

Proof. Let C be the set of all equivalence classes e of Ulm sequences with $w(e, G) \neq 0$. For every $e \in C$ let $0 \neq x + G(\overline{\alpha}^*) \in G(\overline{\alpha})/G(\overline{\alpha}^*)$, where $\overline{\alpha}$ is any element of e. Clearly, x has infinite order, and $\|x\|_p \in e$ since $x \notin G(\overline{\alpha}^*)$.

In the case where G is torsion-free, the submodule $\langle x\rangle^0$ is a torsion-free rank 1 group. Therefore, it is a simply presented $\mathbb{Z}_p$-module by the proof of Lemma 2.6.1, and we let $H(e) = \bigoplus_{w(e,G)}\langle x\rangle^0$. Suppose that G is not torsion-free. Then, by Corollary 2.7.5, there is a simply presented $\mathbb{Z}_p$-module H_x of torsion-free rank 1 containing an element of infinite order with p-Ulm sequence $\|x\|_p$, and we let $H(e) = \bigoplus_{w(e,G)} H_x$. In either case, we define $H' = \bigoplus_{e\in C} H(e)$. Thus, G and H' have the same Warfield invariants.

Let D be the set of all $\alpha \in \mathrm{Ord}_\infty$, where $u(\alpha, G) \neq 0$ or $u(\alpha, H') \neq 0$. For $\alpha \in D$, let P_α be defined as in Theorem 2.7.4 if α is an ordinal, and let $P_\alpha = \mathbb{Z}(p^\infty)$ whenever $\alpha = \infty$. Letting $T = \bigoplus_{\mathfrak{m}} \bigoplus_{\alpha\in D} P_\alpha$, where $\mathfrak{m}$ is an infinite cardinal greater than both $|G|$ and $|H'|$, we have

$$u(\alpha, G \oplus T) = u(\alpha, T) = u(\alpha, H' \oplus T)$$

for all $\alpha \in D$. Then $G \oplus T$ and $H = H' \oplus T$ have the same Ulm and Warfield invariants, and both T and H are simply presented $\mathbb{Z}_p$-modules. If G is torsion-free, then $T = 0$ since $D = \emptyset$; thus, H is completely decomposable. □

We have the following uniqueness result on the class of Warfield modules:

Theorem 3.3.6 (Hunter–Richman [61], Stanton [140]). *The class of Warfield modules is the unique class of $\mathbb{Z}_p$-modules that* (1) *contains all simply presented $\mathbb{Z}_p$-modules, is closed under* (2) *finite direct sums and* (3) *direct summands, and in which* (4) *the Ulm and Warfield invariants distinguish between nonisomorphic groups.*

Proof. There exists a class of $\mathbb{Z}_p$-modules satisfying (1–4), namely the class D of Warfield modules (see Theorems 3.2.8 and 3.3.1). If C is any class of $\mathbb{Z}_p$-modules satisfying (1–4), then C contains D by (1) and (3) since Warfield modules are direct summands of simply presented $\mathbb{Z}_p$-modules. Let $G \in \mathrm{C}$. By Theorem 3.3.5, there are simply presented $\mathbb{Z}_p$-modules H and T such that $G \oplus T$ and H have the same Ulm and Warfield invariants. By (1), H and T are in C; hence, $H \oplus T \in \mathrm{C}$ by (2). But then (4) shows that $G \oplus T \cong H$. It follows that $G \oplus T$ is simply presented; hence, G is a Warfield module. Therefore, $G \in \mathrm{D}$. □

Exercises.

1. (Hill–Megibben [49]) A p-nice submodule A of a $\mathbb{Z}_p$-module G is quasi-sequentially nice if and only if for every $x \in G$ there is an $n < \omega$ and an $a \in A$ such that $A \oplus \langle p^n x + a\rangle$ is a valuated coproduct in G.
2. (Hill–Megibben [49]) Knice submodules are quasi-sequentially nice.
3. Give an example of an element of a $\mathbb{Z}_p$-module that is of infinite order and not primitive.

3.4 The p-localization of a group

To prove the global classification theorem for Warfield groups, we need to consider p-localizations of a given group G. This will allow us to apply what we learned in the local cases to the more general global. Recall from Section 1.17 that localization is accomplished using the localization map,

$$\begin{array}{rccc} v_p: & G & \to & G_p = G \otimes \mathbb{Z}_p \\ & x & \mapsto & x \otimes 1. \end{array}$$

Recall from Lemma 1.17.7 that localization has several useful properties, in particular, v_p is p-height-preserving.

Now we see that v_p preserves other key group characteristics.

Lemma 3.4.1. *Let $v_p : G \to G_p$ be the natural map. Then the following hold:*

1. *If $A = \bigoplus_{i\in I} A_i$ is a valuated coproduct in G, then $A_p = \sum_{i\in I}(A_i)_p$ is a valuated coproduct in G_p.*
2. *If X is a decomposition set in G, then $v_p(X)$ is a decomposition set in G_p and we have $\langle X\rangle_p = \langle v_p(X)\rangle$.*

Proof. To show that the sum $\sum_{i\in I}(A_i)_p$ is direct, suppose $x_1 + \cdots + x_r = 0$ for some $x_1, \dots, x_r \in (A_{i_1})_p, \dots, (A_{i_r})_p$, respectively, where $i_1, \dots, i_r \in I$ are distinct. We may write each x_i as $a_i \otimes \frac{1}{n_i}$, where $a_i \in A_i$ and n_i is a positive integer such that $p \nmid n_i$. Let $k = \prod_{i=1}^r n_i$, and $k_i = \frac{k}{n_i}$ for each i. Then $0 = k(x_1 + \cdots + x_r) = k_1a_1 \otimes 1 + \cdots + k_ra_r \otimes 1 = (k_1a_1+\cdots+k_ra_r)\otimes 1$; hence, $k_1a_1+\cdots+k_ra_r \in \ker v_p$. By Lemma 1.17.8(3), there is a positive integer m, relatively prime to p, such that $m(k_1a_1+\cdots+k_ra_r) = 0$. Let $i \in \{1, \dots, r\}$. Since $\bigoplus_{i\in I} A_i$ is a direct sum, we have $mk_ia_i = 0$, so $0 = mk_ia_i \otimes 1 = mkx_i$. Since $p \nmid mk$, and $x_i \in G_p$, it follows that $x_i = 0$. Thus, the sum $\sum_{i\in I}(A_i)_p$ is direct.

To prove that $\bigoplus_{i\in I}(A_i)_p$ is a valuated coproduct in G_p, let $\sum_{i=1}^r(a_i \otimes \frac{1}{n_i})$ be an arbitrary element of $\bigoplus_{i\in I}(A_i)_p$, as before. Let $k = \prod_{i=1}^r n_i$, and let $k_i = \frac{k}{n_i}$ for each i. We have $|\sum_{i=1}^r a_i \otimes \frac{1}{n_i}|_p^{G_p} = |k(\sum_{i=1}^r a_i \otimes \frac{1}{n_i})|_p^{G_p} = |\sum_{i=1}^r k_ia_i \otimes 1|_p^{G_p} = |(\sum_{i=1}^r k_ia_i) \otimes 1|_p^{G_p} = |\sum_{i=1}^r k_ia_i|_p^G$, since v_p is p-height-preserving, and this then equals $|k_1a_1|_p^G \wedge \cdots \wedge |k_ra_r|_p^G$. For each $i \in I$, we have $|k_ia_i|_p^G = |a_i|_p^G = |a_i \otimes 1|_p^{G_p} = |k_i(a_i \otimes 1)|_p^{G_p} = |k(a_i \otimes \frac{1}{n_i})|_p^{G_p} = |a_i \otimes \frac{1}{n_i}|_p^{G_p}$. Hence, $|\sum_{i=1}^r a_i \otimes \frac{1}{n_i}|_p^{G_p} = |a_1 \otimes \frac{1}{n_1}|_p^{G_p} \wedge \cdots \wedge |a_r \otimes \frac{1}{n_r}|_p^{G_p}$; thus, $\bigoplus_{i\in I}(A_i)_p$ is a valuated coproduct in G_p.

Finally, suppose that X is a decomposition set in G. Then by (1), $\langle X\rangle_p = \bigoplus_{x\in X}\langle x\rangle_p$ is a valuated coproduct in G_p, and we have $\langle v_p(X)\rangle = \sum_{x\in X}\langle v_p(x)\rangle = \sum_{x\in X}\langle x\rangle_p = \bigoplus_{x\in X}\langle x\rangle_p = \bigoplus_{x\in X}\langle v_p(x)\rangle$. Therefore, $v_p(X)$ is a decomposition set in G_p and $\langle v_p(X)\rangle = \langle X\rangle_p$. □

As an application of the result above, we obtain the global versions of Lemmas 3.2.5 and 3.2.7. First, we need a definition. A subgroup A of a group G is called **locally nice** if for every prime p, A_p is p-nice in G_p. Then a decomposition basis X of G is called a **nice decomposition basis** if $\langle X\rangle$ is a locally nice subgroup of G.

Lemma 3.4.2 (Hunter–Richman [61]). *Let G be a group with decomposition basis X. If Y is a finite subset of X, then $\langle Y\rangle$ is locally nice in G.*

Proof. Let p be a prime and $v_p : G \to G_p$ the localization map. By Lemma 3.4.1(2), $v_p(X)$ is a decomposition basis for G_p. Therefore, since $Y_p = v_p(Y)$ is a finite subset of X_p, $\langle Y\rangle_p = \langle Y_p\rangle$ is p-nice in G_p by Lemma 3.2.5. □

Lemma 3.4.3. *Suppose G is a group with a nice decomposition basis X, and let X' be a subordinate of X. Then the following hold:*
1. *X' is a nice decomposition basis for G.*
2. *If $G/\langle X\rangle$ is simply presented, then so is $G/\langle X'\rangle$.*

Proof. Again, let p be a prime and $v_p : G \to G_p$ the localization map. By Lemma 3.4.1(2), $X_p = v_p(X)$ is a decomposition basis for G_p, and we have $\langle X_p\rangle = \langle X\rangle_p$; thus, X_p is nice. Since $X'_p = v_p(X')$ is a subordinate of X_p, $\langle X'\rangle_p = \langle X'_p\rangle$ is nice in G_p by Lemma 3.2.7(1), and (1) follows. If $G/\langle X\rangle$ is simply presented, then $G_p/\langle X_p\rangle \cong (G/\langle X\rangle)_p$ is simply presented (see Lemma 2.8.1); thus, Lemma 3.2.7(2) shows that $t_p(G/\langle X'\rangle) \cong (G/\langle X'\rangle)_p \cong G_p/\langle X'_p\rangle$ is a simply presented p-group, and the second assertion follows. □

Lemma 3.4.4. *Let G and H be groups, A and B subgroups of G and H, respectively, and let p be a prime. Suppose that $f : A \to B$ is a homomorphism, and let $f_p = f \otimes 1_{\mathbb{Z}_p} : A_p \to B_p$. Then we have the following:*
1. *If f is p-height-preserving, then so is f_p.*
2. *If f is an isomorphism, then so is f_p.*

Proof. Suppose f is p-height-preserving. Let $x \in A_p$ and write $x = a \otimes \frac{1}{n}$ for some $a \in A$ and positive integer n with $p \nmid n$. Then $|f_p(x)|_p^{H_p} = |nf_p(x)|_p^{H_p} = |f_p(nx)|_p^{H_p} = |f_p(a\otimes 1)|_p^{H_p} = |f(a)\otimes 1|_p^{H_p}$. By Lemma 1.17.8(5), we have $|f(a)\otimes 1|_p^{H_p} = |f(a)|_p^{H}$, which coincides with $|a|_p^G$ since f is p-height-preserving. Now $|a|_p^G = |a \otimes 1|_p^{G_p} = |n(a \otimes \frac{1}{n})|_p^{G_p} = |a \otimes \frac{1}{n}|_p^{G_p}$; hence, $|f_p(x)|_p^{H_p} = |x|_p^{G_p}$. Therefore, f_p is p-height-preserving.

To prove (2), note that if f is an isomorphism, then $0 \to A \xrightarrow{f} B \to 0$ is exact. Then $0 \to A_p \xrightarrow{f_p} B_p \to 0$ is exact by Theorem 1.17.9, and f_p is an isomorphism. □

Warfield's local-global lemma will be needed:

Lemma 3.4.5 (Warfield [152]). *Let G and H be groups and A and B subgroups of G and H, respectively, such that G/A and H/B are torsion. Suppose $f : A \to B$ is an isomorphism such that for every prime p, the induced homomorphism $f_p : A_p \to B_p$ extends to a homomorphism $g(p) : G_p \to H_p$. Then f extends to a homomorphism $g : G \to H$ such that $g_p = g(p)$ for all primes p. If every map $g(p)$ is injective (bijective), then so is g.*

Proof. Let $v_p : G \to G_p$ and $\mu_p : H \to H_p$ denote the natural maps. Then by Lemma 1.17.8(3), the induced maps $v' : G \to v'(G) \subseteq G' = \prod_{p\in\mathbb{P}} G_p$ and $\mu' : H \to \mu'(H) \subseteq H' = \prod_{p\in\mathbb{P}} H_p$ are injective.

Next we will show that $G'/v'(G)$ is torsion-free. Assume that there is an $x = (x_q)_{q\in\mathbb{P}} \in G'$ such that $px \in v'(G)$ for some prime p, say $px = v'(y)$, where $y \in G$. Since $px_p \in v_p(G)$ and $G_p/v_p(G)$ has no p-torsion (Lemma 1.17.8(1)), we have $x_p = v_p(z)$ for some $z \in G$. Then $v_p(pz - y) = pv_p(z) - v_p(y) = 0$. Hence, $pz - y \in \ker v_p$. By Lemma 1.17.8(3) $\ker v_p$ is p-divisible, so there is an $s \in \ker v_p$ such that $pz - y = ps$.

Letting $w = z - s$, we obtain

$$v_p(w) = v_p(z) - v_p(s) = x_p.$$

Let $q \neq p$ be a prime. Then $pv_q(w) = v_q(pw) = v_q(pz - ps) = v_q(y) = px_q$. Since G_q has no p-torsion it follows that $v_q(w) = x_q$. Thus, $x = v'(w) \in v'(G)$ and, therefore, $G'/v'(G)$ is torsion-free.

To verify that $v'(G) = \{x \in G' : nx \in v'(A) \text{ for some integer } n \neq 0\}$, let $x = v'(c)$, where $c \in G$. Then $nc \in A$ for some nonzero integer n; hence, $nx = v'(nc) \in v'(A)$. Conversely, suppose that $x \in G'$ such that $nx \in v'(A)$ for some $n \neq 0$. Since $G'/v'(G)$ is torsion-free and $nx \in v'(G)$, we obtain $x \in v'(G)$. Likewise, we have $\mu'(H) = \{x \in H' : nx \in \mu'(B) \text{ for some integer } n \neq 0\}$.

By assumption, each homomorphism $f_p : A_p \to B_p$ extends to $g(p) : G_p \to H_p$. This gives rise to a homomorphism $g' : G' \to H'$ by sending $(x_p)_{p\in\mathbb{P}}$ onto $(g(p)(x_p))_{p\in\mathbb{P}}$. We will show that g' extends the isomorphism

$$\begin{array}{ccc} v'(A) & \to & \mu'(B) \\ v'(a) & \mapsto & \mu'(f(a)) \end{array}$$

and maps $v'(G)$ into $\mu'(H)$. First, let $a \in A$. Since $v'(a) = (a \otimes 1, a \otimes 1, \dots)$, we have $g'(v'(a)) = (g(p)(a \otimes 1))_{p\in\mathbb{P}} = (f_p(a \otimes 1))_{p\in\mathbb{P}} = (f(a) \otimes 1, f(a) \otimes 1, \dots) = (\mu_p(f(a)))_{p\in\mathbb{P}} = \mu'(f(a))$, as claimed. To prove the second assertion, let $x \in v'(G)$. Then $nx \in v'(A)$ for some nonzero integer n. Hence, $ng'(x) = g'(nx) \in g'(v'(A)) = \mu'(B)$, and it follows that $g'(x) \in \mu'(H)$.

Let $g'' = g' \upharpoonright v'(G) : v'(G) \to \mu'(H)$. By the paragraph above, the map

$$g = \mu'^{-1} g'' v' : G \to H$$

extends $f : A \to B$. Now fix a prime p. To prove that $g_p = g(p)$, let $x \in G$ and $r \in \mathbb{Z}_p$. Since g' maps $v'(G)$ into $\mu'(H)$, there is an $h \in H$ such that $(g(q)(x \otimes 1))_{q\in\mathbb{P}} = g'(v'(x)) = (h \otimes 1, h \otimes 1, \dots)$. In particular, this yields $g(p)(x \otimes 1) = h \otimes 1$. Since $g(x) = \mu'^{-1} g'' v'(x) = h$, we conclude that $g_p(x \otimes r) = h \otimes r = r(h \otimes 1) = rg(p)(x \otimes 1) = g(p)(x \otimes r)$; hence, $g_p = g(p)$.

If each map $g(p)$ is injective (or bijective), then so is g'. Assume that each map $g(p)$ is bijective. In order to show that g is bijective, it suffices to verify that g'' is surjective, so let $y \in \mu'(H)$. Then $ny \in \mu'(B)$ for some $n \neq 0$. Since g' is bijective, we may define $x \in G$ such that $x = g'^{-1}(y)$, and we have $nx = g'^{-1}(ny) \in g'^{-1}\mu'(B) = v'(A)$; hence, $x \in v'(G)$. Thus, $g''(x) = y$ as desired. □

Exercises.

1. Let $v_p : G \to G_p$ denote the localization map. If X is a nice decomposition basis of G such that $G/\langle X \rangle$ is simply presented, then $v_p(X)$ is a nice decomposition basis of G_p and $G_p/\langle v_p(X) \rangle$ is simply presented.
2. If A is a p-isotype subgroup of G, then A_p is a p-isotype subgroup of G_p.

3.5 Compatible Ulm matrices

We might expect, by an analogue of the local classification, that we may classify general groups using an invariant based on equivalence of Ulm matrices (see Section 1.10), where M and N are **equivalent** if $nM = mN$ for some nonzero integers m and n. This is not the case. Warfield [155] cites an example of four countable groups A_1, A_2, A_3, and A_4 of torsion-free rank 1 such that $\|x_i\|$ are mutually inequivalent, where $x_i \in A_i$ is an element of infinite order for $1 \le i \le 4$, but $A_1 \oplus A_2 \cong A_3 \oplus A_4$ (see Exercise 2). Thus, if we try to define an invariant based on equivalence as we did in the local case, we will find that is not invariant under $\cong$. This problem was faced in generalizing to the global case the classification of Warfield modules up to isomorphism. This was solved by Stanton [140] with the introduction of new invariants based on the concept of compatibility.

If $M = [m_{p,i}]$, and $N = [n_{p,i}]$ are Ulm matrices, we write $M \ge N$ whenever $m_{p,i} \ge n_{p,i}$ for all (p, i). Recall that by $M \wedge N$, we mean the Ulm matrix $[m_{p,i} \wedge n_{p,i}]$, and that the Ulm matrix of an element x of a group G is the Ulm matrix

$$\|x\| = [|p^i x|_p],$$

where the rows are the p-Ulm sequences $\|x\|_p$. Notice that $\|x + y\| \ge \|x\| \wedge \|y\|$ whenever $x, y \in G$. To emphasize the group in which the Ulm matrix of x is computed, we sometimes write $\|x\|_G$ instead of $\|x\|$. For a positive integer n, we define $nM = [m_{p,i+|n|_p}]$, where $|n|_p$ denotes the p-height of n in $\mathbb{Z}$. It is clear that $\|nx\| = n\|x\|$ for all positive integers n and $x \in G$. We call two Ulm matrices M and N **compatible**, and write $M \sim N$, if there are positive integers m and n such that $mM \ge N$ and $nN \ge M$. This yields an equivalence relation, and the resulting equivalence classes are called **compatibility classes.** We write $[M]$ to indicate the compatibility class containing M. If c is a compatibility class, and $c = [\|x\|]$ for some x, we will say that x **is a representative of** c. If c_1 and c_2 are compatibility classes of Ulm matrices, then we write $c_1 \le c_2$ if there are $M_i \in c_i$ $(i = 1, 2)$ such that $M_1 \le M_2$. The following properties hold for all Ulm matrices M, M', N, and positive integer n:

(a) If $M \sim N$, then $M_p = N_p$ for almost all primes p.
(b) $n(M \wedge N) = nM \wedge nN$.
(c) If $M \sim M'$, then $M \wedge N \sim M' \wedge N$.
(d) If $A = \bigoplus_{i \in I} \langle x_i \rangle$ is a valuated coproduct in a group G such that $\|x_i\| \sim M$ for all $i \in I$, and $0 \ne x \in A$, then $\|x\| \sim M$.

The next lemma will be useful:

Lemma 3.5.1. *Let A and C be subgroups of G such that $A \subseteq C$. Suppose that the torsion-free rank of A is finite and that C satisfies the following property: For every finite subset S of C, there is a finite decomposition set X in G such that $X \subseteq C$ and $\langle S, X\rangle / \langle X\rangle$ is finite. Then every increasing sequence of compatibility classes of Ulm matrices realized by elements of A becomes stationary.*

Proof. All heights in this proof are computed in G. Since the Ulm matrix M of any torsion element satisfies $M \sim \overline{\infty}$, we may assume that A is not torsion. Let S be a maximal independent set of elements of A of infinite order. Since S is a finite subset of C, there is a decomposition set $X = \{x_1, \dots, x_n\}$ in G such that $X \subseteq C$ and $\langle S, X\rangle/\langle X\rangle$ is finite. Let c be a compatibility class realized by an element of A, say $x \in A$ is a nonzero element of infinite order with $c = [\|x\|]$. By the maximality of S, we have $x \in \langle S\rangle^0$; thus, $n'x \in \langle X\rangle$ for some positive integer n'. Since $\|n'x\| \sim \|x\|$, we may assume that the representative x of c is in $A \cap \langle X\rangle$. We write

$$x = a_{i_1}x_{i_1} + \cdots + a_{i_m}x_{i_m} \quad (0 \neq a_{i_k} \in \mathbb{Z}, k = 1, \dots, m),$$

where $1 \leq m \leq n$, and let $I_x = \{i_1, \dots, i_m\}$. Now suppose that $c_1 \lneqq c_2$ are compatibility classes with representatives x and y, respectively. To prove that $I_x \neq I_y$, we assume to the contrary that $I_x = I_y$. Since $c_1 \leq c_2$, there is a positive integer l such that $\|x\| \leq \|ly\|$, and we can write

$$ly = b_{i_1}x_{i_1} + \cdots + b_{i_m}x_{i_m} \quad (0 \neq b_{i_k} \in \mathbb{Z}, k = 1, \dots, m).$$

Let $r = b_{i_1} \cdots b_{i_m}$. Then $c_1 \neq c_2$ yields $r\|x\| \ngeq \|ly\|$; thus, $|rp^tx|_p \lneqq |p^tly|_p$ for some prime p and $t < \omega$. But then we obtain

$$|p^tly|_p = \min_{k=1,\dots,m}\{|p^tb_{i_k}x_{i_k}|_p\} \leq \min_{k=1,\dots,m}\{|p^tra_{i_k}x_{i_k}|_p\} = |p^trx|_p \lneqq |p^tly|_p,$$

a contradiction. Therefore, $I_x \neq I_y$. Since $\{1, \dots, n\}$ has only 2^n subsets, it follows that it is impossible to have compatibility classes $c_1 \lneqq c_2 \lneqq \cdots \lneqq c_{2^n+1}$ with representatives in A. Therefore, every increasing sequence $\{c_i\}_{i=1}^{\infty}$ with representatives in A becomes stationary. □

Exercises.

1. Prove (a–d).
2. (Warfield [152]) For an Ulm matrix M, let H_a be a group of torsion-free rank 1 containing an element a of infinite order with $\|a\| = M$ (see Corollary 2.7.7). Show that there are groups H_a, H_b, H_c, H_d such that $H_a \oplus H_b \cong H_c \oplus H_d$, but $\|a\|, \|b\|, \|c\|, \|d\|$ are mutually inequivalent Ulm matrices. [Hint: Define the Ulm matrix $\|a\|$ by the following: For $p = 2, 3$ let $|p^na|_p = 2n$ if n is even and $|p^na|_p = 2n - 1$ if n is odd. For $q \in \mathbb{P} \setminus \{2, 3\}$ let $|q^na|_q = n$ for all $n < \omega$. Then define the Ulm matrix $\|b\|$ as follows: If $p = 2, 3$ let $|p^nb|_p = 2n$ and for $q \in \mathbb{P} \setminus \{2, 3\}$ let $|q^nb|_q = n$ $(n < \omega)$. Finally, in the group $H_a \oplus H_b$, let $c = 9a + b$ and $d = 8a + b$.]

3.6 The subgroups $G(\overline{\alpha}^*, p)$, $G(M^*)$, $G(M^*, p)$ and $*$-valued coproducts

Let p be a prime, $\overline{\alpha} = (\alpha_i)_{i<\omega}$ an Ulm sequence, and M an Ulm matrix. Following Hill and Megibben [50], we define the subgroups

$$G(\overline{\alpha},p)=\{x\in G:\|x\|_p\geq\overline{\alpha}\},$$
$$G(M)=\{x\in G:\|x\|\geq M\}$$

and, using the sets

$$G(\overline{\alpha}^*,p)^\diamond=\{x\in G(\overline{\alpha},p):|p^ix|_p\neq\alpha_i \text{ for infinitely many } i<\omega\},$$
$$G(M^*)^\diamond=\{x\in G(M):\|x\|\nsim M \text{ or } x \text{ is torsion}\},$$
$$G(M^*,p)^\diamond=G(M^*)^\diamond\cup G(M_p^*,p)^\diamond,$$

the subgroups

$$G(\overline{\alpha}^*,p)=\langle G(\overline{\alpha}^*,p)^\diamond\rangle,$$
$$G(M^*)=\langle G(M^*)^\diamond\rangle,$$
$$G(M^*,p)=G(M)\cap\langle G(M^*,p)^\diamond\rangle.$$

Notice that if $F\in\{G(\overline{\alpha}^*,p),G(M^*),G(M^*,p)\}$ and $x\in F$, then $x=z_1+\cdots+z_n$ for some $z_1,\ldots,z_n\in F^\diamond$. We have the following basic properties:

(a) If $M\sim\overline{\infty}$, then $G(M^*)=t(G(M))$. Otherwise, $G(M^*)=\langle x\in G(M):\|x\|\nsim M\rangle$.
(b) If $N\geq M$, then $G(N)\subseteq G(M)$, $G(N^*)\subseteq G(M^*)$ and $G(N^*,p)\subseteq G(M^*,p)$.
(c) If $n>0$, then $G(nM)=n(G(M))$, $G((nM)^*)=n(G(M^*))$ and $G((nM)^*,p)=n(G(M^*,p))$.
(d) If $x\in G(M)$ and $nx\in n(G(M^*,p))$ for some $n>0$, then $x\in G(M^*,p)$.

Clearly, a direct sum $A=\bigoplus_{i\in I}A_i\subseteq G$ is a valuated coproduct in G exactly if $A\cap G(M)=\bigoplus_{i\in I}(A_i\cap G(M))$ for all Ulm matrices M. Following Hill and Megibben [50], we call the valuated coproduct $A=\bigoplus_{i\in I}A_i$ in G a **$*$-valuated coproduct in** G if for each Ulm matrix M, Ulm sequence $\overline{\alpha}$, prime p, and group $F\in\{G(\overline{\alpha}^*,p),G(M^*),G(M^*,p)\}$ the following holds:

$$A\cap F=\bigoplus_{i\in I}(A_i\cap F).$$

Observe that if $A=\bigoplus_{i\in I}A_i$ and $A_i=\bigoplus_{j\in J}A_{ij}$ are $*$-valuated coproducts in G ($i\in I$), then $A=\bigoplus_{i\in I,j\in J}A_{ij}$ is a $*$-valuated coproduct in G. The following observation can be found in [50] and is easily verified:

Lemma 3.6.1. *If $A=\bigoplus_{i\in I}A_i$ is a $*$-valuated coproduct in G and $B=\bigoplus_{i\in I}B_i$ is a valuated coproduct in G such that $A_i\subseteq B_i$ and B_i/A_i is torsion for all $i\in I$, then $B=\bigoplus_{i\in I}B_i$ is a $*$-valuated coproduct in G.*

Lemma 3.6.2. *Let $\phi:G\to H$ be a homomorphism, M an Ulm matrix, $\overline{\alpha}$ an Ulm sequence, and p a prime. Then the following hold:*

1. *$\phi(G(M))\subseteq H(M)$, $\phi(G(M^*)^\diamond)\subseteq H(M^*)^\diamond$, $\phi(G(\overline{\alpha}^*,p)^\diamond)\subseteq H(\overline{\alpha}^*,p)^\diamond$, and $\phi(G(M^*,p)^\diamond)\subseteq H(M^*,p)^\diamond$.*
2. *$\phi(G(M^*))\subseteq H(M^*)$, $\phi(G(\overline{\alpha}^*,p))\subseteq H(\overline{\alpha}^*,p)$, and $\phi(G(M^*,p))\subseteq H(M^*,p)$.*

Proof. The proof of (1) is immediate because $|\phi(x)|_q^H \geq |x|_q^G$ for all $x \in G$ and primes q. (2) follows from (1). □

To prove that a decomposition basis X of a group gives rise to a $*$-valuated coproduct $\langle X\rangle = \bigoplus_{x\in X}\langle x\rangle$, we need the following lemma:

Lemma 3.6.3. *Let G be a group with decomposition basis X, and $\{x_1, \ldots, x_n\} \subseteq X$. Suppose $y = a_1x_1 + \cdots + a_nx_n$, where $a_1, \ldots, a_n$ are nonzero integers, and $y = z_1 + \cdots + z_m$ for some $z_1, \ldots, z_m \in G$, and let p be a prime. Then for every $1 \leq j \leq n$, there is $1 \leq i_0 \leq m$ such that z_{i_0} is not torsion, and the following hold:*
1. *For any Ulm sequence $\overline{\alpha}$, if $y \in G(\overline{\alpha}, p)$ and $z_{i_0} \in G(\overline{\alpha}^*, p)^\diamond$, then $a_jx_j \in G(\overline{\alpha}^*, p)^\diamond$;*
2. *For every Ulm matrix M, if $y \in G(M)$ and $z_{i_0} \in G(M^*)^\diamond$, then $a_jx_j \in G(M^*)^\diamond$.*

Proof. Let $r > 0$ such that $rz_i \in \langle X\rangle$ for all i. In fact, we may choose r such that $rz_i = 0$ whenever z_i is torsion. Then we may write

$$rz_i = c_{i,1}x_1 + \cdots + c_{i,n}x_n + c_{i,n+1}x_{n+1} + \cdots + c_{i,l}x_l$$

for some integers $l, c_{i,1}, \ldots, c_{i,l}$ and $x_{n+1}, \ldots, x_l \in X \setminus \{x_1, \ldots, x_n\}$. Now $rz_1 + \cdots + rz_m = ry = ra_1x_1 + \cdots + ra_nx_n$. Let $1 \leq j \leq n$. Equating the x_j terms, we get

$$\sum_{i=1}^{m} c_{i,j}x_j = ra_jx_j.$$

We note that $c_{i,j}$ cannot be 0 for all i, since then ra_jx_j would be 0, a contradiction. This means that we may choose $1 \leq i_0 \leq n$ such that $c_{i_0,j} \neq 0$ and $|c_{i_0,j}|_p$ is minimal as a p-height in $\mathbb{Z}$. In particular, since $c_{i_0,j} \neq 0$, z_{i_0} is not torsion.

To prove (1), suppose for some Ulm sequence $\overline{\alpha}$, $y \in G(\overline{\alpha}, p)$, and $z_{i_0} \in G(\overline{\alpha}, p)^\diamond$. Then $z_{i_0} \in G(\overline{\alpha}, p)$, and $|p^iz_{i_0}|_p \neq \alpha_i$ for infinitely many $i < \omega$. Since X is a decomposition set, $\|a_jx_j\|_p \geq \|y\|_p \geq \overline{\alpha}$. Now suppose $a_jx_j \notin G(\overline{\alpha}^*, p)^\diamond$, which means $|p^ia_jx_j|_p = \alpha_i$ for almost all $i < \omega$. Let $\tilde{k} = |ra_j|_p \geq \min\{|c_{1,j}|_p, \ldots, |c_{m,j}|_p\} = |c_{i_0,j}|_p$ with p-heights computed in $\mathbb{Z}$. Then, for each i, $|p^{i+\tilde{k}}z_{i_0}|_p = |p^ira_jz_{i_0}|_p = \min\{|p^ia_jc_{i_0,1}x_1|_p, \ldots, |p^ia_jc_{i_0,l}x_l|_p\} \leq |p^ia_jc_{i_0,j}x_j|_p \leq |p^{i+\tilde{k}}a_jx_j|_p$, the latter since $\tilde{k} \geq |c_{i_0,j}|_p$. Then $|p^iz_{i_0}|_p \leq |p^ia_jx_j|_p$ for all $i \geq \tilde{k}$. We may choose k such that $|p^ia_jx_j|_p = \alpha_i$ for all $i \geq k$, and—in fact—we may take $k \geq \tilde{k}$. Then for all $i \geq k$, $\alpha_i \leq |p^iz_{i_0}|_p \leq |p^ia_jx_j|_p = \alpha_i$, contradicting $|p^iz_{i_0}| \neq \alpha_i$ for infinitely many i.

To prove (2), suppose $y \in G(M)$, $z_{i_0} \in G(M)$, and $\|z_{i_0}\| \nsim M$. Then, since X is a decomposition set, we get $\|a_jx_j\| \geq \|y\| \geq M$, and $\|rz_{i_0}\| \leq \|c_{i_0,j}x_j\|$. Now suppose $\|a_jx_j\| \sim M$, so $r'M \geq \|a_jx_j\|$ for some $r' > 0$. Then, if we let $|c_{i_0,j}|$ represent the absolute value of $c_{i_0,j}$ in $\mathbb{Z}$, we get $M \leq \|z_{i_0}\| \leq \|rz_{i_0}\| \leq \|c_{i_0,j}x_j\| \leq \|c_{i_0,j}a_jx_j\| = |c_{i_0,j}|\|a_jx_j\| \leq |c_{i_0,j}|r'M$. But this contradicts the assumption that $\|z_{i_0}\| \nsim M$. Therefore, $\|a_jx_j\| \nsim M$. □

Any decomposition basis gives rise to a $*$-valuated coproduct:

Lemma 3.6.4. *Suppose G is a group with decomposition basis X.*

1. *Let $x_1,\dots,x_n \in X$, $a_1,\dots,a_n$ nonzero integers, M an Ulm matrix, $\overline{\alpha}$ an Ulm sequence, p a prime, and $F \in \{G(\overline{\alpha}^*,p), G(M^*), G(M^*,p)\}$. If $a_1x_1+\cdots+a_nx_n \in F$, then $a_ix_i \in F^\diamond$ for all $1 \le i \le n$.*
2. *$\langle X\rangle = \bigoplus_{x\in X}\langle x\rangle$ is a $*$-valuated coproduct in G.*

Proof. Suppose $y = a_1x_1+\cdots+a_nx_n \in F$, say $y = z_1+\cdots+z_m$, where each $z_i \in F^\diamond$. Let $1 \le j \le n$. Apply Lemma 3.6.3 to get a z_{i_0}. Suppose first that $F = G(\overline{\alpha}^*,p)$. Then by Lemma 3.6.3(1), since $z_{i_0} \in F^\diamond$, $a_jx_j \in G(\overline{\alpha}^*,p)^\diamond = F^\diamond$. If $F = G(M^*)$, then by Lemma 3.6.3(2), $a_jx_j \in G(M^*)^\diamond = F^\diamond$. Finally suppose $F = G(M^*,p)$. Then $z_{i_0} \in G(M)$ and either $z_{i_0} \in G(M_p^*,p)^\diamond$, or $z_{i_0} \in G(M^*)^\diamond$. If $z_{i_0} \in G(M_p^*,p)^\diamond$, then, since $y \in G(M)$ implies $\|y\|_p \ge M_p$, by Lemma 3.6.3(1), $a_jx_j \in G(M_p^*,p)^\diamond \subseteq F^\diamond$. If $z_{i_0} \in G(M^*)$, then by Lemma 3.6.3(2), $a_jx_j \in G(M^*)^\diamond \subseteq F^\diamond$.

Clearly, (2) follows from (1). □

Next we prove Hill and Megibben's Seesaw lemma:

Proposition 3.6.5 (Hill–Megibben [50]). *Suppose that A and B are subgroups of a group G such that $A \subseteq B$ and B/A is finite. If $y_1,\dots,y_n \in G$ have infinite order and $A\oplus\langle y_1\rangle\oplus\cdots\oplus\langle y_n\rangle$ is a valuated coproduct in G, then $B\oplus\langle y_1'\rangle\oplus\cdots\oplus\langle y_n'\rangle$ is a valuated coproduct in G for some nonzero multiples y_i' of y_i $(i = 1,\dots,n)$. The statement remains valid if "valuated coproduct" is replaced by "$*$-valuated coproduct".*

Proof. We may assume that $|B/A| = p$ for some prime p, and write $B = \langle A,x\rangle$. Suppose that $B\oplus\langle y_1\rangle\oplus\cdots\oplus\langle y_n\rangle$ is not a valuated coproduct. Since $|x|_q = |px|_q$ for all primes $q \ne p$, there is an element $u = x + a + \sum_{i=1}^m t_iy_i$ ($a \in A$, $t_i \in \mathbb{Z}$ for $i = 1,\dots,m$, where $m \le n$) such that

$$|u|_p > |t_1y_1|_p \wedge\cdots\wedge |t_my_m|_p,$$

and we can assume that $t_i \ne 0$ for all $i = 1,\dots,m$. Then define $y_i' = pt_iy_i$ if $1 \le i \le m$, and $y_i' = y_i$ if $m < i \le n$. To show that $B\oplus\langle y_1'\rangle\oplus\cdots\oplus\langle y_n'\rangle$ is a valuated coproduct in G, let $w = rx + a' + s_1y_1' + \cdots + s_ny_n'$ with $r \in \{1,\dots,p-1\}$, $a' \in A$, and $s_1,\dots,s_n \in \mathbb{Z}$. Again, it suffices to consider p-heights. Since $A\oplus\langle y_1\rangle\oplus\cdots\oplus\langle y_n\rangle$ is a valuated coproduct, we have

$$|w-ru|_p \le |t_1(ps_1-r)y_1|_p \wedge\cdots\wedge |t_m(ps_m-r)y_m|_p \wedge |s_{m+1}y_{m+1}|_p \wedge\cdots\wedge |s_ny_n|_p.$$

Thus, $|w-ru|_p \le |t_1y_1|_p\wedge\cdots\wedge|t_my_m|_p < |u|_p$, and we obtain $|w|_p = |w-ru|_p \wedge |u|_p \le |s_1y_1'|_p\wedge\cdots\wedge|s_ny_n'|_p$. Therefore, $B\oplus\langle y_1'\rangle\oplus\cdots\oplus\langle y_n'\rangle$ is a valuated coproduct. The second assertion is immediate by Lemma 3.6.1. □

Corollary 3.6.6. *Suppose $\langle ay_1\rangle\oplus\langle y_2\rangle\oplus\cdots\oplus\langle y_n\rangle$ is a valuated (or $*$-valuated) coproduct in G for some $y_1,\dots,y_n \in G$ of infinite order and $a \in \mathbb{Z}\setminus\{0\}$. Then $\langle y_1\rangle\oplus\langle y_2'\rangle\oplus\cdots\oplus\langle y_n'\rangle$ is a valuated (or $*$-valuated) coproduct in G for some nonzero multiples y_i' of y_i $(i = 1,\dots,n)$.*

Exercises.

1. Prove (a) and (b).
2. Prove (c) and (d).
3. If $x \in G(M)$ and $nx \in n(G(M^*))$ for some $n > 0$, then $x \in G(M^*)$.

3.7 The global classification theorem

The following lemma will be useful:

Lemma 3.7.1. *Suppose that G is a group, M an Ulm matrix, and p a prime. Then we have the following:*

1. *If $M_p \nsim (\infty, \infty, \dots)$, then $G(M)/G(M^*, p)$ is a vector space over $\mathbb{Z}/p\mathbb{Z}$.*
2. *If $M_p \sim (\infty, \infty, \dots)$, then $G(M)/G(M^*, p)$ has no p-torsion.*

Proof. Let $x \in G(M)$. If $M_p \nsim (\infty, \infty, \dots)$, then $px \in G(M_p^*, p)^\diamond \subseteq G(M^*, p)$; hence, (1) is true. To prove (2), assume that $M_p \sim (\infty, \infty, \dots)$ and $px \in G(M^*, p)$. Then

$$px = a_1x_1 + \cdots + a_rx_r,$$

where each a_i is an integer and $x_i \in G(M^*, p)^\diamond = G(M^*)^\diamond$. Let $i \in \{1, \dots, r\}$. Since $p^n(M_p) = (\infty, \infty, \dots)$ for some $n < \omega$, there is a $z_i \in G$ such that $|z_i|_p = \infty$ and $p^n x_i = p^{n+1} z_i$. Since $x_i \in G(M)$, we have $z_i \in G(M)$. If $\|x_i\| \nsim M$, then $\|z_i\| \sim \|p^{n+1}z_i\| = \|p^n x_i\| \sim \|x_i\| \nsim M$, and if x_i is torsion, then so is z_i. Therefore, $z_i \in G(M^*)$. We have $p^{n+1}x = \sum a_i p^n x_i = \sum a_i p^{n+1} z_i$; hence,

$$p^{n+1}\left(x - \sum a_i z_i\right) = 0.$$

Then $x - \sum a_i z_i \in t(G(M)) \subseteq G(M^*)$; hence, $x \in G(M^*) = G(M^*, p)$. Therefore, $G(M)/G(M^*, p)$ has no p-torsion. □

For a compatibility class c of Ulm matrices, prime p, and equivalence class e of Ulm sequences, the **(global) Stanton invariant of** G is defined by

$$s(c, p, e, G) = \sup\{r_{p,0}(G(M)/G(M^*, p)) : M \in c \text{ and } M_p \in e\},$$

where $r_{p,0}$ denotes the sum of the p-rank r_p and the torsion-free rank r_0 (see Section 1.4). Then by Lemma 3.7.1, $r_{p,0} = r_p$ if $(\infty, \infty, \dots) \notin e$, and $r_{p,0} = r_0$ otherwise.

Lemma 3.7.2. *Let G be a group, p a prime, and M an Ulm matrix. Then the map*

$$f : G(M)/G(M^*, p) \to G(pM)/G((pM)^*, p)$$

defined by $x + G(M^*, p) \mapsto px + G((pM)^*, p)$ is an isomorphism.

Proof. Let $x \in G(M)$ and write $M = [m_{q,i}]$. If $x \in G(M^*,p)$, then $x = z_1 + \cdots + z_m$, where $z_1,\ldots,z_m \in G(M^*,p)^\diamond$. Let $j \in \{1,\ldots,m\}$ and $z = z_j$. If z is a torsion element, then so is pz. If $\|z\| \nsim M$, then $\|pz\| \sim \|z\| \nsim M \sim pM$, and if $z \in G(M_p^*,p)^\diamond$, then $|p^i pz|_p \neq m_{p,i+1}$ for infinitely many i; hence, $pz \in G((pM)_p^*,p)^\diamond$. In any case, we have $pz \in G((pM)^*,p)^\diamond$; thus, $px \in G((pM)^*,p)$. Thus, the map f is well-defined. Clearly, f is a homomorphism.

To show that f is injective, suppose that $px \in G((pM)^*,p)$, say $px = z_1' + \cdots + z_n'$, where $z_1',\ldots,z_n' \in G((pM)^*,p)^\diamond$. Let $k \in \{1,\ldots,n\}$, and $z' = z_k'$. Then $\|z'\|_p \geq (pM)_p$; hence, $|z'|_p \geq m_{p,1} > m_{p,0}$. Therefore, there is a $z \in G$ such that $pz = z'$ and $|z|_p \geq m_{p,0}$. Then $|p^{i+1}z|_p = |p^i z'|_p \geq m_{p,i+1}$ for all i, so $\|z\|_p \geq M_p$. If $z' \in G((pM)_p^*,p)^\diamond$, then $|p^i z'|_p \neq m_{p,i+1}$ for infinitely many i. Hence, $|p^i pz|_p \neq m_{p,i+1}$ for infinitely many i, and we have $z \in G(M_p^*,p)^\diamond$. Now suppose that $z' \in G((pM)^*)^\diamond$. Since $\|z\|_q = \|z'\|_q$ for all primes $q \neq p$, we have $z \in G(M)$. If z' is torsion, then so is z, and if $\|z'\| \nsim pM$, then $\|z\| \sim \|pz\| \nsim pM \sim M$, yielding $z \in G(M^*)^\diamond$. Hence, in any case, we have $z \in G(M^*,p)^\diamond$. Letting

$$y = z_1 + \cdots + z_n,$$

we have $py = z_1' + \cdots + z_n' = px$. Then $y \in G(M)$, since $\|y\|_p \geq M_p$ and $\|y\|_q = \|x\|_q$ for all primes $q \neq p$. Consequently, $y \in G(M^*,p)$. Then $x - y \in t(G(M))$, which yields $x = y + (x - y) \in G(M^*,p)$. Thus, f is injective.

To prove that f is surjective, let $y \in G(pM)$. Then $|y|_p \geq m_{p,1} > m_{p,0}$. Therefore, there is an $x \in G$ such that $|x|_p \geq m_{p,0}$ and $px = y$. Now f maps $x + G(M^*,p)$ onto $y + G((pM)^*,p)$. □

If G has a decomposition basis X, the **(global) Warfield invariant of** G is given by

$$w_X(c,p,e,G) = |\{x \in X : \|x\| \in c \text{ and } \|x\|_p \in e\}|.$$

As expected, Stanton invariants and Warfield invariants are identical for a group with decomposition basis:

Theorem 3.7.3 (Stanton [140]). *If G is a group with decomposition basis X, then $s(c,p,e,G) = w_X(c,p,e,G)$ for all c,p,e.*

Proof. The proof is similar to the proof of its local version (Theorem 3.2.3). We fix a prime p, a compatibility class c of Ulm matrices and an equivalence class e of Ulm sequences. Let $X_{c,p,e} = \{x \in X : \|x\| \in c \text{ and } \|x\|_p \in e\}$, and suppose that $M \in c$ such that $M_p \in e$. For each $x \in X_{c,p,e}$, there is a positive integer l such that $\|lx\| \geq M$, so by replacing lx with x we may assume that each element $x \in X_{c,p,e}$ satisfies $\|x\| \geq M$. In addition, we can assume that $|x|_p = \infty$ whenever $M_p \sim (\infty,\infty,\ldots)$, and let $\overline{\alpha} = M_p$.

First, consider the case in which $\overline{\alpha} \sim (\infty,\infty,\ldots)$. If $x \in X_{c,p,e}$, then we have $x \in G(M) \setminus G(M^*,p)^\diamond$. Therefore, by Lemma 3.6.4(1), $x \notin G(M^*,p)$. Since $\bigoplus_{x\in X_{c,p,e}} \langle x\rangle$ is a $*$-valuated coproduct in G (Lemma 3.6.4(2)), we obtain $\sum_{x\in X_{c,p,e}} \langle x + G(M^*,p)\rangle = \bigoplus_{x\in X_{c,p,e}} \langle x + G(M^*,p)\rangle$. Notice that this group is torsion-free by Lemma 3.6.4(1).

Now we have $s(c,p,e,G) \geq r_0(G(M)/G(M^*,p)) \geq |X_{c,p,e}| = w_X(c,p,e,G)$. It remains to show that $s(c,p,e,G) \leq w_X(c,p,e,G)$. Assuming $s(c,p,e,G) > w_X(c,p,e,G)$, we may choose M such that

$$r_0(G(M)/G(M^*,p)) > w_X(c,p,e,G).$$

Since $G(M)/G(M^*,p)$ has no p-torsion by Lemma 3.7.1(2), there is an independent set S in $G(M)/G(M^*,p)$ properly containing $\overline{X}_{c,p,e} = \{x + G(M^*,p) : x \in X_{c,p,e}\}$ and consisting only of elements of infinite order. Then there is an element $\overline{y} = y + G(M^*,p) \in S \setminus \overline{X}_{c,p,e}$.

Since X is a decomposition basis for G, there is a positive integer a such that

$$ay = a_1x_1 + \cdots + a_rx_r + b_1y_1 + \cdots + b_sy_s,$$

where $x_1,\ldots,x_r \in X_{c,p,e}, y_1,\ldots,y_s \in X \setminus X_{c,p,e}$ and $a_1,\ldots,a_r,b_1,\ldots,b_s$ are integers. Since for $1 \leq j \leq s$, $\|b_jy_j\| \geq \|ay\| \geq M$, it follows that $b_jy_j \in G(M^*) \subseteq G(M^*,p)$, and we have $ay + G(M^*,p) = a_1x_1 + \cdots + a_rx_r + G(M^*,p)$ with $ay \notin G(M^*,p)$, contradicting independence.

Now consider the case in which $\overline{\alpha} \not\sim (\infty,\infty,\ldots)$. Then for every $x \in X_{c,p,e}$, there is an integer $m \geq 0$ such that $\|p^mx\|_p = p^n\overline{\alpha}$ for some $n \geq 0$. Therefore, by replacing p^mx with x, we can write $X_{c,p,e} = \bigcup_{i<\omega} X_{p^i\overline{\alpha}}$, where $X_{\overline{\beta}} = \{x \in X : \|x\| \sim M, \|x\| \geq M, \text{ and } \|x\|_p = \overline{\beta}\}$. As in the proof of the local theorem, let $Y_0 = X_{\overline{\alpha}}$, and define $Y_1 = X_{p\overline{\alpha}} \cup \{px : x \in Y_0\}$, $Y_2 = X_{p^2\overline{\alpha}} \cup \{px : x \in Y_1\}$, et cetera. Then

$$|Y_0| \leq |Y_1| \leq \cdots \leq |Y_i| \leq \cdots \quad (i < \omega),$$

and therefore, $w_X(c,p,e,G) = |X_{c,p,e}| = \lim_{i\to\infty} |Y_i|$. Let $i < \omega$. If $x \in Y_i$, then $\|x\| \sim M, \|x\| \geq p^iM$, and $\|x\|_p = p^i\overline{\alpha}$. Hence, $x \in G(p^iM) \setminus G((p^iM)^*,p)^\diamond$. Then, by Lemma 3.6.4(1), we have $x \notin G((p^iM)^*,p)$. Since $\bigoplus_{x\in Y_i}\langle x\rangle$ is a $*$-valuated coproduct in G (Lemma 3.6.4(2)), we obtain $\sum_{x\in Y_i}\langle x + G((p^iM)^*,p)\rangle = \bigoplus_{x\in Y_i}\langle x + G((p^iM)^*,p)\rangle$.

For all $i < \omega$, we have $s(c,p,e,G) \geq r_p(G(p^iM)/G((p^iM)^*,p)) \geq |Y_i|$; hence, $s(c,p,e,G) \geq w_X(c,p,e,G)$.

It remains to show that $s(c,p,e,G) \leq w_X(c,p,e,G)$. Assuming $s(c,p,e,G) > w_X(c,p,e,G)$, we may choose M such that

$$r_p(G(M)/G(M^*,p)) > w_X(c,p,e,G).$$

Now define a subordinate of X and corresponding sets Y_i $(i < \omega)$ in terms of M and M_p as in the first part of this proof. For $i < \omega$, we define $\overline{Y_i} = \{x + G((p^iM)^*,p) : x \in Y_i\}$. Let

$$\phi_i : G(M)/G(M^*,p) \to G(p^iM)/G((p^iM)^*,p)$$

be the isomorphism sending $x + G(M^*,p)$ onto $p^ix + ((p^iM)^*,p)$ (see Lemma 3.7.2). Since

$$\overline{Y_0} = \phi_0^{-1}(\overline{Y_0}) \subseteq \phi_1^{-1}(\overline{Y_1}) \subseteq \phi_2^{-1}(\overline{Y_2}) \subseteq \cdots,$$

the set $S' = \bigcup_{i<\omega} \phi_i^{-1}(\overline{Y_i})$ is independent in $G(M)/G(M^*,p)$, and we have $|S'| = \lim_{i\to\infty} |\phi_i^{-1}(\overline{Y_i})| = \lim_{i\to\infty} |Y_i| = |X_{c,p,e}|$. Consequently, there is an independent set S in $G(M)/G(M^*,p)$ properly containing S'. Then there is an element $\overline{y} = y+G(M^*,p) \in S\backslash S'$, so the subset $\{\phi_i(\overline{y})\} \cup \overline{Y_i}$ of $\phi_i(S)$ is independent for each $i < \omega$.

Since X is a decomposition basis for G, there is a positive integer $a = p^k a'$ ($k < \omega$ and a' an integer with $p \nmid a'$) such that

$$ay = a_1x_1 + \cdots + a_rx_r + b_1y_1 + \cdots + b_sy_s,$$

where $x_1, \ldots, x_r \in \bigcup_{i=0}^k X_{p^i\overline{a}}, y_1, \ldots, y_s \in X \setminus \bigcup_{i=0}^k X_{p^i\overline{a}}$, and $a_1, \ldots, a_r, b_1, \ldots, b_s \in \mathbb{Z}$. Let $i \in \{1, \ldots, r\}$, and $j \in \{1, \ldots, s\}$. Since $ay \in G(p^kM)$, we have $a_ix_i, b_jy_j \in G(p^kM)$ as well. Then $x_i \in X_{p^{t_i}\overline{a}}$ for some $t_i \le k$. Thus, we have $p^{k-t_i}|a_i$, and therefore, a_ix_i is a multiple of $p^{k-t_i}x_i \in Y_k$.

In the case where $y_j \in X_{p^l\overline{a}}$ for some $l > k$, we have $y_j \in G(p^k\overline{a}^*,p)^\diamond$. If y_j is in $X \setminus \bigcup_{i=0}^\infty X_{p^i\overline{a}}$, then $\|y_j\| \nsim M$ or $\|y_j\|_p \notin e$; thus, $b_jy_j \in G(M^*)^\diamond$ or $b_jy_j \in G(p^k\overline{a}^*,p)^\diamond$. In any case, we obtain $b_jy_j \in G((p^kM)^*,p)^\diamond$. But then we have

$$ay + G((p^kM)^*,p) = a_1x_1 + \cdots + a_rx_r + G((p^kM)^*,p)$$

with $ay \notin G((p^kM)^*,p)$, contradicting independence of the set $\{\phi_k(\overline{y})\} \cup \overline{Y_k}$. It follows that $s(c,p,e,G) \le w_X(c,p,e,G)$, as desired. □

As in the local case, this result allows us to drop the subscript X.

Proposition 3.7.4 (Arnold–Hunter–Walker [5]). *Suppose that $A = \langle x_1\rangle \oplus \langle x_2\rangle$ is a valuated coproduct in the group G, where $\|x_1\|$ and $\|x_2\|$ are compatible Ulm matrices. Let p be a prime. Then A coincides with a valuated coproduct $\langle y_1\rangle \oplus \langle y_2\rangle$ in G for some $y_1, y_2 \in G$ such that $\|x_1\|_p = \|y_2\|_p$, $\|x_2\|_p = \|y_1\|_p$, and $\|x_i\|_q = \|y_i\|_q$ for $i = 1,2$ and all primes $q \neq p$.*

Proof. Since $\|x_1\| \sim \|x_2\|$, we have $m\|x_1\| \ge \|x_2\|$ and $m\|x_2\| \ge \|x_1\|$ for some positive integer m. If $p \nmid m$, the elements $y_1 = x_1$ and $y_2 = x_2$ have the desired properties. Now assume that $p|m$, and write $m = p^jn$, where $p \nmid n$. Then there are integer s and t such that $sp^{2j} - tn^2 = 1$, and we let

$$\begin{pmatrix} y_1 \\ y_2 \end{pmatrix} = \begin{pmatrix} sp^j & tn \\ n & p^j \end{pmatrix} \begin{pmatrix} x_1 \\ x_2 \end{pmatrix}.$$

Since the determinant of the square matrix above is 1, its inverse has integer entries as well, giving $\langle y_1\rangle + \langle y_2\rangle = \langle x_1\rangle \oplus \langle x_2\rangle$. Further, we have $\langle y_1\rangle \cap \langle y_2\rangle = 0$ since $k_1y_1 = k_2y_2$ implies

$$\begin{pmatrix} sp^j & n \\ tn & p^j \end{pmatrix} \begin{pmatrix} k_1 \\ -k_2 \end{pmatrix} = \begin{pmatrix} 0 \\ 0 \end{pmatrix},$$

which yields $k_1 = k_2 = 0$. Therefore, $A = \langle x_1\rangle \oplus \langle x_2\rangle$ coincides with $\langle y_1\rangle \oplus \langle y_2\rangle$. We have $\|y_1\|_p = \|sp^jx_1\|_p \wedge \|tnx_2\|_p = \|x_2\|_p$ since $\|sp^jx_1\|_p = \|smx_1\|_p \ge \|sx_2\|_p$ and $\|tnx_2\|_p = \|x_2\|_p$.

Likewise, we have $\|y_2\|_p = \|nx_1\|_p \wedge \|p^j x_2\|_p = \|x_1\|_p$. Now let q be a prime $\neq p$. We show that $\|y_1\|_q = \|x_1\|_q$ and $\|y_2\|_q = \|x_2\|_q$. If $q \nmid m$, then $\|y_1\|_q = \|sp^j x_1\|_q \wedge \|tnx_2\|_q = \|sx_1\|_q \wedge \|tx_2\|_q$. But this is equal to $\|x_1\|_q$ since $\|x_1\|_q = \|x_2\|_q$, and $q \nmid s$ or $q \nmid t$. Similarly, we obtain $\|y_2\|_q = \|x_2\|_q$. Now suppose that $q|m$. Then $q|n$; hence $q \nmid s$, and we have $\|y_1\|_q = \|x_1\|_q \wedge \|tnx_2\|_q$. This equals $\|x_1\|_q$ since $\|tnx_2\|_q = \|tmx_2\|_q \geq \|tx_1\|_q$. Similarly, it can be shown that $\|y_2\|_q = \|x_2\|_q$. Finally, we need to verify that $A = \langle y_1\rangle \oplus \langle y_2\rangle$ is a valuated coproduct in G. Let p' be any prime, and

$$\alpha = |m_1y_1 + m_2y_2|_{p'}, \quad \beta = |m_1y_1|_{p'} \wedge |m_2y_2|_{p'},$$

where m_1, m_2 are integers. Then $\alpha \geq \beta$, and if $|m_1y_1|_{p'} \neq |m_2y_2|_{p'}$ then $\alpha = \beta$. Therefore, assume $|m_1y_1|_{p'} = |m_2y_2|_{p'}$. Letting $c_1 = m_1sp^j + m_2n$ and $c_2 = m_1tn + m_2p^j$, we have $m_1 = p^jc_1 - nc_2$ and $m_2 = -tnc_1 + sp^jc_2$. Computing p'-heights in $\mathbb{Z}$, we let $r = |c_1|_{p'} \wedge |c_2|_{p'}$ and obtain $|m_1|_{p'} = |p^jc_1 - nc_2|_{p'} \geq r$, $|m_2|_{p'} = |-tnc_1 + sp^jc_2|_{p'} \geq r$.

First, we assume that $r = |c_1|_{p'}$. Then

$$\alpha = |m_1y_1 + m_2y_2|_{p'} = \left|(m_1sp^j + m_2n)x_1\right|_{p'} \wedge \left|(m_1tn + m_2p^j)x_2\right|_{p'} \leq |c_1x_1|_{p'}.$$

If $p' = p$, then $\beta = |m_2y_2|_{p'} \geq r|y_2|_{p'} = r|x_1|_{p'} = |c_1x_1|_{p'}$. If $p' \neq p$, then $\beta = |m_1y_1|_{p'} = |m_1x_1|_{p'} \geq r|x_1|_{p'} = |c_1x_1|_{p'}$. In any case, we conclude that $\alpha \leq \beta$.

Finally, assume $r = |c_2|_{p'}$. Using the same argument as above, we conclude that $\alpha \leq r|x_2|_{p'} \leq \beta$. Consequently, $A = \langle y_1\rangle \oplus \langle y_2\rangle$ is a valuated coproduct in G. □

To prove the global version of Theorem 3.2.4, the following lemma is needed:

Lemma 3.7.5 (Richman [133]). *Let I and J be sets. If $\phi_n : I \to J$ are bijections ($n \in \mathbb{N}$), then there exist partitions $I = \bigcup_{\lambda\in\Lambda} I_\lambda$ and $J = \bigcup_{\lambda\in\Lambda} J_\lambda$ such that*

$$\phi_n(I_\lambda) = J_\lambda \quad \textit{and} \quad |I_\lambda| \leq \aleph_0$$

for all $n \in \mathbb{N}$, and $\lambda \in \Lambda$.

Proof. We define an equivalence relation on I by $i_1 \sim i_2$ if there are positive integers $n_1, \ldots, n_{2k}$ such that

$$i_2 = \phi_{n_{2k}}^{-1}\phi_{n_{2k-1}} \cdots \phi_{n_2}^{-1}\phi_{n_1}(i_1).$$

A similar equivalence relation is defined on J, with ϕ_n and ϕ_n^{-1} interchanged. Then each equivalence class $[i]$ is countable since $\mathbb{N}$ is countable, and we have $I = \bigcup_{i\in I}[i]$ and $J = \bigcup_{i\in I}[\phi_1(i)]$. Let $i \in I$ and $n \in \mathbb{N}$. Then the image of $[i]$ under ϕ_n is $[\phi_n(i)]$, which is equal to $[\phi_1(i)]$ since $\phi_n(i) = \phi_n\phi_1^{-1}\phi_1(i)$. □

Theorem 3.7.6 (Hunter–Richman [61]). *Suppose G and H are groups with decomposition bases X and Y, respectively. Then G and H have the same Warfield invariants if and only if there is a height-preserving isomorphism $\langle X'\rangle \to \langle Y'\rangle$ for some subordinates X' of X and Y' of Y.*

Proof. Only necessity needs to be proved. Therefore, suppose G and H have the same Warfield invariants. We write $X = \{x_i : i \in I\}$ and $Y = \{y_j : j \in J\}$. For a compatibility class c of Ulm matrices with $w(c, p, e, G) \neq 0$ for some p and e, we define $I_c = \{i \in I : \|x_i\| \in c\}$ and J_c accordingly. This yields partitions $I = \bigcup_c I_c$, and $J = \bigcup_c J_c$, and a bijection $I \to J$ that maps I_c onto J_c for all c. Therefore, we may assume that $I = I_c$ and $J = J_c$ for some fixed c. For a prime p and an Ulm sequence $\overline{\alpha}$, let

$$I(p, \overline{\alpha}) = \{i \in I : \|x_i\|_p \sim \overline{\alpha}\},$$

and define $J(p, \overline{\alpha})$ in the same way. Then for each prime p, we have partitions

$$I = \bigcup_{\overline{\alpha}} I(p, \overline{\alpha}), \quad J = \bigcup_{\overline{\alpha}} J(p, \overline{\alpha})$$

and a bijection $\phi_p : I \to J$ that maps $I(p, \overline{\alpha})$ onto $J(p, \overline{\alpha})$ for all Ulm sequences $\overline{\alpha}$. By Lemma 3.7.5, we may assume that $|I| = |J| \leq \aleph_0$, say, $I = J = \{1, 2, 3, \dots\}$. Next we will construct for each prime p a bijection $\psi_p : I \to J$ that maps $I(p, \overline{\alpha})$ onto $J(p, \overline{\alpha})$ for all $\overline{\alpha}$ and has the additional property that for each $i \in I$, $\psi_p(i) = i$ for almost all primes p. The maps ψ_p will then be used to construct the desired subordinates of X and Y.

Let $T = \{(i, p, j) \in I \times \mathbb{P} \times J : \|x_i\|_p \sim \|y_j\|_p\}$. For any $i \in I$, we have $\|x_i\| \sim \|y_i\|$. Hence, for almost all primes p, $\|x_i\|_p$ coincides with $\|y_i\|_p$, and therefore, $(i, p, i) \in T$. We define $A_1 = \{(i, p, j) \in T : i = j = 1\}$, $P_1 = \{p \in \mathbb{P} : (1, p, 1) \notin A_1\}$, $B_1 = \{(1, p, \phi_p(1)), (\phi_p^{-1}(1), p, 1) : p \in P_1\}$, and $S_1 = A_1 \cup B_1$. Letting $A_2 = \{(i, p, j) \in T : i = j = 2\}$, and $P_2 = \{p \in \mathbb{P} : (2, p, 2) \notin A_2\}$, we define B_2 to be the union of the sets

$$\{(2, p, j) \in T : j = \phi_p(2), p \in P_2 \text{ and } (2, p, n) \notin S_1 \text{ for all } n \in J\},$$

and

$$\{(i, p, 2) \in T : i = \phi_p^{-1}(2), p \in P_2 \text{ and } (m, p, 2) \notin S_1 \text{ for all } m \in I\},$$

and let $S_2 = A_2 \cup B_2 \cup S_1$, et cetera. This yields an increasing sequence of sets

$$S_1 \subseteq S_2 \subseteq \cdots \subseteq S_n \subseteq \cdots \quad (n \in \mathbb{N}),$$

whose union is denoted by S. Let p be a fixed prime. Then by construction of S, for every $i \in I$ (or $j \in J$), there is a unique $j \in J$ (or $i \in I$) such that $(i, p, j) \in S$. This yields a bijection $\psi_p : I \to J$, defined by $\psi_p(i) = j$ if and only if $(i, p, j) \in S$. Notice that $(i, p, \phi_p(i)) \in S$ if $\|x_i\|_p \neq \|y_i\|_p$, and $(i, p, i) \in S$ otherwise. Therefore, ψ_p maps each set $I(p, \overline{\alpha})$ onto $J(p, \overline{\alpha})$, and for every $i \in I$, we have $\psi_p(i) = i$ for almost all primes p.

For each $(i, p, j) \in S$, there are smallest positive integers $m_{p,i}$ and $n_{p,j}$ such that $\|p^{m_{p,i}} x_i\|_p = \|p^{n_{p,j}} y_j\|_p$. Let $i \in I$. Then for almost all primes p, we have $\|x_i\|_p = \|y_i\|_p$, and therefore, $m_{p,i} = 1$, which yields the unique positive integer $a_i = \prod_p p^{m_{p,i}}$. By symmetry, we obtain for each $j \in J$ the unique positive integer $b_j = \prod_p p^{n_{pj}}$. Since ψ_p maps each $I(p, \overline{\alpha})$ onto $J(p, \overline{\alpha})$, we have

$$|\{i \in I : \|a_i x_i\|_p = \overline{\alpha}\}| = |\{j \in J : \|b_j y_j\|_p = \overline{\alpha}\}|$$

for all p and $\overline{\alpha}$. This yields subordinates $X' = \{a_i x_i : i \in I\}$ of X and $Y' = \{b_j y_j : j \in J\}$ of Y such that each ψ_p induces a p-height preserving isomorphism $\langle X' \rangle \to \langle Y' \rangle$. Now replace all elements $a_i x_i$ by x_i and $b_j y_j$ by y_j. Since $\|x_1\|_p = \|y_1\|_p$ for almost all primes p, there is a positive integer n such that $\psi_p(1) \in \{1, \dots, n\}$ for all primes p. Applying Proposition 3.7.4 repeatedly, we obtain

$$\langle y_1 \rangle \oplus \langle y_2 \rangle \oplus \cdots \oplus \langle y_n \rangle = \langle y_1' \rangle \oplus \langle y_2' \rangle \oplus \cdots \oplus \langle y_n' \rangle,$$

where both direct sums are valuated coproducts in H such that $\|x_1\| = \|y_1'\|$, $\|y_j\|_p = \|y_j'\|_p$ for all $j \in \{1, \dots, n\}$ whenever $\psi_p(1) = 1$, $\|y_1\|_p = \|y_k'\|_p$ and $\|y_k\|_p = \|y_1'\|_p$ whenever $k = \psi_p(1) > 1$, and $\|y_j\|_p = \|y_j'\|_p$ if $\psi_p(1) > 1$ but $\psi_p(1) \neq j > 1$.

Let $j \in \{2, \dots, n\}$ and p a prime. If $\|x_1\|_p = \|y_1\|_p$, then $\psi_p(1) = 1$; hence,

$$\|y_j'\|_p = \|x_i\|_p \text{ for some integer } i > 1. \tag{$*$}$$

Now suppose $\|x_1\|_p \neq \|y_1\|_p$. If $\|y_j'\|_p = \|y_1\|_p$, then $(*)$ follows since $i = \psi_p^{-1}(1) > 1$. If $\|y_j'\|_p \neq \|y_1\|_p$, then $j = \psi_p(1) > 1$ is clearly impossible. Hence, $\psi_p(1) \neq j$, and we have $\|y_j\|_p = \|y_j'\|_p$, so $(*)$ holds again. Thus, there is a positive integer m such that for every $j \in \{2, \dots, n\}$ and prime p, there is an $i \in \{2, \dots, m\}$ such that $\|y_j'\|_p = \|x_i\|_p$. Using Proposition 3.7.4 repeatedly again, we obtain

$$\langle x_2 \rangle \oplus \langle x_3 \rangle \oplus \cdots \oplus \langle x_m \rangle = \langle x_2' \rangle \oplus \langle x_3' \rangle \oplus \cdots \oplus \langle x_m' \rangle,$$

and $\|y_i'\| = \|x_i'\|$ for $i = 2, 3, \dots, n$. Going back and forth, we obtain decomposition bases $X'' = \{x_1' = x_1, x_2', \dots\}$ and $Y'' = \{y_1', y_2', \dots\}$ for G and H such that $\langle X' \rangle = \langle X'' \rangle$, $\langle Y' \rangle = \langle Y'' \rangle$, and $\|x_i'\| = \|y_i'\|$ for all $i \in I$. Then $x_i' \mapsto y_i'$ induces the desired height-preserving isomorphism $\langle X' \rangle \to \langle Y' \rangle$. □

Now we are ready to prove the global classification theorem, which we will see classifies Warfield groups:

Theorem 3.7.7 (Hunter–Richman [61], Stanton [140]). *Let G and H be groups with nice decomposition bases X and Y, respectively, such that $G/\langle X \rangle$ and $H/\langle Y \rangle$ are simply presented. Then $G \cong H$ if and only if for all c, p, e, and $\alpha \in \mathrm{Ord}_\infty$, we have $u_p(\alpha, G) = u_p(\alpha, H)$ and $w(c, p, e, G) = w(c, p, e, H)$.*

Proof. Suppose G and H have the same Ulm and Warfield invariants. Using Lemmas 3.1.6 and 3.2.4, we replace the decomposition bases X and Y by suitable lower decomposition bases X' and Y', for which there exists a height-preserving isomorphism $f : \langle X' \rangle \to \langle Y' \rangle$. Then by Lemma 3.4.4, the induced map $f_p : \langle X' \rangle_p \to \langle Y' \rangle_p$ is a p-height-preserving isomorphism. For every ordinal α, we have

$$u_{\alpha,p}(G, \langle X' \rangle) = u_{\alpha,p}(H, \langle Y' \rangle)$$

(see the proof of Theorem 3.2.8). By Lemma 3.4.3, X' and Y' are nice, and $G/\langle X' \rangle$ and $H/\langle Y' \rangle$ are simply presented. Then

$$u_{\alpha,p}(G_p, \langle X' \rangle_p) = u_{\alpha,p}(H_p, \langle Y' \rangle_p)$$

by Lemma 1.17.11. The p-groups $G_p/\langle X' \rangle_p$ and $H_p/\langle Y' \rangle_p$ are simply presented, so by Theorem 2.3.2, the p-height-preserving isomorphism

$$f_p : \langle X' \rangle_p \to \langle Y' \rangle_p$$

can be extended to an isomorphism $G_p \to H_p$. Finally, we apply Warfield's Lemma 3.4.5 and obtain an isomorphism $G \to H$. The converse of the statement is immediate. □

Exercises.

1. It is possible that $G(M)/G(M^*, p)$ is neither p-torsion nor divisible. [Hint: Consider $G = \mathbb{Z}_q$ for $q \neq p$.]

3.8 Nice subgroups

It was pointed out in [50] that the subgroup $A = 2\mathbb{Z}$ of $G = \mathbb{Z}$ fails to satisfy

$$p^\alpha(G/A) = (p^\alpha G + A)/A \text{ for all primes } p \text{ and ordinals } \alpha. \qquad (*)$$

In fact, for any prime $p \neq 2$, we have $p^\omega(\mathbb{Z}/2\mathbb{Z}) = \mathbb{Z}/2\mathbb{Z}$, but $(p^\omega\mathbb{Z} + 2\mathbb{Z})/2\mathbb{Z} = 0$. Therefore, a global definition of niceness requiring $(*)$ is not suitable if one wishes to include cyclic subgroups. Following Hill and Megibben [50], we call a subgroup A of a group G **nice** if

$$\frac{p^\alpha(G/A)}{(p^\alpha G + A)/A}[p] = 0$$

for all primes p and ordinals α. In other words, A is nice in G if and only if the group $p^\alpha(G/A)/((p^\alpha G + A)/A)$ has no p-torsion for all primes p and ordinals α. As we will see, this extends the definition of local niceness (Proposition 3.8.2), and cyclic subgroups will always be nice (Corollary 3.8.13). The following facts are easily verified:

(a) If G/A is a p-group, then A is nice in G exactly if A is p-nice in G. In particular, for p-groups this definition of niceness is equivalent to Hill's original definition of niceness; that is, a subgroup A of a p-group G is nice if and only if $p^\alpha(G/A) = (p^\alpha G + A)/A$ for all ordinals α.

(b) A is nice in G if and only if $x + A \in p^\alpha(G/A)$ and $px \in p^\alpha G + A$ yield $x \in p^\alpha G + A$ for all $x \in G$, primes p, and ordinals α.
(c) If A_i are nice subgroups of groups G_i $(i \in I)$, then $\bigoplus_{i\in I} A_i$ is nice in $\bigoplus_{i\in I} G_i$.

Since finite subgroups are p-nice for every prime p, we have the following:

Proposition 3.8.1 (Hill–Megibben [50]). *Finite subgroups are nice.*

A subgroup is nice whenever it is locally nice:

Proposition 3.8.2 (Hill–Megibben [50]). *Locally nice subgroups are nice.*

Proof. Suppose that A is locally nice in G. Let p be a prime, α an ordinal, and $x + A \in p^\alpha(G/A)$. We will show that there is a positive integer n such that $p \nmid n$ and $nx \in p^\alpha G{+}A$. This implies that $p^\alpha(G/A)/((p^\alpha G + A)/A)$ has no p-torsion, which proves that A is nice in G. We may identify $(G/A)_p$ with G_p/A_p. Then the localization map

$$G/A \to (G/A)_p$$

sends $x + A$ onto $(x \otimes 1) + A_p$. Thus, $(x \otimes 1) + A_p \in p^\alpha(G_p/A_p) = (p^\alpha(G_p) + A_p)/A_p$, and we can write $x \otimes 1 = y + z$ with $y \in A_p$ and $z \in p^\alpha(G_p)$. Letting v_p denote the localization map from G to G_p, Lemma 1.17.8(1) shows that $G_p/v_p(G)$ is torsion with no p-torsion. Therefore, there is a positive integer k_1 with $p \nmid k_1$ such that $k_1 y = v_p(a)$ and $k_1 z = v_p(b)$ for some $a \in A$ and $b \in G$. Notice that $b \in p^\alpha G$ since v_p is p-height-preserving by Lemma 1.17.8(5). Since

$$v_p(k_1 x) = k_1(x \otimes 1) = k_1 y + k_1 z = v_p(a) + v_p(b),$$

we have $k_1 x - a - b \in \ker v_p$. Therefore, by Lemma 1.17.8(3), there is a positive integer k_2 such that $p \nmid k_2$ and $k_2(k_1 x - a - b) = 0$. Hence, $k_2 k_1 x = k_2 a + k_2 b \in p^\alpha G + A$, as desired. □

To show that niceness carries over to p-localizations, the following result is needed:

Lemma 3.8.3. *Let A be a subgroup of G and p a prime. Then*

$$\frac{p^\alpha(G_p/A_p)}{(p^\alpha(G_p) + A_p)/A_p} \cong \left(\frac{p^\alpha(G/A)}{(p^\alpha G + A)/A}\right)_p.$$

Proof. By Lemma 1.17.10, the isomorphism $G_p/A_p \to (G/A)_p$ maps $p^\alpha(G_p/A_p)$ onto $(p^\alpha(G/A))_p$. Using Lemma 1.17.10 again, we obtain $(p^\alpha(G_p){+}A_p)/A_p = ((p^\alpha G)_p{+}A_p)/A_p = (p^\alpha G + A)_p/A_p$, which is mapped onto $((p^\alpha G + A)/A)_p$. □

Lemma 3.8.4. *Let A be a subgroup of G, p a prime, and α an ordinal. Then we have the following:*

$$\textit{If } \frac{p^\alpha(G/A)}{(p^\alpha G + A)/A}[p] = 0, \quad \textit{then } \frac{p^\alpha(G_p/A_p)}{(p^\alpha(G_p) + A_p)/A_p}[p] = 0.$$

Proof. Letting $C = p^\alpha(G/A)/((p^\alpha G+A)/A)$, we have $C_p[p] \cong C[p] = 0$ by Lemma 1.17.8(4). Since $C_p \cong p^\alpha(G_p/A_p)/((p^\alpha(G_p) + A_p)/A_p)$ by Lemma 3.8.3, the assertion follows. □

Corollary 3.8.5 (Hill–Megibben [50]). *Let A be a nice subgroup of G such that G/A is torsion. Then A is locally nice in G.*

Proof. Let p be a prime. Then $G_p/A_p \cong t_p(G/A)$; hence, A_p is p-nice in G_p by Lemma 3.8.4. □

Recall that a decomposition basis X is called nice whenever $\langle X\rangle$ is a locally nice subgroup of G. Now Corollary 3.8.5 and Proposition 3.8.2 show the following:

Corollary 3.8.6. *Suppose that G has a decomposition basis X. Then X is nice if and only if $\langle X\rangle$ is nice.*

Lemma 3.8.7. *Suppose that A is a subgroup of G, and that p is a prime, such that $(p^\alpha(G/A))/((p^\alpha G + A)/A)[p] = 0$ for all ordinals α. If $x \in G$ such that $px \in A$, then the coset $x + A$ has an element x' such that $|x'|_p^G = |x + A|_p^{G/A}$.*

Proof. Let $\alpha = |x + A|_p^{G/A}$. Since $px \in p^\alpha G + A$, we have $x \in p^\alpha G + A$. Therefore, there is an $x' \in x + A$ such that $|x'|_p^G \geq \alpha$; thus, $|x'|_p^G = |x + A|_p^{G/A}$. □

The order of an element $x \in G$ is denoted by $o(x)$. We will need the following basic fact:

Lemma 3.8.8. *Let G be a torsion group and p a prime. If $x \in G$, then $x = y + pz$ for some $y, z \in G$ such that $o(y) = p^n$ for some $n < \omega$ and $p \nmid o(z)$.*

Proof. Write $o(x) = p^n m$, where $p \nmid m$. Then there are integers k and l such that $km + lp^{n+1} = 1$; thus, we have $x = kmx + p(lp^n)x$. Now let $y = kmx$ and $z = lp^n x$. □

Proposition 3.8.9 (Hill–Megibben [50]). *Let A and B be subgroups of G such that $A \subseteq B$. Suppose that A is nice in G, and B/A is nice in G/A. Then we have:*
1. *If α is an ordinal, p is a prime, and $x \in G$ such that $x+B \in p^\alpha(G/B)$ and $px \in p^\alpha G+B$, then $x + b + A \in p^\alpha(G/A)$ for some $b \in B$.*
2. *If B/A is torsion, then B is nice in G.*

Proof. Assume that $x + B \in p^\alpha(G/B)$, where $px \in p^\alpha G + B$. Since $G/B \cong (G/A)/(B/A)$, this yields $(x + A) + (B/A) \in p^\alpha((G/A)/(B/A))$ and $px + A \in (p^\alpha G + B)/A \subseteq p^\alpha(G/A) + B/A$. Since B/A is nice in G/A, we obtain $x + A \in p^\alpha(G/A) + B/A$, so there is a $b \in B$ such that $x + b + A \in p^\alpha(G/A)$, as claimed in (1).

Now suppose that in addition, B/A is torsion. To prove that B is nice, we need to show that $x \in p^\alpha G + B$. Since $p(x + b) \in p^\alpha G + B$, we can write $p(x + b) = b_1 + z$ for some $b_1 \in B$ and $z \in p^\alpha G$. By Lemma 3.8.8, there are $b_2, b_3 \in B$ such that $b_1 + A = b_2 + pb_3 + A$, where $p^n b_2 \in A$ for some $n < \omega$, and p does not divide the order k of $b_3 + A$. Letting $x' + A = x + b - b_3 + A$, we have $kx' + A \in p^\alpha(G/A)$. Thus, $x' + A \in p^\alpha(G/A)$ by

Lemma 1.10.2, and $p(x+b) - z + A = b_1 + A = b_2 + pb_3 + A$, so there is an $a \in A$ such that $p(x + b - b_3) = b_2 + z + a$. Then

$$p^{n+1}x' + A = p^n(b_2 + z + a) + A = p^n z + A \in (p^\alpha G + A)/A.$$

Since A is nice, it follows that $x+b-b_3+A = x'+A \in (p^\alpha G+A)/A$; hence, $x \in p^\alpha G+B$. □

Since finite subgroups are nice, Proposition 3.8.9(2) yields

Theorem 3.8.10 (Hill–Megibben [50]). *Finite extensions of nice subgroups are nice.*

Lemma 3.8.11 (Hill–Megibben [50]). *Suppose that A is a subgroup of G that fails to be nice. If p is a prime and β is the least ordinal such that*

$$(p^\beta(G/A)/(p^\beta G + A)/A)[p] \neq 0,$$

then β is a limit ordinal. Further, if $x + A \in p^\beta(G/A)$ and $p^n(x+A) \in (p^\beta G+A)/A$ for some $n < \omega$, then $x \in p^\alpha G + A$ for all $\alpha < \beta$.

Proof. Let p, β, x, and n be as in the statement. Suppose that $\alpha < \beta$. Then there is an $x'+A \in p^\alpha(G/A)$ such that $x+A = p(x'+A)$. Hence, $p^{n+1}(x'+A) = p^n(x+A) \in (p^\alpha G+A)/A$. By the minimality of β, we have $x' + A \in (p^\alpha G+A)/A$. Thus, $x' \in p^\alpha G + A$, and it follows that $x \in p^\alpha G + A$. This proves the second part of the lemma.

It remains to show that β is a limit ordinal. Assume that $\beta - 1$ exists, and that $x+A \notin (p^\beta G+A)/A$. Letting $\alpha = \beta-1$, we have $x' \in p^\alpha G+A$. Thus, $x \in p^{\alpha+1}G+A = p^\beta G+A$, which is impossible. Since $\beta \neq 0$, the assertion follows. □

Theorem 3.8.12 (Hill–Megibben [50]). *Suppose that $B = A\oplus\langle x_1\rangle\oplus\cdots\oplus\langle x_n\rangle$ is a valuated coproduct in G such that A is nice. Then B is nice.*

Proof. It will be enough to consider the case $n = 1$. If B/A is finite, the assertion follows from Theorem 3.8.10, so we suppose that $x = x_1$ has infinite order. Assume that B is not nice in G. Then there exists a prime p and a smallest ordinal β such that $g + B \in p^\beta(G/B) \setminus (p^\beta G + B)/B$ with $pg \in p^\beta G + B$ for some $g \in G$. Therefore, we can write $pg = a + tx + z$ with $a \in A$, $t \in \mathbb{Z}$, and $z \in p^\beta G$. Let $\alpha < \beta$. By Lemma 3.8.11, β is a limit ordinal and $g \in p^\alpha G + B$. It follows that

$$g + a_\alpha + t_\alpha x \in p^\alpha G \setminus p^\beta G$$

for some $a_\alpha \in A$ and $t_\alpha \in \mathbb{Z}$. Thus,

$$|tx + pt_\alpha x|_p \geq |tx + pt_\alpha x + a + pa_\alpha|_p \geq |pg + pa_\alpha + pt_\alpha x|_p \wedge |z|_p \geq \alpha.$$

Claim: $p^\beta G \cap \langle x\rangle \neq 0$.

Proof: If $p \nmid t$, then for any $\alpha < \beta$, $|x|_p = |tx + pt_\alpha x|_p \geq \alpha$, which implies $x \in p^\beta G$ since β is a limit ordinal. Suppose that $p|t$, say $t = pt'$ with $t' \in \mathbb{Z}$. Letting $g' = g - t'x$ and $t'_\alpha = t' + t_\alpha$, we have

$$g' + a_\alpha + t'_\alpha x = g + a_\alpha + t_\alpha x \in p^\alpha G \quad\text{and}\quad |pt'_\alpha x|_p \geq \alpha$$

for all $\alpha < \beta$. If $g' + A \in p^{\beta}(G/A)$, then $pg' = pg - tx = a + z \in p^{\beta}G + A$ shows that $g' \in p^{\beta}G+A$ since A is nice, contradicting $g \notin p^{\beta}G+B$. Therefore, $g'+A \notin p^{\beta}(G/A)$. Thus, there is a $\mu < \beta$ such that $g' \notin p^{\mu}G + A$, and we obtain $t'_{\mu}x \notin p^{\mu}G$. Since $|pt'_{\mu}x|_p \geq \mu$ and $t'_{\mu} \neq 0$, there is a smallest $s(\mu) < \omega$ such that $|p^{s(\mu)}x|_p \geq \mu$, and it follows that $p^{s(\mu)} \nmid t'_{\mu}$. Then for any $\mu < \alpha < \beta$, we have

$$|(t'_{\alpha} - t'_{\mu})x|_p \geq |g' + a_{\alpha} + t'_{\alpha}x|_p \wedge |g' + a_{\mu} + t'_{\mu}x|_p \geq \mu.$$

Thus, $p^{s(\mu)} | (t'_{\alpha} - t'_{\mu})$, and we write $t'_{\alpha} = t'_{\mu} + l_{\alpha}p^{s(\mu)}$. Since $p^{s(\mu)} \nmid t'_{\mu}$, we have

$$\alpha \leq |pt'_{\alpha}x|_p = |(pt'_{\mu} + l_{\alpha}p^{s(\mu)+1})x|_p = |pt'_{\mu}x|_p \wedge |l_{\alpha}p^{s(\mu)+1}x|_p = |pt'_{\mu}x|_p$$

for all $\mu < \alpha < \beta$. Hence, $pt'_{\mu}x \in p^{\beta}G$, and we obtain $p^{\beta}G \cap \langle x \rangle \neq 0$.

Consequently, there is an $s < \omega$ such that $p^s x \in p^{\beta}G$, and for each $\alpha < \beta$, there is a nonnegative integer $t^*_{\alpha} < p^s$ with $t^*_{\alpha} \equiv t_{\alpha} (\mathrm{mod}\, p^s)$. Then $t^*_{\alpha} = t_{\alpha} + rp^s$ for some $r \in \mathbb{Z}$. Therefore, we let $a^*_{\alpha} = a_{\alpha} - rp^s x$, and obtain $g + a^*_{\alpha} + t^*_{\alpha}x = g + a_{\alpha} + t_{\alpha}x \in p^{\alpha}G \setminus p^{\beta}G$. Since the set $\{t^*_{\alpha} : \alpha < \beta\}$ is finite, there exists a cofinal subset Λ of $\beta = \sup\{|g + a^*_{\alpha} + t^*_{\alpha}x|_p : \alpha < \beta\}$, and a $t^* \in \mathbb{Z}$ such that

$$g + a^*_{\lambda} + t^*x \in p^{\lambda}G \setminus p^{\beta}G$$

for all $\lambda \in \Lambda$. But then it follows that $g + t^*x + A \in p^{\beta}(G/A)$. Since $p^{s+1}(g + t^*x) = p^s(a + tx + z) + p^{s+1}t^*x \in p^{\beta}G + A$, and A is nice, we obtain $g + t^*x \in p^{\beta}G + A$. Thus, $g \in p^{\beta}G + B$, which is impossible. Therefore, B is nice. □

As a corollary of this result, we obtain:

Corollary 3.8.13 (Hill–Megibben [50]). *Cyclic subgroups are nice.*

Proposition 3.8.14 (Hill–Megibben [50]). *Suppose that $C = A \oplus B$ is a valuated coproduct in G. If C is nice, then so are A and B.*

Proof. Assume that A is not nice. By Lemma 3.8.11, there is an $x \in G$, a prime p, and a limit ordinal β such that $x + A \in p^{\beta}(G/A) \setminus (p^{\beta}G + A)/A$ with $px \in p^{\beta}G + A$ and $x \in p^{\alpha}G + A$ for all $\alpha < \beta$. It follows that $x + C \in p^{\beta}(G/C)$ and $px + C \in (p^{\beta}G + C)/C$. This yields $x \in p^{\beta}G + C$ since C is nice. Therefore, we can write $x = a + b + z$ with $a \in A$, $b \in B$, and $z \in p^{\beta}G$. For every $\alpha < \beta$, there is an $a_{\alpha} \in A$ such that $x + a_{\alpha} \in p^{\alpha}G$. Thus, $a + a_{\alpha} + b = x + a_{\alpha} - z \in p^{\alpha}G$, which implies that $b \in p^{\alpha}G$ since $A \oplus B$ is a valuated coproduct in G. But then $b \in p^{\beta}G$, which yields $x = a + b + z \in p^{\beta}G + A$, a contradiction. Therefore, A is nice. □

Proposition 3.8.15 (Hill–Megibben [50]). *Suppose that A is a nice subgroup of G, and $A \oplus \langle x \rangle$ is a valuated coproduct in G. Then $\|x\|_G = \|x + A\|_{G/A}$.*

Proof. Assume that $|p^n x + A|_p^{G/A} \neq |p^n x|_p^G$ for some prime p and $n < \omega$. Let $y = p^n x$, and $\alpha = |y|_p^G$. Then $\alpha \neq \infty$ since $|y + A|_p^{G/A} \ngeq |y|_p^G$. Therefore, there is a $g \in G$ such that $y + A = p(g + A)$ with $g + A \in p^\alpha(G/A)$. By Theorem 3.8.12, $B = A \oplus \langle y \rangle$ is nice in G; thus, $g + B \in p^\alpha(G/B)$ and $pg \in p^\alpha G + B$ yield $g \in p^\alpha G + B$. We write $g = a + my + z$ ($a \in A$, $m \in \mathbb{Z}$, $z \in p^\alpha G$) and $pg = y + a'$ ($a' \in A$). Then

$$a' - pa + (1 - pm)y = pg - pa - pmy = pz \in p^{\alpha+1}G.$$

Thus, $y \in p^{\alpha+1}G$ since $A \oplus \langle y \rangle$ is a valuated coproduct in G, contradicting $|y|_p^G = \alpha$. Consequently, we have $\|x\|_G = \|x + A\|_{G/A}$. □

Corollary 3.8.16. *Let A be a nice subgroup of a group G and X a decomposition set in G such that $A \oplus \langle X \rangle$ is a valuated coproduct in G. Then $\{x + A : x \in X\}$ is a decomposition set in G/A.*

Proof. Clearly, the sum $\sum_{x \in X} \langle x + A \rangle$ is direct. Let M be an Ulm matrix, and let $z = r_1 x_1 + \cdots + r_m x_m$ with $r_i \in \mathbb{Z}$, and $x_i \in X$ for $i = 1, \ldots, m$ such that $z + A \in (G/A)(M)$. Since A is nice in G, and $A \oplus \langle z \rangle$ is a valuated coproduct in G, Proposition 3.8.15 yields $\|z\|_G = \|z + A\|_{G/A} \geq M$. By assumption, $\langle x_1 \rangle \oplus \cdots \oplus \langle x_m \rangle$ is a valuated coproduct in G, so we obtain $\|r_i x_i + A\|_{G/A} = \|r_i x_i\|_G \geq M$ for all $i = 1, \ldots, m$. Consequently, $\{x + A : x \in X\}$ is a decomposition set in G/A. □

The following useful result on nice subgroups can be found in the proof of [53, Theorem 3.2]:

Corollary 3.8.17. *Let A be a nice subgroup of a group G with decomposition basis X, and let $Y_A = X \setminus (A \cap X)$. If $A \oplus \langle Y_A \rangle$ is a valuated coproduct in G, then $\{x + A : x \in Y_A\}$ is a decomposition basis for G/A.*

Lemma 3.8.18 (Hill–Megibben [50]). *Suppose that $C = A \oplus B$ is a nice subgroup of $G = H \oplus K$, where $A \subseteq H$ and $B \subseteq K$. Then A is nice in H.*

Proof. By Proposition 3.8.14, A is nice in G. Since the groups $p^\alpha(G/A)/((p^\alpha G + A)/A)$ and $p^\alpha(H/A)/((p^\alpha H + A)/A)$ are isomorphic for all primes p and ordinals α, it follows that A is nice in H. □

By Lemma 3.8.18 and Section 3.8(c), we have:

Corollary 3.8.19 (Hill–Megibben [50]). *If A_i are subgroups of groups G_i ($i \in I$), then $\bigoplus_{i \in I} A_i$ is nice in $\bigoplus_{i \in I} G_i$ if and only if each A_i is nice in G_i.*

Now Theorem 2.2.8 can be applied to obtain the following result:

Corollary 3.8.20 (Hill [47]). *The class of groups having an $H(\aleph_0)$-family of nice subgroups is closed under arbitrary direct sums and direct summands.*

Exercises.
1. Prove (a–c).
2. (Hill–Megibben [50]) If G is a group of torsion-free rank 1, then every finitely generated subgroup of G is nice.

3.9 Quasi-sequentially nice subgroups

A nice subgroup A of a group G is called **quasi-sequentially nice** if for each $x \in G$ there is a positive integer n and an $a \in A$ such that $\|nx + a\|_G = \|nx + A\|_{G/A}$ (see Loth [102]). The following facts are easily verified:
(a) Every nice subgroup of a torsion group is quasi-sequentially nice.
(b) If A_i are quasi-sequentially nice subgroups of groups G_i $(i \in I)$, then $\bigoplus_{i \in I} A_i$ is quasi-sequentially nice in $\bigoplus_{i \in I} G_i$.

The next lemma follows from the proof of [50, Theorem 3.1]:

Lemma 3.9.1. *Suppose that $C = A \oplus B$ is a quasi-sequentially nice subgroup of $G = H \oplus K$, where $A \subseteq H$ and $B \subseteq K$. Then A is quasi-sequentially nice in H.*

Proof. Notice that $G/C \cong H/A \oplus K/B$. By Lemma 3.8.18, A is nice in G. To show that A is quasi-sequentially nice in H, let $x \in H$. Then there is a positive integer n and a $c \in C$ such that $\|nx + c\|_G = \|nx + C\|_{G/C} = \|nx + A\|_{H/A}$. We write $c = a + b$ with $a \in A$, $b \in B$, and have $\|nx + a\|_H = \|nx + a\|_G \geq \|nx + a\|_G \wedge \|b\|_G = \|nx + c\|_G$. It follows that $\|nx + a\|_H = \|nx + A\|_{H/A}$. Therefore, A is quasi-sequentially nice in H. □

By (b) and the above result, we have the following:

Corollary 3.9.2. *If A_i are subgroups of groups G_i $(i \in I)$, then $\bigoplus_{i \in I} A_i$ is quasi-sequentially nice in $\bigoplus_{i \in I} G_i$ if and only if each A_i is quasi-sequentially nice in G_i.*

We apply Theorem 2.2.8 and obtain the following fact:

Corollary 3.9.3. *The class of groups having an $H(\aleph_0)$-family of quasi-sequentially nice subgroups is closed under arbitrary direct sums and direct summands.*

The next result is contained in the proof of [53, Corollary 1.10]:

Lemma 3.9.4 (Hill–Megibben [53]). *Suppose that A is a pure quasi-sequentially nice subgroup of a group G. Then A is p-nice in G for all primes p.*

Proof. Let p be a prime and α an ordinal. It suffices to show that $p^\alpha(G/A) \subseteq (p^\alpha G + A)/A$, so let $x + A \in p^\alpha(G/A)$. Then $nx \in np^\alpha G + A$ for some positive integer n; hence, there is a $y \in p^\alpha G$ such that

$$nx - ny \in A \cap nG = nA.$$

It follows that there exists an $a \in A$ such that $x' = x - y - a \in G[n]$. We write $n = p^r n'$, where $p \nmid n'$, and have $p^r x' \in G[n']$. Since $G[n']$ is torsion with no p-torsion, it is p-divisible, and is, therefore, a subset of $p^\alpha G$. But then $p^r x + A \in (p^\alpha G + A)/A$. Since A is nice in G, we conclude that $x + A \in (p^\alpha G + A)/A$. □

Corollary 3.9.5. *Suppose that A and B are subgroups of a group G such that $A \subseteq B$. If B is a pure quasi-sequentially nice subgroup of G, then B/A is a pure quasi-sequentially nice subgroup of G/A.*

Proof. By Lemma 1.7.2, B/A is pure in G/A. For any prime p, B is p-nice in G by Lemma 3.9.4; thus, B/A is p-nice in G/A by Lemma 2.1.4(1). Consequently, B/A is nice in G/A. Letting $x + A \in G/A$, we obtain $\|nx + b\|_G = \|nx + B\|_{G/B}$ for some positive integer n and element $b \in B$. On the other hand, we have

$$\|nx + B\|_{G/B} = \|(nx + A) + B/A\|_{(G/A)/(B/A)} \geq \|(nx + b) + A\|_{G/A} \geq \|nx + b\|_G.$$

But then $\|n(x + A) + (b + A)\|_{G/A} = \|n(x + A) + B/A\|_{(G/A)/(B/A)}$; hence, B/A is quasi-sequentially nice in G/A. □

Quasi-sequentially nice subgroups are characterized as follows:

Lemma 3.9.6. *The following conditions are equivalent for a nice subgroup A of a group G:*
1. *A is quasi-sequentially nice;*
2. *For every $x \in G$, there is a positive integer n such that $x + A \in (G/A)(M)$ implies $nx \in G(nM) + A$ for any Ulm matrix M;*
3. *For every $x \in G$, there is a positive integer n and an $a \in A$ such that $A \oplus \langle nx + a\rangle$ is a valuated coproduct in G;*
4. *For every infinite cyclic subgroup B of G, there is a positive integer n such that $(G/A)(M) \cap (nB + A)/A = (G(M) + A)/A \cap (nB + A)/A$ for each Ulm matrix M.*

Proof. (1) implies (2). Assuming that A is quasi-sequentially nice, let $x \in G$. Then $\|nx + a\|_G = \|nx + A\|_{G/A}$ for some positive integer n and $a \in A$. If M is an Ulm matrix such that $x + A \in (G/A)(M)$, then $\|nx + a\|_G = \|nx + A\|_{G/A} \geq nM$; thus, $nx \in G(nM) + A$.

(2) implies (3). Let $x \in G$. If $\langle x\rangle \cap A \neq 0$, then $nx \in A$ for some positive integer n, and we let $a = -nx$. If $\langle x\rangle \cap A = 0$, let n be a positive integer as in (2). Since $x + A \in (G/A)(M)$ with $M = \|x + A\|_{G/A}$, we obtain $nx + a \in G(nM)$ for some $a \in A$. Then the sum $A + \langle nx + a\rangle$ is direct, and we have $\|nx + A\|_{G/A} \geq \|nx + a\|_G \geq nM$. Thus, $\|nx + a\|_G = \|nx + A\|_{G/A}$. Letting p be a prime, $y \in A$, and $m \in \mathbb{Z}$, we have

$$|m(nx + a)|_p = |mnx + A|_p \geq |y + m(nx + a)|_p.$$

Hence, $|y|_p \geq |y + m(nx + a)|_p \wedge |m(nx + a)|_p = |y + m(nx + a)|_p$. But then we obtain $|y + m(nx + a)|_p \leq |y|_p \wedge |m(nx + a)|_p$; thus, $A \oplus \langle nx + a\rangle$ is a valuated coproduct.

(3) implies (4). Let $B = \langle x \rangle$ be an infinite cyclic subgroup of G. By (2), there is an $n \in \mathbb{N}$ such that $A \oplus \langle nx + a \rangle$ is a valuated coproduct in G for some $a \in A$. Let M be an Ulm matrix. It suffices to prove that $(G/A)(M) \cap (nB+A)/A \subseteq (G(M)+A)/A \cap (nB+A)/A$. Let $r \in \mathbb{Z}$. If $rnx + A \in (G/A)(M)$, then we have $\|r(nx + a)\|_G = \|rnx + ra + A\|_{G/A} = \|rnx + A\|_{G/A} \geq M$ by Proposition 3.8.15. Thus, $rnx + A = r(nx + a) + A \in (G(M) + A)/A$, as desired.

(4) implies (1). Let $x \in G$. If the order of x is finite, simply let $n = \text{order}(x)$ and $a = 0$. Assume that x has infinite order. Then there is an $n \in \mathbb{N}$ such that for $M = \|nx + A\|_{G/A}$, we have $nx + A = x' + A$ for some $x' \in G(M)$. It follows that $\|x'\|_G \geq \|nx + A\|_{G/A} \geq \|x'\|_G$; thus, A is quasi-sequentially nice in G. □

The following observation will be useful:

Lemma 3.9.7. *Suppose that G has a composition series $\{N_\alpha\}_{\alpha\leq\lambda}$ such that if $N_{\alpha+1}/N_\alpha$ is infinite, then $N_{\alpha+1} = N_\alpha \oplus \langle x_\alpha \rangle$ is a valuated coproduct in G for some $x_\alpha \in G$. Then the elements x_α form a decomposition basis for G.*

Proof. Let X be the set of all elements x_α as described in the statement. Then $\langle X \rangle = \bigoplus_{x\in X} \langle x \rangle$ is a valuated coproduct in G. Let $g \in G$, and

$$\gamma = \min\{\alpha \leq \lambda : ng \in N_\alpha + \langle X \rangle \text{ for some positive integer } n\}.$$

Then $\gamma = 0$; hence, $G/\langle X \rangle$ is torsion. □

Theorem 3.9.8 (Loth [102]). *The following conditions are equivalent for a group G:*
1. *G has a composition series of quasi-sequentially nice subgroups;*
2. *G has a composition series $\{N_\alpha\}_{\alpha\leq\lambda}$ of nice subgroups such that if $N_{\alpha+1}/N_\alpha$ is infinite, there is an $x_\alpha \in G$ such that $N_{\alpha+1} = N_\alpha \oplus \langle x_\alpha \rangle$ is a valuated coproduct in G.*

In fact, every composition series as described in (1) *can be refined to a composition series as in* (2), *and each member of a composition series as in* (2) *is a quasi-sequentially nice subgroup of G.*

Proof. Suppose that $\{N_\alpha\}_{\alpha\leq\lambda}$ is a composition series of quasi-sequentially nice subgroups of G. If $N_{\alpha+1}/N_\alpha \cong \mathbb{Z}$, then $N_{\alpha+1} = \langle N_\alpha, x \rangle$ for some element $x \in G$ of infinite order. By Lemma 3.9.6(3), there is a $y \in N_\alpha$ such that $N_\alpha \oplus \langle nx + y \rangle$ is a valuated coproduct in G for some positive integer n. Let $x_\alpha = nx + y$. Then $B_1 = N_\alpha \oplus \langle x_\alpha \rangle$ is nice in G by Theorem 3.8.12, and $N_{\alpha+1}/B_1$ is a cyclic torsion group. Using Theorem 3.8.10, we have nice subgroups

$$B_1 \subseteq B_2 \subseteq \cdots \subseteq B_n = N_{\alpha+1}$$

of G such that each quotient B_{i+1}/B_i is cyclic of prime order. Thus, we obtain a suitable refined composition series, as desired.

Conversely, assume that $\{N_\alpha\}_{\alpha\leq\lambda}$ is a composition series of nice subgroups as in (2). By Lemma 3.9.7, the set X of all elements x_α is a decomposition basis for G. Now let $g \in G$, $\alpha < \lambda$, and define $Z = X \setminus (X \cap N_\alpha)$. Then there exists a positive integer n such that $ng = y + z$ for some $y \in N_\alpha$ and $z \in \langle Z\rangle$. Since $N_\alpha \oplus \langle Z\rangle$ is a valuated coproduct in G, N_α is quasi-sequentially nice by Lemma 3.9.6. □

The following result will be needed:

Lemma 3.9.9. *Suppose that $\{N_\alpha\}_{\alpha\leq\lambda}$ is a smooth chain of subgroups of G. If p is a prime, then $\{(N_\alpha)_p\}_{\alpha\leq\lambda}$ is a smooth chain of submodules of G_p.*

Proof. Let $\beta \leq \lambda$ be a limit ordinal. To prove that $(N_\beta)_p \subseteq \sum_{\alpha<\beta}(N_\alpha)_p$, let $x \in (N_\beta)_p$, say $x = a \otimes r$, where $a \in N_\beta$ and $r \in \mathbb{Z}_p$. Then $a \in N_\gamma$ for some $\gamma < \beta$; thus, $x \in (N_\gamma)_p \subseteq \sum_{\alpha<\beta}(N_\alpha)_p$. Therefore, $\{(N_\alpha)_p\}_{\alpha\leq\lambda}$ is a smooth chain of submodules of G_p. □

Hunter and Richman [61] proved that Warfield groups are exactly those groups possessing a nice decomposition basis with simply presented cokernel. Then Hill and Megibben [53] established a variety of additional characterizations of Warfield groups, including condition (2) of Theorem 3.9.8. We will prove the equivalence of these various characterizations of Warfield groups using Theorem 3.9.8. A key link is the following result extending Moore's Theorem 3.3.2 to the global case (see Loth [102]). Our proof generalizes Moore's proof techniques in [120, Theorem].

Theorem 3.9.10. *The following conditions are equivalent for a group G:*
1. *G has a nice decomposition basis X such that $G/\langle X\rangle$ is simply presented;*
2. *G has a composition series of quasi-sequentially nice subgroups.*

Proof. Suppose that $X = \{x_\alpha\}_{\alpha<\lambda}$ is a nice decomposition basis for G such that $G/\langle X\rangle$ is simply presented. Since $t_p(G/\langle X\rangle) \cong (G/\langle X\rangle)_p$ is a simply presented p-group for every prime p, Theorem 2.10.1(6) yields a smooth chain for $(G/\langle X\rangle)_p$. Since $G/\langle X\rangle = t(G/\langle X\rangle) = \bigoplus_{p\in\mathbb{P}} t_p(G/\langle X\rangle)$, this then gives us a smooth chain

$$0 = N_0/\langle X\rangle \subseteq \cdots \subseteq N_\beta/\langle X\rangle \subseteq \cdots \subseteq N_\mu/\langle X\rangle = G/\langle X\rangle$$

of locally nice subgroups of $G/\langle X\rangle$ with $(N_{\beta+1}/\langle X\rangle)/(N_\beta/\langle X\rangle) \cong N_{\beta+1}/N_\beta$ cyclic of prime order for all $\beta < \mu$. Notice that by Proposition 3.8.2, $\langle X\rangle$ is nice in G, and each quotient $N_\beta/\langle X\rangle$ is nice in $G/\langle X\rangle$. Thus, N_β is nice in G by Proposition 3.8.9(2). Letting $M_\alpha = \bigoplus_{\gamma<\alpha}\langle x_\gamma\rangle$ for all $\alpha < \lambda$, we have

$$0 = M_0 \subseteq \cdots \subseteq M_\alpha \subseteq \cdots \subseteq M_\lambda = \langle X\rangle,$$

and $\langle X\rangle = M_\alpha \oplus \bigoplus_{\gamma\geq\alpha}\langle x_\gamma\rangle$ is a valuated coproduct in G. Then Proposition 3.8.14 shows that M_α is nice in G. For every $\beta \leq \mu$, we define $M_{\lambda+\beta} = N_\beta$, obtaining a smooth chain

$$0 = M_0 \subseteq \cdots \subseteq M_\lambda \subseteq M_{\lambda+1} \subseteq \cdots \subseteq M_{\lambda+\mu} = G$$

of nice subgroups of G, such that $M_{\alpha+1} = M_\alpha \oplus \langle x_\alpha \rangle$ is a valuated coproduct if $\alpha < \lambda$ and $M_{\alpha+1}/M_\alpha$ has prime order whenever $\lambda \le \alpha < \lambda + \mu$. By Theorem 3.9.8, $\{M_\alpha\}_{\alpha \le \lambda+\mu}$ is a composition series of quasi-sequentially nice subgroups of G; hence, (2) follows.

To prove that (2) implies (1), suppose that G has a composition series of quasi-sequentially nice subgroups. By Theorem 3.9.8, G has a composition series $\{N_\alpha\}_{\alpha \le \lambda}$ of nice subgroups such that if $N_{\alpha+1}/N_\alpha$ is infinite, there is an $x_\alpha \in G$ such that $N_{\alpha+1} = N_\alpha \oplus \langle x_\alpha \rangle$ is a valuated coproduct in G. Then Lemma 3.9.7 shows that the elements x_α form a decomposition basis X for G. For every prime p, we have a smooth chain

$$0 = (N_0)_p \subseteq \cdots \subseteq (N_\alpha)_p \subseteq \cdots \subseteq (N_\lambda)_p = G_p$$

by Lemma 3.9.9, and we can assume that for each $\alpha < \lambda$, we have $(N_{\alpha+1})_p/(N_\alpha)_p \cong \mathbb{Z}/p\mathbb{Z}$, or else there is an $x_\alpha \in X$ such that $(N_{\alpha+1})_p = (N_\alpha)_p \oplus \langle x_\alpha \rangle_p$ is a valuated coproduct in G_p (Lemma 3.4.1(1)). Now fix a prime p and an ordinal $\alpha < \lambda$. To prove that

$$A = (N_\alpha)_p + \langle X \rangle_p$$

is p-nice in G_p, we will show the following:

Claim: If $\alpha < \beta \le \lambda$, then every coset $g + A$ with $g \in (N_\beta)_p$ has an element of maximal p-height.

Proof: Since $|0|_p = \infty$, we can assume that $g \notin A$. For the induction base, assume that $\beta = \alpha + 1$. Since $g \notin A$, there is no $x_\alpha \in X$ such that $(N_{\alpha+1})_p = (N_\alpha)_p \oplus \langle x_\alpha \rangle$. Thus, $pg \in (N_\alpha)_p$. Letting $X_1 = X \cap N_\alpha$ and $X_2 = X \setminus N_{\alpha+1}$, we have $\langle X_1 \rangle_p \subseteq (N_\alpha)_p$, and $(N_{\alpha+1})_p \oplus \langle X_2 \rangle_p$ is a valuated coproduct in G_p. Since N_α is nice in G, $(N_\alpha)_p$ is nice in G_p by Lemma 3.8.4; hence, Lemma 3.8.7 yields an element $x' \in (N_\alpha)_p$ such that

$$|g + x'|_p \ge |g + z|_p \quad \text{for all } z \in (N_\alpha)_p.$$

Let $a \in A$, say $a = h + y_1 + y_2$ with $h \in (N_\alpha)_p$, $y_1 \in \langle X_1 \rangle_p$, and $y_2 \in \langle X_2 \rangle_p$. Since $y_1 \in (N_\alpha)_p$ and $(N_{\alpha+1})_p \oplus \langle X_2 \rangle_p$ is a valuated coproduct, we obtain $|g + x'|_p \ge |g + h + y_1|_p \ge |g + a|_p$. Thus, $|g + x'|_p$ is an element of maximal p-height in the coset $g + A$.

Now let $\alpha + 1 < \beta \le \lambda$, and assume that the claim holds for all ordinals γ with $\alpha < \gamma < \beta$. If β is a limit ordinal, then $g \in (N_\gamma)_p$ for some $\gamma < \beta$. Therefore, by the induction hypothesis, there exists an $x' \in A$ such that $|g + x'|_p \ge |g + a|_p$ for all $a \in A$. If $\beta - 1$ exists, every coset $b + A$ with $b \in (N_{\beta-1})_p$ has an element of maximal p-height by the induction hypothesis. As the base case shows, every coset $c + (N_{\beta-1})_p + \langle X \rangle_p$ with $c \in (N_\beta)_p$ has an element of maximal p-height. Since $g \in (N_\beta)_p$, Lemma 2.1.5 shows that the coset $g + A$ has an element of maximal p-height. This completes the induction and the proof of the claim.

Consequently, $A = (N_\alpha)_p + \langle X \rangle_p$ is p-nice in G_p. In particular, it follows that $\langle X \rangle_p$ is p-nice, which implies that X is a nice decomposition basis for G since p is an arbitrary prime. Moreover, $M_\alpha = ((N_\alpha)_p + \langle X\rangle_p)/\langle X\rangle_p$ is p-nice in $G_p/\langle X\rangle_p$ by Lemma 2.1.4(1); thus, we obtain a smooth chain

$$0 = M_0 \subseteq \cdots \subseteq M_\alpha \subseteq \cdots \subseteq M_\lambda = G_p/\langle X\rangle_p$$

with each quotient $M_{\alpha+1}/M_\alpha \cong ((N_{\alpha+1})_p + \langle X\rangle_p)/((N_\alpha)_p + \langle X\rangle_p)$ either trivial or cyclic of order p. Thus, $t_p(G/\langle X\rangle) \cong (G/\langle X\rangle)_p \cong G_p/\langle X\rangle_p$ has a nice composition series. Therefore, by Theorem 2.10.1 it is a simply presented p-group. Then $G/\langle X\rangle = \bigoplus_{p\in\mathbb{P}} t_p(G/\langle X\rangle)$ is simply presented, as desired. □

Exercises.

1. Prove (a) and (b).
2. For a submodule A of a $\mathbb{Z}_p$-module G, the definitions of "quasi-sequentially nice" in Sections 3.3 and 3.9 are equivalent.

3.10 Primitive elements and $*$-decomposition sets

Let G be any group, M an Ulm matrix, $\overline{\alpha} = \{\alpha_i\}_{i<\omega}$ an Ulm sequence, and p a prime. Recall from Section 3.6 that $G(M^*)^\diamond = \{x \in G(M) : \|x\| \nsim M$, or x is torsion$\}$ and that $G(\overline{\alpha}^*, p)^\diamond = \{x \in G(\overline{\alpha}, p) : |p^i x|_p \neq \alpha_i$ for infinitely many $i < \omega\}$. Then

$$G(M^*) = \langle G(M^*)^\diamond\rangle \quad \text{and} \quad G(M^*, p) = G(M) \cap \langle G(M^*, p)^\diamond\rangle,$$

where $G(M^*, p)^\diamond = G(M^*)^\diamond \cup G(M_p^*, p)^\diamond$. Following Hill and Megibben [50], we call an element $x \in G$ **primitive** if for any Ulm matrix M, prime p, and positive integer n,

$$nx \in G(M^*, p) \quad \text{implies that } \|x\| \nsim M \text{ or } nx \in G(M_p^*, p)^\diamond.$$

Notice that primitive elements have infinite order since any torsion element x satisfies $\|x\| \sim \overline{\infty}$. In particular, 0 is not a primitive element. A decomposition set X in G is called a **$*$-decomposition set** if X consists entirely of primitive elements such that $\langle X\rangle = \bigoplus_{x\in X}\langle x\rangle$ is a $*$-valuated coproduct in G (see Section 3.6). We note the following basic facts:

(a) An element $x \in G$ is primitive if and only if x has infinite order and for every nonzero multiple x' of x, Ulm matrix M, and prime p, we have $x' \in G(M^*, p)^\diamond$ whenever $x' \in G(M^*, p)$.

(b) If $x \in G$ has infinite order and $\|x\| \sim \overline{\infty}$, then x is primitive.

(c) If X is a $*$-decomposition set in $G = H \oplus K$, and if $X \subseteq H$, then X is a $*$-decomposition set in H.

(d) If $\{x_1, \ldots, x_n\}$ is a $*$-decomposition set, then $\{a_1x_1, \ldots, a_nx_n\}$ is a $*$-decomposition set for any $a_1, \ldots, a_n \in \mathbb{Z} \setminus \{0\}$.

As an immediate consequence of Lemma 3.6.4, we have:

Corollary 3.10.1 (Hill–Megibben [50]). *If G has a decomposition basis X, then X is a $*$-decomposition set.*

Proposition 3.10.2 (Hill–Megibben [50]). *If n is a positive integer, then $x \in G$ is primitive if and only if nx is primitive.*

Proof. By definition, only sufficiency needs to be proved. Assume that nx is primitive, and that $mx \in G(M^*, p)$ for some positive integer m, Ulm matrix M, and prime p. Then $mnx \in G((nM)^*, p)$ by Section 3.6(c); hence, $\|x\| \sim \|nx\| \nsim nM \sim M$, or else $mnx \in G((nM)^*_p, p)^\diamond$, in which case, we have $|p^i mnx|_p \neq m_{p,i+|n|_p}$ for infinitely many $i < \omega$. Thus, $mx \in G(M^*_p, p)^\diamond$. Therefore, x is primitive. □

A primitive element is in $G(M^*, p)^\diamond$ whenever it is in $G(M^*, p)$. Moreover, we have the following:

Proposition 3.10.3 (Hill–Megibben [50]). *Let x be primitive in G. Then the following hold:*

1. *If $x \in G(M^*)$ for some Ulm matrix M, then $\|x\| \nsim M$.*
2. *If $x \in G(\overline{\alpha}^*, p)$ for some Ulm sequence $\overline{\alpha}$ and prime p, then $x \in G(\overline{\alpha}^*, p)^\diamond$.*

Proof. To prove (1), let $x \in G(M^*)$, and assume that $\|x\| \sim M$. Then $\|x\|_p = M_p$ for only finitely many primes p. Therefore, if p is such a prime, then $x \notin G(M^*_p, p)^\diamond$. Since x is primitive, and $x \in G(M^*) \subseteq G(M^*, p)$, it follows that $\|x\| \nsim M$, a contradiction.

Now suppose that $x \in G(\overline{\alpha}^*, p)$. If $x \notin G(\overline{\alpha}^*, p)^\diamond$, then there is a $j < \omega$ such that $|p^i x|_p = \alpha_i$ for all $i \geq j$. Let M be the Ulm matrix, where $M_p = \overline{\alpha}$ and $M_q = \|x\|_q$ for all $q \neq p$. Then $x \in G(M^*_p, p) \subseteq G(M^*, p)$. Thus, $\|x\| \sim M$ and $|p^i x|_p = m_{p,i}$, for all $j \geq i$, which is impossible since x is primitive. Therefore, (2) follows. □

Corollary 3.10.4. *Suppose that X is a $*$-decomposition set in G, and M is an Ulm matrix such that $\|x\| \sim M$ for all $x \in X$. If $B = \langle X \rangle \oplus A$ is a $*$-valuated coproduct in G with $A \subseteq G(M^*)$, then $G(M^*) \cap B = A$.*

Proof. We have $G(M^*) \cap B = \bigoplus_{x \in X}(G(M^*) \cap \langle x \rangle) \oplus A$. If $G(M^*) \cap \langle x \rangle \neq 0$ for some $x \in X$, then there is a positive integer n such that $nx \in G(M^*)$. But then Proposition 3.10.3(1) yields $\|x\| \sim \|nx\| \nsim M$, a contradiction. □

Proposition 3.10.5 (Hill–Megibben [50]). *Let x be primitive in G with $M = \|x\|$. If $y \in x + G(M^*)$, then y is primitive in G, and we have $\|ny\| = \|nx\|$ for some positive integer n.*

Proof. We have $y \in G(M)$. Therefore, since $x \notin G(M^*)$ by Proposition 3.10.3(1), we have $y \notin G(M^*)$; hence, $\|y\| \sim M$. Assume that y is not primitive. Then there is a positive integer n, an Ulm matrix N, a prime p, and a $j < \omega$ such that

$$ny \in G(N^*, p) \text{ with } \|y\| \sim N \text{ and } |p^i ny|_p = n_{p,i} \text{ for all } i \geq j.$$

Now let $y = x + z$ with $z \in G(M^*)$. Then we obtain $nx = ny - nz \in G(N^*, p) + G((nM)^*) \subseteq G((N \wedge nM)^*, p)$ by Section 3.6(b). Since $y \in G(M)$, we have $ny \in G(nM)$; therefore,

$n_{p,i} \geq m_{p,i+|n|_p}$ for all $i \geq j$. But then $|p^i nx|_p = m_{p,i+|n|_p} = m_{p,i+|n|_p} \wedge n_{p,i}$ for all $i \geq j$, which is impossible since x is primitive. Therefore, y is primitive.

We claim that, for every prime p, there is a $k < \omega$ such that $|p^i y|_p = m_{p,i}$ for all $i \geq k$. Assume that this fails for some prime p. Then $y \in G(M_p^*, p)$, which yields $x = y + z \in G(M) \cap (G(M_p^*, p) + G(M^*)) = G(M^*, p)$, contradicting primitivity of x. Now the last assertion follows since $\|y\|_p = M_p$ for almost all primes p. □

Proposition 3.10.6 (Hill–Megibben [50]). *Suppose that $B = A \oplus \langle x \rangle$ is a ∗-valuated coproduct in G, where x is primitive in G. If $y = a + x$ with $\|y\| = \|x\|$ and $a \in A$, then $B = A \oplus \langle y \rangle$ is a ∗-valuated coproduct in G and y is primitive in G.*

Proof. Let M be an Ulm matrix, $\overline{\alpha}$ an Ulm sequence, p a prime, and n a nonzero integer. If $ny \in G(M^*, p)$, then $nx \in G(M^*, p)$; thus, $\|y\| = \|x\| \sim M$, or else $x \in G(M_p^*, p)^\diamond$, in which case, we have $y \in G(M_p^*, p)^\diamond$. Therefore, y is primitive. To prove that $B = A \oplus \langle y \rangle$ is a ∗-valuated coproduct, let $b = a' + ny$, where $a' \in A$. Then $b = a' + na + nx$. Therefore, if $b \in G(M)$, then $nx \in G(M)$. Thus, $ny \in G(M)$. If $b \in G(M^*)$, then $nx \in G(M^*)$. Since x is primitive, it follows that $\|ny\| = \|nx\| \nsim M$ by Proposition 3.10.3(1); hence, $ny \in G(M^*)$. Similarly, if $b \in G(\overline{\alpha}^*, p)$, then $nx \in G(\overline{\alpha}^*, p)$, and therefore, $nx \in G(\overline{\alpha}^*, p)^\diamond$ (Proposition 3.10.3(2)); thus, $ny \in G(\overline{\alpha}^*, p)^\diamond$. Finally, suppose that $b \in G(M^*, p)$. Then $nx \in G(M^*, p)$; hence, $\|y\| = \|x\| \nsim M$, or $nx \in G(M_p^*, p)^\diamond$. But then $ny \in G(M^*, p)$, as desired. □

The following result is contained in the proof of [50, Lemma 3.6]:

Proposition 3.10.7. *Let A and B be subgroups of G such that $A \subseteq B$. Suppose that $B/A \oplus \langle y_1 + A \rangle \oplus \cdots \oplus \langle y_n + A \rangle$ is a valuated coproduct in G/A such that $y_i + A$ is primitive in G/A and $\|y_i + A\|_{G/A} = \|y_i\|_G$ for $i = 1, \ldots, n$. Then the following hold:*
1. *$B \oplus \langle y_1 \rangle \oplus \cdots \oplus \langle y_n \rangle$ is a valuated coproduct in G, and each y_i is primitive in G.*
2. *If $B/A \oplus \langle y_1 + A \rangle \oplus \cdots \oplus \langle y_n + A \rangle$ is a ∗-valuated coproduct in G/A, then $B \oplus \langle y_1 \rangle \oplus \cdots \oplus \langle y_n \rangle$ is a ∗-valuated coproduct in G.*

Proof. To show that the sum $B + \langle y_1 \rangle + \cdots + \langle y_n \rangle$ is direct, assume that $b + r_1 y_1 + \cdots + r_n y_n = 0$, where $b \in B$ and $r_1, \ldots, r_n \in \mathbb{Z}$. If $i \in \{1, \ldots, n\}$, then $r_i y_i + A = 0$. Hence, $r_i = 0$ since primitive elements have infinite order.

Now let $y \in \{y_1, \ldots, y_n\}$, and assume that y is not primitive. Then we can find an Ulm matrix M, a prime p, a positive integer k, and $j < \omega$ such that $ky \in G(M^*, p)$, $\|y\|_G \sim M$, and $|p^i ky|_p = m_{p,i}$ for all $i \geq j$. But then $ky + A \in (G/A)(M^*, p)$ by Lemma 3.6.2, $\|y + A\|_{G/A} = \|y\|_G \sim M$, and $ky + A \notin (G/A)(M_p^*, p)^\diamond$ since $|p^i ky + A|_p^{G/A} = |p^i ky|_p^G = m_{p,i}$ for all $i \geq j$, contradicting primitivity of $y + A$ in G/A. Therefore, each element y_i is primitive in G.

Let $x = x_0 + x_1 + \cdots + x_n$ with $x_0 \in B$, and $x_i \in \langle y_i \rangle$ for $i = 1, \ldots, n$, and let M be an Ulm matrix. If $x \in G(M)$, then $\|x + A\|_{G/A} \geq \|x\|_G \geq M$. Hence, $\|x_i\|_G = \|x_i + A\|_{G/A} \geq M$ for all $i = 1, \ldots, n$. Therefore, $x_0, \ldots, x_n \in G(M)$, so $B \oplus \langle y_1 \rangle \oplus \cdots \oplus \langle y_n \rangle$ is a valuated coproduct in G.

Assuming that $B/A \oplus \langle y_1 + A\rangle \oplus \cdots \oplus \langle y_n + A\rangle$ is a $*$-valuated coproduct in G/A, let $F \in \{G(\overline{\alpha}^*, p), G(M^*), G(M^*, p)\}$, where M is an Ulm matrix, $\overline{\alpha}$ is an Ulm sequence, and p is a prime. Suppose that $x = x_0 + x_1 + \cdots + x_n \in F$, and let $i \in \{1, \ldots, n\}$. First, we let $F = G(M^*, p)$. Then $x + A \in (G/A)(M^*, p)$; hence, $x_i + A \in (G/A)(M^*, p)$. In particular, this yields $x_i \in G(M)$ because $x_i + A \in (G/A)(M)$. Since $x_i + A$ is primitive in G/A, it is an element of $(G/A)(M^*, p)^\diamond$; hence, $\|x_i\|_G = \|x_i + A\|_{G/A} \nsim M$, or else $|p^j x_i|_p^G = |p^j x_i + A|_p^{G/A} \neq m_{p,j}$ for infinitely many $j < \omega$. It follows that $x_i \in F^\diamond$; thus, $x_0, \ldots, x_n \in F$.

If $F = G(M^*)$, then $x + A \in (G/A)(M^*)$ by Lemma 3.6.2. Hence, $x_i + A \in (G/A)(M^*)$. Again, $x_i \in G(M)$, and since $x_i + A$ is primitive in G/A, we have $\|x_i\|_G = \|x_i + A\|_{G/A} \nsim M$ by Proposition 3.10.3(1). Thus, $x_i \in F^\diamond$, and we obtain $x_0, \ldots, x_n \in F$.

Likewise, if $F = G(\overline{\alpha}^*, p)$, then $x + A \in (G/A)(\overline{\alpha}^*, p)$, and we have $x_i + A \in (G/A)(\overline{\alpha}^*, p)$. Since $\|x_i\|_p^G = \|x_i + A\|_p^{G/A} \geq \overline{\alpha}$ and $|p^j x_i|_p^G = |p^j x_i + A|_p^{G/A} \neq \alpha_j$ for infinitely many $j < \omega$ (cf. Proposition 3.10.3(2)), we obtain $x_i \in F^\diamond$, and again it follows that $x_0, \ldots, x_n \in F$. Therefore, $B \oplus \langle y_1\rangle \oplus \cdots \oplus \langle y_n\rangle$ is a $*$-valuated coproduct in G. □

Corollary 3.10.8. *Let A and B be subgroups of a group G such that $A \subseteq B$ and A is quasi-sequentially nice in G. Suppose that $B/A \oplus \langle y_1 + A\rangle \oplus \cdots \oplus \langle y_n + A\rangle$ is a valuated coproduct in G/A, where $y_1 + A, \ldots, y_n + A$ are primitive in G/A. Then the following hold:*
1. *There exist positive integers m_i and elements $y_i' \in m_i y_i + A$ $(i = 1, \ldots, n)$ such that $B \oplus \langle y_1'\rangle \oplus \cdots \oplus \langle y_n'\rangle$ is a valuated coproduct in G and $y_1', \ldots, y_n'$ are primitive in G.*
2. *If $B/A \oplus \langle y_1 + A\rangle \oplus \cdots \oplus \langle y_n + A\rangle$ is a $*$-valuated coproduct in G/A, then $B \oplus \langle y_1'\rangle \oplus \cdots \oplus \langle y_n'\rangle$ in (1) is a $*$-valuated coproduct in G.*

Proof. For every $i \in \{1, \ldots, n\}$, there is a positive integer m_i and an element $y_i' \in m_i y_i + A$ such that $\|y_i'\|_G = \|y_i' + A\|_{G/A}$. Then by Proposition 3.10.7, $B \oplus \langle y_1'\rangle \oplus \cdots \oplus \langle y_n'\rangle$ is a valuated coproduct in G and y_i' is primitive in G, as claimed. (2) follows from (1) and Proposition 3.10.7. □

Corollary 3.10.9. *Let A be a subgroup of G, $x \in G$, n a positive integer, and $c \in nx + A$ such that $\|c\|_G = \|nx + A\|_{G/A}$. If $x + A$ is primitive in G/A, then c is primitive in G.*

Lemma 3.10.10 (Hill–Megibben [50]). *Suppose that $m_1, \ldots, m_n$ are positive integers that are pairwise relatively prime. Let $M = \prod_{i=1}^n m_i$ and $M_i = \frac{M}{m_i}$ for $i = 1, \ldots, n$. Then there are nonzero integers $\lambda_2, \ldots, \lambda_n, \mu_2, \ldots, \mu_n$, satisfying the following conditions:*
1. $\det A = 1$, *where*

$$A = \begin{pmatrix} 1 & \lambda_2 M_2 & \lambda_3 M_3 & \cdots & \lambda_n M_n \\ 1 & \mu_2 m_2 & 0 & \cdots & 0 \\ 1 & 0 & \mu_3 m_3 & \cdots & 0 \\ \vdots & \vdots & \vdots & & \vdots \\ 1 & 0 & 0 & \cdots & \mu_n m_n \end{pmatrix}.$$

2. $\gcd(\lambda_k, \mu_k) = 1$ *for all* $k = 2, \ldots, n$, *and* $\gcd(\mu_k, m_i) = 1$ *whenever* $i \neq k$.

Proof. For $k = 2, \dots, n$, we will construct inductively integers λ_k, μ_k satisfying property (2) such that $\det A_k = 1$, where A_k is the $k \times k$-submatrix in the upper left corner of the matrix A, as shown in the lemma. Notice that m_k and M_k are relatively prime. Then for $k = 2$, there are integers λ_2 and μ_2 such that $\mu_2 m_2 - \lambda_2 M_2 = 1$. Thus, $\det A_2 = 1$ and λ_2, μ_2 satisfy property (2). Now let $2 \le k < n$, and assume that we already constructed integers $\lambda_2, \dots, \lambda_k, \mu_2, \dots, \mu_k$ with $\det A_k = 1$, satisfying conditions (2). Letting $\delta_{k+1} = M_{k+1} \prod_{i=2}^{k} \mu_i m_i$, we have $\gcd(m_{k+1}, \delta_{k+1}) = 1$ by the induction hypothesis. Therefore, there are integers λ_{k+1} and μ_{k+1} such that $\mu_{k+1} m_{k+1} - \lambda_{k+1}\delta_{k+1} = 1$. Computing the determinant of A_{k+1} by expansion along the last column of A_{k+1}, we obtain $\det A_{k+1} = (-1)^k \lambda_{k+1} M_{k+1} \prod_{i=2}^{k} (-1)^{k+1} \mu_i m_i + \mu_{k+1} m_{k+1} \det A_k$. Thus,

$$\det A_{k+1} = -\lambda_{k+1}\delta_{k+1} + \mu_{k+1} m_{k+1} \det A_k = 1,$$

as desired. □

The following two results can be found in [50]:

Lemma 3.10.11. *Suppose that $\{x_1, \dots, x_n\}$ is a ∗-decomposition set in G such that $\|x_1\| \sim \dots \sim \|x_n\|$. Then $y = x_1 + \dots + x_n$ is primitive in G exactly if for every prime p there is an $s < \omega$ and an $i \in \{1, \dots, n\}$ such that $\|p^s y\|_p = \|p^s x_i\|_p$.*

Proof. Suppose that y is primitive, and assume that there is a prime p such that $\|p^s y\|_p \ne \|p^s x_i\|_p$ for all $s < \omega$ and $i = 1, \dots, n$. Letting $\overline{\alpha} = \|y\|_p$, we have $\|x_i\|_p \ge \overline{\alpha}$; hence, $x_i \in G(\overline{\alpha}^*, p)$ for all $i = 1, \dots, n$. But then $y \in G(\overline{\alpha}^*, p)$, which is impossible by Proposition 3.10.3(2). Thus, the assertion follows.

To prove the converse, assume that y is not primitive. Then there is an Ulm matrix M, a prime p, and positive integers j, r such that

$$ry \in G(M^*, p), \quad \|y\| \sim M \quad \text{and} \quad |p^k ry|_p = m_{p,k} \quad \text{for all } k \ge j.$$

It follows that for all $i = 1, \dots, n$, we obtain $rx_i \in G(M^*, p)$, and $\|x_i\| \sim M$ by Section 3.5(d). Thus, $rx_i \in G(M_p^*, p)^\diamond$ since x_i is primitive. But then we cannot have $\|p^s x_i\|_p = \|p^s y\|_p$ for any $s < \omega$. This completes the proof. □

Lemma 3.10.12. *Suppose that $X = \{x_1, \dots, x_n\}$ is a ∗-decomposition set in G with $\|x_1\| \sim \dots \sim \|x_n\|$ such that $y = x_1 + \dots + x_n$ is primitive in G. Let P denote the set of all primes p such that $\|x_1\|_p = \dots = \|x_n\|_p$, and let $Q = \mathbb{P} \setminus P$. Then there is a nonzero multiple y' of y and a subset X' of $\langle X\rangle$ containing elements $x_1', \dots, x_r'$ such that the following conditions are satisfied:*

1. $y' = x_1' + \dots + x_r'$;
2. *X' is a ∗-decomposition set in G with $\langle X\rangle / \langle X'\rangle$ finite;*
3. *The sets $P_i' = \{p \in Q : \|y'\|_p = \|x_i'\|_p\}$ $(i = 1, \dots, r)$ form a partition of Q.*

Proof. Notice that the set Q is finite. By Lemma 3.10.11, for every prime p, there exists an $i \le n$ and an $s < \omega$ such that $\|p^s y\|_p = \|p^s x_i\|_p$. Hence, $\|ay\|_p = \|ax_i\|_p$, where

$a = \prod_{p\in Q} p^s$. Replacing y by ay and x_i by ax_i ($i = 1,\dots,n$), we may assume that for every prime p, there exists an $i \le n$ such that

$$\|y\|_p = \|x_i\|_p.$$

Replacing y and all x_i by a nonzero multiple once more, we can assume that $|p^k y|_p \ne \infty$ for all $k < \omega$ whenever $p \in Q$ since $\langle x_1\rangle \oplus \cdots \oplus \langle x_n\rangle$ is a valuated coproduct. Now write $Q = P_1 \cup P_2 \cup \cdots \cup P_n$, where

$$P_i = \{p \in Q : \|y\|_p = \|x_i\|_p\}.$$

To construct the desired partition of Q, suppose that $P_i \cap P_j \ne \emptyset$ for some $i \ne j$, say $P_1 \cap P_2 \ne \emptyset$. Since $\|x_1\| \sim \|x_2\|$, there is a positive integer k such that

$$\|kx_2\|_p \ge \|x_1\|_p \quad \text{for all } p \in Q \setminus (P_1 \cap P_2),$$

and since $\|x_1\|_p = \|x_2\|_p$ for all $p \in P \cup (P_1 \cap P_2)$, we may assume that the prime divisors of k are exactly the primes in $Q \setminus (P_1 \cap P_2)$. Letting m be the product of all primes in $P_1 \cap P_2$, we can find integers λ and μ such that $\lambda m + \mu k = 1$. Let

$$x_1' = x_1 + \mu k x_2 \quad \text{and} \quad x_2' = \lambda m x_2.$$

Then $\|x_1'\| = \|x_1\| \wedge \|\mu k x_2\| = \|x_1\|$; thus, x_1' is primitive by Proposition 3.10.6. Note that we have $x_1' + x_2' = x_1 + x_2$, and $\langle x_1'\rangle + \langle x_2'\rangle = \langle x_1'\rangle \oplus \langle x_2'\rangle \subseteq \langle x_1\rangle \oplus \langle x_2\rangle$. Let $P_i' = \{p \in Q : \|y\|_p = \|x_i'\|_p\}$ for $i = 1, 2$. Clearly, we have $P_1' = P_1$. Suppose that $p \in P_2'$. Assuming $p \in P_1 \cap P_2$, we have $\|x_1\|_p = \|y\|_p = \|x_2\|_p$; thus, $\|y\|_p \ne \|x_2'\|_p$ since $p|m$, contradicting the fact that $p \in P_2'$; it follows that $p \in Q \setminus (P_1 \cap P_2)$. But then $p|k$, which implies that $p \nmid \lambda m$, and we obtain $\|y\|_p = \|x_2'\|_p = \|x_2\|_p$; thus, $p \in P_2 \setminus (P_1 \cap P_2)$. Conversely, suppose that $p \in P_2 \setminus (P_1 \cap P_2)$. Then $\|y\|_p = \|x_2\|_p$, and since $P_2 \setminus (P_1 \cap P_2) \subseteq Q \setminus (P_1 \cap P_2)$, we have $p \nmid \lambda m$. Hence, $\|y\|_p = \|x_2\|_p = \|x_2'\|_p$, which shows that $p \in P_2'$. Therefore, we obtain $P_2' = P_2 \setminus (P_1 \cap P_2)$; thus,

$$y = x_1' + x_2' + x_3 + \cdots + x_n \quad \text{and} \quad P_1' \cap P_2' = \emptyset.$$

To show that $\langle x_1'\rangle \oplus \langle x_2'\rangle$ is a $*$-valuated coproduct, let $z = t_1 x_1' + t_2 x_2'$ with $t_1, t_2 \in \mathbb{Z} \setminus \{0\}$, M an Ulm matrix, p a prime, and $\overline{\alpha}$ an Ulm sequence. If $z \in G(M)$, then $t_1 x_1 \in G(M)$ since $z = t_1 x_1 + (t_1 \mu k + t_2 \lambda m) x_2$ and $\langle x_1\rangle \oplus \langle x_2\rangle$ is a valuated coproduct. Hence, $\|x_1'\| = \|x_1\|$ shows that $t_1 x_1' \in G(M)$. Thus, $\langle x_1'\rangle \oplus \langle x_2'\rangle$ is a valuated coproduct.

If $z \in G(M^*)$, then $t_1 x_1 \in G(M^*)$, and since x_1 is primitive, it follows that $\|x_1'\| = \|x_1\| \nleq M$ (Proposition 3.10.3(1)). Thus, $t_1 x_1' \in G(M^*)$. If $z \in G(\overline{\alpha}^*, p)$, then $t_1 x_1 \in G(\overline{\alpha}^*, p)^\diamond$ by Proposition 3.10.3(2). Hence, $t_1 x_1' \in G(\overline{\alpha}^*, p)^\diamond$. Finally, if $z \in G(M^*, p)$, then $t_1 x_1 \in G(M^*, p)$. Therefore, since x_1 is primitive, we have $\|x_1\| \nleq M$ or $t_1 x_1 \in G(M_p^*, p)^\diamond$, and it follows that $\|x_1'\| \nleq M$, or $t_1 x_1' \in G(M_p^*, p)^\diamond$; thus, $t_1 x_1' \in G(M_p^*, p)$. Therefore, $\langle x_1'\rangle \oplus \langle x_2'\rangle$ is a $*$-valuated coproduct, as desired.

Since y' and $\{x_1', x_2', x_3, \dots, x_n\}$ satisfy the same conditions as the original y and $\{x_1, \dots, x_n\}$, and since $P_1' \subseteq P_1$ and $P_2' \subseteq P_2$, we may apply this procedure again as necessary to x_1', x_3, then x_2', x_3', and so on. This gives us Q as a union of pairwise disjoint sets P_i $(i = 1, \dots, n)$.

Now suppose that some P_i, say P_n, is an empty set. For every $i = 1, \dots, n-1$, there is a positive integer m_i whose prime divisors are exactly the primes in $Q \setminus P_i$ such that $\|m_i x_n\|_p \geq \|x_i\|_p$ for all $p \in Q \setminus P_i$. Since

$$(Q \setminus P_1) \cap (Q \setminus P_2) \cap \dots \cap (Q \setminus P_{n-1}) = \bigcap_{i=1}^{n-1} \Big(\bigcup_{j \neq i} P_j \Big) = P_n = \emptyset,$$

we have $\gcd(m_1, \dots, m_{n-1}) = 1$. Therefore, there are integers $\lambda_1, \dots, \lambda_{n-1}$ such that $\lambda_1 m_1 + \dots + \lambda_{n-1} m_{n-1} = 1$. For $i = 1, \dots, n-1$, we define $x_i' = x_i + \lambda_i m_i x_n$ and have $y = x_1' + \dots + x_{n-1}'$. It is clear that the sum $\langle x_1' \rangle + \dots + \langle x_{n-1}' \rangle + \langle x_n \rangle$ is direct. We claim that $X' = \{x_1', \dots, x_{n-1}', x_n\}$ is a $*$-decomposition set in G. For all $i = 1, \dots, n-1$, we have $\|\lambda_i m_i x_n\|_p \geq \|x_i\|_p$ for all $p \in Q \setminus P_i$ and $\|x_n\|_p \geq \|y\|_p = \|x_i\|_p$ for all $p \in P_i$. Thus,

$$\|x_i\| = \|x_i\| \wedge \|\lambda_i m_i x_n\| = \|x_i'\|,$$

which shows that x_i' is primitive (Proposition 3.10.6). Now let $w = \sum_{i=1}^{n-1} t_i x_i' + t_n x_n$, where $t_1, \dots, t_{n-1}, t_n \in \mathbb{Z}$. Then $w = \sum_{i=1}^{n-1} t_i x_i + r x_n$ for some integer r. Let $i \in \{1, \dots, n-1\}$. If $w \in G(M)$, then $t_i x_i \in G(M)$ since $\{x_1, \dots, x_n\}$ is a decomposition set in G. Since $\|x_i'\| = \|x_i\|$, it follows that $t_i x_i' \in G(M)$, and therefore, $t_n x_n = w - \sum_{i=1}^{n-1} t_i x_i' \in G(M)$. Thus, X' is a decomposition set in G. Apply this process repeatedly for all i such that $P_i = \emptyset$.

Using similar arguments as before, it is easily verified that $w \in G(M^*)$ (or $G(\overline{\alpha}^*, p)$, $G(M^*, p)$) implies that $t_i x_i \in G(M^*)$ (or $G(\overline{\alpha}^*, p), G(M^*, p)$). Therefore, X' is a $*$-decomposition set in G. Repeated application of this process yields the desired $*$-decomposition set and partition of Q. □

Now we can prove Hill and Megibben's exchange lemma:

Proposition 3.10.13 (Hill–Megibben [50]). *Suppose that $X = \{x_1, \dots, x_m\}$ is a $*$-decomposition set in G such that $\|x_1\| \sim \dots \sim \|x_m\|$. If y_1 is primitive in G, where $y_1 \in C = \langle X \rangle$, then there are $y_2, \dots, y_m \in C$ such that $Y = \{y_1, y_2, \dots, y_m\}$ is a $*$-decomposition set in G with $C/\langle Y \rangle$ finite.*

Proof. Let $P = \{p \in \mathbb{P} : \|x_1\|_p = \dots = \|x_m\|_p\}$, and $Q = \mathbb{P} \setminus P$. We may assume that $y_1 = x_1 + \dots + x_n$ for some $n \leq m$. Notice that by Corollary 3.6.6, we can replace y_1 by any nonzero multiple. Therefore, by Lemma 3.10.12, we may assume that the sets $P_i = \{p \in Q : \|y_1\|_p = \|x_i\|_p\}$ form a partition of Q. Observe that for any $p \in P_i$, we have $\|x_i\|_p = \|y_1\|_p \leq \|x_j\|_p$ for all $j = 1, \dots, n$. Since $\|x_1\| \sim \dots \sim \|x_n\|$, we have $\|m_i x_i\|_p \geq \|p x_k\|_p$ $(k = 1, \dots, n)$ for some integers m_i whose prime divisors are exactly the primes in P_i. It follows that

$$\|m_i x_j\|_p \geq \|m_i x_i\|_p \geq \|p x_k\|_p \quad \text{for all } j,k = 1,\dots,n \text{ whenever } p \in P_i. \qquad (*)$$

For $i = 2,\dots,n$, we let M_i, λ_i, and μ_i be the integers as in Lemma 3.10.10, and define

$$y_i = \lambda_i M_i x_1 + \mu_i m_i x_i.$$

If $p \in P_j$ for $j \neq i$, then $\|\lambda_i M_i x_1\|_p = \|\lambda_i m_j x_1\|_p \geq \|\lambda_i p x_i\|_p$ and $\|\mu_i m_i x_i\|_p = \|x_i\|_p$, and if $p \in P_i$, then $\|\lambda_i M_i x_1\|_p = \|x_1\|_p$ and $\|\mu_i m_i x_i\|_p \geq \|\mu_i p x_1\|_p$. It follows that for all $i = 2,\dots,n$, we have

$$\|y_i\|_p = \begin{cases} \|x_i\|_p & \text{if } p \in \mathbb{P} \setminus P_i, \\ \|x_1\|_p & \text{if } p \in P_i. \end{cases}$$

Note that $(x_1,\dots,x_n)A = (y_1,\dots,y_n)$, where A is the integer matrix in Lemma 3.10.10, and that $\langle y_i \rangle \cap \sum_{j\neq i}\langle y_j \rangle = 0$ for all $i = 1,\dots,n$. Since A has determinant 1, A^{-1} is an integer matrix, and we obtain $C = \langle x_1 \rangle \oplus \cdots \oplus \langle x_n \rangle = \langle y_1 \rangle \oplus \cdots \oplus \langle y_n \rangle$.

It remains to show that $\{y_1,\dots,y_n\}$ is a $*$-decomposition set in G. Clearly, the elements of Y are primitive by Lemma 3.10.11. Now let $0 \neq y = t_1 y_1 + \cdots + t_n y_n \in C$ ($t_1,\dots,t_n \in \mathbb{Z}$). Then $y = s_1 x_1 + \cdots + s_n x_n$ with $s_1 = t_1 + t_2\lambda_2 M_2 + \cdots + t_n \lambda_n M_n$, and $s_i = t_1 + t_i \mu_i m_i$ ($i = 2,\dots,n$). Let p be a prime, and put $\sigma = |t_1 y_1|_p \wedge \cdots \wedge |t_n y_n|_p$. To prove that $\{y_1,\dots,y_n\}$ is a decomposition set, we need to show that $|y|_p \leq \sigma$. First, suppose that $p \in P$. Then $\|y_i\|_p = \|x_i\|_p = \|x_1\|_p$ for all $i = 1,\dots,n$. Hence, $\sigma = |p^t y_1|_p = |p^t x_1|_p$ with $t = |t_1|_p \wedge \cdots \wedge |t_n|_p$, and $|y|_p = |p^s x_1|_p$ with $s = |s_1|_p \wedge \cdots \wedge |s_n|_p$. Then

$$A\begin{pmatrix} t_1 \\ \vdots \\ t_n \end{pmatrix} = \begin{pmatrix} s_1 \\ \vdots \\ s_n \end{pmatrix}$$

implies that

$$\begin{pmatrix} t_1 \\ \vdots \\ t_n \end{pmatrix} = A^{-1}\begin{pmatrix} s_1 \\ \vdots \\ s_n \end{pmatrix}.$$

Hence, $p^s | t_i$ for all $i = 1,\dots,n$, and it follows that $s \leq t$; thus, $|y|_p \leq \sigma$. Now suppose that $p \notin P$. Then $p \in P_k$ for some $k \in \{1,\dots,n\}$. Let i be the smallest integer $\leq n$ with $\sigma = |t_i y_i|_p$. In the case where $|t_l y_l|_p \neq |t_i y_i|_p$ for all $l \neq i$, we have $|y|_p = \sigma$. Therefore, we assume that $|t_l y_l|_p = |t_i y_i|_p$ for some $l \neq i$. Since $\{x_1,\dots,x_n\}$ is a decomposition set, it suffices to show the following:

Claim (A): There exists a $j \in \{1,\dots,n\}$ such that $s_j \neq 0$ and $|s_j x_j|_p = \sigma$.

Proof: If $\sigma = \infty$, then $|y|_p = \infty$, and therefore, $|s_jx_j|_p = \infty$ for all $j = 1,\dots,n$. Since $y \neq 0$, the assertion follows. Now assume that σ is an ordinal. In the following, we will use both $(*)$ and the fact that $p \in P_k$ repeatedly.

Case I: $i = 1$ and $k \neq 1$. Then

$$|t_km_kx_k|_p \geq |t_kpx_1|_p > |t_kx_1|_p = |t_ky_k|_p \geq |t_1y_1|_p = |t_1x_k|_p;$$

thus, $|s_kx_k|_p = |t_1x_k|_p \wedge |t_k\mu_km_kx_k|_p = |t_1x_k|_p = |t_1y_1|_p = \sigma$.

Case II: $i = k = 1$. For any $j \neq 1$, we have

$$|t_jM_jx_1|_p = |t_jm_1x_1|_p \geq |t_jpx_j|_p > |t_jx_j|_p = |t_jy_j|_p \geq |t_1y_1|_p = |t_1x_1|_p;$$

hence, $|s_1x_1|_p = |t_1x_1|_p \wedge |t_2\lambda_2M_2x_1|_p \wedge \cdots \wedge |t_n\lambda_nM_nx_1|_p = |t_1x_1|_p = |t_1y_1|_p = \sigma$.

Case III: $i \neq 1$ and $k \neq i$. Then $p \notin P_i$, so the minimality of i and $\|y_1\|_p \leq \|x_i\|_p$ yield $|t_ix_i|_p = |t_iy_i|_p < |t_1y_1|_p \leq |t_1x_i|_p$. Since $p \nmid \mu_im_i$, we, therefore, have $|s_ix_i|_p = |t_1x_i|_p \wedge |t_i\mu_im_ix_i|_p = |t_ix_i|_p = |t_iy_i|_p = \sigma$.

Case IV: $i \neq 1$ and $k = i$. By assumption, there is a $j \neq i$ such that $|t_jy_j|_p = |t_iy_i|_p$, and we have $j \neq 1$ by the minimality of i. Then $|t_jx_j|_p = |t_jy_j|_p$ since $p \notin P_j$. Thus, $|t_jx_j|_p < |t_1y_1|_p = |t_1x_k|_p \leq |t_1x_j|_p$ since $p \in P_k$. Now $k \neq j$ yields $p \nmid \mu_jm_j$; hence, $|s_jx_j|_p = |t_1x_j|_p \wedge |t_j\mu_jm_jx_j|_p = |t_jx_j|_p = \sigma$.

Therefore, $\{y_1,\dots,y_n\}$ is a decomposition set in G. Now let M be an Ulm matrix, $\overline{\alpha}$ an Ulm sequence, and p a prime. If $y \in G(M^*)$, then $s_ix_i \in G(M^*)$ for all i. We have $s_j \neq 0$ for some j, so since x_j is primitive, it follows that $\|x_j\| \nsim M$ by Proposition 3.10.3(1). But then for each $i = 1,\dots,n$, we have $\|y_i\| \sim \|x_j\| \nsim M$ by Section 3.5(d); hence, $t_iy_i \in G(M^*)$.

Suppose that $y \in G(\overline{\alpha}^*,p)$, and notice that $s_jx_j \in G(\overline{\alpha}^*,p)^\diamond$ whenever $s_j \neq 0$ (Proposition 3.10.3(2)). Let $i \in \{1,\dots,n\}$ with $t_iy_i \neq 0$, and assume first that $p \in P$. Since $s \leq t$, as shown earlier, we have $|s_j|_p \leq |t_i|_p$ for some j. Then $\|y_i\|_p = \|x_i\|_p$ shows that $\|t_iy_i\|_p \geq \|s_jy_i\|_p = \|s_jx_i\|_p = \|s_jx_j\|_p$, and it follows that $t_iy_i \in G(\overline{\alpha}^*,p)^\diamond$. Now suppose that $p \notin P$, say $p \in P_k$ for $k \in \{1,\dots,n\}$, and assume that $t_iy_i \notin G(\overline{\alpha}^*,p)$. Since $y = t_1y_1 + \cdots + t_ny_n \in G(\overline{\alpha}^*,p)$, we have $t_ly_l \notin G(\overline{\alpha}^*,p)$ for some $l \neq i$, so since $\{y_1,\dots,y_n\}$ is a decomposition set in G, we have

$$\|p^rt_ly_l\|_p = \|p^rt_iy_i\|_p = (\alpha_{r+j})_{j<\omega}$$

for some $r < \omega$. Letting $m \geq r$, we have $\alpha_m = |p^mt_1y_1|_p \wedge \cdots \wedge |p^mt_ny_n|_p$. Then by Claim (A), there is a $j \in \{1,\dots,n\}$ such that $s_j \neq 0$ and $|p^ms_jx_j|_p = \alpha_m$. But then there is a $j_0 \in \{1,\dots,n\}$ with $s_{j_0} \neq 0$ and $|p^ms_{j_0}x_{j_0}|_p = \alpha_m$ for infinitely many $m \geq r$, contradicting $s_{j_0}x_{j_0} \in G(\overline{\alpha}^*,p)^\diamond$. Therefore, we have $t_iy_i \in G(\overline{\alpha}^*,p)$.

Finally, suppose that $y \in G(M^*,p)$. Then $t_iy_i \in G(M)$ for all $i = 1,\dots,n$, and we have $0 \neq s_jx_j \in G(M^*,p)$ for some j. Since x_j is primitive, it follows that $\|x_j\| \nsim M$ or $s_jx_j \in G(M_p^*,p)^\diamond$. If $\|x_j\| \nsim M$, then for all $i = 1,\dots,n$, we have $\|y_i\| \nsim M$, Thus, $t_iy_i \in G(M^*) \subseteq G(M^*,p)$. It remains to consider the case in which $s_jx_j \in G(M_p^*,p)^\diamond$ for all j with $s_j \neq 0$. Then $y \in G(M_p^*,p)$, so by the paragraph above, we have $t_iy_i \in$

$G(M_p^*, p) \subseteq G(M^*, p)$ for all $i = 1, \dots, n$. Consequently, $\{y_1, \dots, y_n\}$ is a $*$-decomposition set in G. Let $Y = \{y_1, \dots, y_n, x_{n+1}, \dots, x_m\}$. Then since $\langle y_1\rangle \oplus \cdots \oplus \langle y_n\rangle = \langle x_1\rangle \oplus \cdots \oplus \langle x_n\rangle$, Y is a $*$-decomposition set in G. □

Theorem 3.10.14 (Hill–Megibben [50]). *Suppose that $A \oplus \langle x_1\rangle$ and $B = A \oplus \langle y_1\rangle \oplus \cdots \oplus \langle y_n\rangle$ are $*$-valuated coproducts in G, where $x_1, y_1, \dots, y_n$ are primitive in G. If $x_1 \in B$, then there are primitive elements $x_2, \dots, x_n$ in G such that $\{x_2, \dots, x_n\} \subseteq B$ and $B' = A \oplus \langle x_1\rangle \oplus \langle x_2\rangle \oplus \cdots \oplus \langle x_n\rangle$ is a $*$-valuated coproduct in G with B/B' finite.*

Proof. We may assume that $x_1 = a + y_1 + \cdots + y_m$, where $a \in A$ and $m \le n$. In addition, we can assume that for some $k \le m$, we have $\|y_i\| \sim \|x_1\|$ if $i \le k$ and $\|y_i\| \nsim \|x_1\|$ whenever $i > k$. Then $x_1 = a + y + g$ with $y = \sum_{i=1}^{k} y_i$, and $g = \sum_{i=k+1}^{m} y_i$. Since $A \oplus \langle x_1\rangle$ and $B = A \oplus \langle y_1\rangle \oplus \cdots \oplus \langle y_m\rangle$ are valuated coproducts, we have $\|x_1\| \le \|a\|$ and $\|x_1 - a\| = \|x_1\| \wedge \|a\| = \|x_1\|$. Suppose that there is a positive integer r and an $i_0 < \omega$ such that $r(x_1 - a) \in G(M^*, p)$ with $M \sim \|x_1 - a\|$ and $|p^i r(x_1 - a)|_p = m_{p,i}$ for all $i \ge i_0$. Then $rx_1 \notin G(M_p^*, p)^\diamond$, contradicting the fact that x_1 is primitive. Therefore, $x_1 - a$ is primitive. Since $y \in (x_1 - a) + G(\|x_1\|^*)$, y is primitive and $\|sy\| = \|sx_1\|$ for some positive integer s by Proposition 3.10.5. By Proposition 3.10.13, there are $z_2, \dots, z_k \in C = \langle y_1, \dots, y_k\rangle$ such that $Z = \{sy, z_2, \dots, z_k\}$ is a $*$-decomposition set in G with $C/\langle Z\rangle$ finite. Then

$$B''' = A \oplus \langle sy\rangle \oplus \langle z_2\rangle \oplus \cdots \oplus \langle z_k\rangle \oplus \langle y_{k+1}\rangle \oplus \cdots \oplus \langle y_m\rangle$$

is a $*$-valuated coproduct in G with B/B''' finite. Therefore, by Proposition 3.10.6,

$$B''' = A \oplus \langle sx_1\rangle \oplus \langle z_2\rangle \oplus \cdots \oplus \langle z_k\rangle \oplus \langle y_{k+1}\rangle \oplus \cdots \oplus \langle y_m\rangle$$

is a $*$-valuated coproduct in G. Then Proposition 3.6.5 yields a $*$-valuated coproduct

$$B'' = (A + \langle x_1\rangle) \oplus \langle x_2\rangle \oplus \cdots \oplus \langle x_k\rangle \oplus \langle x_{k+1}\rangle \oplus \cdots \oplus \langle x_m\rangle,$$

where x_i is a nonzero multiple of z_i $(i = 2, \dots, k)$, and x_j is a nonzero multiple of y_j $(j = k+1, \dots, m)$. Now let $B' = B'' \oplus \langle y_{m+1}\rangle \oplus \cdots \oplus \langle y_n\rangle$. It is clear that B/B' is finite, so the proof is complete. □

Corollary 3.10.15 (Hill–Megibben [53]). *Suppose that X and Y are finite $*$-decomposition sets in G. Let $C = \langle Y\rangle$. If $X \subseteq C$, then there is a finite $*$-decomposition set Z in G such that $X \subseteq Z \subseteq C$ and $C/\langle Z\rangle$ is finite.*

Proof. We write $X = \{x_1, \dots, x_n\}$. For $n = 1$, the claim follows from Theorem 3.10.14 with $A = 0$. Therefore, let $n > 1$, and assume that the assertion holds for $n - 1$. Then there is a finite $*$-decomposition set Z' such that $\{x_1, \dots, x_{n-1}\} \subseteq Z' \subseteq C$ with $C/\langle Z'\rangle$ finite. Let $A = \langle x_1, \dots, x_{n-1}\rangle$, $B = \langle Z'\rangle$, and write $Z' \setminus X = \{y_1, \dots, y_m\}$. Then $A \oplus \langle x_n\rangle$ and $B = A \oplus \langle y_1\rangle \oplus \cdots \oplus \langle y_m\rangle$ are $*$-valuated coproducts in G such that $x_n' \in B$, where x_n' is some nonzero multiple of x_n. Using Theorem 3.10.14 again, we obtain elements $z_2, \dots, z_m \in B$

such that $Z'' = \{x_1, \ldots, x_{n-1}, x_n', z_2, \ldots, z_m\}$ is a $*$-decomposition set in G with $B/\langle Z''\rangle$ finite. By Proposition 3.6.5, there are nonzero multiples z_i' of z_i ($i = 2, \ldots, m$) such that

$$Z = \{x_1, \ldots, x_{n-1}, x_n, z_2', \ldots, z_m'\}$$

is a $*$-decomposition set in G. Since $C/\langle Z\rangle$ is finite, the assertion follows. □

Exercises.

1. Prove (a) and (b).
2. Prove (c) and (d).
3. In Lemma 3.10.10, the integers $\lambda_2, \ldots, \lambda_n, \mu_2, \ldots, \mu_n$ can be chosen to be positive.

3.11 *k*-Subgroups and *k*-basic subgroups

A subgroup A of a group G is called a ***k*-subgroup**, if for every finite subset S of A there is a finite (possibly empty) $*$-decomposition set X in G such that $X \subseteq A$ and $\langle S, X\rangle/\langle X\rangle$ is finite. If in addition for every primitive element x in G with $x \in A$ there is a subgroup B of A such that $A' = \langle x\rangle \oplus B$ is a $*$-valuated coproduct in G with A/A' finite, then A is called ***k*-basic in** G (see Megibben–Ullery [117], Hill–Megibben [53]). As a corollary to Theorem 3.10.14 and Corollary 3.10.1, we have:

Corollary 3.11.1. *If X is a $*$-decomposition set in a group G, then $\langle X\rangle$ is a k-basic subgroup of G. In particular, every subset of a decomposition basis generates a k-basic subgroup.*

The next proposition contains some basic facts on k-basic subgroups, which can be found in Hill–Megibben [53]:

Proposition 3.11.2. *Suppose that A is a k-subgroup of a group G. Then we have the following:*

1. *Assume that A' is a subgroup of A with A/A' torsion. Then A' is a k-subgroup of G and A' is k-basic in G whenever A is k-basic in G.*
2. *Let X be a finite $*$-decomposition set in G that is contained in A. If S is a finite subset of A, then there is a finite $*$-decomposition set Z in G such that $X \subseteq Z \subseteq A$ and $\langle S, Z\rangle/\langle Z\rangle$ is finite.*
3. *Suppose that $A' = B \oplus \langle x_1\rangle \oplus \cdots \oplus \langle x_n\rangle$ is a $*$-valuated coproduct in G, where each element x_i is primitive in G and $A' \subseteq A$ with A/A' torsion. Then B is a k-subgroup of G, and B is k-basic in G whenever A is k-basic in G.*

Proof. The first part of (1) is obvious. To prove the second part, assume that A is k-basic in G, and let $x \in A'$ be primitive in G. Then there is a $*$-valuated coproduct $C = \langle x\rangle \oplus B$ in G such that $B \subseteq A$ and A/C is finite. Letting $B' = B \cap A'$, the group $C' = \langle x\rangle \oplus B'$ coincides with $(\langle x\rangle \oplus B) \cap A'$; thus, A'/C' is finite. Thus, A' is k-basic in G.

To show (2), we write $X = \{x_1, \ldots, x_n\}$, and let S be a finite subset of A. Then there is a finite $*$-decomposition set Y in G such that $Y \subseteq A$ and $k\langle S \cup X\rangle \subseteq \langle Y\rangle$ for some positive integer k. Since $\langle Y\rangle$ contains the elements $x_i' = kx_i$ ($i = 1, \ldots, n$), Corollary 3.10.15 shows that there is a finite $*$-decomposition set Y' in G containing $X' = \{x_1', \ldots, x_n'\}$ such that $Y' \subseteq \langle Y\rangle$ with $\langle Y\rangle/\langle Y'\rangle$ finite. Write $Y' \setminus X' = \{y_1, \ldots, y_m\}$. Then by Proposition 3.6.5, there are nonzero multiples y_i' of y_i ($i = 1, \ldots, m$) such that $Z = \{x_1, \ldots, x_n, y_1', \ldots, y_m'\}$ is a $*$-decomposition set in G, and we have $\langle S, Z\rangle/\langle Z\rangle$ finite, as desired.

Finally, assume that $A' = B \oplus \langle x_1\rangle \oplus \cdots \oplus \langle x_n\rangle \subseteq A$ is a $*$-valuated coproduct in G as in (3), and let S be a finite subset of B. By (2), there is a finite $*$-decomposition set Z' containing $X = \{x_1, \ldots, x_n\}$ such that $Z' \subseteq A$ and $k'\langle S\rangle \subseteq \langle Z'\rangle$ for some positive integer k'. Write $Z' \setminus X = \{y_1, \ldots, y_m\}$, and let $i \in \{1, \ldots, m\}$. Then there is a nonzero multiple y_i' of y_i such that $y_i' \in A'$, say $y_i' = b_i + w_i$ with $b_i \in B$, $w_i \in \langle X\rangle$, and we obtain $\|y_i'\| = \|b_i\|$ since $A' = B \oplus \langle X\rangle$ and $C = \langle X\rangle \oplus \langle y_1'\rangle \oplus \cdots \oplus \langle y_m'\rangle$ are valuated coproducts in G. Then by Proposition 3.10.6, $C = \langle X\rangle \oplus \langle b_1\rangle \oplus \cdots \oplus \langle b_m\rangle$ is a $*$-valuated coproduct in G with each b_i primitive in G. Letting $Z = \{b_1, \ldots, b_m\}$, we have $k\langle S\rangle \subseteq (\langle X\rangle \oplus \langle Z\rangle) \cap B = \langle Z\rangle$ for some positive integer k. Thus, B is a k-subgroup of G. Finally, assume that A is k-basic in G, and let $b \in B$ be a primitive element in G. Then there is a $*$-valuated coproduct $\langle b\rangle \oplus D \subseteq A$ in G with $A/(\langle b\rangle \oplus D)$ finite. Thus, $B/(\langle b\rangle \oplus D) \cap B = B/(\langle b\rangle \oplus (B \cap D))$ is finite. Consequently, B is k-basic in G. □

The following two results are contained in the proof of [53, Theorem 2.3]. We will need these results to prove Theorem 3.11.7, a necessary step in proving a key result about closure under direct summands.

Lemma 3.11.3. *Let $x_1, \ldots, x_n$ be any elements of a group G and p a prime. Then there is a nonempty subset J of $I = \{1, \ldots, n\}$ and a $\gamma < \omega$ such that $\|p^\gamma x_l\|_p = \|p^\gamma x_r\|_p$ for all $l, r \in J$ and $|p^\alpha x_i|_p \gneqq |p^\alpha x_l|_p$ for infinitely many $\alpha < \omega$ if $l \in J$ and $i \in I \setminus J$.*

Proof. For elements $x, y \in G$, we write:

$$\begin{aligned} x >_f y &\quad \text{if } |p^\alpha x|_p \gneqq |p^\alpha y|_p \text{ for only finitely many } \alpha < \omega; \\ x >_i y &\quad \text{if } |p^\alpha x|_p \gneqq |p^\alpha y|_p \text{ for infinitely many } \alpha < \omega; \\ x \geq y &\quad \text{if } \|p^\beta x\|_p \geq \|p^\beta y\|_p \text{ for some } \beta < \omega. \end{aligned}$$

It suffices to prove that there is an $m \in I$ such that $x_k \geq x_m$ or $x_k >_i x_m$ for all $k \in I$ since we can then take J to be the set of $k \in I$ such that $x_k >_i x_m$ fails. The case $n = 1$ being trivial, let $n > 1$ and assume that the assertion is true for $n - 1$. Then there is an $m' \in I' = \{1, \ldots, n-1\}$ such that $x_k \geq x_{m'}$ or $x_k >_i x_{m'}$ for all $k \in I'$. If $x_n >_i x_{m'}$, let $m = m'$. If $x_n >_f x_{m'}$, then there is a positive integer r such that $|p^\alpha r x_n|_p \gneqq |p^\alpha r x_{m'}|_p$ is not true for any $\alpha < \omega$, which implies that $x_{m'} \geq x_n$. Then for any $k \in I'$, we have $x_k \geq x_{m'} \geq x_n$ or $x_k >_i x_{m'} \geq x_n$. Thus, $m = n$ yields $x_k \geq x_m$ or $x_k >_i x_m$ for all $k \in I$, as desired. □

Lemma 3.11.4. *Let C be a subgroup of G, and let $x \in C$ be primitive in G. Assume that $C = A \oplus B$ and $F = \langle x_1 \rangle \oplus \cdots \oplus \langle x_{m'} \rangle \oplus W$ are $*$-valuated coproducts in G such that $F \subseteq C$ and $W \subseteq G(M^*)$, where $\|x_1\| \sim \cdots \sim \|x_{m'}\|$, $x_1, \ldots, x_{m'}$ are primitive in G, $x_1 = rx$ for some $r \in \mathbb{Z}$ with $M = \|a_0\|$ for some $a_0 \in A$. Suppose that*

(a) $\delta a_0 = x_1 + \cdots + x_m + w_0$ $(\delta \in \mathbb{Z}, m \leq m', w_0 \in W)$;

(b) $x_i = a_i + b_i$ $(a_i \in A, b_i \in B, i = 1, \ldots, m)$ and $ka_i \in F$ $(i = 1, \ldots, m)$ for some positive integer k.

Then there is a primitive element g in G such that $g \in A + W$ and $\|g\| \sim \|x\|$. In fact, if Λ is any finite (possibly empty) set of primes that do not divide r such that $\|x\|_p \leq \|a_0\|_p$ for all $p \in \Lambda$, then after replacing x by a suitable nonzero multiple of itself, the primitive element g satisfies the following condition $(*)$ with respect to Λ and x:

$(*)$ $g = t(\delta a_0 - w_0) + g' = c_1 x_1 + \cdots + c_{m'} x_{m'}$ $(t, c_1, \ldots, c_{m'} \in \mathbb{Z}, c_1 \neq 0, g' \in A + W)$, and for all $p \in \Lambda$, we have $p \nmid t$, $\|g'\|_p \geq \|pg\|_p$, and $\|g\|_p = \|\delta a_0\|_p = \|x\|_p = \|c_1 x_1\|_p$.

If n is a positive integer and g is any element satisfying condition $(*)$ with respect to Λ and x, then ng satisfies condition $(*)$ with respect to Λ and nx.

Proof. Since $C = A \oplus B$ is a $*$-valuated coproduct in G, and $\delta a_0 - (x_1 + \cdots + x_m) = w_0 \in C \cap G(M^*)$, we have $b_1 + \cdots + b_m \in G(M^*)$. By (b), we can write

$$ka_i = \sum_{j=1}^{m'} c_{ij} x_j + w_i \quad (c_{ij} \in \mathbb{Z}, \ w_i \in W)$$

for $i = 1, \ldots, m$. Then $k(b_1 + \cdots + b_m) \in G(M^*) \cap F = W$ by Corollary 3.10.4, and we obtain $k(x_1 + \cdots + x_m) = \sum_{i=1}^{m} (\sum_{j=1}^{m'} c_{ij} x_j + w_i + kb_i) = \sum_{i=1}^{m} \sum_{j=1}^{m'} c_{ij} x_j + w$ for some $w \in W$. Then a comparison of coefficients of x_j yields

$$k = c_{1j} + \cdots + c_{mj}$$

for $j = 1, \ldots, m$. Now let Λ be a finite set of primes not dividing r such that $\|x\|_p \leq \|a_0\|_p$ for all $p \in \Lambda$. Clearly, $g = \delta a_0 - w_0$ satisfies condition $(*)$, but g is not necessarily primitive.

Let P be the set of all primes p such that $\|p^\alpha x_1\|_p = \cdots = \|p^\alpha x_{m'}\|_p$ for some $\alpha < \omega$, and notice that $\mathbb{P} \setminus P$ is a finite set. Then for any element

$$g = t(\delta a_0 - w_0) + g' = c_1 x_1 + \cdots + c_{m'} x_{m'} \quad (t, c_1, \ldots, c_{m'} \in \mathbb{Z}, \ c_1 \neq 0, g' \in A + W)$$

satisfying condition $(*)$, let

$$Q(g) = \{p \in (\mathbb{P} \setminus P) \cup \Lambda : \|g\|_p = \|c_i x_i\|_p \text{ for some } i \leq m'\}.$$

Notice that $Q(g)$ contains the set Λ, and that for any positive integer n,

$$ng = t(\delta n a_0 - n w_0) + ng' = c_1 n x_1 + \cdots + c_{m'} n x_{m'}$$

satisfies condition $(*)$ with respect to Λ and nx since $\|ng'\|_p \geq \|png\|_p$ and $\|ng\|_p = \|\delta na_0\|_p = \|nx\|_p = \|c_1nx_1\|_p$ for all $p \in \Lambda$. Thus, the last statement in the lemma follows. Therefore, we have $Q(g) \subseteq Q(ng)$, and we can replace x by a nonzero multiple if necessary.

If $\mathbb{P} \setminus P \subseteq Q(g)$, then g is primitive (Lemma 3.10.11) as desired. Therefore, assume that $\mathbb{P} \setminus P \nsubseteq Q(g)$. Letting $q \in \mathbb{P} \setminus P$ with $q \notin Q(g)$, we will show that there exists a $\overline{g}$ in G, satisfying condition $(*)$, such that $\{q\} \cup Q(g) \subseteq Q(\overline{g})$. Then finitely many applications of this result will produce an element $g^* \in G$ satisfying condition $(*)$, such that $\mathbb{P} \setminus P \subseteq Q(g^*)$. Hence, g^* is primitive.

Since $q \notin Q(g)$, we have $\|g\|_q \neq \|c_ix_i\|_q$ for $i = 1, \dots, m'$. Therefore, by Lemma 3.11.3, there exists a nonempty subset J of $I = \{1, \dots, m\}$ and a $\gamma < \omega$ such that $\|q^\gamma x_l\|_q = \|q^\gamma x_r\|_q$ for all $l, r \in J$ and $|q^\alpha x_i|_q \ngeq |q^\alpha x_l|_q$ for infinitely many $\alpha < \omega$ whenever $l \in J$ and $i \in I \setminus J$. By replacing x with $q^\gamma x$, we may assume that $\|x_l\|_q = \|x_r\|_q$ for all $l, r \in J$. Let $k = q^j k'$ with $q \nmid k'$. Then we have

$$q^{j+1} | c_{il} \quad \text{whenever } l \in J \text{ and } i \in I \setminus J$$

since $q^{j+1} \nmid c_{il}$ yields $|q^{\alpha+j}x_l|_q \geq |q^\alpha c_{il}x_l|_q \geq |q^\alpha ka_i|_q = |q^{\alpha+j}a_i|_q \geq |q^{\alpha+j}x_i|_q$ for all $\alpha < \omega$, which is impossible. For all $l, r \in J$, we have

$$\|ka_l\|_q \leq \|c_{lr}x_r\|_q = \|c_{lr}x_l\|_q \leq \|c_{lr}a_l\|_q;$$

hence, $q^j | c_{lr}$. Now fix an $r \in J$. Since $k = \sum_{i=1}^m c_{ir}$ and $q^{j+1} | c_{ir}$ for all $i \in I \setminus J$, there exists an $l \in J$ such that $q^{j+1} \nmid c_{lr}$. Then $\|q^j x_l\|_q = \|kx_l\|_q \leq \|ka_l\|_q \leq \|ka_l - w_l\|_q \leq \|c_{lr}x_r\|_q = \|q^j x_r\|_q = \|q^j x_l\|_q$. Thus,

$$\|kx_l\|_q = \|ka_l\|_q = \|ka_l - w_l\|_q = \|c_{lr}x_r\|_q.$$

Since $c_{lr} \neq 0$, we have $\|ka_l - w_l\| \sim \|x\| \sim \|g\|$. Therefore, there are positive integers c and β with $q \nmid c$ such that

$$\|q^\beta g\|_q \geq \|q(ka_l - w_l)\|_q \quad \text{and} \quad \|c(ka_l - w_l)\|_p \geq \|pg\|_p \text{ for all } p \in Q(g).$$

Let $\overline{g} = q^\beta g + c(ka_l - w_l) = \overline{c}_1x_1 + \cdots + \overline{c}_{m'}x_{m'}$ with $\overline{c_i} = q^\beta c_i + cc_{li}$ $(i = 1, \dots, m')$. It is clear that β can be replaced by a larger integer, so since $c_1 \neq 0$, we can assume that $\overline{c}_1 \neq 0$ and have $\|\overline{g}\| \sim \|x\|$. Since $\|q^\beta g\|_q > \|c(ka_l - w_l)\|_q$, $\|q^\beta c_r x_r\|_q \geq \|q^\beta g\|_q$, and $\|cc_{lr}x_r\|_q = \|ka_l - w_l\|_q < \|q(ka_l - w_l)\|_q \leq \|q^\beta g\|_q$, we obtain

$$\|\overline{g}\|_q = \|ka_l - w_l\|_q = \|c_{lr}x_r\|_q = \|q^\beta c_r x_r\|_q \wedge \|cc_{lr}x_r\|_q = \|\overline{c}_r x_r\|_q.$$

To verify that $\overline{g}$ satisfies condition $(*)$, we write

$$\overline{g} = q^\beta t(\delta a_0 - w_0) + \overline{g}' = \overline{c}_1x_1 + \cdots + \overline{c}_{m'}x_{m'}$$

with $\overline{g}' = q^\beta g' + c(ka_l - w_l) \in A + W$, and assume that $p \in \Lambda$. Then $p \nmid q^\beta t$, and

$$\|\overline{g}'\|_p \geq \|q^\beta g'\|_p \wedge \|c(ka_l - w_l)\|_p \geq \|pg\|_p = \|pq^\beta g\|_p \wedge \|pc(ka_l - w_l)\|_p = \|p\overline{g}\|_p.$$

By assumption, we have $\|g\|_p = \|\delta a_0\|_p = \|x\|_p = \|c_1 x_1\|_p$; hence,

$$\|\overline{g}\|_p = \|q^\beta g\|_p \wedge \|c(ka_l - w_l)\|_p = \|g\|_p$$

and $\|cc_{l1}x_1\|_p \geq \|\sum_{i=1}^{m'} cc_{li}x_i\|_p = \|c(ka_l - w_l)\|_p \geq \|pg\|_p$ yield

$$\|\overline{g}\|_p = \|\delta a_0\|_p = \|x\|_p = \|c_1 x_1\|_p = \|q^\beta c_1 x_1\|_p \wedge \|cc_{l1}x_1\|_p = \|\overline{c}_1 x_1\|_p.$$

Consequently, $\overline{g}$ satisfies condition $(*)$. It remains to show that $\{q\} \cup Q(g) \subseteq Q(\overline{g})$. Clearly, $q \in Q(\overline{g})$ since $\|\overline{g}\|_q = \|\overline{c}_r x_r\|_q$. If $p \in Q(g)$, then $\|g\|_p = \|c_i x_i\|_p$ for some $i \leq m'$. Therefore, since $\|cc_{li}x_i\|_p \geq \|c(ka_l - w_l)\|_p \geq \|pg\|_p$, we have

$$\|\overline{g}\|_p = \|g\|_p = \|c_i x_i\|_p = \|c_i x_i\|_p \wedge \|cc_{li}x_i\|_p = \|\overline{c}_i x_i\|_p.$$

Therefore, $p \in Q(\overline{g})$ as required. □

Corollary 3.11.5. *Let C be a k-basic subgroup of G such that $C = A \oplus B$ is a $*$-valuated coproduct in G, and let $a_0 \in A$ with $M = \|a_0\|$. Assume that*

$$\delta a_0 = x_1 + \cdots + x_m + w_0$$

for some positive integer δ, where $\langle x_1\rangle \oplus \cdots \oplus \langle x_m\rangle \oplus \langle w_0\rangle \subseteq C$ is a $*$-valuated coproduct in G with $\|x_1\| \sim \cdots \sim \|x_m\| \sim M$, $x_1, \ldots, x_m$ primitive in G, and $w_0 \in G(M^*)$. Then there is a primitive element g in G such that $\|g\| \sim M$ and $g \in (A + G(M^*)) \cap C$. If g is any such element in G, then $g = a + b'$ ($a \in A, b' \in B \cap G(M^*)$) and a is primitive in G with $\|na\| = \|ng\|$ for some positive integer n.

Proof. For $i \in \{1, \ldots, m\}$, we write $x_i = a_i + b_i$ ($a_i \in A$, $b_i \in B$), and have $\|a_i\| \geq \|x_i\| \geq \|\delta a_0\| \geq M$. By Proposition 3.11.2(2), there is a positive integer k such that $k\langle a_1, \ldots, a_m\rangle \subseteq \langle Z\rangle$, where $Z = X \dot\cup Y \subseteq C$ is a finite $*$-decomposition set in G satisfying $\{x_1, \ldots, x_m\} \subseteq X$, $\|x\| \sim M$ for all $x \in X$ and $Y \subseteq G(M^*)$. Using Proposition 3.11.2(2) again, we obtain a finite $*$-decomposition set Z' in G with $Z \subseteq Z' \subseteq C$ and containing a set $Y' = \{y_1, \ldots, y_s\}$ with $lw_0 = l_1 y_1 + \cdots + l_s y_s$ for some nonzero integers $l, l_1, \ldots, l_s$. Let $i \in \{1, \ldots, s\}$. Since $lw_0 \in G(M^*)$, we have $l_i y_i \in G(M^*)$. Thus, $\|y_i\| \nsim M$ since y_i is primitive (Proposition 3.10.3(1)). Then $W = \langle Y \cup Y'\rangle$ is contained in $G(M^*)$, and we obtain the $*$-valuated coproduct $F = \langle X\rangle \oplus W$ with $F \subseteq C$, $lw_0 \in W$,

$$\delta l a_0 = lx_1 + \cdots + lx_m + lw_0 \in F,$$

and $k\langle la_1, \ldots, la_m\rangle \subseteq F$. By Lemma 3.11.4 applied to $lx_1, \ldots, lx_m$, there is a primitive element $g \in A + W \subseteq (A + G(M^*)) \cap C$ with $\|g\| \sim \|lx_1\| \sim M$. This proves such a g exists.

Now let g be such an element of G, and write $g = a+b'$ with $a \in A$ and $b' \in B$. Then b' is an element of $A + G(M^*)$ since g is, so we have $b' = a' + g'$ for some $a' \in A$ and $g' \in G(M^*)$; hence, $b' \in G(M^*)$. Since C is k-basic, there is a finite $*$-decomposition set X' in G such that $X' \subseteq C$ and $rb' \in \langle X' \rangle$ for some positive integer r. If $rb' = 0$, then $rg = ra$. Hence, a is primitive in G, and the assertion follows. Assuming that b' has infinite order, we obtain $rb' = r_1z_1+\dots+r_tz_t$ for some $z_1,\dots,z_t \in X'$ and nonzero integers $r_1,\dots,r_t$. Letting $z \in \{r_1z_1,\dots,r_tz_t\}$, we have $z \in G(M^*)$. Hence, $\|z\| \nsim M$ since z is primitive. Then $\|z\| \geq \|rb'\| \geq \|rg\|$ and $\|z\| \nsim \|g\|$ yield $z \in G(\|rg\|^*)$; thus, $rb' \in G(\|rg\|^*)$. By Proposition 3.10.5, $ra \in rg + G(\|rg\|^*)$ is primitive in G and $\|na\| = \|ng\|$ for some positive integer n. □

Corollary 3.11.6 (Hill–Megibben [53]). *Let $A \oplus B$ be a $*$-valuated coproduct in a group G such that $C = A \oplus B$ is a k-basic subgroup of G, and A has finite torsion-free rank. Then there is a finite $*$-decomposition set X in G such that $X \subseteq A$ and $A/\langle X \rangle$ is torsion.*

Proof. We show the assertion by induction on $n = r_0(A)$. For $n = 0$, simply let $X = \emptyset$. Now suppose that $n > 0$, and that the claim holds for all $k < n$. Since C is a k-subgroup of G, Lemma 3.5.1 shows that every increasing sequence of compatibility classes of Ulm matrices realized by nonzero elements of A of infinite order becomes stationary. Thus, there is an Ulm matrix M and a nonzero element $a_0 \in A$ of infinite order such that $\|a_0\| = M$ and $A \cap G(M^*)$ is torsion; hence, $a_0 \notin G(M^*)$. Since C is k-basic in G, we have

$$\delta a_0 = x_1 + \dots + x_m + w_0$$

for some positive integer δ, where $\langle x_1 \rangle \oplus \dots \oplus \langle x_m \rangle \oplus \langle w_0 \rangle \subseteq C$ is a $*$-valuated coproduct in G, each element x_i is primitive in G with $\|x_i\| \sim M$ and $w_0 \in G(M^*)$. By Corollary 3.11.5, there is a primitive element g in G such that $g \in C$ and if $g = a + b'$ $(a \in A, b' \in B)$, then a is primitive in G. Since C is k-basic in G, there is a $*$-valuated coproduct

$$C_1 = \langle a \rangle \oplus D_1$$

in G with C/C_1 finite. Let $A' = A \cap D_1$. Then $C' = \langle a \rangle \oplus A' \oplus B$ is a $*$-valuated coproduct in G with C/C' finite, and by Proposition 3.11.2(1), C' is k-basic in G. We have $r_0(A') < r_0(A)$. Therefore, by the induction hypothesis, there is a finite $*$-decomposition set X' in G such that $X' \subseteq A'$ with $A/\langle X' \rangle$ torsion. Then $X = \{a\} \cup X'$ is a $*$-decomposition set in G, and $A/\langle X \rangle$ is torsion, as desired. □

Theorem 3.11.7 (Hill–Megibben [53]). *Let $A \oplus B$ be a $*$-valuated coproduct in a group G such that $C = A \oplus B$ is a k-basic subgroup of G. Suppose that $x \in C$ is a primitive element in G. Then there is a $*$-valuated coproduct $C' = \langle x' \rangle \oplus \langle y \rangle \oplus A' \oplus B'$ in G with $A' \subseteq A$, $B' \subseteq B$ and $C' \subseteq C$ such that*

1. *$r_0(A/A') \leq 1$, $r_0(B/B') \leq 1$, and C/C' is finite;*
2. *x' is a nonzero multiple of x, and either $y = 0$ or y is primitive in G with $\|x\| \sim \|y\|$.*

Proof. We write $x = a_0 + b_0$ with $a_0 \in A$ and $b_0 \in B$, and let $M = \|a_0\|$. By Proposition 3.11.2(2), there is a positive integer δ and an $r \in \mathbb{Z}$ such that

$$\delta a_0 = rx + x_2 + \cdots + x_m + w_0,$$

where $\langle x\rangle \oplus \langle x_2\rangle \oplus \cdots \oplus \langle x_m\rangle \oplus \langle w_0\rangle \subseteq C$ is a $*$-valuated coproduct in G with each x_i primitive, $\|x_i\| \sim \|x\|$, and $w_0 \in G(M^*)$. Letting $s = \delta - r$, we have

$$\delta b_0 = sx - (x_2 + \cdots + x_m + w_0).$$

Case I: $s = 0$. Then $\delta = r$; therefore, we have $\delta a_0 = \delta x + x_2 + \cdots + x_m + w_0$. Now $\|\delta a_0\| \leq \|\delta x\|$ and $\|x\| \leq \|a_0\|$ yield $\|\delta a_0\| = \|\delta x\|$. Hence, Proposition 3.10.6 shows that δa_0 is primitive in G. Since C is k-basic in G, there is a $*$-valuated coproduct $C_1 = \langle \delta a_0\rangle \oplus C_2$ in G with $C_1 \subseteq C$ and C/C_1 finite. Let $A' = A \cap C_2$. Then $A \cap C_1 = \langle \delta a_0\rangle \oplus A' \subseteq \langle \delta a_0\rangle \oplus C_2$ is a $*$-valuated coproduct in G; thus, $C' = \langle \delta a_0\rangle \oplus A' \oplus B$ is a $*$-valuated coproduct in G. Since $A/(A \cap C_1) \cong (A + C_1)/C_1 \subseteq C/C_1$ is finite, $C/C' = (A \oplus B)/(\langle \delta a_0\rangle \oplus A' \oplus B)$ is finite, and we have $r_0(A/A') = r_0(C/(A' \oplus B)) = r_0(C'/(A' \oplus B)) + r_0(C/C') = 1$. We use Proposition 3.10.6 again and obtain the $*$-valuated coproduct $C' = \langle \delta x\rangle \oplus A' \oplus B$. Letting $y = 0$ and $B' = B$, the assertion follows.

Case II: $r = 0$. Then $\delta = s$; hence, $\delta b_0 = \delta x - (x_2 + \cdots + x_m + w_0)$, and we have $\|\delta b_0\| = \|\delta x\|$. Using a similar argument as above, we obtain a subgroup B' of B with $r_0(B/B') = 1$ such that $C' = \langle \delta x\rangle \oplus A \oplus B'$ is a $*$-valuated coproduct in G with C/C' finite, and we let $y = 0$ and $A' = A$.

Case III: $r \neq 0$ and $s \neq 0$. Then $\|x\| \leq M \leq \|rx\|$ shows that $\|x\| \sim M$. If $M \sim \overline{\infty}$, we can proceed as in Case I since a_0 is primitive by Section 3.10(b) and $\|nx\| = \|na_0\| = \overline{\infty}$ for some positive integer n. Now suppose that $M \nsim \overline{\infty}$. By replacing x with a nonzero multiple, we may assume that $\gcd(\delta, r) = 1$, and therefore, $\gcd(r, s) = 1$. Let $x_1 = rx$, and write $x_i = a_i + b_i$ ($a_i \in A$, $b_i \in B$) for $i = 1, \ldots, m$. Then $\|a_i\| \geq \|x_i\| \geq M$ for all $i = 1, \ldots, m$. By Proposition 3.11.2(2), there is a positive integer k such that $k\langle a_1, \ldots, a_m\rangle \subseteq \langle Z\rangle$, where $Z = X \dot{\cup} Y \subseteq C$ is a finite $*$-decomposition set in G with $X = \{x_1, \ldots, x_{m'}\}$ for some $m' \geq m$, $\|x_i\| \sim M$ for all $i = 1, \ldots, m'$, and $Y \subseteq G(M^*)$. Using Proposition 3.11.2(2) again, we can find a $*$-valuated coproduct $F = \langle x_1\rangle \oplus \cdots \oplus \langle x_{m'}\rangle \oplus W$ in G such that $F \subseteq C$, $Y \subseteq W \subseteq G(M^*)$, and $lw_0 \in W$ for some positive integer l (see the first part of the proof of Corollary 3.11.5). Then

$$\delta l a_0 = lx_1 + \cdots + lx_m + lw_0 \in F,$$

and by replacing x with lx, we may assume that $l = 1$. Now let Λ be any finite set of primes not dividing r. Then by Lemma 3.11.4, after replacing x by a suitable nonzero multiple, there is a primitive element

$$g = t(\delta a_0 - w_0) + g'$$

with $g' \in A + W$ and $\|g\| \sim \|x\|$ such that $p \nmid t$, $\|g\|_p = \|x\|_p$, and $\|g'\|_p \geq \|pg\|_p$ for all $p \in \Lambda$. Since C is k-basic in G, we have a $*$-valued coproduct $C_1 = \langle g \rangle \oplus D_1$ in G with $C_1 \subseteq C$ and C/C_1 finite, so there are integers $d \neq 0$ and λ such that

$$d\delta a_0 = \lambda g + u \quad (u \in D_1),$$

and $dw_0 \in C_1$. Replacing x by a nonzero multiple, we can assume that $\gcd(d, \lambda) = 1$ (see the last statement in Lemma 3.11.4), and have $t\lambda g + dg' = t(d\delta a_0 - u) + dg - dt(\delta a_0 - w_0) = dg + t(dw_0 - u)$. If $dw_0 = d'g + y$ for some nonzero integer d' and $y \in D_1$, then $d'g \in G(M^*)$, which is impossible since $d'g$ is primitive (Proposition 3.10.3(1)). Thus, $dw_0 \in D_1$. Then for any $p \in \Lambda$ with $p|\lambda$, we obtain $\|x\|_p = \|g\|_p = \|dg\|_p \geq \|dg\|_p \wedge \|t(dw_0 - u)\|_p$. Thus,

$$\|x\|_p \geq \|t\lambda g + dg'\|_p \geq \|t\lambda g\|_p \wedge \|dg'\|_p \geq \|pg\|_p = \|px\|_p,$$

which shows that $|x|_p = \infty$. Consequently, we have

$$p \nmid \lambda \text{ and } \|x\|_p = \|g\|_p \quad \text{whenever } p \in \Lambda \text{ and } |x|_p \neq \infty.$$

Likewise, for any finite set Λ' of primes not dividing s, there exists a primitive element

$$h = t'(\delta b_0 - w_0) + h'$$

with $h' \in B + W'$ for some $W' \subseteq G(\|b_0\|^*)$ and $\|h\| \sim \|x\|$ such that $p \nmid t'$, $\|h\|_p = \|x\|_p$, and $\|h'\|_p \geq \|ph\|_p$ for all $p \in \Lambda'$. Then there is a $*$-valuated coproduct $C_2 = \langle h \rangle \oplus D_2$ in G with $C_2 \subseteq C$ and C/C_2 finite. Therefore, there are integers $e \neq 0$ and μ such that

$$e\delta b_0 = \mu h + v \quad (v \in D_2),$$

and after replacing x with a nonzero multiple if necessary, we have

$$p \nmid \mu \text{ and } \|x\|_p = \|h\|_p \quad \text{whenever } p \in \Lambda' \text{ and } |x|_p \neq \infty.$$

Now let $\Lambda = \{p \in \mathbb{P} : p \mid s\}$, and

$$\Lambda' = \{p \in \mathbb{P} : p \mid \lambda \text{ or } \|g\|_p \neq \|x\|_p\} \cap \{p \in \mathbb{P} : |x|_p \neq \infty\}.$$

Notice that the primes in Λ do not divide r. Clearly, Λ and Λ' are disjoint; thus, the primes in Λ' do not divide s. If $\lambda \neq 0$, then Λ' is finite since $\|g\| \sim \|x\|$. If $\lambda = 0$, we write $\mathbb{P} = \{p_i : i < \omega\}$ and, if possible, replace Λ by $\Lambda_n = \Lambda \cup \{p_i : p_i \nmid r \text{ and } i \leq n\}$ for sufficiently large $n < \omega$ such that the resulting primitive element g_n satisfies

$$d_n\delta a_0 = \lambda_n g_n + u_n \quad (d_n \in \mathbb{Z} \setminus \{0\}, u_n \in D_1)$$

with $\lambda_n \neq 0$. In this case, we replace λ by λ_n and g by g_n to obtain a finite Λ'. If $\lambda_n = 0$ for all $n < \omega$, we have $|x|_p = \infty$ for all $p \in \Lambda_n$ $(n < \omega)$, that is, for all primes p with $p \nmid r$. But then Λ need not be replaced since Λ' is finite.

Next we will show that

$$\|de\delta x\|_p = \begin{cases} \|e\lambda g\|_p & \text{if } p \notin \Lambda', \\ \|d\mu h\|_p & \text{if } p \in \Lambda'. \end{cases}$$

Since $\|de\delta x\| = \|de\delta a_0\| \wedge \|de\delta b_0\| \le \|e\lambda g\| \wedge \|d\mu h\|$, the claim follows for primes p with $|x|_p = \infty$. Suppose that $|x|_p \neq \infty$. If $p \in \Lambda'$, then $p \nmid \mu$ and $\|x\|_p = \|h\|_p$; thus, $\|de\delta x\|_p \le \|d\mu h\|_p = \|dx\|_p \le \|de\delta x\|_p$. If $p \notin \Lambda'$, then $p \nmid \lambda$ and $\|g\|_p = \|x\|_p$, which implies that $\|de\delta x\|_p \le \|e\lambda g\|_p = \|eg\|_p \le \|de\delta x\|_p$. Therefore, the claim holds in both cases.

Now we write $g = a + b'$ for $a \in A$ and $b' \in B$. By Corollary 3.11.5, a is primitive in G, $b' \in G(M^*)$, and $\|na\| = \|ng\|$ for some positive integer n. By replacing x with a nonzero multiple and using the last statement in Lemma 3.11.4, we may assume that $\|a\| = \|g\|$. Similarly, if $h = a' + b$ ($a' \in A$, $b \in B$), then b is primitive in G, and we can assume that $\|b\| = \|h\|$. By replacing x with a nonzero multiple, we can assume that $b' \in C_1$. Since $C_1 = \langle g\rangle \oplus D_1$ is a $*$-valuated coproduct, and g is primitive with $\|g\| \sim M$, we have $b' \in D_1$. Likewise, we can assume that $a' \in D_2$. Since $a = g - b'$ with $\|a\| = \|g\|$ and $b = h - a'$ with $\|b\| = \|h\|$, we obtain $*$-valuated coproducts $C_1 = \langle a\rangle \oplus D_1$ and $C_2 = \langle b\rangle \oplus D_2$ by Proposition 3.10.6. Now let $A' = A \cap D_1$ and $B' = B \cap D_2$. Then

$$C_0 = \langle a\rangle \oplus \langle b\rangle \oplus A' \oplus B'$$

is a $*$-valuated coproduct in G, and since the groups $C/C_1 = (A \oplus B)/(\langle a\rangle \oplus D_1)$ and $C/C_2 = (A \oplus B)/(\langle b\rangle \oplus D_2)$ are finite, $C/C_0 \cong A/(\langle a\rangle \oplus A') \oplus B/(\langle b\rangle \oplus B')$ is finite. Further, we have $r_0(A/A') = r_0((\langle a\rangle \oplus A')/A') + r_0(A/(\langle a\rangle \oplus A')) = 1$, and similarly, $r_0(B/B') = 1$.

Since $d\delta a_0 = \lambda g + u = \lambda(a + b') + u$, we can write $d\delta a_0 = \lambda a + u'$ with $u' = \lambda b' + u \in D_1 \cap A = A'$. Likewise, we have $e\delta b_0 = \mu b + v'$ with $v' = \mu a' + v \in B'$. Then

$$de\delta x = e(d\delta a_0) + d(e\delta b_0) = e\lambda a + d\mu b + (eu' + dv').$$

Thus, for any prime p, we have $\|e\lambda a + d\mu b\|_p \ge \|de\delta x\|_p$. If $p \notin \Lambda'$, then $\|de\delta x\|_p = \|e\lambda g\|_p = \|e\lambda a\|_p$, and if $p \in \Lambda'$, we have $\|de\delta x\|_p = \|d\mu h\|_p = \|d\mu b\|_p$. In any case, we obtain $\|e\lambda a + d\mu b\|_p \ge \|de\delta x\|_p \ge \|e\lambda a + d\mu b\|_p$. Thus,

$$\|de\delta x\| = \|e\lambda a + d\mu b\|.$$

Letting $z = e\lambda a + d\mu b$, we have $\|z\| = \|de\delta x\|$; thus, $\|z\|_p = \|e\lambda a\|_p$ whenever $p \notin \Lambda'$, and $\|z\|_p = \|d\mu b\|_p$ if $p \in \Lambda'$. Notice that $\|x\| \sim M \nsim \overline{\infty}$ implies that $|x|_p \neq \infty$ for some prime p. In the case where $\lambda = 0$, we have $p \in \Lambda'$; thus $p \nmid \mu$, which shows that $\mu \neq 0$. Consequently, $z \neq 0$ in any case. Then Lemma 3.10.11 shows that z is primitive in G since $\|a\| \sim \|b\|$. By Theorem 3.10.14, there is a primitive element y in G such that $y \in \langle a\rangle \oplus \langle b\rangle$ and $\langle z\rangle \oplus \langle y\rangle$ is a $*$-valuated coproduct in G with $(\langle a\rangle \oplus \langle b\rangle)/(\langle z\rangle \oplus \langle y\rangle)$ finite. Then $\|y\| \sim \|a\| \sim \|x\|$ and $C' = \langle z\rangle \oplus \langle y\rangle \oplus A' \oplus B'$ is a $*$-valuated coproduct in G with C/C' finite. Letting $x' = de\delta x$, we have $\|x'\| = \|z\|$ and $x' = z + (eu' + dv')$; hence, $C' = \langle x'\rangle \oplus \langle y\rangle \oplus A' \oplus B'$ is a $*$-valuated coproduct in G by Proposition 3.10.6. This completes the proof. □

The next result follows from the proof of [53, Corollary 2.5]:

Corollary 3.11.8. *Let $A \oplus B$ be a $*$-valuated coproduct in a group G, and $X = \{x_1, x_2, \dots\}$ a countable $*$-decomposition set in G. Let $C = \langle X \rangle$, $A_0 = A \cap C$, $B_0 = B \cap C$, and $C_0 = A_0 \oplus B_0$. If C/C_0 is torsion, then for every positive integer n, there exists a nonzero multiple x_n' of x_n and subgroups A_n, B_n, and C_n of C such that*
1. $C_n = \langle x_1' \rangle \oplus \dots \oplus \langle x_n' \rangle \oplus A_n \oplus B_n$ *is a $*$-valuated coproduct in G;*
2. $C_n \subseteq C_{n-1}$ *and C_0/C_n is finite;*
3. $A_n \subseteq A_{n-1}$ *and A_0/A_n has torsion-free rank $\le n$.*

Proof. By Corollary 3.11.1 and Proposition 3.11.2(1), $C_0 = A_0 \oplus B_0$ is k-basic in G. We will show the assertion by induction, so first assume that $n = 1$. Then Theorem 3.11.7 yields a $*$-valuated coproduct $C_1 = \langle x_1' \rangle \oplus A_1 \oplus B_1$ in G (with $B_1 = \langle y \rangle \oplus B'$ as in Theorem 3.11.7) such that x_1' is a nonzero multiple of x_1, $C_1 \subseteq C_0$ with C_0/C_1 finite, and $A_1 \subseteq A_0$ with $r_0(A_0/A_1) \le 1$.

Now let n be an integer > 1. Assuming that the claim holds for all $k < n$, there is a $*$-valuated coproduct $C_{n-1} = \langle x_1' \rangle \oplus \dots \oplus \langle x_{n-1}' \rangle \oplus A_{n-1} \oplus B_{n-1}$ in G with nonzero multiples x_i' of x_i ($i = 1, \dots, n-1$) such that C_0/C_{n-1} is finite and $r_0(A_0/A_{n-1}) \le n-1$. By Proposition 3.11.2(3), $D = A_{n-1} \oplus B_{n-1}$ is k-basic in G. Since C_{n-1} contains a nonzero multiple $\tilde{x}_n$ of x_n, we can write $\tilde{x}_n = y + z$, where $y \in \langle x_1' \rangle \oplus \dots \oplus \langle x_{n-1}' \rangle$ and $z \in D$. Hence, $\|\tilde{x}_n\| \le \|y\|$. Then $\|z\| = \|\tilde{x}_n\| \wedge \|y\| = \|\tilde{x}_n\|$. Thus, z is primitive in G by Proposition 3.10.6. Using Theorem 3.11.7, we obtain a positive integer m and a $*$-valuated coproduct $D' = \langle mz \rangle \oplus A_n \oplus B_n$ in G such that $A_n \subseteq A_{n-1}$ with $r_0(A_{n-1}/A_n) \le 1$, and $D' \subseteq D$ with D/D' finite. Now let $x_n' = m\tilde{x}_n$. Then $x_n' = my + mz \in \langle x_1' \rangle \oplus \dots \oplus \langle x_{n-1}' \rangle \oplus A_n \oplus B_n \oplus \langle mz \rangle$, and we have $\|x_n'\| = \|mz\|$. Thus Proposition 3.10.6 shows that

$$C_n = \langle x_1' \rangle \oplus \dots \oplus \langle x_{n-1}' \rangle \oplus \langle x_n' \rangle \oplus A_n \oplus B_n$$

is a $*$-valuated coproduct in G. Notice that $C_n = \langle x_1' \rangle \oplus \dots \oplus \langle x_{n-1}' \rangle \oplus D'$. Hence, $C_n \subseteq C_{n-1}$ and C_{n-1}/C_n is finite, which implies that C_0/C_n is finite. Finally, we have $r_0(A_0/A_n) = r_0(A_{n-1}/A_n) + r_0(A_0/A_{n-1}) \le n$, as desired. This completes the induction. □

Exercises.
1. (Megibben–Ullery [117]) The set of all k-subgroups of a group G is closed under ascending unions.
2. (Megibben–Ullery [117]) Let A be a subgroup of G such that $|A| \le \kappa$. If C is a k-subgroup of G containing A, then there is a k-subgroup B of G such that $A \subseteq B \subseteq C$ and $|B| \le \kappa$.

3.12 Knice subgroups and k-groups

Let G be a group. Then a nice subgroup A of G is called **knice** if for every finite subset S of G there is a finite (possibly empty) set of primitive elements $x_1, \dots, x_n \in G$ such that $B = A \oplus \langle x_1 \rangle \oplus \cdots \oplus \langle x_n \rangle$ is a $*$-valuated coproduct in G with $\langle S, B \rangle / B$ finite. If 0 is a knice subgroup of G, then G is called a **k-group**. Clearly, we have:

(a) A nice subgroup A of G is knice exactly if for every finite subset S of G there is a finite $*$-decomposition set X in G such that $A \oplus \langle X \rangle$ is a $*$-valuated coproduct in G containing a nonzero multiple of $\langle S \rangle$.

(b) If A_i are knice subgroups of groups G_i $(i \in I)$, then $\bigoplus_{i \in I} A_i$ is knice in $\bigoplus_{i \in I} G_i$.

Theorem 3.12.1 (Hill–Megibben [50]). *Finite extensions of knice subgroups are knice.*

Proof. Suppose that A and B are subgroups of G such that $A \subseteq B$, B/A is finite, and A is knice in G. Then B is nice in G by Theorem 3.8.10. If S is a finite subset of G, then there is a $*$-valuated coproduct $A \oplus \langle x_1 \rangle \oplus \cdots \oplus \langle x_n \rangle$ in G containing a nonzero multiple of $\langle S \rangle$ such that $x_1, \ldots, x_n$ are primitive in G. By Proposition 3.6.5, there are nonzero multiples x_i' of x_i $(i = 1, \ldots, n)$ such that $C = B \oplus \langle x_1' \rangle \oplus \cdots \oplus \langle x_n' \rangle$ is a $*$-valuated coproduct in G. Since $x_1', \ldots, x_n'$ are primitive in G, and $\langle S, C \rangle / C$ is finite, it follows that B is knice in G. □

Proposition 3.12.2 (Hill–Megibben [50]). *Suppose that A and B are subgroups of G such that $A \subseteq B$. If A is knice in G and B/A is nice in G/A, then B is nice in G.*

Proof. Let p be a prime, α an ordinal, and $x + B \in p^\alpha(G/B)$. Assuming that $px + B \in (p^\alpha G + B)/B$, we have $x + b + A \in p^\alpha(G/A)$ for some $b \in B$ by Proposition 3.8.9(1). Since A is knice in G, there is a $c \in G$ such that $A \oplus \langle c \rangle$ is a valuated coproduct in G and $n(x + b) = a + c$ for some positive integer n and $a \in A$. Then by Proposition 3.8.15, we have

$$|c|_p^G = |c + A|_p^{G/A} \geq \alpha;$$

hence, $n(x+b) = a+c \in p^\alpha G + A$. We write $n = p^r n'$, where $p \nmid n'$. Since $x+b+A \in p^\alpha(G/A)$ and A is nice in G, this yields $n'(x + b) \in p^\alpha G + A$; thus, $n'x \in p^\alpha G + B$. Then there are integers k and l such that $kn' + lp = 1$; therefore, $x = kn'x + lpx \in p^\alpha G + B$. Consequently, B is nice in G. □

Theorem 3.12.3 (Hill–Megibben [50]). *Suppose that $B = A \oplus \langle x_1 \rangle \oplus \cdots \oplus \langle x_n \rangle$ is a $*$-valuated coproduct in a group G such that A is knice in G, and each element x_i is primitive. Then B is knice in G.*

Proof. By induction, it suffices to show the assertion for $n = 1$. Let S be a finite subset of G. Then there are primitive elements $y_1, \ldots, y_m$ in G such that $C = A \oplus \langle y_1 \rangle \oplus \cdots \oplus \langle y_m \rangle$ is a $*$-valuated coproduct in G containing $k\langle S, x_1 \rangle$ for some positive integer k. Since $x_1' = kx_1 \in C$, Theorem 3.10.14 shows that there is a $*$-valuated coproduct $C' = A \oplus \langle x_1' \rangle \oplus \langle x_2 \rangle \oplus \cdots \oplus \langle x_m \rangle$ in G with $x_2, \ldots, x_m$ primitive in G, $C' \subseteq C$, and C/C' finite. Now $B/(A \oplus \langle x_1' \rangle)$ is finite; therefore, by Proposition 3.6.5 there are nonzero multiples x_i' of x_i $(i = 2, \ldots, m)$ such that $B \oplus \langle x_2' \rangle \oplus \cdots \oplus \langle x_m' \rangle$ is a $*$-valuated coproduct in G. Clearly, $\langle S, B \rangle / B$ is finite. Since B is nice in G (Theorem 3.8.12), it follows that B is knice in G. □

Corollary 3.12.4 (Hill–Megibben [50]). *Let A be a knice subgroup of G. If S is a finite subset of G, then there is a knice subgroup C of G containing A and S such that C/A is finitely generated. In particular, if G is a k-group, then every finite subset is contained in some finitely generated knice subgroup of G.*

Proof. By the definition of knice and Theorem 3.12.3, there is a knice subgroup B of G containing A such that B/A is finitely generated and $\langle S, B\rangle/B$ is finite. Now let $C = \langle S, B\rangle$, and apply Theorem 3.12.1. □

The next result follows from the proof of [53, Proposition 1.7]:

Lemma 3.12.5. *Suppose that A is quasi-sequentially nice in a group G, M is an Ulm matrix and p a prime. Then we have:*
1. *If $x + A \in (G/A)(M^*, p)$, then $nx \in G((nM)^*, p) + A$ for some positive integer n.*
2. *If $x + A \in (G/A)(M^*)$, then $nx \in G((nM)^*) + A$ for some positive integer n.*
3. *If $x + A \in (G/A)(\overline{\alpha}^*, p)$, then $nx \in G((p^r\overline{\alpha})^*, p) + A$ for some positive integer $n = p^r n'$ with $p \nmid n'$.*

Proof. To prove (1), suppose that $x + A \in (G/A)(M^*, p)$. Then $x + A \in (G/A)(M)$, and we can write $x + A = z_1 + \cdots + z_m + A$, where $z_i + A \in (G/A)(M^*, p)^\diamond$ for all $i = 1, \ldots, m$. Then there is a positive integer n and $a, a_1, \ldots, a_m \in A$ such that $\|nx + a\|_G = \|nx + A\|_{G/A}$, and for $i = 1, \ldots, m$, $\|nz_i + a_i\|_G = \|nz_i + A\|_{G/A}$.

Let $x' = nx + a$ and $x_i = nz_i + a_i$. Notice that

$$\|x'\|_G = \|x' + A\|_{G/A} = \|nx + A\|_{G/A} \geq nM$$

by Proposition 3.8.15. Let $i \in \{1, \ldots, m\}$. If $z_i + A \in (G/A)(M^*)^\diamond$, then—as before— we have $\|x_i\|_G = \|x_i + A\|_{G/A}$. Hence, $\|x_i\|_G \geq nM$ and $\|x_i\|_G \not\sim nM$. Similarly, if $z_i + A \in (G/A)(M_p^*, p)^\diamond$, then we have $\|x_i\|_p^G = \|x_i + A\|_p^{G/A} = \|nz_i + A\|_p^{G/A} \geq (nM)_p$ and $|p^j x_i|_p^G = |p^j x_i + A|_p^{G/A} = |p^j nz_i + A|_p^{G/A} \neq m_{p,j+|n|_p}$ for infinitely many $j < \omega$; thus, $x_i \in G((nM)_p^*, p)^\diamond$. Consequently, we have $x_1, \ldots, x_m \in G((nM)^*, p)^\diamond$. Since $nx + A = nz_1 + \cdots + nz_m + A = x_1 + \cdots + x_n + A \subseteq G((nM)^*, p) + A$, we have $nx \in G((nM)^*, p) + A$, as desired.

(2) and (3) follow from the proof of (1). □

The proof of [53, Proposition 1.7] shows the following result:

Lemma 3.12.6. *Let A be a quasi-sequentially nice subgroup of a group G and X a $*$-decomposition set in G such that $A \oplus \langle X\rangle$ is a $*$-valuated coproduct in G. Then $\{x + A : x \in X\}$ is a $*$-decomposition set in G/A.*

Proof. Let $x \in X$, and notice that $\|x\|_G = \|x + A\|_{G/A}$ by Proposition 3.8.15. Let M be an Ulm matrix and p a prime. To prove that $x + A$ is primitive in G/A, assume that $kx + A \in (G/A)(M^*, p)$ for some positive integer k. Then by Lemma 3.12.5(2), there is a positive integer n such that $nkx \in G((nM)^*, p) + A$. By assumption, $A \oplus \langle x\rangle$ is a $*$-valuated coproduct in G; thus, $nkx \in G((nM)^*, p)$. Since x is primitive in G, we have $\|x + A\|_{G/A} =$

$\|x\|_G \not\sim nM \sim M$, or else $nkx \in G((nM)^*_p, p)^\diamond$; that is, $\|nkx + A\|_p^{G/A} = \|nkx\|_p^G \geq (nM)_p$, and $|p^i nkx + A|_p^{G/A} = |p^i nkx|_p^G \neq m_{p,i+|n|_p}$ for infinitely many $i < \omega$. It follows that $\|x + A\|_{G/A} \not\sim M$, or else $kx + A \in (G/A)((nM)^*_p, p)^\diamond$; hence, $x + A$ is primitive in G/A.

It remains to show that $\bigoplus_{x\in X}\langle x + A\rangle$ is a $*$-valuated coproduct. Note that it is at least a valuated coproduct in G/A by Corollary 3.8.16. Let $z = r_1x_1 + \cdots + r_mx_m$, where $0 \neq r_i \in \mathbb{Z}$, $x_i \in X$ $(i = 1, \ldots, m)$, and let $i \in \{1, \ldots, m\}$.

First, we assume that $z+A \in (G/A)(M^*, p)$ and use Lemma 3.12.5(2) again to obtain $nz \in G((nM)^*, p)+A$ for some positive integer n. Since $A\oplus\langle x_1\rangle\oplus\cdots\oplus\langle x_m\rangle$ is a $*$-valuated coproduct in G, we have $nr_ix_i \in G((nM)^*, p)$; thus, $nr_ix_i + A \in (G/A)((nM)^*, p)$ by Lemma 3.6.2(2). Since r_ix_i+A is primitive in G/A, we have $\|r_ix_i+A\|_{G/A} \not\sim nM \sim M$, or else $\|nr_ix_i+A\|_p^{G/A} \geq (nM)_p$ and $|p^j nr_ix_i+A|_p^G \neq m_{p,j+|n|_p}$ for infinitely many $j < \omega$. On the other hand, we have $r_ix_i+A \in (G/A)(M)$ since $z+A \in (G/A)(M)$ and $\bigoplus_{x\in X}\langle x+A\rangle$ is a valuated coproduct in G/A. It follows that $r_ix_i + A \in (G/A)(M) \cap (G/A)(M^*, p)^\diamond \subseteq (G/A)(M^*, p)$, as desired.

Next we assume that $z + A \in (G/A)(M^*)$. By Lemma 3.12.5(3), we have $nz \in G((nM)^*) + A$ for some positive integer n. Then $nr_ix_i \in G((nM)^*)$; hence, $nr_ix_i + A \in (G/A)((nM)^*)$ (Lemma 3.6.2(2)). By Proposition 3.10.3(1) we obtain $\|nr_ix_i + A\|_{G/A} \not\sim nM$. Since $r_ix_i + A \in (G/A)(M)$, this yields $r_ix_i + A \in (G/A)(M^*)$.

Finally, suppose that $z + A \in (G/A)(\overline{\alpha}^*, p)$. Then Lemma 3.12.5(4) shows that $nx \in G((p^r\overline{\alpha})^*, p) + A$ for some positive integer $n = p^rn'$ with $p \nmid n'$. It follows that $nr_ix_i \in G((p^r\overline{\alpha})^*, p)$, and therefore, $nr_ix_i + A \in (G/A)((p^r\overline{\alpha})^*, p)$ (Lemma 3.6.2(2)). By Proposition 3.10.3(2), we have $nr_ix_i + A \in (G/A)((p^r\overline{\alpha})^*, p)^\diamond$. Then $r_ix_i + A \in (G/A)(\overline{\alpha}, p)$ shows that $r_ix_i + A \in (G/A)(\overline{\alpha}^*, p)$. This completes the proof. □

Knice subgroups are characterized as follows:

Proposition 3.12.7 (Hill–Megibben [53]). *Let A be a subgroup of a group G. Then the following conditions are equivalent:*
1. *A is knice in G;*
2. *A is quasi-sequentially nice in G, and G/A is a k-group.*

Proof. Suppose that A is knice in G. Let $x \in G$. Then there is a finite $*$-decomposition set X such that $A \oplus \langle X\rangle$ is a $*$-valuated coproduct containing nx for some nonzero integer n. Thus, for some $a \in A$, $nx + a \in \langle X\rangle$ and $A \oplus \langle nx + a\rangle$ is a valuated coproduct. Then by Lemma 3.9.6(3), A is quasi-sequentially nice.

To prove that G/A is a k-group, let S be a finite subset of G. Since A is knice, there is a finite $*$-decomposition set X in G such that $A \oplus \langle X\rangle$ is a $*$-valuated coproduct in G containing a nonzero multiple of $\langle S\rangle$. Then $\{x + A : x \in X\}$ is a $*$-decomposition set in G/A by Lemma 3.12.6; hence, G/A is a k-group.

To prove the converse, assume that A is quasi-sequentially nice in G, and that G/A is a k-group. Let S be a finite subset of G. Then there is a finite $*$-decomposition set $\{x_1 + A, \ldots, x_n + A\}$ in G/A such that $\langle x_1 + A\rangle \oplus \cdots \oplus \langle x_n + A\rangle$ contains a nonzero multiple of $\langle S, A\rangle/A$. By Corollary 3.10.8 with $B = A$, there are positive integers m_i and elements $x_i' \in m_i x_i + A$ $(i = 1, \ldots, n)$ such that $C = A \oplus \langle x_1'\rangle \oplus \cdots \oplus \langle x_n'\rangle$ is a $*$-valuated coproduct in G and $x_1', \ldots, x_n'$ are primitive in G. Since $\langle S, C\rangle/C$ is finite, it follows that A is knice in G. □

Corollary 3.12.8 (Megibben–Ullery [116]). *Suppose that B is a pure knice subgroup of a group G. If A is a subgroup of B, then B/A is a pure knice subgroup of G/A.*

Proof. By Proposition 3.12.7, B is quasi-sequentially nice in G and G/B is a k-group. Then Corollary 3.9.5 shows that B/A is pure and quasi-sequentially nice in G/A. Since $(G/A)/(B/A) \cong G/B$ is a k-group, B/A is knice in G/A by Proposition 3.12.7. □

Theorem 3.12.9 (Hill–Megibben [50]). *Let A and B be subgroups of a group G such that $A \subseteq B$. If A is knice in G and B/A is knice in G/A, then B is knice in G.*

Proof. Let S be a finite subset of G. Then there is a $*$-valuated coproduct $B/A \oplus \langle x_1 + A\rangle \oplus \cdots \oplus \langle x_n + A\rangle$ in G/A containing $m(\langle S, A\rangle/A)$ for some positive integer m, where each element $x_i + A$ is primitive in G/A. Let $i \in \{1, \ldots, n\}$. By Proposition 3.12.7, A is quasi-sequentially nice in G. Therefore, by Corollary 3.10.8, there is a positive integer m_i such that $x_i' = m_i x_i$ is primitive in G, and $C = B \oplus \langle x_1'\rangle \oplus \cdots \oplus \langle x_n'\rangle$ is a $*$-valuated coproduct in G. Then $\langle S, C\rangle/C$ is finite. Since B is a nice subgroup of G by Proposition 3.12.2, it follows that B is knice in G. □

The following result was extracted from the proof of [116, Theorem 2.1]:

Proposition 3.12.10. *Let $A \oplus B$ be a $*$-valuated coproduct in a group G, and let $F = \langle X\rangle$, where X is a countable $*$-decomposition set in G. Suppose that S is a finite subset of A with $\langle S, F\rangle/F$ finite, and that $F/((F \cap A) \oplus (F \cap B))$ is torsion. Then there is a finite $*$-decomposition set Y in G such that $Y \subseteq A$ and $\langle S, Y\rangle/\langle Y\rangle$ is finite.*

Proof. Since $\langle S, F\rangle/F$ is finite, there exists a finite subset $X_0 = \{x_1, \ldots, x_n\}$ of X such that $\langle S, F_0\rangle/F_0$ is finite, where $F_0 = \langle X_0\rangle$. Then, by Corollary 3.11.8, there are subgroups A' and B' of F and nonzero multiples x_i' of x_i $(i = 1, \ldots, n)$ such that

1. $F' = \langle x_1'\rangle \oplus \cdots \oplus \langle x_n'\rangle \oplus A' \oplus B'$ is a $*$-valuated coproduct in G;
2. F/F' is torsion;
3. $A' \subseteq A$ and $(F \cap A)/A'$ has torsion-free rank $\leq n$.

Let $F_0' = \langle x_1'\rangle \oplus \cdots \oplus \langle x_n'\rangle$, and notice that $\langle S, F_0'\rangle/F_0'$ is finite since F_0/F_0' and $\langle S, F_0\rangle/F_0$ are finite. Letting $A'' = A \cap (F_0' \oplus B')$, we have

$$F' \cap A = (F_0' \oplus A' \oplus B') \cap A = A' \oplus (A \cap (F_0' \oplus B')) = A' \oplus A''.$$

By Corollary 3.11.1 and Proposition 3.11.2(1), F' is k-basic in G since F/F' is torsion, and $A' \oplus A'' \oplus (F' \cap B) = (F' \cap A) \oplus (F' \cap B)$ is k-basic in G since $F'/((F' \cap A) \oplus (F' \cap B))$ is torsion. Now $A'' \cong (F' \cap A)/A'$ has torsion-free rank $\leq n$, and $A' \oplus A'' \oplus (F' \cap B)$ is a $*$-valuated coproduct in G. Therefore, by Corollary 3.11.6, there exists a finite $*$-decomposition set Y in G such that $Y \subseteq A''$ and $A''/\langle Y\rangle$ is torsion. Since

$$\langle S, A''\rangle/A'' \cong \langle S\rangle/(\langle S\rangle \cap A'') = \langle S\rangle/(\langle S\rangle \cap (F_0' \oplus B')) \cong \langle S, F_0' \oplus B'\rangle/F_0' \oplus B'$$

is finite, we conclude that the finitely generated group $\langle S, Y\rangle/\langle Y\rangle$ is torsion, hence finite. Since $Y \subseteq A$, the proof is complete. □

The class of k-groups is closed under direct summands:

Theorem 3.12.11 (Hill–Megibben [53]). *A direct summand of a k-group is a k-group.*

Proof. Suppose that $G = H \oplus K$ is a k-group, and let S be a finite subset of H. Then there is a $*$-decomposition set $T_0 = \{x_1, \ldots, x_{n_0}\}$ in G such that $\langle S, F_0\rangle/F_0$ is finite, where $F_0 = \langle T_0\rangle$. We write $x_i = h_i + k_i$ ($h_i \in H, k_i \in K$) for $i = 1, \ldots, n_0$, and let $S_0 = \{h_1, \ldots, h_{n_0}, k_1, \ldots, k_{n_0}\}$. By Theorem 3.12.3, F_0 is knice in G. Therefore, T_0 is contained in some $*$-decomposition set $T_1 = \{x_1, \ldots, x_{n_0}, x_{n_0+1}, \ldots, x_{n_1}\}$ in G such that $\langle S_0, F_1\rangle/F_1$ is finite, where $F_1 = \langle T_1\rangle$. Write $x_i = h_i + k_i$ ($h_i \in H, k_i \in K$) for $i = n_0 + 1, \ldots, n_1$, and let $S_1 = \{h_1, \ldots, h_{n_1}, k_1, \ldots, k_{n_1}\}$. Again, we can find a $*$-decomposition set $T_2 = \{x_1, \ldots, x_{n_2}\}$ in G containing T_1 such that $\langle S_1, F_2\rangle/F_2$ is finite with $F_2 = \langle T_2\rangle$, et cetera. This gives us ascending sequences $\{T_n\}_{n<\omega}$ and $\{S_n\}_{n<\omega}$ of finite subsets of G, and a sequence $\{F_n = \langle T_n\rangle\}_{n<\omega}$ of subgroups of G such that $F = \bigcup_{n<\omega} F_n = \langle T\rangle$, where $T = \bigcup_{n<\omega} T_n$ is a $*$-decomposition set in G, and $F/((F \cap H) \oplus (F \cap K))$ is torsion. By Proposition 3.12.10, there is a finite $*$-decomposition set Y in G such that $Y \subseteq H$ and $\langle S, Y\rangle/\langle Y\rangle$ is finite. Since Y is a $*$-decomposition set in H by Section 3.10(c), H is a k-group. □

Lemma 3.12.12 (Hill–Megibben [53]). *Let $C = A \oplus B$ be a knice subgroup of $G = H \oplus K$, where $A \subseteq H$ and $B \subseteq K$. Then A is knice in H.*

Proof. Since C is knice in G, C is quasi-sequentially nice in G and G/C is a k-group by Proposition 3.12.7. By Lemma 3.9.1, A is quasi-sequentially nice in H. Now $G/C \cong H/A \oplus K/B$ and Theorem 3.12.11 show that H/A is a k-group, so by Proposition 3.12.7, A is knice in H. □

By the above result and (b), we have the following:

Corollary 3.12.13 (Hill–Megibben [53]). *If A_i are subgroups of groups G_i ($i \in I$), then $\bigoplus_{i\in I} A_i$ is knice in $\bigoplus_{i\in I} G_i$ if and only if each A_i is knice in G_i.*

As in the case of nice and quasi-sequentially nice subgroups (see Corollaries 3.8.20 and 3.9.3), we apply Theorem 2.2.8 once more to obtain the following result:

Corollary 3.12.14 (Hill–Megibben [53]). *The class of groups having an $H(\aleph_0)$-family of knice subgroups is closed under arbitrary direct sums and direct summands.*

In order to prove that a group possessing a composition series of quasi-sequentially nice subgroups has also a composition series of knice subgroups, the following fact will be needed. It is based on an observation of its p-local case made by Hill and Megibben [49]; the proof carries over to the global case:

Lemma 3.12.15. *Suppose that G is a group with a composition series $\{N_\alpha\}_{\alpha\leq\lambda}$ such that if $N_{\alpha+1}/N_\alpha$ is infinite, there is an $x_\alpha \in G$ such that $N_{\alpha+1} = N_\alpha \oplus \langle x_\alpha\rangle$ is a valuated coproduct in G. Let X be the set of all such elements x_α. Then we have the following:*
1. *X is a $*$-decomposition set.*
2. *Each valuated coproduct $N_{\alpha+1} = N_\alpha \oplus \langle x_\alpha\rangle$ is a $*$-valuated coproduct in G.*
3. *If N_α is nice in G for some $\alpha < \lambda$, then N_α is knice in G.*

Proof. By Lemma 3.9.7, X is a decomposition basis for G; hence, X is a $*$-decomposition set (Corollary 3.10.1). For $\alpha < \lambda$, let $Y = X \cap N_\alpha$ and $Z = X \setminus Y$. Then $N_\alpha \oplus \langle Z\rangle$ is a valuated coproduct in G. Since $\langle Y\rangle \oplus \langle Z\rangle$ is a $*$-valuated coproduct and $N_\alpha/\langle Y\rangle$ is torsion, it follows from Lemma 3.6.1 that $N_\alpha \oplus \langle Z\rangle$ is a $*$-valuated coproduct in G; thus, we have (2) and (3). □

Corollary 3.12.16. *The following conditions are equivalent for a group G:*
1. *G has a composition series $\{N_\alpha\}_{\alpha\leq\lambda}$ of nice subgroups such that if $N_{\alpha+1}/N_\alpha$ is infinite, there is an $x_\alpha \in G$ such that $N_{\alpha+1} = N_\alpha \oplus \langle x_\alpha\rangle$ is a valuated coproduct in G;*
2. *G has a composition series $\{N_\alpha\}_{\alpha\leq\lambda}$ of knice subgroups such that if $N_{\alpha+1}/N_\alpha$ is infinite, there is a primitive element $x_\alpha \in G$ such that $N_{\alpha+1} = N_\alpha \oplus \langle x_\alpha\rangle$ is a $*$-valuated coproduct in G.*

Corollary 3.12.17. *The following conditions are equivalent for a group G:*
1. *G has a composition series of quasi-sequentially nice subgroups;*
2. *G has a composition series of knice subgroups.*

Proof. (1) implies (2) by Theorem 3.9.8 and Lemma 3.12.15. The converse follows from the fact that knice subgroups are quasi-sequentially nice (Proposition 3.12.7). □

Exercises.
1. A k-group of finite torsion-free rank has a decomposition basis.
2. Let G be a torsion-free group of finite rank. Then G is a k-group if and only if G is completely decomposable. [Hint: Exercise 1 and Theorem 3.1.2.]
3. (Ullery [145]) Hill and Ullery [54] call a system $\mathcal{C}$ of subgroups of G a **weak Axiom 3 system** for G if the following hold:
   – (1) $0 \in \mathcal{C}$;
   – (2) $\mathcal{C}$ is closed under ascending unions;
   – (3) for every countable subgroup A of G, there is a countable $C \in \mathcal{C}$ such that $A \subseteq C$.

   Show that a group G is a k-group if and only if the set of all k-subgroups of G is a weak Axiom 3 system for G. [Hint: Use Section 3.11, Exercises 1, 2.]

3.13 k-Subgroups and strong k-subgroups

Recall that a subgroup A of G is called a **k-subgroup of** G if for every finite subset S of A there is a finite (possibly empty) $*$-decomposition set X in G such that $X \subseteq A$ and $\langle S,X\rangle/\langle X\rangle$ is finite (see Section 3.11). We call A a **strong k-subgroup of** G, if for every countable subset S of A there is a countable (possibly empty) $*$-decomposition set X in G such that $X \subseteq A$ and $\langle S,X\rangle/\langle X\rangle$ is torsion. Clearly, strong k-subgroups are k-subgroups.

Theorem 3.13.1 (Megibben–Ullery [116]). *Suppose that A is a subgroup of a k-group G. Then the following hold:*

1. *If A is a knice subgroup of G, then A is a k-subgroup of G.*
2. *If A is a pure knice subgroup of G, then A is a strong k-subgroup of G.*

Proof. Suppose that A is a knice subgroup of G, and let S be a finite subset of A. Since 0 is knice in G, there is a $*$-decomposition set $T_0 = \{x_1,\dots,x_{k_0}\}$ in G such that $\langle S,F_0\rangle/F_0$ is finite, where $F_0 = \langle T_0\rangle$. Since A is knice, there is a $*$-valuated coproduct $A \oplus B_0$ in G containing a nonzero multiple of F_0, where $B_0 = \langle Y_0\rangle$ for some finite $*$-decomposition set Y_0 in G. By replacing the elements of T_0 by suitable nonzero multiples, we may assume that $F_0 \subseteq A\oplus B_0$. Then we write $x_i = a_i + b_i$ ($a_i \in A, b_i \in B_0$) for $i = 1,\dots,k_0$, and let $S_0 = \{a_1,\dots,a_{k_0}, b_1,\dots,b_{k_0}\}$. Since F_0 is knice in G (Theorem 3.12.3), T_0 is contained in some $*$-decomposition set $T_1 = \{x_1,\dots,x_{k_0},x_{k_0+1,\dots,}x_{k_1}\}$ in G such that $\langle S_0,F_1\rangle/F_1$ is finite with $F_1 = \langle T_1\rangle$. By Theorem 3.12.3, $A \oplus B_0$ is knice in G. Therefore, some nonzero multiple of F_1 is contained in a $*$-valuated coproduct $A \oplus B_0 \oplus B_1$ in G, where $B_1 = \langle Y_1\rangle$ for some finite $*$-decomposition set Y_1 in G. Replacing elements of $T_1 \setminus T_0$ by nonzero multiples if necessary, we may assume that $F_1 \subseteq A \oplus B_0 \oplus B_1$. We write $x_i = a_i + b_i$ ($a_i \in A, b_i \in B_0 \oplus B_1$) for $i = k_0 + 1,\dots,k_1$, and let $S_1 = \{a_1,\dots,a_{k_1}, b_1,\dots,b_{k_1}\}$. Since F_1 is knice in G, we can find a $*$-decomposition set $T_2 = \{x_1,\dots,x_{k_2}\}$ in G containing T_1 such that $\langle S_1,F_2\rangle/F_2$ is finite with $F_2 = \langle T_2\rangle$, et cetera.

This yields ascending sequences $\{T_n\}_{n<\omega}$ and $\{S_n\}_{n<\omega}$ of finite subsets of G, and sequences $\{F_n = \langle T_n\rangle\}_{n<\omega}$ and $\{B_n\}_{n<\omega}$ of subgroups of G such that

$$F = \bigcup_{n<\omega} F_n \quad \text{and} \quad B = \bigoplus_{n<\omega} B_n$$

satisfy the following conditions: $F = \langle T\rangle$, where T is a countable $*$-decomposition set in G, $A \oplus B$ is a $*$-valuated coproduct in G and $F/((F \cap A) \oplus (F \cap B))$ is torsion. By Proposition 3.12.10, there is a finite $*$-decomposition set Y in G such that $Y \subseteq A$ and $\langle S, Y\rangle/\langle Y\rangle$ is finite; thus, A is a k-subgroup of G.

To prove (2), suppose that A is a pure knice subgroup of G, and let S be a countable subset of A. Then $S = \bigcup_{n<\omega} S_n$, where

$$S_0 \subseteq S_1 \subseteq \cdots \subseteq S_n \subseteq \cdots$$

are finite subsets of G. We will construct inductively an ascending sequence $\{Y_n\}_{n<\omega}$ of finite $*$-decomposition sets in G such that $Y_n \subseteq A$ and $\langle S_n, Y_n\rangle/\langle Y_n\rangle$ is finite for all $n < \omega$. Then $Y = \bigcup_{n<\omega} Y_n$ is a countable $*$-decomposition set in G with $Y \subseteq A$ and $\langle S, Y\rangle/\langle Y\rangle$ torsion; hence, A is a strong subgroup of G.

Since A is a k-subgroup by (1), there is a $*$-decomposition set Y_0 in G such that $Y_0 \subseteq A$ and $\langle S_0, Y_0\rangle/\langle Y_0\rangle$ is finite. Let $n < \omega$, and suppose that Y_n has already been constructed. Let $T = S_{n+1} \setminus S_n$, $N_n = \langle Y_n\rangle$, and $\overline{T} = \{t + N_n : t \in T\}$. Since N_n is knice, G/N_n is a k-group (Proposition 3.12.7) and contains A/N_n as a knice subgroup (Corollary 3.12.8). Then (1) shows that A/N_n is a k-subgroup of G/N_n. Therefore, there exists a $*$-decomposition set $\{y_1 + N_n, \dots, y_s + N_n\}$ $(y_1, \dots, y_s \in A)$ such that $\langle \overline{T}, \overline{N}\rangle/\overline{N}$ is finite, where $\overline{N} = \bigoplus_{i=1}^{s}\langle y_i + N_n\rangle$. By Proposition 3.12.7, N_n is quasi-sequentially nice in G; hence, Corollary 3.10.8 yields positive integers m_i and elements $y_i' \in m_i y_i + N_n$ $(i = 1, \dots, s)$ such that $\{y_1', \dots, y_s'\}$ is a $*$-decomposition set in G and $\langle Y_n\rangle \oplus \langle y_1'\rangle \oplus \cdots \oplus \langle y_s'\rangle$ is a $*$-valuated coproduct in G. Then $Y_{n+1} = Y_n \cup \{y_1', \dots, y_s'\}$ is a $*$-decomposition set in G, $Y_{n+1} \subseteq A$, and $\langle S_{n+1}, Y_{n+1}\rangle/\langle Y_{n+1}\rangle$ is finite, as required. This completes the induction. □

Theorem 3.13.2 (Megibben–Ullery [116]). *Suppose that A and B are pure knice subgroups of a k-group G such that $A \subseteq B$ and $|B/A| \le \aleph_0$. Then there is an ordinal $\mu \le \omega$ and an ascending sequence*

$$A = N_0 \subseteq N_1 \subseteq \cdots \subseteq N_i \subseteq \cdots \subseteq N_\mu = B$$

of knice subgroups of G such that each quotient N_{i+1}/N_i is cyclic of prime or infinite order. In the latter case, $N_{i+1} = N_i \oplus \langle x_i\rangle$ is a $*$-valuated coproduct in G for some primitive element $x_i \in G$.

Proof. First, we assume that $A = 0$. By Theorem 3.13.1(2), there is a countable $*$-decomposition set Y in G such that $Y \subseteq B$ and $B/\langle Y\rangle$ is torsion. If Y is finite, we write $Y = \{x_1, \dots, x_r\}$ and define $N_0 = 0$, and $N_k = \bigoplus_{i=1}^{k}\langle x_i\rangle$ for $k = 1, \dots, r$. Then by Theorem 3.12.3, each subgroup N_k is knice. Since B/N_r is a countable torsion group, we can find an ordinal $r \le \mu \le \omega$ and a sequence

$$N_r \subseteq N_{r+1} \subseteq \cdots \subseteq N_{r+i} \subseteq \cdots \subseteq N_\mu = B$$

such that each quotient N_{r+i+1}/N_{r+i} is cyclic of prime power. As a finite extension of a knice subgroup, each subgroup N_{r+i} is knice in G, as required.

Now suppose that the set Y is infinite, say $Y = \{y_1, y_2, \dots\}$. Since B is countable, it is the union of an ascending sequence of subgroups

$$0 = B_0 \subseteq B_1 \subseteq \cdots \subseteq B_n \subseteq \cdots \quad (n < \omega)$$

such that B_n is a finite extension of $A_n = \langle y_1, \dots, y_n\rangle$ for all $n \ge 1$. Since A_n is knice by Theorem 3.12.3, B_n is knice as well by Theorem 3.12.1. Then by Proposition 3.6.5, there is a nonzero multiple x_n of y_n such that $B_{n-1} \oplus \langle x_n \rangle$ is a $*$-valuated coproduct in G, and x_n is primitive. Since $B_1/\langle x_1 \rangle$ is finite, there is a finite sequence

$$0 = N_0 \subseteq N_1 = \langle x_1 \rangle \subseteq N_2 \subseteq \cdots \subseteq N_{k_1} = B_1$$

of subgroups such that for each $i \geq 1$, the quotient N_{i+1}/N_i is cyclic of prime order. Continuing in this manner, we construct a finite sequence

$$B_1 \subseteq N_{k_1+1} = B_1 \oplus \langle x_2 \rangle \subseteq N_{k_1+2} \subseteq \cdots \subseteq N_{k_2} = B_2,$$

et cetera. This yields a suitable ascending sequence $\{N_i\}_{i<\omega}$ of knice subgroups of G, whose union is B.

Finally, assume that $A \neq 0$. We need to construct a suitable ascending sequence $A = N_0 \subseteq N_1 \subseteq \cdots \subseteq N_i \subseteq \cdots \subseteq N_\mu = B$ of knice subgroups of G. By Corollary 3.12.8, B/A is pure and knice in G/A, which is a k-group by Proposition 3.12.7. Then 0 is knice in G/A. Therefore, as in the first part of this proof, we can find an ordinal $\lambda \leq \omega$ and an ascending sequence

$$0 = N_0/A \subseteq N_1/A \subseteq \cdots \subseteq N_\lambda/A = B/A$$

of knice subgroups of G/A such that for each $i < \lambda$ we have the following:
(i) $(N_{i+1}/A)/(N_i/A)$ is cyclic of prime order, or
(ii) $N_{i+1}/A = N_i/A \oplus \langle x_i + A \rangle$ is a $*$-valuated coproduct in G/A for some primitive element $x_i + A$ in G/A.

Let $i < \mu$. Then by Theorem 3.12.9, N_i is knice in G, and in case (ii), there is a positive integer m_i and an element $x_i' \in m_i x_i + A$ such that x_i' is primitive in G and $N' = N_i \oplus \langle x_i' \rangle$ is a $*$-valuated coproduct in G (see Proposition 3.12.7 and Corollary 3.10.8(1)). Since N_{i+1}/N' is finite, we have a finite sequence $N' = N_0' \subseteq \cdots \subseteq N_k' = N_{i+1}$ of knice subgroups of G such that each quotient N_{j+1}'/N_j' is cyclic of prime order. Thus, we obtain an ascending sequence

$$A = N_0 \subseteq N_1 \subseteq \cdots \subseteq N_i \subseteq \cdots \subseteq N_\mu = B$$

of knice subgroups of G having the desired properties. □

Theorem 3.13.3 (Megibben–Ullery [116]). *Suppose that G is a group with an $F(\aleph_0)$-family $\mathcal{C}$ of pure knice subgroups. Then $\mathcal{C}$ can be refined to a composition series*

$$0 = N_0 \subseteq N_1 \subseteq \cdots \subseteq N_\beta \subseteq \cdots \subseteq N_\lambda = G$$

of knice subgroups of G such that if $N_{\beta+1}/N_\beta$ is infinite, then $N_{\beta+1} = N_\beta \oplus \langle x_\beta \rangle$ is a $*$-valuated coproduct in G for some primitive element $x_\beta \in G$.

Proof. Let $0 = M_0 \subseteq M_1 \subseteq \cdots \subseteq M_\alpha \subseteq \cdots \subseteq M_\lambda = G$ be an $F(\aleph_0)$-family of pure knice subgroups of G. We may assume that G is infinite, and that λ is the smallest ordinal of cardinality $|G|$. For $\lambda = \aleph_0$, the assertion follows directly from Theorem 3.13.2, so assume that $\lambda > \aleph_0$.

By Theorem 3.13.2, for every $\alpha < \lambda$, there is a $\mu_\alpha \leq \omega$ and an ascending chain

$$M_\alpha = N_{\alpha,0} \subseteq N_{\alpha,1} \subseteq \cdots \subseteq N_{\alpha,i} \subseteq \cdots \subseteq N_{\alpha,\mu_\alpha} = M_{\alpha+1}$$

of knice subgroups of G such that each quotient $N_{\alpha,i+1}/N_{\alpha,i}$ $(i < \mu_\alpha)$ is cyclic of prime order, or $N_{\alpha,i+1} = N_{\alpha,i} \oplus \langle x_{\alpha,i} \rangle$ is a $*$-valuated coproduct in G for some primitive element $x_{\alpha,i} \in G$.

We will construct inductively a smooth ascending chain

$$0 = N_0 \subseteq N_1 \subseteq \cdots \subseteq N_\beta \subseteq \cdots \subseteq N_\lambda = G$$

of knice subgroups of G such that
(i) for every $\beta < \lambda$, there is an ordinal $\alpha < \lambda$ and $\mu \leq \omega$ such that $N_\beta = N_{\alpha,\mu}$.
(ii) if $\beta \leq \lambda$ is a limit ordinal, then $\bigcup_{\gamma<\beta} N_\gamma = M_\alpha$ for some $\alpha \leq \lambda$.

Since $|M_\alpha| < |G|$ for all $\alpha < \lambda$, this will yield $\bigcup_{\beta<\lambda} N_\beta = M_\lambda = G$, and we obtain a composition series as desired.

Starting with $N_0 = 0$, let $0 < \beta \leq \lambda$, and suppose that N_γ has been constructed for all $\gamma < \beta$. If $\beta - 1$ exists, then by the induction hypothesis $N_{\beta-1} = N_{\alpha,\mu}$ for some α, and some $\mu \leq \mu_\alpha$. If $\mu < \mu_\alpha$, which means it is finite, choose $N_\beta = N_{\alpha,\mu+1}$. Otherwise, let $N_\beta = N_{\alpha+1,1}$.

If β is a limit ordinal, we define $N_\beta = \bigcup_{\gamma<\beta} N_\gamma$. Then for $\gamma < \beta$, if we let α_γ be defined such that $N_\gamma = N_{\alpha_\gamma,\mu}$ for some $\mu \leq \omega$, we have $M_{\alpha_\gamma+1} \subseteq N_\beta \subseteq \bigcup_{\gamma<\beta} M_{\alpha_\gamma+1}$; hence, $N_\beta = M_\alpha$ for some $\alpha \leq \lambda$. This completes the induction. □

Exercises.
1. (Hill–Megibben [53]) If A is a pure knice subgroup of G, then A is p-nice in G for all primes p.

3.14 Sequentially-pure-projective groups

Following Warfield [152], we call a short exact sequence $E : 0 \to A \to B \xrightarrow{\nu} C \to 0$ of groups **sequentially pure** if the induced sequence

$$0 \to A(M) \to B(M) \to C(M) \to 0$$

is exact for every Ulm matrix M. It is clear that E is sequentially pure exactly if $A(M) = B(M) \cap A$ and $\nu(B(M)) = C(M)$ for all Ulm matrices M.

Lemma 3.14.1 (Warfield [152, 154]). *Let $E : 0 \to A \to B \xrightarrow{\nu} C \to 0$ be an exact sequence of groups such that $\nu(B(M)) = C(M)$ for all Ulm matrices M. Then the following conditions are equivalent:*

1. *E is sequentially pure;*
2. *A is isotype in B;*
3. *$\nu((p^\alpha B)[p]) = (p^\alpha C)[p]$ for all primes p and ordinals α.*

Proof. Notice that A is isotype in B if and only if $A(M) = B(M) \cap A$ for all Ulm matrices M. Therefore, (1) and (2) are equivalent.

To show that (2) implies (3), assume that A is isotype in B, let p be a prime and α an ordinal. Assume that $x \in (p^\alpha C)[p]$. Then $x = \nu(y)$ for some $y \in p^\alpha B$; hence, $\nu(py) = px = 0$, and we have $py \in \ker \nu \cap p^{\alpha+1}B = A \cap p^{\alpha+1}B = p^{\alpha+1}A$. Then there is a $z \in p^\alpha A$ such that $py = pz$; thus, $y - z \in (p^\alpha B)[p]$, and $\nu(y - z) = \nu(y) = x$. Therefore, (3) follows.

Conversely, assume that ν maps $(p^\alpha B)[p]$ onto $(p^\alpha C)[p]$ for all ordinals α and a fixed prime p. We will show inductively that

$$A \cap p^\alpha B \subseteq p^\alpha A$$

for all ordinals α, proving that (3) implies (2). The claim being true for $\alpha = 0$, let $\alpha > 0$, pick an element $x \in A \cap p^\alpha B$, and assume that the assertion holds for all $\beta < \alpha$. If α is a limit ordinal, then $x \in A \cap p^\beta B \subseteq p^\beta A$ for all $\beta < \alpha$ by the induction hypothesis; hence, $x \in p^\alpha A$. If $\alpha - 1$ exists, then $x = py$ for some $y \in p^{\alpha-1}B$. Then $\nu(y) \in p^{\alpha-1}C$, and $p\nu(y) = \nu(x) = 0$ since $x \in A$. Thus, $\nu(y) \in (p^{\alpha-1}C)[p]$. Therefore, by assumption, there is a $z \in (p^{\alpha-1}B)[p]$ such that $\nu(y) = \nu(z)$. Therefore, $y - z \in \ker \nu \cap p^{\alpha-1}B = A \cap p^{\alpha-1}B$, which is a subset of $p^{\alpha-1}A$ by the induction hypothesis. Consequently, we have $x = py = p(y - z) \in p^\alpha A$, and the induction is complete. □

Following Warfield [152], we call a group G **sequentially-pure-projective** if G has the projective property relative to every sequentially pure sequence; in other words, if for every sequentially pure sequence $0 \to A \to B \xrightarrow{\nu} C \to 0$ and homomorphism $\gamma : G \to C$, there is a homomorphism $\tau : G \to B$ such that the diagram

$$\begin{array}{ccccccccc} & & & & & & G & & \\ & & & & & \tau\swarrow & \downarrow\gamma & & \\ 0 & \to & A & \to & B & \xrightarrow{\nu} & C & \to & 0 \end{array}$$

is commutative.

Now we are ready to prove a variety of characterizations of Warfield groups. The equivalence of (1) and (7) is due to Warfield [152]. Hill and Megibben [53] proved characterizations in terms of $H(\aleph_0)$- and $G(\aleph_0)$-families and composition series; then Megibben and Ullery [116] established a characterization in terms of an $F(\aleph_0)$-family. Let us call a group an **Axiom 3 group** if it has an $H(\aleph_0)$-family of knice subgroups. We follow the proofs in [53] and [116].

Theorem 3.14.2 (Hill–Megibben [53], Megibben–Ullery [116], Warfield [152]). *The following conditions are equivalent for a group G:*

1. *G is a Warfield group;*
2. *G is an Axiom 3 group;*
3. *G has a $G(\aleph_0)$-family of knice subgroups;*
4. *G has a $G(\aleph_0)$-family of pure knice subgroups;*
5. *G has an $F(\aleph_0)$-family of pure knice subgroups;*
6. *G has a composition series $\{N_\alpha\}_{\alpha\leq\lambda}$ of knice subgroups such that if $N_{\alpha+1}/N_\alpha$ is infinite, there is a primitive element $x_\alpha \in G$ such that $N_{\alpha+1} = N_\alpha \oplus \langle x_\alpha\rangle$ is a $*$-valuated coproduct in G;*
7. *G has a composition series $\{N_\alpha\}_{\alpha\leq\lambda}$ of nice subgroups such that if $N_{\alpha+1}/N_\alpha$ is infinite, there is an $x_\alpha \in G$ such that $N_{\alpha+1} = N_\alpha \oplus \langle x_\alpha\rangle$ is a valuated coproduct in G;*
8. *G is sequentially-pure-projective.*

Proof. (1) implies (2). Suppose that G is a direct summand of a simply presented group G'. By Theorem 2.5.4, G' is a direct sum of a simply presented torsion group and simply presented groups H_i of torsion-free rank 1, so let us assume that H is one of the groups H_i. Let x be one of the elements of infinite order that is used for a simple presentation of H. Then $H/\langle x\rangle$ is a simply presented torsion group. Therefore, by Theorem 2.10.1(1)⇒(4), it has an $H(\aleph_0)$-family $\mathcal{C}_{H/\langle x\rangle}$ of nice subgroups. Since $\{x\}$ is a decomposition basis for H, x is primitive (Corollary 3.10.1); hence, 0 is a knice subgroup of H. Now let $\mathcal{C}$ consist of the trivial subgroup 0 of H and all subgroups A of H such that $A/\langle x\rangle \in \mathcal{C}_{H/\langle x\rangle}$. Then $A/\langle x\rangle$ is nice in $H/\langle x\rangle$, and $\langle x\rangle$ is nice in H by Corollary 3.8.13. Thus, $\langle x\rangle$ is knice in H, which implies that A is knice in H (Lemma 3.12.9). Therefore, $\mathcal{C}$ is an $H(\aleph_0)$-family of knice subgroups of H, so by Corollary 3.12.14, G has an $H(\aleph_0)$-family of knice subgroups.

Clearly (2) implies (3), and we have (3) ⇒ (4) ⇒ (5) by Lemmas 2.2.2 and 2.2.3 (see the remark after [116, Theorem 2.5]). Theorem 3.13.3 shows that (5) implies (6), and it is obvious that (7) follows from (6).

(7) implies (8). Suppose that $\{N_\alpha\}_{\alpha\leq\lambda}$ is a composition series for G as in (7). Let $E : 0 \to A \to B \xrightarrow{\nu} C \to 0$ be a sequentially pure sequence, and $\phi : G \to C$ a homomorphism. We construct inductively an ascending chain $\{\tau_\alpha : N_\alpha \to B\}_{\alpha\leq\lambda}$ of homomorphisms satisfying

$$\nu\tau_\alpha = \phi \quad \text{and} \quad \|\tau_\alpha(x)\|_B \geq \|x\|_G$$

for all $x \in N_\alpha$, yielding a map $\tau : G \to B$ satisfying $\nu\tau = \phi$. Assume that for $\alpha < \lambda$, the map τ_α has already been constructed.

First, consider the case where $N_{\alpha+1}/N_\alpha$ is infinite and $N_{\alpha+1} = N_\alpha \oplus \langle x_\alpha\rangle$ is a valuated coproduct in G. Let $M = \|x_\alpha\|_G$. Since E is sequentially pure, we can find an element $y \in B(M)$ such that $\nu(y) = \phi(x_\alpha)$. For every element $z \in N_{\alpha+1}$, there are unique elements

$g \in N_\alpha$ and $r \in \mathbb{Z}$ such that $z = g + rx_\alpha$; hence, the homomorphism

$$\begin{aligned} \tau_{\alpha+1}: \quad N_{\alpha+1} &\to \quad B \\ z \quad &\mapsto \quad \tau_\alpha(g) + ry \end{aligned}$$

extends τ_α. Then we have $\nu\tau_{\alpha+1}(z) = \nu(\tau_\alpha(g) + ry) = \phi(g) + r\phi(x_\alpha) = \phi(z)$, and

$$\|\tau_{\alpha+1}(z)\|_B \geq \|\tau_\alpha(g)\|_B \wedge \|ry\|_B \geq \|g\|_G \wedge \|rx_\alpha\|_G = \|g + rx_\alpha\|_G = \|z\|_G.$$

Now assume that $N_{\alpha+1}/N_\alpha$ is cyclic of prime order p. Then there exists an $x_\alpha \in G$ such that $N_{\alpha+1} = \langle N_\alpha, x_\alpha\rangle$ and $|x_\alpha|_p^G = |x_\alpha + N_\alpha|_p^{G/N_\alpha}$ (see Lemma 3.8.7). Again, let $M = \|x_\alpha\|_G$. There is a $b \in B(M)$ such that $\nu(b) = \phi(x_\alpha)$ since E is sequentially pure. Since $\|\tau_\alpha(px_\alpha)\|_B \geq \|px_\alpha\|_G$, this yields $pb - \tau_\alpha(px_\alpha) \in \ker \nu \cap B(pM) = A(pM)$. Therefore, there is an $a \in A(M)$ such that $pa = pb - \tau_\alpha(px_\alpha)$, and we let $y = b - a$. Then $\nu(y) = \nu(b) - \nu(a) = \phi(x_\alpha)$, and $\|y\|_B \geq M$. For every $z \in N_{\alpha+1}$, there are unique elements $g \in N_\alpha$ and $r \in \{0, 1, \ldots, p-1\}$ such that $z = g + rx_\alpha$; thus,

$$\begin{aligned} \tau_{\alpha+1}: \quad N_{\alpha+1} &\to \quad B \\ z \quad &\mapsto \quad \tau_\alpha(g) + ry \end{aligned}$$

defines a homomorphism extending the map τ_α, and we have $\nu\tau_{\alpha+1}(z) = \phi(g) + r\nu(y) = \phi(g) + r\phi(x_\alpha) = \phi(z)$. Notice that $|rx_\alpha|_p^G = |rx_\alpha + N_\alpha|_p^{G/N_\alpha}$; hence, rx_α is proper with respect to N_α, and we obtain

$$|\tau_{\alpha+1}(z)|_p^B \geq |\tau_\alpha(g)|_p^B \wedge |ry|_p^B \geq |g|_p^G \wedge |rx_\alpha|_p^G = |g + rx_\alpha|_p^G = |z|_p^G.$$

If q is any prime $\neq p$, then $\|\tau_{\alpha+1}(z)\|_q^B = \|\tau_{\alpha+1}(pz)\|_q^B = \|\tau_\alpha(pz)\|_q^B \geq \|pz\|_q^G = \|z\|_q^G$. It follows that $\|\tau_{\alpha+1}(z)\|_B \geq \|z\|_G$, as required. Since $\{N_\alpha\}_{\alpha\leq\lambda}$ is a smooth chain, the induction is complete.

(8) implies (1). Let G be a sequentially-pure-projective group. We will show that there exists a sequentially pure sequence $0 \to A \to B \to G \to 0$, where B is simply presented. Then by assumption, the sequence splits, and (1) follows. Let X be the set of all elements of G, F the free group on X, and R the set of all elements $mx - ny \in F$ (with $x, y \in X$ and $m, n \in \mathbb{Z}$) such that $mx = ny$ is true in G. Then $B = F/\langle R\rangle$ is a simply presented group. For $x \in X$ let $\overline{x} = x + \langle R\rangle$. It is clear that $\overline{x} \mapsto x$ gives rise to a surjective homomorphism $\phi : B \to G$.

Now let p be a prime. We will prove inductively that if α is an ordinal and $x \in p^\alpha G$, then $\overline{x} \in p^\alpha B$. The case $\alpha = 0$ being evident, let $\alpha > 0$, and suppose the claim holds for all $\beta < \alpha$. Assume that $x \in p^\alpha G$. If α is a limit ordinal, then for all $\beta < \alpha$, we have $x \in p^\beta G$. Hence, $\overline{x} \in p^\beta B$ by the induction hypothesis, and we have $\overline{x} \in p^\alpha B$. If $\alpha - 1$ exists, then $x = py$ for some $y \in p^{\alpha-1}G$; thus, $\overline{y} \in p^{\alpha-1}B$. Consequently, $\overline{x} = \overline{py} = py + \langle R\rangle = p\overline{y} \in p^\alpha B$, and the induction is complete. Since $px = 0$ implies $px \in R$, and therefore, $p\overline{x} = \overline{px} = 0$, this result shows that

$$\phi((p^\alpha B)[p]) = (p^\alpha G)[p]$$

for all ordinals α. In addition, it follows that $|x|_p^G = |\overline{x}|_p^B$ for all $x \in G$. Since $\overline{px} = p\overline{x}$, this yields $\|x\|_G = \|\overline{x}\|_B$ for all $x \in G$; hence,

$$\phi(B(M)) = G(M)$$

for all Ulm matrices M. By Lemma 3.14.1, the sequence $0 \to \ker\phi \to B \to G \to 0$ is sequentially pure.

Now apply the definition of sequentially-pure-projective to the sequence and the identity map on G. Then by Corollary 1.16.9, the sequence splits, giving $B \cong G \oplus \ker\phi$. □

Now we have what we need to prove a key characterization of Warfield groups:

Corollary 3.14.3. *A group G is a Warfield group if and only if it has a nice decomposition basis X such that $G/\langle X\rangle$ is simply presented.*

Proof. By Theorem 3.14.2, G is a Warfield group if and only if it satisfies condition (7) of that theorem. By Theorems 3.9.10 and 3.9.8, condition (7) holds if and only if G has a nice decomposition basis X such that $G/\langle X\rangle$ is simply presented. □

Exercises.

1. (Hill–Megibben [53]) A subgroup A of G is called **sequentially pure** if the exact sequence $0 \to A \to G \to G/A \to 0$ is sequentially pure. Then we have the following:
   (a) A is sequentially pure in G if and only if A is isotype in G and $(G/A)(M) = (G(M) + A)/A$ for all Ulm matrices M.
   (b) A is isotype and knice in G if and only if G/A is a k-group and A is sequentially pure in G.
2. (Warfield [152]) Let $E : 0 \to A \to B \to C \to 0$ be any exact sequence of groups. If every simply presented group has the projective property relative to E, then E is sequentially pure.

3.15 Locally compatible and separable subgroups

Hill and Megibben [53] defined two subgroups A and B of a group G to be **locally compatible**, written $A\|B$, if

$$(A + B) \cap p^\alpha G = (A \cap p^\alpha G) + (B \cap p^\alpha G)$$

for every prime p and ordinal α. We have the following:

(a) If $A \oplus B$ is a valuated coproduct in G, then $A\|B$.
(b) If F is a subgroup of G, and $\{A_i\}_{i\in I}$ is an ascending chain of subgroups of G such that $A_i\|F$ for all $i \in I$, then $(\bigcup_{i\in I} A_i)\|F$.

Suppose G is a group with decomposition basis X. For a subgroup A of G, we define $X_A = X \cap A$ and $Y_A = X \setminus X_A$. The following result is taken from the proof of [53, Theorem 3.2]:

Lemma 3.15.1 (Hill–Megibben [53]). *Let A be a subgroup of a group G with decomposition basis X. If $A \| \langle X \rangle$ and $A/\langle X_A \rangle$ is torsion, then $A \oplus \langle Y_A \rangle$ is a $*$-valuated coproduct in G.*

Proof. Clearly, the sum $A + \langle Y_A \rangle$ is direct. To show that it is a valuated coproduct in G, let p be a prime and α an ordinal. Assume that $a \in A$ and $y \in \langle Y_A \rangle$ such that $a+y \in p^\alpha G$. Since $A \| \langle X \rangle$, we have $a+y = a'+x+y'$ for some elements $a' \in A, x \in \langle X_A \rangle$, and $y' \in \langle Y_A \rangle$, each of which is in $p^\alpha G$. Since $A/\langle X_A \rangle$ is torsion, there is a positive integer n such that na and na' are elements of $\langle X_A \rangle$. Then $na + ny = na' + nx + ny'$; hence, $ny = ny'$, and therefore $y = y' \in p^\alpha G$. This shows that $a \in p^\alpha G$; thus, $A \oplus \langle Y_A \rangle$ is a valuated coproduct in G. But then Corollary 3.10.1 and Lemma 3.6.1, applied to $\langle X_A \rangle \oplus \langle Y_A \rangle$ and $A \oplus \langle Y_A \rangle$, show that $A \oplus \langle Y_A \rangle$ is a $*$-valuated coproduct in G. □

The next result follows from an observation made in the proof of [53, Theorem 3.2]:

Lemma 3.15.2. *Suppose that A is a pure quasi-sequentially nice subgroup of a group G. Let B and F be subgroups of G such that B contains A. If $A \| F$ and $B/A \| (F+A)/A$, then $B \| F$.*

Proof. Let p be a prime, and α an ordinal. Using Lemma 3.9.4, we obtain $p^\alpha(G/A) = (p^\alpha G + A)/A$. Then

$$\frac{B+F}{A} \cap \frac{p^\alpha G + A}{A} = \frac{B \cap (p^\alpha G + A)}{A} + \frac{(F+A) \cap (p^\alpha G + A)}{A}$$

yields $A + ((B+F) \cap p^\alpha G) = A + (B \cap p^\alpha G) + (F \cap p^\alpha G)$. Let $x \in (B+F) \cap p^\alpha G$. Then there are elements $a \in A$, $b \in B \cap p^\alpha G$, and $f \in F \cap p^\alpha G$, such that $x = a + b + f$; hence, $a = x - b - f \in p^\alpha G$. This shows that $x \in (B \cap p^\alpha G) + (F \cap p^\alpha G)$. It follows that $B \| F$, as desired. □

A subgroup A of a group G is called **separable,** if for every $x \in G$ there is a countable subset $\{a_n : n < \omega\}$ of A such that the following is true: For every $a \in A$, there exists an $n < \omega$ such that $|x + a_n|_p \geq |x + a|_p$ for all primes p.

Proposition 3.15.3 (Hill–Megibben [53]). *Let G be a group with decomposition basis X. Then $\langle X \rangle$ is separable.*

Proof. We write $A = \langle X \rangle = \bigoplus_{i \in I} \langle x_i \rangle$ and assume that A is not separable. Then there exists an $x \in G$ such that for every countable subset C of A, there is an $a \in A$ satisfying the following: For every $c \in C$, we have

$$|x + c|_p \lneqq |x + a|_p$$

for some prime p. First, we will construct inductively an ascending sequence $\{I_n\}_{n<\omega}$ of finite subsets of I such that the subgroups $C_n = \bigoplus_{i\in I_n}\langle x_i\rangle$ satisfy the following condition: If $c_n \in C_n$, and $|x+c_n|_p \ngeq |x+a|_p$ for some $a \in A$ and prime p, then this inequality holds for some $a \in C_{n+1}$. Starting with $I_0 = \emptyset$, assume that C_n has already been constructed. Then there is an $a_n \in A$, such that for every $c \in C_n$ we have $|x+c|_p \ngeq |x+a_n|_p$ for some prime p. We have $a_n \in \bigoplus_{i\in I_n'}\langle x_i\rangle$ for some finite subset I_n' of I, and define $I_{n+1} = I_n + I_n'$. Since $a_n \in C_{n+1} = \bigoplus_{i\in I_{n+1}}\langle x_i\rangle$, the construction is complete.

Let $C = \bigcup_{n<\omega} C_n$ and $D = \bigoplus_{i\in I\setminus\bigcup_{n<\omega} I_n}\langle x_i\rangle$. Then C is a countable subgroup of A. Therefore, there is an $a = c + d \in A = C \oplus D$ (with $c \in C$ and $d \in D$) such that $|x+c|_p \ngeq |x+a|_p$ for some prime p. Then $c \in C_n$ for some $n < \omega$; hence, there is a $c' \in C_{n+1}$ such that

$$|x+c|_p \ngeq |x+c'|_p;$$

thus, $|a-c'|_p \geq |x+a|_p \wedge |x+c'|_p \gneq |x+c|_p \geq |x+c|_p \wedge |x+c'|_p = |c-c'|_p$. On the other hand, $A = C \oplus D$ is a valuated coproduct in G, so we have

$$|c-c'|_p \geq |c-c'|_p \wedge |a-c|_p = |a-c'|_p,$$

a contradiction. Therefore, A is separable. □

Lemma 3.15.4 (Hill–Megibben [51]). *Let A be a separable subgroup of G. Then every countable subset S of G is contained in some countable subgroup C of G such that $A\|C$.*

Proof. We will construct inductively an ascending sequence $\{C_n\}_{n<\omega}$ of countable subgroups of G satisfying the following property: For each $n < \omega$ and $(a, c) \in A \times C_n$, there exists an $a' \in A \cap C_{n+1}$ such that

$$|a'+c|_p \geq |a+c|_p$$

for all primes p. Starting with $C_0 = \langle S\rangle$, let $n < \omega$, and assume that the subgroups C_k ($k \leq n$) are already constructed. We write $C_n = \{c_i : i < \omega\}$. Then, since A is separable, for every $i < \omega$, there is a countable subset $\{a_{i,j} : j < \omega\}$ of A, such that for every $a \in A$ there exists a $j_a < \omega$ such that

$$|a_{i,j_a} + c_i|_p \geq |a + c_i|_p$$

for all primes p. Now define $C_{n+1} = \langle C_n \cup \{a_{i,j} : i,j < \omega\}\rangle$. If $(a, c_i) \in A \times C_n$, we let $a' = a_{i,j_a}$. Then $a' \in A \cap C_{n+1}$ and $|a' + c_i|_p \geq |a + c_i|_p$ for all primes p. This completes the induction.

Let $C = \bigcup_{n<\omega} C_n$. Clearly, C is a countable group containing S. To verify that $A\|C$, let p be a prime, α an ordinal, $a \in A$, and $c \in C$ such that $a + c \in p^\alpha G$. Then $c \in C_n$ for some $n < \omega$. Therefore, there is an $a' \in A \cap C_{n+1}$ such that $|a' + c|_p \geq \alpha$, and it follows that $a + c = (a' + c) + (a - a') \in (C \cap p^\alpha G) + (A \cap p^\alpha G)$. Thus, we have $A\|C$. □

Theorem 3.15.5 (Hill–Megibben [53]). *The following conditions are equivalent for a group G:*
1. *G has an $H(\aleph_0)$-family of nice subgroups and possesses a decomposition basis;*
2. *G has a $G(\aleph_0)$-family of nice subgroups and possesses a decomposition basis;*
3. *G has a $G(\aleph_0)$-family of knice subgroups.*

Proof. Clearly, (1) implies (2). To prove that (2) implies (3), suppose that G has a $G(\aleph_0)$-family $\mathcal{C}$ of nice subgroups, and that X is a decomposition basis for G. Let $F = \langle X \rangle$, and for a subgroup A of G, let $X_A = X \cap A$ and $Y_A = X \setminus X_A$. We will show that

$$\mathcal{C}' = \{A \in \mathcal{C} : A \text{ is pure in } G, A \| F, \text{ and } A/\langle X_A \rangle \text{ is torsion}\}$$

is a $G(\aleph_0)$-family of knice subgroups of G. By Lemma 3.15.1, each member of $\mathcal{C}'$ is knice in G. It is clear that $\mathcal{C}'$ contains 0 and is closed under ascending unions of subgroups. To prove that $\mathcal{C}'$ satisfies (H3) (see Section 2.2), let S be a countable subset of G and $A \in \mathcal{C}'$. Then Lemma 3.15.1 shows that $A \oplus \langle Y_A \rangle$ is a valuated coproduct in G. Therefore, the set $\{x + A : x \in Y_A\}$ is a decomposition basis for G/A by Corollary 3.8.17. By Proposition 3.15.3, $(F + A)/A = \langle x + A : x \in Y_A \rangle$ is separable.

There is an $A_1 \in \mathcal{C}$ containing $\langle S, A \rangle$ such that $|A_1/A| \leq \aleph_0$. Since $(F + A)/A$ is separable, there is a subgroup B_1 of G such that $A_1 \subseteq B_1$, $|B_1/A| \leq \aleph_0$ and $B_1/A \| (F+A)/A$ (Lemma 3.15.4 applied to $(F + A)/A$ and A_1/A). Recall that the knice subgroup A is automatically quasi-sequentially nice (Proposition 3.12.7); thus, $B_1 \| F$ by Lemma 3.15.2. By Proposition 1.7.3, there is a countable pure subgroup C_1/A of G/A containing B_1/A. Then C_1 is pure in G by Lemma 1.7.2(3). There exists an $A_2 \in \mathcal{C}$ containing C_1 such that $|A_2/A| \leq \aleph_0$. Then there is a subgroup B_2 of G such that $A_2 \subseteq B_2$, $|B_2/A| \leq \aleph_0$, and $B_2 \| F$. Again, there is a countable pure subgroup C_2/A of G/A containing B_2/A; hence, C_2 is pure in G, et cetera. This yields an ascending sequence of subgroups

$$A_1 \subseteq B_1 \subseteq C_1 \subseteq A_2 \subseteq B_2 \subseteq C_2 \subseteq \cdots,$$

and we let $H = \bigcup_{i=1}^{\infty} A_i$. Then $H \in \mathcal{C}$, $H/\langle X_H \rangle$ is torsion, H is pure in G, and $H \| F$. Therefore, $H \in \mathcal{C}'$. It is clear that $\langle S, A \rangle \subseteq H$ and $|H/A| \leq \aleph_0$; hence, $\mathcal{C}'$ satisfies (H3). Consequently, $\mathcal{C}'$ is a $G(\aleph_0)$-family of knice subgroups of G, as desired.

(3) implies (1). Assume that G has a $G(\aleph_0)$-family of knice subgroups. Then G has an $H(\aleph_0)$-family of nice subgroups by Theorem 3.14.2(3) ⇒ (2) and possesses a decomposition basis (Theorem 3.14.2(3) ⇒ (7) and Lemma 3.9.7). □

Remark. Condition (2) of Theorem 3.15.5 does not appear explicitly in [53, Theorem 3.2], but follows from the proof.

Exercises.
1. Prove (a) and (b).
2. (Megibben–Ullery [116]) A subgroup A of G is called **strongly separable** in G, if for every $x \in G$ there is a countable subset $\{a_n : n < \omega\}$ of A such that the following

holds: For every $a \in A$, there is an $n < \omega$ such that $\|x + a_n\| \geq \|x + a\|$. Show: If A is an isotype subgroup of G such that A possesses a decomposition basis X, then $\langle X\rangle$ is a strongly separable subgroup of G.

3.16 The class of global Warfield groups

We summarize the various characterizations of global Warfield groups (see Theorems 3.9.8, 3.9.10, 3.14.2, 3.15.5 and Corollaries 3.12.16, 3.12.17):

Theorem 3.16.1. *Let G be a group. Then the following conditions are equivalent:*
1. *G is a Warfield group;*
2. *G is sequentially-pure-projective;*
3. *G has a nice decomposition basis X such that $G/\langle X\rangle$ is simply presented;*
4. *G is an Axiom 3 group;*
5. *G has a $G(\aleph_0)$-family of knice subgroups;*
6. *G has a $G(\aleph_0)$-family of pure knice subgroups;*
7. *G has an $F(\aleph_0)$-family of pure knice subgroups;*
8. *G has a composition series of knice subgroups;*
9. *G has a composition series of quasi-sequentially nice subgroups;*
10. *G has a composition series $\{N_\alpha\}_{\alpha\leq\lambda}$ of knice subgroups such that if $N_{\alpha+1}/N_\alpha$ is infinite, there is a primitive element $x_\alpha \in G$ such that $N_{\alpha+1} = N_\alpha \oplus \langle x_\alpha\rangle$ is a $*$-valuated coproduct in G;*
11. *G has a composition series $\{N_\alpha\}_{\alpha\leq\lambda}$ of nice subgroups such that if $N_{\alpha+1}/N_\alpha$ is infinite, there is an $x_\alpha \in G$ such that $N_{\alpha+1} = N_\alpha \oplus \langle x_\alpha\rangle$ is a valuated coproduct in G;*
12. *G has an $H(\aleph_0)$-family of nice subgroups and possesses a decomposition basis;*
13. *G has a $G(\aleph_0)$-family of nice subgroups and possesses a decomposition basis.*

The next result is taken from [73]. In terms of an affable (instead of a simply presented) group H, it can be found in [140, p. 55].

Theorem 3.16.2. *Suppose that G is a group. Then there is a simply presented group H such that $G \oplus H$ and H have the same Ulm and Warfield invariants. If G is torsion-free, we may take H to be completely decomposable.*

Proof. Let $I = \{(c,p,e) : w(c,p,e,G) \neq 0\}$. Then for every $(c,p,e) \in I$, there is a nonzero coset $x + G(M^*,p) \in G(M)/G(M^*,p)$ for some $M \in c$ with $M_p \in e$. Then $\|x\| \sim M$, and x has infinite order since $x \notin G(M^*)$. Further, $x \notin G(M_p^*,p)$ shows that $|p^ix|_p \neq m_{p,i}$ for at most finitely many i; hence, $\|x\|_p \sim M_p \in e$. Let $\mathfrak{m} = |G| \cdot \aleph_0$. In the case where G is torsion-free, the group $\langle x\rangle^0$ is simply presented since it is a torsion-free rank 1 group, and we define $H(c,p,e) = \bigoplus_{\mathfrak{m}} \langle x\rangle^0$. Now suppose that G is not torsion-free. By Corollary 2.7.7, there is a simply presented group H_y of torsion-free rank 1 containing an element of infinite order with Ulm matrix M, and we let $H(c,p,e) = \bigoplus_{\mathfrak{m}} H_y$. In either case, let $H' = \bigoplus_{(c,p,e)\in I} H(c,p,e)$. Then $G \oplus H'$ and H' have the same Warfield invariants, since the Warfield invariant of a direct sum is the sum of the invariants of the summands, and $\mathfrak{m} + w(c, p, e, G) = \mathfrak{m}$ for all $(c, p, e) \in I$.

Let

$$J = \{(p, \alpha) : p \text{ is a prime}, \alpha \in \mathrm{Ord}_{\infty}, u_p(\alpha, G) \neq 0\},$$

and pick $(p, \alpha) \in J$. If α is an ordinal, let $T_{p,\alpha}$ be the simply presented p-group P_α, as in Theorem 2.7.4. If $\alpha = \infty$, let $T_{p,\alpha} = \mathbb{Z}(p^\infty)$. We define $T = \bigoplus_{(p,\alpha)\in J} \bigoplus_{\mathfrak{m}} T_{p,\alpha}$ and $H = H' \oplus T$. Then it follows that

$$u_p(\alpha, G \oplus H) = u_p(\alpha, H)$$

for all $(p, \alpha) \in J$. Consequently, $G \oplus H$ and H have the same Ulm and Warfield invariants, and H is simply presented. If G is torsion-free, then H is torsion-free and completely decomposable by construction of H'. □

Finally, we establish the uniqueness theorem for the class of global Warfield groups, extending its p-local version (Theorem 3.3.6):

Theorem 3.16.3 (Hunter–Richman [61], Stanton [140]). *The class of Warfield groups is the unique class of groups that* (1) *contains all simply presented groups, is closed under* (2) *finite direct sums and* (3) *direct summands, and in which* (4) *the Ulm and Warfield invariants distinguish between nonisomorphic groups.*

Proof. The class D of Warfield groups satisfies (1) and (3) by definition, and (2) since the class of simply presented groups is closed under finite direct sums. If we characterize Warfield groups using Theorem 3.16.1(3), then by Theorem 3.7.7, the class satisfies (4). Since Warfield groups are direct summands of simply presented groups, any class C of groups satisfying (1–4) contains D by (1) and (3). Let $G \in$ C. By Theorem 3.16.2, there is a simply presented group H such that $G \oplus H$ and H have the same Ulm and Warfield invariants. By (1), H is in C; hence, $G \oplus H \in$ C by (2). Then (4) shows that $G \oplus H \cong H$; hence, $G \oplus H$ is simply presented. Therefore, G is a Warfield group, and we have $G \in$ D. □

Theorem 3.16.4. *A torsion group is a Warfield group if and only if it is simply presented. A torsion-free group is a Warfield group if and only if it is completely decomposable.*

Proof. For the first assertion, see Corollary 2.8.21. The second claim follows from Theorems 3.16.1(1) ⇒ (12), 3.1.2 and Corollary 2.6.2. □

Corollary 3.16.5 (Baer [6], Kulikov [90], Kaplansky [79]). *The class of completely decomposable torsion-free groups is closed under direct summands.*

3.17 Going further

The history of abelian group classification has been one of successive expansion of the class of groups considered. There seems to be a limit to this expansion. In fact, there may not be a classification theory for even the class of torsion-free groups of finite rank. There is current research investigating the limits of classification, determining the point at which the invariants become as complex as the isomorphisms themselves. See Hjorth [56], Thomas [144], and Downey–Montalbán [20].

Hunter, Richman, and Walker [65] studied Warfield modules as valuated extensions, and extended Warfield's Theorem 3.2.8 to valuated $\mathbb{Z}_p$-modules G having a quasi-sequentially nice valuated submodule A such that G/A is a Warfield module (see also Stanton [141]). In [66], they established necessary and sufficient conditions for the existence of Warfield modules with prescribed Ulm and Warfield invariants. Richman [130] generalized the notion of a Warfield module in terms of valuated modules other than direct sums of cyclics and proved a classification theorem. Lane [92] extended the theory of Warfield modules to SW-**modules** (= direct sums of Warfield modules and S-groups). The torsion part of a Warfield module is an S-group (see Hunter [60], Hunter–Walker [62]). Hill [43, 44] investigated and classified A-**modules,** which are certain isotype submodules of Warfield modules. In [52], Hill and Megibben investigated IW-**groups** (= the isotype submodules of Warfield modules). Richman [132] developed the constructive theory of countably generated Warfield modules. Files [22] proved that Warfield modules are determined (up to isomorphism) by their endomorphism rings.

In [61], Hunter and Richman proved existence theorems of (global) Warfield groups. It is shown in [61] that a Warfield group is a direct summand of a simply presented group with a simply presented torsion complement. Stanton [142] studied Warfield groups in terms of **affable groups** (= direct sums of Warfield groups of torsion-free rank 1) and **almost affable groups**, defined to be groups G such that $G \oplus T$ is an affable group with T a simply presented torsion group. Hunter [60] investigated sequentially pure exact sequences (calling them *balanced exact*; not to be confused with "balanced exact" as defined in Section 2.11) and established classification results for projectives and injectives relative to them. Richman [131] generalized the notion of a (global) Warfield group and proved a classification theorem.

Problems.

1. (Fuchs [29, Problem 15.1]) Using the theory of totally projective p-groups and Ulm matrices, develop a theory of mixed groups G such that tG is totally projective and G/tG is completely decomposable (or divisible).
2. Classify and characterize isotype subgroups of Warfield groups.
3. Characterize the groups that can be embedded in Warfield groups.
4. Is there an Axiom 3 characterization of k-groups?
5. Is there an Axiom 3 characterization of torsion parts of Warfield groups?
6. Characterize those groups possessing an $H(\aleph_0)$-family of quasi-sequentially nice subgroups.

4 Infinitary logic looks at groups

We explore expanding the class to be classified by easing the isomorphism requirement. Partial isomorphism is a weakened version of isomorphism, which has an interesting model theoretic interpretation. Specifically, anything that we can state in the language $L_{\infty\omega}$ will be true in a group if and only if it is true in any partially isomorphic group. Barwise and Eklof made use of this approach to classify all torsion groups up to partial isomorphism.

4.1 Partial isomorphisms

Up to this point, we have been pushing the limits of the classes of groups that we can classify up to isomorphism. By weakening the isomorphism requirement, we can broaden the class of groups that we can classify. Here we look at groups that are "not quite isomorphic", but are "locally isomorphic", in the sense that they are covered by isomorphic subgroups that may be extended to finitely many arbitrary elements in either group.

Let G and H be groups. If S and T are any subgroups of G and H, respectively, then we call an isomorphism $f : S \to T$ a **partial isomorphism (from G to H)**. We say G is **partially isomorphic** to H, written $G \cong_p H$, if there is a nonempty set I of partial isomorphisms from G to H with the **back-and-forth property**: For any $f \in I$ and $a \in G$ (or $b \in H$), there is a $g \in I$ extending f, written $f \subseteq g$, such that $a \in \text{domain}(g)$ (or $b \in \text{range}(g)$). We will sometimes write

$$I : G \cong_p H.$$

Such a set of isomorphisms is called a **Karp system.** The following proof shows how this property allows us to go back and forth between two partially isomorphic groups to build a countable isomorphism.

Theorem 4.1.1. *If G and H are countable groups, and $G \cong_p H$, then $G \cong H$.*

Proof. Suppose that $I : G \cong_p H$. Enumerate the elements of G and H, say $G = \{x_0, x_1, \dots\}$, and $H = \{y_0, y_1, \dots\}$. Since I is not empty, we may choose some $f_0 \in I$. We define f_n by induction on $n > 0$. If n is odd, say $n = 2m + 1$, choose $f_n \in I$ such that $f_{n-1} \subseteq f_n$ and $y_m \in \text{range}(f_n)$. If n is even, say $n = 2m$, choose $f_n \in I$ such that $f_{n-1} \subseteq f_n$ and $x_{m-1} \in \text{domain}(f_n)$. Now let $f = \bigcup_n f_n$. If $x \in G$, then for some $n \geq 0$, $x \in \text{domain}(f_n) \subseteq \text{domain}(f)$, so domain$(f) = G$, and similarly, range$(f) = H$. Since each f_n is an isomorphism, so is f. □

As we have seen, the heights of elements of groups are essential to classifying the group. Both the Ulm and the Warfield invariants are defined in terms of heights. Partial isomorphisms preserve this key attribute.

https://doi.org/10.1515/9783110427684-004

Theorem 4.1.2. *Suppose $I : G \cong_p H$. If $f \in I$ and $x \in$ domain(f), then $|f(x)|_p^H = |x|_p^G$ for every prime p.*

Proof. By transfinite induction on the ordinal α, we claim $x \in p^\alpha G$ implies $f(x) \in p^\alpha H$. If $\alpha - 1$ exists, then $x = py$ for some $y \in p^{\alpha-1}G$. By the induction hypothesis, $f(y) \in p^{\alpha-1}H$, and so $f(x) = pf(y) \in p^\alpha H$. If α is a limit ordinal, then for every $\beta < \alpha$, $x \in p^\beta G$, and hence $f(x) \in p^\beta H$, so $f(x) \in p^\alpha H$. Applying the same argument to f^{-1} shows that $f(x) \in p^\alpha H$ implies $x \in p^\alpha G$. □

Just in case the reader may be starting to think that partially isomorphic groups are as good as isomorphic, here is an example of partially isomorphic groups that are of different cardinalities. On one side, we have the group of all finite sequences of integers (countable), and on the other, the group of all infinite sequences of integers (uncountable). Furthermore, one has a decomposition basis and the other does not.

The groups we examine are $\bigoplus \mathbb{Z} = \bigoplus_{i<\omega} Z_i$ and $\prod \mathbb{Z} = \prod_{i<\omega} Z_i$, where $Z_i = \mathbb{Z}$ for all $i < \omega$. First, we need some notation and a lemma. If G is $\bigoplus \mathbb{Z}$ or $\prod \mathbb{Z}$ and $i_1, \ldots, i_k$ are integers, we define $G_{i_1,\ldots,i_k}$ to be the subgroup of all $x = (x_i) \in G$ such that $x_i = 0$ for all $i \in \{i_1, \ldots, i_k\}$.

Lemma 4.1.3. *If $G = \bigoplus \mathbb{Z}$ or $G = \prod \mathbb{Z}$, and $d \in G_{i_1,\ldots,i_k}$ for some positive integers $i_1, \ldots, i_k$, with $d = (d_i)$ and $|d_j| = 1$ for a given $j < \omega$ with $j \notin \{i_1, \ldots, i_k\}$, then we have $G_{i_1,\ldots,i_k} = \langle d \rangle \oplus G_{i_1,\ldots,i_k,j}$.*

Proof. Let $(a_i)_{i<\omega}$ be an arbitrary element of $G_{i_1,\ldots,i_k}$. Then either $(a_i) - a_j d$ or $(a_i) + a_j d$ is in $G_{i_1,\ldots,i_k,j}$ (depending on the sign of d_j); therefore, $G_{i_1,\ldots,i_k} = \langle d \rangle + G_{i_1,\ldots,i_k,j}$. The sum is direct since if $ad \in G_{i_1,\ldots,i_k,j}$, then $ad_j = 0$; hence, $a = 0$. □

Theorem 4.1.4. $\bigoplus \mathbb{Z} \cong_p \prod \mathbb{Z}$.

Specifically, let I be the set of all isomorphisms f from some $\langle x_1, \ldots, x_n \rangle \subseteq \bigoplus \mathbb{Z}$ onto a subgroup of $\prod \mathbb{Z}$ such that

$$\bigoplus \mathbb{Z} = \langle x_1 \rangle \oplus \cdots \oplus \langle x_n \rangle \oplus \Big(\bigoplus \mathbb{Z}\Big)_{i_1,\ldots,i_n}$$

for some positive integers $i_1, \ldots, i_n$ and

$$\prod \mathbb{Z} = \langle f(x_1) \rangle \oplus \cdots \oplus \langle f(x_n) \rangle \oplus \Big(\prod \mathbb{Z}\Big)_{j_1,\ldots,j_n}$$

for some positive integers $j_1, \ldots, j_n$. Then I is a Karp system.

Proof. First, note that I is not empty since it contains the zero function. Let $x \in \bigoplus \mathbb{Z}$, and $f \in I$ with domain $\langle x_1 \rangle \oplus \cdots \oplus \langle x_n \rangle$ and associated $i_1, \ldots, i_n$ and $j_1, \ldots, j_n$. We may assume that $x \notin$ domain(f). We may write $x = c + d$, where $c \in \langle x_1 \rangle \oplus \cdots \oplus \langle x_n \rangle$ and $d \in (\bigoplus \mathbb{Z})_{i_1,\ldots,i_n}$. Let $(d_i)_{i<\omega} = d$, and $m = \min_{d_i \neq 0}\{|d_i|\}$, say $|d_j| = m$. We will prove that f extends to x by induction on m.

Suppose $m = 1$. By the lemma, we get $\bigoplus \mathbb{Z} = \langle x_1 \rangle \oplus \cdots \oplus \langle x_n \rangle \oplus \langle d \rangle \oplus (\bigoplus \mathbb{Z})_{i_1,\ldots,i_n,j}$, and $x = c + d \in \langle x_1, \ldots, x_n, d \rangle$. Now extend f to d by letting $f(d)$ be some element y of $\prod \mathbb{Z}$ with $|y_k| = 1$, for some index k distinct from $j_1, \ldots, j_n$. By the lemma again, $\prod \mathbb{Z} = \langle f(x_1) \rangle \oplus \cdots \oplus \langle f(x_n) \rangle \oplus \langle f(d) \rangle \oplus (\prod \mathbb{Z})_{j_1,\ldots,j_n,k}$. Since f is an isomorphism on the summands, f is an isomorphism.

Now suppose $m > 1$. For each $i < \omega$, write $d_i = q_i m + r_i$ for some $q_i \in \mathbb{Z}$ and $0 \le r_i < m$. Let $q = (q_i)$, and $r = (r_i)$. Since $|d_j| = m$, $|q_j| = 1$, so by the $m = 1$ case, we may extend f to q. By the induction hypothesis, f may be further extended to r, and hence to $x = c + mq + r$. The other direction is the same. □

We will see in Section 5.7 that $\prod \mathbb{Z}$ has no decomposition basis. Clearly, $\bigoplus \mathbb{Z}$ has a decomposition basis.

Exercises.

1. Suppose $G \cong_p H$. Prove or provide a counterexample:
   (a) If G is torsion, then so is H.
   (b) If G is divisible then so is H.
   (c) If G is finitely generated, then so is H.
   (d) If G is completely decomposable, then so is H.
2. Without using results from the following sections, prove that $\cong_p$ is an equivalence relation.

4.2 The language $L_{\infty\omega}$ and $L^{\lambda}_{\infty\omega}$

So far we have been taking an algebraic view of things. Now we look at groups from the view of logic and model theory. We can compare groups by the logical statements they agree on. If they agree on all possible statements in a language, they are alike in some sense. The richer the language, the more alike they are. We will look at the algebraic implications of equivalence in certain languages.

We start with the **signature** L **of groups**, which consists of the following **nonlogical symbols:** the constant symbol 0, the binary function symbol +, and the unary function symbol − for the inverse. Rather than the clumsy $+(x, y)$, we will write the more familiar $x + y$. We can put these together to form terms, such as $x + 0$, $y + (-y)$ and $x + (y + z)$. The **atomic formulas** of the signature $L = (0, +, -)$ are certain expressions in terms of variables $v_1, v_2, v_3, \ldots$ and the elements of L, and can be written as

$$\sum_{i=1}^{k} n_i v_i = 0,$$

where $n_1, \ldots, n_k \in \mathbb{Z}$. To keep the notation simple and familiar, sometimes we will use x, y and z as names of variables rather than the formal v_1, v_2, v_3. We wrap the atomic formulas with the familiar language of first-order logic with equality. This provides the following symbols: $\wedge$ (and), $\vee$ (or), $\neg$ (not), $\exists$ (there exists an element...), $\forall$ (for all elements...), $\rightarrow$ (implies), and parentheses. Some may quibble that some of these are redundant, that $\rightarrow$ and $\forall$ may be written in terms of the others, but we include them here to make the formulas we write easier to read. Formally, the **(finitary) first-order language** $L_{\omega\omega}$ is defined to be the smallest class Y of elements, called **formulas**, containing the atomic formulas such that

1. if $\varphi, \psi \in Y$, then $\neg\varphi$, $\varphi \wedge \psi$, $\varphi \vee \psi \in Y$;
2. if $\varphi \in Y$ and v is a variable, then $\forall v\varphi$, $\exists v\varphi \in Y$.

Variables that are not under the control of $\forall$ or $\exists$ are called **free variables.** For example, in the formula $\exists x(x + y = z)$, the variables y and z are free. We may include the free variables in the name of the formula, such as $\varphi(y, z)$.

Sentences are formulas with no free variables. The language $L_{\omega\omega}$ allows us to write sentences such as these:

Associativity: $\forall x\forall y\forall z(x + (y + z) = (x + y) + z)$
Identity: $\forall x((x + 0 = x) \wedge (0 + x = x))$
Inverse: $\forall x((x + (-x) = 0) \wedge ((-x) + x = 0))$

For talking about groups, we define a model to consist of a nonempty set G, a binary function $+$ on the set, an element 0 in the set, and a function $-$ on the set, written $(G, +, -, 0)$. We say $(G, +, -, 0)$ is a **model** of a sentence φ, written $(G, +, -, 0) \models \varphi$, if the sentence is true in G. For formal definitions, see Hodges [57]. Generally, when we talk about a group G, the $+$ and $-$ operators, and the 0 are implicit, and so we will write simply $G \models \varphi$. More generally, we may replace the free variables with elements of the group under discussion and write $G \models \varphi[a_1, \dots, a_n]$ to mean that the statement that results from substituting the elements $a_1, \dots, a_n$, respectively, for the free variables holds in the group G. We can also say a group G **models** a set of sentences T, $G \models T$.

The three example sentences above are clearly the definition of a group. Therefore, if T is the set of those three sentences, $G \models T$ if and only if G is a group. In other words, this language is rich enough to say, "G is a group" using only finitely many sentences.

We want to talk about abelian groups, so we add the following sentence:

Commutativity: $\forall x\forall y(x + y = y + x)$

We can combine the four sentences together to get a single sentence:

$\forall x\forall y\forall z((x + (y + z) = (x + y) + z) \wedge (x + 0 = x) \wedge (x + (-x) = 0) \wedge (x + y = y + x))$,
which says, "This is an abelian group."

We say two models are **equivalent** in the language if they satisfy the same sentences. Hence, if G and H are equivalent in $L_{\omega\omega}$, and one is a group, then so is the other, and if one is abelian, then so is the other.

But if we try to say anything more complex, we run into the limits of this finite language. We cannot express concepts such as divisible, torsion, or even infinite. Algebraists will naturally want to say "for every integer," but this language will not support that. The quantifiers apply only to elements of the group.

To prove that there is no way of expressing these concepts, we turn to one form of the Compactness theorem. For a proof, see Hodges [57]. We say a sentence φ is a **consequence** of a set of sentences T if every model of T is a model of φ.

Theorem 4.2.1 (Compactness theorem). *If T is a set of sentences of first-order logic, φ a sentence of first-order logic, and φ is a consequence of T, then φ is a consequence of U for some finite $U \subseteq T$.*

The compactness theorem tells us that there is no statement of first-order logic that says, "This group is divisible." For suppose there were some such statement φ. We write ny as an abbreviation for $y + \cdots + y$ with n copies of y. We can prove, by induction, that for any n, there is such an expression in our first-order language. Let T be the set of statements $\{\varphi_n : n \in \mathbb{Z}^+\}$, where φ_n is $\forall x \exists y(ny = x)$. Then any group that satisfies all the statements in T is divisible, and so satisfies φ. By compactness, φ is a consequence of some finite $U \subseteq T$. Let n be the largest positive integer such that $\varphi_n \in U$. Choose some prime $p > n$. Then $\mathbb{Z}_p$ satisfies all the statements of U, but is not divisible, a contradiction.

We can get a language that allows us to talk about some of these key properties of groups if we consider $L_{\infty\omega}$, which allows infinitely many conjunctions and disjunctions. This is a language in the formal sense, since the sentences may be infinitely long, and so cannot be written out precisely. However, we will adopt abbreviations that will allow us to talk about the sentences. Specifically, once we have proven that an algebraic expression can be written in the language, we will use it in quotation marks to represent its formula in more complex formulas.

The ∞ in $L_{\infty\omega}$ tells us that we may use infinitely many conjunctions and disjunctions, whereas the ω tells us that we may have only finitely many quantifiers in a row. We first define $L^{\lambda}_{\infty\omega}$ for λ an ordinal.

The atomic formulas are the quantifier-free, well-formed formulas of group theory, such as $(x + y) + z = 0$ or $x + y = y + x$.

Now for each ordinal λ we define, by induction on λ, $L^{\lambda}_{\infty\omega}$ to be the smallest class Y containing the atomic formulas and such that

(i) if $\varphi \in Y$, then $\neg\varphi \in Y$;
(ii) if Φ is a set and $\Phi \subseteq Y$, then $\wedge\Phi \in Y$ and $\vee\Phi \in Y$;
(iii) if $\varphi \in L^{\beta}_{\infty\omega}$ for some $\beta < \lambda$, and if v is a variable, then $\forall v\varphi \in Y$ and $\exists v\varphi \in Y$.

Then we let $L_{\infty\omega} = \bigcup_{\lambda \in \mathrm{Ord}} L^{\lambda}_{\infty\omega}$. We define free and bound variables as before and say that φ is a **formula** of $L_{\infty\omega}$ if $\varphi \in L_{\infty\omega}$ and φ has at most a finite number of free variables. An **existential formula** is one of the form $\exists v\varphi$. Again, **sentences** are formulas with no free variables.

Now we can build some of those sentences that allow us to talk about some of the characteristics of infinite groups. As an example, we will build a sentence to say, "This group is torsion."

As before, we let "nx" be the abbreviation for "$x + \cdots + x$", a representation of the term that comes from successively adding the variable x n times. Then "$nx = 0$" is an atomic formula, call it $\varphi_n(x)$. As an atomic formula, $\varphi_n(x)$ is in $L^0_{\infty\omega}$. Let Φ be the set of formulas $\{\varphi_n(x) : n \in \mathbb{N}\}$. Then by (ii), $\vee\Phi(x)$ is a formula of $L^0_{\infty\omega}$. This formula essentially says, "For some positive integer n, $nx = 0$." Finally, by (iii), $\forall x(\vee\Phi(x))$ is a formula of $L^1_{\infty\omega}$. Note that, for a group G, we have $G \models \forall x(\vee\Phi(x))$ if and only if G is torsion. We may write this formula informally as $\forall x \bigvee_{n\in\mathbb{N}}(\text{“}nx = 0\text{”})$.

The minimal ordinal λ such that a formula φ is in $L^\lambda_{\infty\omega}$ is called the **quantifier rank** of φ, written $qr(\varphi)$. This is a measure of the complexity of the formula, and indicates how deeply nested the quantifiers are. The example formula has quantifier rank of only 1, but as we explore the invariants, related to heights of elements, we will find ourselves building much more complex formulas.

We may use equivalences in these languages to compare groups, specifically, for G and H groups, we write

$$G \equiv_\infty H$$

and say that G **is equivalent to** H **in** $L_{\infty\omega}$, if G and H satisfy the same sentences of $L_{\infty\omega}$. Likewise,

$$G \equiv_\lambda H$$

means G and H satisfy the same sentences of $L^\lambda_{\infty\omega}$.

Exercises.

1. Prove that "torsion-free" is not expressible in the first-order language.
2. Write a formula of $L_{\infty\omega}$ to express each of the following. You may use abbreviations as long as you make it clear that the abbreviated formula is in the language. Find the quantifier rank of each formula.
   (a) Divisible
   (b) Torsion-free
   (c) Reduced
   (d) Finitely generated
   (e) Completely decomposable
   (f) Infinite
3. Suppose $G \equiv_1 H$ for some groups G and H. Prove that if G is a mixed group (neither torsion nor torsion-free), then so is H.

4.3 Karp's theorem

Now we have the algebraic view of groups: Groups are exactly the same if they are isomorphic, and are exactly the same in regions if they are partially isomorphic. We saw that groups that are partially isomorphic share many of the key characteristics that distinguish groups. We also have the logical view: While two structures that are equivalent in $L_{\omega\omega}$ share only some basic properties, such as being a group, groups that are equivalent in $L_{\infty\omega}$ share many more properties.

This gives us two very different ways to look at groups that are equivalent in some sense. Remarkably, Karp proved that these two views are exactly the same. Whereas Karp proved this for general models, we state it here specifically for groups.

Karp's theorem can be generalized to cover equivalence in $L^{\lambda}_{\infty\omega}$. We will prove that first, since the standard theorem follows from it. This more general theorem will be useful later.

Theorem 4.3.1 (Karp [80]). *Let G and H be R-modules, where $R = \mathbb{Z}$ or $R = \mathbb{Z}_p$ for some prime p, and let $\lambda \in \mathrm{Ord}_\infty$. Then the following are equivalent:*
1. $G \equiv_\lambda H$;
2. *For each ordinal $\nu \le \lambda$, there is a nonempty set I_ν of isomorphisms on finitely generated submodules of G into H such that*
   (a) *if $\nu \le \mu$, then $I_\mu \subseteq I_\nu$, and if $\lambda = \infty$, then $I_\mu = I_\nu$;*
   (b) *if $\nu < \lambda$, $f \in I_{\nu+1}$, and $x \in G$ (or $y \in H$), then f extends to a map $f' \in I_\nu$ such that $x \in \mathrm{domain}(f')$ (or $y \in \mathrm{range}(f')$).*

In that case, if $\varphi(x_1,\dots,x_n)$ is a formula with $qr(\varphi) = \alpha \le \lambda$, $f \in I_\alpha$, $a_1,\dots,a_n \in$ domain(f), then $G \models \varphi[a_1,\dots,a_n]$ if and only if $H \models \varphi[f(a_1),\dots,f(a_n)]$.

Proof. Suppose for every $\nu \le \lambda$, there is a set I_ν as in (2). Then we prove, by induction on $\nu \le \lambda$, that for any $f \in I_\nu$, $a_1,\dots,a_n \in$ domain(f), and $\varphi(x_1,\dots,x_n)$ a formula of quantifier rank ν, $G \models \varphi[a_1,\dots,a_n]$ if and only if $H \models \varphi[f(a_1),\dots,f(a_n)]$. Consider a formula φ of the form $\exists x_{n+1}\psi(x_1,\dots,x_{n+1})$, since the other cases are obvious or can be derived from this. Suppose $G \models \varphi[a_1,\dots,a_n]$. Then $G \models \psi[a_1,\dots,a_n,a]$ for some $a \in G$. If $\beta = qr(\psi)$, then $\beta + 1 = \nu$; hence, $f \in I_{\beta+1}$. Extend f to some $g \in I_\beta$ with $a \in$ domain(g). Then, by the induction hypothesis, $H \models \varphi[g(a_1),\dots,g(a_n),g(a)]$; hence, $H \models \psi[f(a_1),\dots,f(a_n),b]$ for some $b \in H$, which gives us $H \models \varphi[f(a_1),\dots,f(a_n)]$. The other direction is the same. Note that, in particular, if φ is a sentence, then $G \models \varphi$ if and only if $H \models \varphi$.

Now suppose $G \equiv_\lambda H$. Let I_ν be the set of all isomorphisms f between some S and some T, where S and T are finitely generated submodules of G and H, respectively, such that for all $a_1,\dots,a_n \in S$ and all $\varphi(x_1,\dots,x_n) \in L^{\nu}_{\infty\omega}$, $G \models \varphi[a_1,\dots,a_n]$ if and only if $H \models \varphi[f(a_1),\dots,f(a_n)]$. Since the zero function is in I_ν, I_ν is not empty.

Now suppose $f \in I_{\nu+1}$ and let $a \in G$. We wish to show that f can be extended to a. We may suppose $a \notin$ domain(f), since otherwise we are done. Suppose domain(f) =

$\langle a_1,\dots,a_n\rangle$. We want to find a $b \in H$ such that for all formulas $\varphi(x_1,\dots,x_{n+1})$ of quantifier rank $\le \nu$, if $G \models \varphi[a_1,\dots,a_n,a]$, then $H \models \varphi[f(a_1),\dots,f(a_n),b]$. If this were not possible, then for each $b \in H$ there would be a formula φ_b of quantifier rank $\le \nu$, for which this fails, specifically, $G \models \varphi_b[a_1,\dots,a_n,a]$, and $H \models \neg\varphi_b[f(a_1),\dots,f(a_n),b]$. Now let $\psi(x_1,\dots,x_{n+1}) = \bigwedge\{\varphi_b(x_1,\dots,x_{n+1}) : b \in H\}$, which is a formula in $L^{\nu}_{\infty\omega}$. Then we have $G \models \exists x_{n+1}\psi(x_{n+1})[a_1,\dots,a_n]$, whereas $H \models \neg\exists x_{n+1}\psi(x_{n+1})[f(a_1),\dots,f(a_n)]$. But this contradicts the definition of $I_{\nu+1}$ since $f \in I_{\nu+1}$. Thus, there is a $b \in H$ such that we may extend f to a $g \in I_\nu$ by letting $g(a) = b$. Any formula involving elements of $\langle a_1,\dots,a_n\rangle$ may be converted into one involving $a_1,\dots,a_n$, so g preserves all formulas. □

A system $\{I_\nu : \nu \le \lambda\}$ as in Theorem 4.3.1(2) is called a **Karp system.** Now it follows, by taking $\lambda = \infty$, that partial isomorphism and equivalence in $L_{\infty\omega}$ are two different views of the same group comparison:

Theorem 4.3.2 (Karp [80]). *Let G and H be groups. Then $G \equiv_\infty H$ if and only if $G \cong_p H$. In that case, we may choose $I : G \cong_p H$ such that every $f \in I$ has finitely generated domain and range.*

Note that I is the set of all isomorphisms f between some S and some T, where S and T are finitely generated subgroups of G and H, respectively, such that for all $a_1,\dots,a_n \in S$ and all $\varphi(x_1,\dots,x_n) \in L_{\infty\omega}$, we have $G \models \varphi[a_1,\dots,a_n]$ if and only if $H \models \varphi[f(a_1),\dots,f(a_n)]$.

In the following and in Chapter 5, we will classify groups up to partial isomorphism, and it will be handy to be able to use both of these very different viewpoints.

Exercises.

1. Prove that $\cong_p$ is an equivalence relation.
2. Prove that "uncountable" cannot be expressed by a sentence of $L_{\infty\omega}$.
3. Suppose $G \equiv_\infty H$, and let G_0 be a countable subgroup of G. Prove that H has a subgroup H_0 such that $G_0 \cong H_0$.
4. (Barwise [8]) Let $G_j \cong_p H_j$ be groups for each j in some index set J. Prove that
$$\prod_{j\in J} G_j \cong_p \prod_{j\in J} H_j$$
and
$$\bigoplus_{j\in J} G_j \cong_p \bigoplus_{j\in J} H_j.$$
5. (Barwise–Eklof [9]) Let I and J be infinite index sets, and suppose for some group G, $G_i = G = H_j$ for all $i \in I$ and $j \in J$. Then $\bigoplus_{i\in I} G_i \cong_p \bigoplus_{j\in J} H_j$.

4.4 Expressibility in $L_{\infty\omega}$

Barwise and Eklof [9] examined torsion groups from the viewpoint of $L_{\infty\omega}$ and in the process found that many of the key characteristics of groups are expressible in $L_{\infty\omega}$. Their work culminated in a generalization of Ulm's theorem. We first note that height is expressible. Whereas Barwise and Eklof proved this for p-groups, the proof works just as well for arbitrary groups.

Lemma 4.4.1 (Barwise–Eklof [9]). *Fix a prime p, and let $\alpha = \omega\delta + n$ be an ordinal, where $n < \omega$. Then there is a formula $\theta_\alpha(x)$, existential if $\alpha \neq 0$, whose quantifier rank is at most*

$$\begin{cases} \delta & \text{if } n = 0, \\ \delta + 1 & \text{if } n > 0, \end{cases}$$

such that for any group G and any $a \in G$, we have $G \models \theta_\alpha[a]$ if and only if $a \in p^\alpha G$.

Proof. The proof is by transfinite induction on α. For $\theta_0(x)$, we may use $x = x$. Assume that there is such a formula $\theta_\beta(x)$ for all $\beta < \alpha$. First, consider the case of $n = 0$. Let $\theta_\alpha(x) = \bigwedge_{\beta<\omega\delta} \theta_\beta(x)$. Then, for any group G, $a \in p^{\omega\delta}G$ if and only if for all $\beta < \omega\delta$, $a \in p^\beta G$, if and only if $G \models \theta_\alpha[a]$. For any $\beta < \alpha = \omega\delta$, the induction hypothesis tells us that the quantifier rank of $\theta_\beta(x)$ is at most δ, and so the quantifier rank of θ_α is at most δ.

Now suppose $n > 0$. By the induction hypothesis, there is a formula $\theta_{\omega\delta}(x)$ of quantifier rank no more than δ. Then $a \in p^\alpha G$, if and only if there is a $b \in p^{\omega\delta}G$ such that $a = p^n b$. Let $\theta_\alpha(x) = \exists y(\theta_{\omega\delta}(y) \wedge x = p^n y)$. The quantifier rank of this formula is at most $\delta + 1$. □

As we have seen, the notion of height is central to the study of mixed groups. This result allows us to write formulas for analogues of rank and Ulm invariants. We will henceforth adopt the convention of using an algebraic formula in quotation marks to represent the corresponding expression in $L_{\infty\omega}$. For example, the above lemma may be written as

"$x \in p^{\omega\delta+n}G$" is expressible by a formula of quantifier rank $\leq \delta$ if $n = 0$, $\leq \delta + 1$ if $n > 0$.

Alternatively, we may write

"$x \in p^{\omega\delta+n}G$" $\in L^\delta_{\infty\omega}$ if $n = 0$ and in $L^{\delta+1}_{\infty\omega}$ if $n > 0$.

Then we may use these expressions to represent the corresponding formula within formulas of $L_{\infty\omega}$. For example,

"$|x|_p = \alpha$", which may be written as "$x \in p^\alpha G$" $\wedge \neg$"$x \in p^{\alpha+1}G$", is in $L^{\delta+2}_{\infty\omega}$, where $\alpha = \omega\delta + n$.

Note that "x has infinite order" and "$x \in t_pG$" are both expressible by formulas of quantifier rank 0.

Lemma 4.4.2. *Suppose that H is a subgroup of a group G and "$x \in H$" is in $L^{\delta}_{\infty\omega}$. Then "$x_1, \dots, x_m$ are independent modulo H" is also in $L^{\delta}_{\infty\omega}$.*

Proof. $x_1, \dots, x_n$ are independent modulo H if and only if for every $a_1, \dots, a_n \in \mathbb{Z}$, $a_1x_1 + \cdots + a_nx_n \in H$ implies $a_ix_i \in H$ for all $1 \le i \le n$. This may be written as $\bigvee_{a_1\in\mathbb{Z}} \cdots \bigvee_{a_n\in\mathbb{Z}}$("$a_1x_1 + \cdots + a_nx_n \in H$" $\to \bigvee_{i=1}^{n}$ "$a_ix_i \in H$"). Since each subexpression has quantifier rank $\le \delta$ and no quantifiers are introduced, this expression has quantifier rank $\le \delta$. □

Corollary 4.4.3. *Let G be a group, p a prime, and $\alpha = \omega\delta + n$ with $n < \omega$. Then "$x_1, \dots, x_m$ are independent modulo $p^{\alpha+1}G$" may be written as an expression in $L^{\delta+1}_{\infty\omega}$, and "$x_1, \dots, x_n$ are independent" has an expression in $L^{0}_{\infty\omega}$.*

We may also express the p-rank r_p, the torsion-free rank r_0, and the ranks $r_{p,0}$ and r of a group, for finite ranks.

Theorem 4.4.4 (Barwise–Eklof [9]). *Let G be a group, p a prime, and $\alpha = \omega\delta + n$ with $n < \omega$. If $m < \omega$, then "$r_p(p^{\alpha}G) \ge m$", "$r_0(p^{\alpha}G) \ge m$", "$r_{p,0}(p^{\alpha}G) \ge m$", and "$r(p^{\alpha}G) \ge m$" can be expressed in $L^{\delta+m}_{\infty\omega}$.*

Proof. We note that $r_p(p^{\alpha}G) \ge m$ if and only if there are m independent elements of p-height $\ge \omega\delta + n$, if and only if there are m elements $x_1, \dots, x_m$ of p-height $\ge \omega\delta$ such that $p^nx_1, \dots, p^nx_m$ are independent, if and only if $G \models \exists x_1 \dots x_m(\bigwedge_{i=1}^{m}$ "$x_i \in p^{\omega\delta}G$" $\wedge$ "$p^nx_1, \dots, p^nx_m$ are independent" $\wedge$ "$x_1, \dots, x_n \in t_pG$"). The proof is similar for r_0, and the result follows for $r_{p,0}$ and r. □

For any cardinal-valued function g, we define $\hat{g}$, g **hat**, as

$$\hat{g}(x) = \begin{cases} g(x) & \text{if } g(x) \text{ is finite,} \\ \omega & \text{if } g(x) \text{ is infinite.} \end{cases}$$

In particular, we have the modified Ulm invariant $\hat{u}$, the modified Warfield invariant $\hat{w}$ (which we will use in the next chapter), and the modified ranks $\hat{r}_0$, $\hat{r}_p$, and $\hat{r}$. The next theorem tells us that $\hat{u}$ is expressible.

Theorem 4.4.5 (Barwise–Eklof [9]). *Let p be a prime. For every $m < \omega$ and ordinal $\alpha = \omega\delta + n$ with $n < \omega$, there is a statement of $L^{\delta+m+1}_{\infty\omega}$ that says $u_p(\alpha, G) \ge m$.*

Proof. Given $\alpha = \omega\delta + n$, and $m < \omega$, we have $u_p(\alpha, G) \ge m$ if and only if $G \models \exists x_1 \dots x_m[\bigwedge_{i=1}^{m}$("$x_i \in p^{\alpha}G$" $\wedge\ px_i = 0) \wedge$ "$x_1, \dots, x_m$ are independent modulo $p^{\alpha+1}G$"]. Since the expressions in quotes have quantifier rank $\le \delta + 1$, and we have introduced m quantifiers, the quantifier rank of the expression is $\le \delta + m + 1$. □

Corollary 4.4.6. *If $G \equiv_p H$ for some groups G and H, then $\hat{u}_p(\alpha, G) = \hat{u}_p(\alpha, H)$ for all $\alpha \in \mathrm{Ord}_{\infty}$ and primes p.*

Proof. The result follows if α is an ordinal. If $\alpha = \infty$, note that $x \in p^{\infty}G$ if and only if $x \in p^{\mathrm{length}(G)}G$, so we may similarly construct a sentence that says "$\hat{u}_p(\infty, G) \ge m$". □

This gives us everything we need in order to talk about torsion groups in $L_{\infty\omega}$. Looking ahead, we examine expressibility for mixed groups. As we will see, not all of the key definitions are expressible in $L_{\infty\omega}$. In particular, "decomposition set" is not (Theorem 5.9.1). This means that if we wish to examine Warfield invariants in $L^{\alpha}_{\infty\omega}$, as we will in the next chapter, we will need to construct a related invariant. First, we need some definitions of familiar concepts modified for considering groups "up to α", which will be useful in the next chapter. Fix an ordinal α. For ordinals β and γ, we write

$$\beta =_{\alpha} \gamma$$

if $\beta \wedge \alpha = \gamma \wedge \alpha$. Likewise, we define $\beta \geq_{\alpha} \gamma$ to mean $\beta \wedge \alpha \geq \gamma \wedge \alpha$. In other words, $\beta \geq_{\alpha} \gamma$ holds exactly if $\beta \geq \gamma$ whenever $\gamma < \alpha$ and $\beta \geq \alpha$ whenever $\gamma \geq \alpha$. By $\beta >_{\alpha} \gamma$, we mean $\beta \geq_{\alpha} \gamma$ and $\beta \neq_{\alpha} \gamma$. Two Ulm sequences $u = (\beta_i)_{i<\omega}$ and $v = (\gamma_i)_{i<\omega}$ are said to be **equal up to** α, written

$$u =_{\alpha} v$$

if $\beta_i \wedge \alpha = \gamma_i \wedge \alpha$ for all $i < \omega$. We define $u \geq_{\alpha} v$ and $u \leq_{\alpha} v$ in the same way. For example, $u \geq_{\alpha} v$ if and only if $\beta_i \wedge \alpha \geq \gamma_i \wedge \alpha$ for all $i < \omega$. We say that u and v are **α-equivalent**, and write

$$u \sim_{\alpha} v$$

if $(\beta_{i+m})_{i<\omega} =_{\alpha} (\gamma_{i+n})_{i<\omega}$ for some $m, n > 0$. If e is an equivalence class of Ulm sequences and u an Ulm sequence, we write $u \sim_{\alpha} e$ if $u \sim_{\alpha} u'$ for some $u' \in e$. We say two Ulm matrices $M = [m_{p,i}]$ and $N = [n_{p,i}]$ are **equal up to** α, and we write

$$M =_{\alpha} N$$

if $m_{p,i} \wedge \alpha = n_{p,i} \wedge \alpha$ for all primes p and $i < \omega$, and $M \geq_{\alpha} N$ and $M \leq_{\alpha} N$ are defined analogously. We say M and N are **α-compatible**, written

$$M \sim_{\alpha} N$$

if there are $m, n > 0$ such that $mM \geq_{\alpha} N$ and $nN \geq_{\alpha} M$. Note that α-compatibility yields an equivalence relation. We say $M \sim_{\alpha} c$ for an Ulm matrix M and a compatibility class c of Ulm matrices if for some $N \in c$, $M \sim_{\alpha} N$.

Let G be a group, p a prime, M an Ulm matrix, and α an ordinal. We define the following:

$G_{\alpha}(M) = \{x \in G : \|x\| \geq_{\alpha} M\}$;
$G_{\alpha}(M^*)^{\diamond} = \{x \in G_{\alpha}(M) : \|x\| \nsim_{\alpha} M \text{ or } x \text{ is torsion}\}$;
$G_{\alpha}(M_p^*, p)^{\diamond} = \{x \in G : \|x\|_p \geq_{\alpha} M_p \text{ and } |p^i x|_p \neq_{\alpha} m_{p,i} \text{ for infinitely many } i\}$;

$$G_\alpha(M^*,p)^\diamond = G_\alpha(M^*)^\diamond \cup G_\alpha(M^*_p,p)^\diamond;$$
$$G_\alpha(M^*,p) = G_\alpha(M) \cap \langle G_\alpha(M^*,p)^\diamond \rangle.$$

Then for c a compatibility class of Ulm matrices and e an equivalence class of Ulm sequences, we modify the Stanton invariant as follows. Let

$$\hat{s}_\alpha(c,p,e,G) = \sup\{\hat{r}_{p,0}(G_\alpha(M)/G_\alpha(M^*,p))\},$$

where the supremum is taken over all Ulm matrices M with $M \sim_\alpha c, M_p \sim_\alpha e$ (see Section 3.7).

In the next chapter, this construct and its expressibility will help us prove a classification theorem.

Lemma 4.4.7 (Jacoby–Loth [70], Göbel–Leistner–Loth–Strüngmann [37]). *The statement "$\hat{s}_\alpha(c,p,e,G) \geq m$" is expressible by a formula of $L_{\infty\omega}$. If we write $\alpha = \omega\delta + n$ with $n < \omega$, the formula has quantifier rank $\leq \delta + \omega + m$.*

Proof. The local case was proved in [37], and we follow that proof here. Suppose $M \in c$, and $M_p \in e$. By Lemma 4.4.1, "$x \in q^\alpha G$" is expressible in a formula of quantifier rank δ or $\delta + 1$ for any q. We will use this to build our desired formula. To write "$x \in G_\alpha(M)$", we note that this is the same as "for all q and i, $q^i x \in q^{m_{q,i} \wedge \alpha} G$". Since "$q^i x \in q^{m_{q,i} \wedge \alpha} G$" is expressible with quantifier rank $\leq \delta + 1$, and since "for all q and i" may be represented using $\bigwedge$, which does not add to the quantifier rank, we see that "$x \in G_\alpha(M)$" is expressible with quantifier rank $\leq \delta + 1$.

Next, "$x \in G_\alpha(M^*)^\diamond$" is true if and only if $x \in G_\alpha(M)$ and either x is torsion or $\|x\| \nsim_\alpha M$. If $x \in G_\alpha(M)$, then $\|x\| \nsim_\alpha M$ if and only if for every $j < \omega$, $jM \ngeq_\alpha \|x\|$. This holds if and only if for every $j < \omega$ there are q and i, for which $|q^i x| > m_{q,i+|j|_q}$ and $m_{q,i+|j|_q} < \alpha$. We may write this so that all subformulas have quantifier rank no more than δ+1: $\bigwedge_j \bigvee_{q,i}($"$q^i x \in q^{(m_{q,i+|j|_q}+1) \wedge \alpha} G$"$\wedge m_{q,i+|j|_q} < \alpha)$. Since "$x \in G_\alpha(M)$" has quantifier rank $\leq \delta + 1$, and "x is torsion" has quantifier rank 0, we find that "$x \in G_\alpha(M^*)^\diamond$" has quantifier rank $\leq \delta + 1$.

The statement "$x \in G_\alpha(M^*_p, p)^\diamond$" holds if and only if $|p^i x|_p \geq m_{p,i} \wedge \alpha$ for all i, and for every i_0, there is an $i \geq i_0$ such that $p^i x \in p^{(m_{p,i}+1) \wedge \alpha} G$ and $p^i x \notin p^\alpha G$. Again, we have a formula of quantifier rank $\leq \delta + 1$.

Combining these, we see that "$x \in G_\alpha(M^*,p)^\diamond$" is expressible with quantifier rank $\leq \delta + 1$. Then "$x \in G_\alpha(M^*,p)$" is true if and only if $x \in G_\alpha(M)$ and there are $n < \omega$ and $z_1, \ldots, z_n \in G_\alpha(M^*,p)^\diamond$ such that $x = z_1 + \cdots + z_n$. The existence statement for the $z_1, \ldots, z_n$ adds n to the quantifier rank for each $n < \omega$, so this formula has quantifier rank $\leq \delta + \omega$.

Next, "$x_1, \ldots, x_m \in G_\alpha(M)$ are independent modulo $G_\alpha(M^*,p)$" has an expression of quantifier rank $\leq \delta + \omega$ by Lemma 4.4.2. Finally, we may write "$\hat{s}_\alpha(c,p,e,G) \geq m$" as

> $\exists x_1, \ldots, \exists x_m$("$x_1, \ldots, x_m \in G_\alpha(M)$, independent modulo $G_\alpha(M^*,p)$, and each x_i is of either p-power or infinite order").

This statement has quantifier rank $\leq \delta + \omega + m$. □

4.5 The classification of torsion groups up to partial isomorphism

Following Barwise and Eklof [9], we will prove that the modified Ulm invariants classify p-groups up to $L^{\lambda}_{\infty\omega}$-equivalence for appropriate λ, or up to partial isomorphism. First, we need some lemmas.

Lemma 4.5.1. *If G is a p-group, S a subgroup of G of finite p-rank, α a limit ordinal $\leq$ length(G), and $\beta < \alpha$, then for any $n < \omega$, there exists a $z \in p^{\beta}G$ such that $p^n z \notin S$ and $p^{n+1}z = 0$.*

Proof. Let $\beta = \omega\delta + n$ with $n < \omega$. By Lemma 1.13.7(2) there are infinitely many γ between β and α with $u(\gamma, G) \neq 0$. We claim that the rank $r_p(p^{\beta}G)$ is infinite. This is immediate if $r_p(p^{\gamma}G)$ is infinite for some such γ, since $r_p(p^{\beta}G) \geq r_p(p^{\gamma}G)$. So suppose for all γ with $\beta < \gamma < \alpha$ and $u(\gamma, G) \neq 0$, $r_p(p^{\gamma}G)$ is finite.

Suppose $x_1, \ldots, x_n$ are independent in $(p^{\gamma+1}G)[p]$. Since $u(\gamma, G) \neq 0$, there is a $y \in (p^{\gamma}G)[p] \setminus (p^{\gamma+1}G)[p]$. Then $x_1, \ldots, x_n, y$ are independent, since otherwise $ny \in \langle x_1, \ldots, x_n \rangle$ for some $0 < n < p$, and then, since n and p are relatively prime, $y \in \langle ny \rangle \subseteq p^{\gamma+1}G$, a contradiction. This implies that $r_p(p^{\gamma}G) > r_p(p^{\gamma+1}G)$. Since there are infinitely many such γ, we conclude that $r_p(p^{\beta}G)$ is infinite.

Now suppose for all $z \in p^{\beta}G$ with $p^{n+1}z = 0$, $p^n z \in S$. By Corollary 1.4.3, we obtain $r_p(S) \geq r_p((p^{\beta+n}G)[p]) = r_p(p^{\beta+n}G)$. Since this is infinite, we get the desired contradiction. □

Let S and T be subsets of R-modules G and H, respectively. Let $p \in R$ be a prime and α an ordinal. Then a map $f : S \to T$ is said to **preserve p-heights up to α** if

$$|f(x)|^H_p \wedge \alpha = |x|^G_p \wedge \alpha$$

for all $x \in S$. We say that f **preserves heights up to α**, and we call f **α-height-preserving**, if f preserves p-heights up to α for all primes p. For any map $f : S \to T$ we have the following:

(a) f preserves p-heights up to α if and only if $|f(x)|^H_p = |x|^G_p$ whenever $|x|^G_p < \alpha$, and $|f(x)|^H_p \geq \alpha$ whenever $|x|^G_p \geq \alpha$.

(b) f preserves p-heights up to α if and only if $|f(x)|^H_p =_{\alpha} |x|^G_p$ for all $x \in S$.

(c) In the case where $R = \mathbb{Z}_p$, f preserves p-heights up to α if and only if f preserves heights up to α.

(d) If f is a bijection and preserves p-heights up to α, then so does f^{-1}.

Lemma 4.5.2 (Barwise–Eklof [9]). *Let G and H be p-groups. Let ν be an ordinal such that $\omega(\nu + 1) \leq$ length(G), and $\omega(\nu + 1) \leq$ length(H). Let $f : S \to T$ be an isomorphism, where S and T are finitely generated subgroups of G and H, respectively, and f is $\omega(\nu + 1)$-height preserving. Then for every $a \in G$ such that $|a|^G_p \geq \omega\nu$, there is a $b \in H$ and an $\omega\nu$-height-preserving isomorphism $f' : \langle S, a \rangle \to \langle T, b \rangle$ such that $f \subseteq f'$ and $f'(a) = b$.*

Proof. If $a \in S$, let $f' = f$, and we are done. So suppose $a \notin S$, and let $r \geq 0$ be least such that $p^{r+1}a \in S$. Since $|p^{r+1}a| \geq \omega\nu + r + 1$, and f preserves heights up to $\omega(\nu + 1)$, $|f(p^{r+1}a)| \geq \omega\nu + r + 1$. Let $b_0 \in p^{\omega\nu}H$ such that $f(p^{r+1}a) = p^{r+1}b_0$. If $p^r b_0 \notin T$, we may let $b = b_0$. So suppose $p^r b_0 \in T$. By Lemma 4.5.1, there is a $b_1 \in p^{\omega\nu}H$ such that $p^r b_1 \notin T$ and $p^{r+1}b_1 = 0$. Let $b = b_0 + b_1$. In either case, we have $|b| \geq \omega\nu$, $p^r b \notin T$, and $p^{r+1}b = f(p^{r+1}a)$.

Repeated application of Lemma 1.6.5 yields an isomorphism $f' : \langle S, a\rangle \to \langle T, b\rangle$, extending f by $a \mapsto b$. We claim that f' is $\omega\nu$-height-preserving. Let $s \in S$ and $n \in \mathbb{Z}$. If $|s + na| \geq \omega\nu$, then $|f(s)| = |s| \geq \omega\nu$ and $|f(s + na)| = |f(s) + nb| \geq \omega\nu$. If $|s + na| < \omega\nu$, then $|f(s)| = |s| < \omega\nu$ and $|f(s + na)| = |f(s) + nb| = |f(s)| = |s| = |s + na|$. □

Let G be an R-module, A a submodule of G, $p \in R$ a prime, and α an ordinal. Recall that in Section 1.13, we defined

$$\phi : D_{\alpha,p}(A) = \frac{\{x \in A \cap p^\alpha G : px \in p^{\alpha+2}G\}}{A \cap p^{\alpha+1}G} \to \frac{(p^\alpha G)[p]}{(p^{\alpha+1}G)[p]}$$

by $\phi(x + A \cap p^{\alpha+1}G) = x - y + (p^{\alpha+1}G)[p]$, where $y \in p^{\alpha+1}G$ and $px = py$. Since ϕ is a monomorphism, we have $\dim D_{\alpha,p}(A) = \dim I_{\alpha,p}(A)$, where $I_{\alpha,p}(A)$ denotes the image of ϕ. Notice that this dimension is finite whenever A is finitely generated. As before, the subscript p is dropped if it is clear which prime p is used.

Lemma 4.5.3. *Let S and T be submodules of R-modules G and H, respectively. Let p be a prime, α an ordinal, and $f : S \to T$ an isomorphism that preserves p-heights up to $\alpha+2$. Then $D_{\alpha,p}(S) \cong D_{\alpha,p}(T)$.*

Proof. The isomorphism f maps the set $\{x \in S \cap p^\alpha G : px \in p^{\alpha+2}G\}$ onto $\{y \in T \cap p^\alpha H : py \in p^{\alpha+2}H\}$, and $S \cap p^{\alpha+1}G$ onto $T \cap p^{\alpha+1}H$. □

The following fact will be useful:

Lemma 4.5.4 (Kaplansky [78]). *Let G be an R-module, A a submodule of G, $p \in R$ a prime, and α an ordinal. Then the following statements are equivalent:*
1. *There is an element $x \in G[p]$ of height α which is proper with respect to A;*
2. $I_{\alpha,p}(A) \neq (p^\alpha G)[p]/(p^{\alpha+1}G)[p]$.

Proof. Kaplansky's proof of (1)⇔(2) for p-groups applies just as well to R-modules. First, suppose there is an element x in $(p^\alpha G)[p]$ of height α that is proper with respect to A. We claim the coset $x + (p^{\alpha+1}G)[p]$ is not in the range of ϕ. Suppose it is the image of the coset of some y. Specifically, $x + (p^{\alpha+1})G[p] = y - z + (p^{\alpha+1}G)[p]$, where $y \in A$, $z \in p^{\alpha+1}G$, and $p(y - z) = 0$. Then $|x - y| > \alpha$, which contradicts the assumption that x is proper with respect to A.

Now suppose that $I_{\alpha,p}(A)$ is not all of $(p^\alpha G)[p]/(p^{\alpha+1}G)[p]$, say $x + (p^{\alpha+1}G)[p] \notin I_{\alpha,p}(A)$, where $px = 0$ and $x \in p^\alpha G$. Then $|x| = \alpha$. We claim that x is proper with respect to A. Suppose to the contrary that there is some $a \in A$ with $|a - x| > \alpha$, say $a - x = py$

for some $y \in p^\alpha G$. Then $pa = px + p^2y = p^2y$, and so $a + A \cap p^{\alpha+1}G$ is in the domain of ϕ. Its image is $a - py + (p^{\alpha+1}G)[p] = x + (p^{\alpha+1}G)[p]$, contradicting the assumption that $x + (p^{\alpha+1}G)[p]$ is not in the range. □

Corollary 4.5.5. *Let G be an R-module, A a finitely generated submodule of G, $p \in R$ a prime, and α an ordinal. Then the following statements are equivalent:*
1. *There is an element $x \in G[p]$ of height α which is proper with respect to A;*
2. $\dim I_{\alpha,p}(A) < u_p(\alpha, G)$*;*
3. $\dim D_{\alpha,p}(A) < u_p(\alpha, G)$*.*

As an important application, we obtain the following:

Lemma 4.5.6 (Barwise–Eklof [9]). *Let G and H be p-groups. Suppose that α is a limit ordinal, $\alpha \le$ length(G), $\alpha \le$ length(H), and $f : S \to T$, where S and T are finitely generated subgroups of G and H, respectively, and f is an α-height-preserving isomorphism. Let a be any element of $G\setminus S$ such that $pa \in S$ and the following hold with $\beta = \max\{|x| : x \in a+S\}$:*
1. $\beta < \alpha$*,*
2. *If $\hat{u}(\beta, G) > \dim D_\beta(S)$, then $\hat{u}(\beta, H) > D_\beta(S)$.*

Under these assumptions, there is an α-height preserving isomorphism f' extending f such that domain$(f') = \langle S, a\rangle$.

Proof. Since S is nice, we may assume a is proper with respect to S, and specifically, $|a| = \beta$. We may also assume that we have chosen a such that $|pa|$ is maximal. We will find an appropriate $b \in H$ to be the image of a.

Case 1: $|pa| = \beta + 1$. Then $|f(pa)| = \beta + 1$ since f preserves heights up to $\alpha > \beta + 1$. Then there is a $b \in p^\beta H$ such that $f(pa) = pb$. Since $|pb| = |pa| = \beta + 1$, we must have $|b| = \beta$. We claim that b is proper with respect to T. Suppose $|b + t| \ge \beta + 1$ for some $t \in T$, say $t = f(s)$ for some $s \in S$. Then $|s| = |t| = \beta$, and so $|a + s| = \beta$ since a is proper. Assuming that $|p(a+s)| = \beta+1$, we have $\beta+1 = |f(pa+ps)| = |pb+pt| > |b+t| \ge \beta+1$, which is impossible. Thus, $|p(a + s)| > \beta + 1$, contradicting our choice of a from the coset with $|pa|$ maximal. Therefore, b is proper with respect to T.

Case 2: $|pa| > \beta + 1$. Then there is an $x \in p^{\beta+1}G$ such that $pa = px$. It follows that $a - x \in (p^\beta G)[p]$ and $a - x \notin S + p^{\beta+1}G$; thus, $a - x$ is proper with respect to S. Then by Corollary 4.5.5, $\dim D_\beta(S) < u(\beta, G)$. Since S is finitely generated, $\dim D_\beta(S)$ is finite, so $\hat{u}(\beta, G) > \dim D_\beta(S)$, and so by the second hypothesis, $\hat{u}(\beta, H) > \dim D_\beta(S)$. Since f is an isomorphism preserving heights up to $\alpha > \beta+2$, Lemma 4.5.3 yields $D_\beta(S) \cong D_\beta(T)$, giving us $\hat{u}(\beta, H) > \dim D_\beta(T)$. Applying Corollary 4.5.5 to H and T, there is an element $b_0 \in H[p]$ of height β that is proper with respect to T. Since $|f(pa)| > \beta + 1$, we may choose $b_1 \in H$ such that $f(pa) = pb_1$ and $|b_1| \ge \beta + 1$. Let $b = b_0 + b_1$. Then $f(pa) = pb$, $|b| = \beta$, and b is proper with respect to T.

In either case, we may define $f' : \langle S, a\rangle \to \langle T, b\rangle$ as the isomorphism extending f by $a \mapsto b$ (Lemma 1.6.5). It may be verified that f' is an α-height-preserving map. □

Theorem 4.5.7 (Barwise–Eklof [9]). *Let G and H be p-groups, and let λ be an ordinal. Suppose the following:*

1. $\hat{u}(\alpha, G) = \hat{u}(\alpha, H)$ *for all ordinals* $\alpha < \omega\lambda$;
2. *If* $\text{length}(G) < \omega\lambda$, *then* $\hat{u}(\infty, G) = \hat{u}(\infty, H)$.

Then $G \equiv_\lambda H$. If λ is a limit ordinal, the converse also holds.

Proof. Case 1: $\text{length}(G) < \omega\lambda$. Then $\text{length}(H) < \omega\lambda$ by Lemma 1.13.7(2); thus, $\hat{u}(\alpha, G) = \hat{u}(\alpha, H)$ for all $\alpha \in \text{Ord}_\infty$. Let I be the set of all height-preserving isomorphisms $f : S \to T$, where S and T are finitely generated subgroups of G and H, respectively. Given such a map $f : S \to T$, Lemmas 1.13.8 and 1.13.9 show that

$$u_\alpha(G, S) \neq 0 \quad \text{if and only if} \quad u_\alpha(H, T) \neq 0$$

for all $\alpha \in \text{Ord}_\infty$. By the proof of Ulm's Theorem 1.14.2, I has the back-and-forth property. It follows that $G \equiv_p H$.

Case 2: $\text{length}(G) \geq \omega\lambda$. Then $\text{length}(H) \geq \omega\lambda$. For each $\nu \leq \lambda$, we define I_ν as $\{f : S \to T : S, T$ finitely generated subgroups of G and H, respectively, f an $\omega\nu$-height-preserving isomorphism$\}$. We will prove that I_ν has the extension property of Karp's Theorem 4.3.1. Clearly, each I_ν is nonempty since it contains the zero function, and $I_\mu \subseteq I_\nu$ if $\nu \leq \mu$. Suppose $f : S \to T$ is in $I_{\nu+1}$, where $\nu + 1 \leq \lambda$, and assume that $x \in G \setminus S$. If $|x| \geq \omega\nu$, f extends to x by Lemma 4.5.2. So suppose $|x| < \omega\nu$. Let $r \geq 0$ be least such that $p^{r+1}x \in S$. Consider the cosets $p^m x + \langle S, p^{m+1}x\rangle$ with $m \leq r$.

Case 2a: For every $m \leq r$ and every $y \in p^m x + \langle S, p^{m+1}x\rangle$, $|y| < \omega\nu$. We claim by induction on $k \geq 0$ that there is a map $f_k \in I_\nu$ such that $p^{r-k+1}x \in \text{domain}(f_k)$ and $f = f_0 \subseteq f_1 \subseteq \cdots \subseteq f_{r+1}$. Suppose we have f_k. For every $y \in p^{r-k}x + \langle S, p^{r-k+1}x\rangle$, we get $|y| < \omega\nu$, so condition (1) of Lemma 4.5.6 is satisfied with $\alpha = \omega\nu$, and we obtain $\beta = \max\{|y| : y \in p^{r-k}x + \langle S, p^{r-k+1}x\rangle\} < \alpha$, since $\langle S, p^{r-k+1}x\rangle$ is finite. Furthermore, $\hat{u}(\beta, G) = \hat{u}(\beta, H)$, so condition (2) of that lemma is satisfied as well. Lemma 4.5.6 applied to $\langle S, p^{r-k+1}x\rangle$ and $p^{r-k}x$ gives us $f_{k+1} \in I_\nu$ with domain $\langle S, p^{r-k}x\rangle$. This completes the induction. Let $f' = f_{r+1}$ with domain $\langle S, x\rangle$.

Case 2b: For some $m \leq r$ and $y \in p^m x + \langle S, p^{m+1}x\rangle$, $|y| \geq \omega\nu$. Choose $m \geq 0$ least and such an element y. By Lemma 4.5.2, f extends to an $\omega\nu$-height-preserving isomorphism f' with domain $\langle S, y\rangle$. Write $y = p^m x + s + ap^{m+1}x$, where $s \in S$ and $a \in \mathbb{Z}$. Let $z = p^m x + ap^{m+1}x = (1 + ap)p^m x$. Since $1 + ap$ and p are relatively prime, $\langle z\rangle = \langle p^m x\rangle$, so $\langle S, y\rangle = \langle S, z\rangle = \langle S, p^m x\rangle$. If $m = 0$, we are done. If $m > 0$, we may successively adjoin $p^{m-1}x, \dots, px, x$ by applying Lemma 4.5.6 as in Case 2a, noting that the condition (1) of that lemma is met since m was chosen to be minimal. Each time we get an isomorphism in I_ν. This completes Case 2.

For the converse, suppose $G \equiv_\lambda H$, where λ is a limit ordinal. Let $\alpha < \omega\lambda$, say $\alpha = \omega\lambda' + n'$, where $\lambda' < \lambda$ and $n' < \omega$. Suppose $\hat{u}(\alpha, G) = m < \omega$. Then $G \models$ "$\hat{u}(\alpha, G) \geq m$" $\wedge \neg$"$\hat{u}(\alpha, G) \geq m + 1$", an expression of quantifier rank $\leq \lambda' + m + 2$ by Lemma 4.4.5. Since λ is a limit ordinal, this is less than λ, and so H models the same expression;

that is, $\hat{u}(\alpha, H) = m$. Next suppose $\hat{u}(\alpha, G) = \omega$. Then $G \models \bigvee_{m<\omega}$ "$\hat{u}(\alpha, G) > m$", of quantifier rank $\lambda' + \omega \leq \lambda$, and so $\hat{u}(\alpha, H) = \omega$. Suppose length$(G) < \omega\lambda$ and $u(\infty, G) = m < \omega$. Let α = length(G). Then, since $p^\infty G = p^\alpha G$ and $\alpha < \omega\lambda$, say $\alpha = \omega\lambda' + n'$, $G \models$ "$r(p^\alpha G) = m$", of quantifier rank $\leq \lambda' + m < \lambda$. It follows that $H \models$ "$r(p^\alpha H) = m$". Now suppose $\hat{u}(\infty, G) = \omega$. Then $G \models \bigvee_{m<\omega}$ "$r(p^\alpha G) \geq m$", an expression of quantifier rank $\leq \lambda' + \omega \leq \lambda$, so $\hat{u}(\infty, H) = \omega$. □

Condition (2) of this theorem was originally stated in terms of rank, $\hat{r}(p^\infty G)$, which is equivalent to $\hat{u}(\infty, G)$. The use of $\hat{u}$ gives us a simple criterion for classification of torsion groups up to partial isomorphism.

Corollary 4.5.8 (Barwise–Eklof [9]). *Let G and H be p-groups. Then $G \cong_p H$ if and only if $\hat{u}(\alpha, G) = \hat{u}(\alpha, H)$ for all $\alpha \in \mathrm{Ord}_\infty$.*

Proof. By Karp's Theorem 4.3.2, $G \cong_p H$ if and only if $G \equiv_\infty H$. Choose a limit ordinal λ such that length$(G) < \omega\lambda$ and length$(H) < \omega\lambda$. Suppose $\hat{u}(\alpha, G) = \hat{u}(\alpha, H)$ for all $\alpha \in \mathrm{Ord}_\infty$. Then $G \equiv_\lambda H$. Since λ was arbitrary, $G \equiv_\infty H$. Next suppose $G \equiv_\infty H$. Then $G \equiv_\lambda H$, and hence $\hat{u}(\infty, G) = \hat{u}(\infty, H)$ and $\hat{u}(\alpha, G) = \hat{u}(\alpha, H)$ for all ordinals $\alpha < \omega\lambda$, and hence, for all ordinals α. □

Theorem 4.5.9. *Every torsion group G is partially isomorphic to a simply presented torsion group.*

Proof. Since a torsion group is a direct sum of its p-torsion parts, we may assume that G is a p-group. By Theorem 2.7.4, for each ordinal α there is a simply presented p-group P_α such that $p^\alpha P_\alpha$ is cyclic of order p, generated by an element of height α. Let x be the generator. Then $x \in (p^\alpha P_\alpha)[p] \setminus (p^{\alpha+1} P_\alpha)[p]$, so $u_p(\alpha, P_\alpha) = 1$. Let $T_\alpha = \bigoplus_{u(\alpha,G)} P_\alpha$, the direct sum of $u_p(\alpha, G)$ copies of P_α. Let $H = \bigoplus_{\alpha<\mathrm{length}(G)} T_\alpha \oplus p^\infty G$. Then T is simply presented. Since H and G have the same Ulm invariants, they are partially isomorphic. □

Exercises.

1. (Barwise–Eklof [9]) Let $G = R \oplus D$ be a p-group, where R is reduced and D is divisible, and let $\omega\alpha \leq$ length(G). Then for any formula $\varphi \in L^\alpha_{\infty\omega}$ and any $a_1, \ldots, a_n \in R$, $G \models \varphi[a_1, \ldots, a_n]$ if and only if $R \models \varphi[a_1, \ldots, a_n]$.
2. (Barwise–Eklof [9]) Given a group G and a set $X \subseteq G$, we say that X is a set of **α-indiscernibles in** G if for each formula $\varphi(v_1, \ldots, v_n)$ of quantifier rank at most α, and for all sequences $x_1, \ldots, x_n, y_1, \ldots, y_n$ from X such that $x_i \neq x_j$ and $y_i \neq y_j$ for all $i \neq j$, we have
$$G \models \varphi[x_1, \ldots, x_n]$$
if and only if $G \models \varphi[y_1, \ldots, y_n]$. Prove that if G is a p-group and X a subset of $(p^{\omega\alpha} G)[p] \setminus \{0\}$ such that for any distinct $x, y \in X$, $|x|^G \neq |y|^G$, then X is a set of α-indiscernibles in G.

4.6 Going further

There is a game-theoretic approach to partial isomorphisms, specifically two groups are partially isomorphic if and only if player ∃ has a winning strategy in the Ehrenfeucht–Fraïssé game. There is also a modification of this game that addresses isomorphism of uncountable groups. See [118] for the details. Mekler and Oikkonen [118] used this approach to prove, assuming the continuum hypothesis, that the Barwise–Eklof results cannot be extended to a larger class of groups without adding more invariants.

An area for exploration is how partial isomorphism differs from full isomorphism. We saw that groups of differing cardinalities may be partially isomorphic, but there are further characteristics that distinguish partially isomorphic groups. Leistner [94] explored the question of how closely related partially isomorphic uncountable p-groups are if they have the same cardinality. She constructed a class of modules over a certain class of rings and proved that for every sufficiently large cardinal λ there are at least 2^λ such modules of cardinality λ that are partially isomorphic, but pairwise nonisomorphic. On the other hand, there may be interesting group properties that are preserved under partial isomorphism, which may give us insights into arbitrary torsion groups.

Here we have considered only $L_{\infty\omega}$, which has a nice algebraic analogue. If we consider $L_{\infty\alpha}$ for some $\alpha > \omega$, it will allow us to write more complex and expressive formulas, which may move us algebraically closer to isomorphism. An area of inquiry is whether there is some α or class of ordinals that gives us a meaningful and useful algebraic analogue. From an algebraic viewpoint, we may seek an α that strengthens the language enough to add further group characteristics without losing the classification of all torsion groups. Alternatively, there may be some α that adds significant algebraic richness, and then the question is to determine the largest class of groups that can be classified in $L_{\infty\alpha}$.

From an algebraic point of view, we have considered what it means for groups to be partially isomorphic, but there are there other types of "almost isomorphic" that appear in the literature, such as quasi-isomorphic and nearly isomorphic; see for instance Arnold [4]: Two finite rank torsion-free groups G and H are called **quasi-isomorphic** if there is a nonzero integer n and homomorphisms $\alpha : G \to H, \beta : H \to G$ such that $\beta\alpha = n1_G$ and $\alpha\beta = n1_H$. The groups G and H are called **nearly isomorphic** if for every nonzero integer m there are n, α and β as described above such that $\gcd(m, n) = 1$.

Problems.

1. Is there some ordinal α or class of ordinals for which $L_{\infty\alpha}$ has a meaningful and useful algebraic analogue?
2. What is the largest α for which torsion groups can be classified by their Ulm invariants in $L_{\infty\alpha}$?

3. Does quasi-isomorphic have a useful model-theoretic counterpart? What about nearly isomorphic?
4. Identify the group characteristics that are maintained under partial isomorphism and those that are not.

5 Groups with partial decomposition bases

We strive to generalize all of the above, by defining a class of groups that includes all torsion groups and all Warfield groups, and classifying it up to partial isomorphism. The vehicle for doing this is the partial decomposition basis, a collection of finite subsets with an extension property. The groups of this class are called PDB groups. They are classified using modified Ulm and Warfield invariants in $L_{\infty\omega}$. We see that PDB groups are exactly the k-groups of Hill and Megibben. We consider $L^{\alpha}_{\infty\omega}$, the statements with quantifier rank (a measure of complexity) less than or equal to α. Further modifications of the invariants classify PDB groups in $L^{\alpha}_{\infty\omega}$. It follows that the class of PDB groups is not definable in $L_{\infty\omega}$, nor is any class that generalizes the Warfield groups.

5.1 Partial decomposition bases

At this point, we have seen two very different generalizations of Ulm's theorem. Warfield extended it to a class of mixed groups that may be classified up to isomorphism with additional invariants. Barwise and Eklof extended it to the classification of all torsion groups up to partial isomorphism. Here we seek a class of mixed groups that includes all torsion groups and all Warfield groups, and such that all groups in the class may be classified up to partial isomorphism. As we will see, this can be done using only modified Ulm and Warfield invariants.

We will see later that a decomposition basis is not maintained under partial isomorphism, since, for example, $\prod_{\omega} \mathbb{Z} \cong_p \bigoplus_{\omega} \mathbb{Z}$, even though the latter has a decomposition basis, whereas the former does not. We will need a generalization of the decomposition basis, and we find it by taking inspiration from the extension property that makes partial isomorphisms go back and forth. We define it for modules over a principal ideal domain R with 1 so that we may use it both in the local case for modules over $\mathbb{Z}_p$ and the global case for modules over $\mathbb{Z}$. In the following, R always denotes an arbitrary principal ideal domain with 1, unless stated otherwise.

For G an R-module, we say a nonempty collection C of subsets of G is a **partial decomposition basis** for G if the following hold:

1. Every $X \in C$ is a finite decomposition set;
2. For every $X \in C$ and $x \in G$, there is a $Y \in C$ such that $X \subseteq Y$ and $x \in \langle Y \rangle^0$.

Note that (2) implies that for any $X \in C$ and finite $A \subseteq G$, there is a $Y \in C$ such that $X \subseteq Y$ and $A \subseteq \langle Y \rangle^0$, a fact we will use repeatedly. Note that any torsion group has a partial decomposition basis, specifically the set containing exactly the empty set. Furthermore, if a group has a decomposition basis, then the set of all finite subsets of the decomposition basis forms a partial decomposition basis. The class of groups with

https://doi.org/10.1515/9783110427684-005

partial decomposition bases thus includes all torsion groups and all Warfield groups. We call a group G with a partial decomposition basis a **PDB group**. In the p-local case (that is, $R = \mathbb{Z}_p$) we call G a **PDB module**. For a given prime p, let $v_p : G \to G_p$ denote the localization map (see Section 1.17). Then we have the following:

Lemma 5.1.1. *Let G be a PDB group and p a prime. Then G_p is a PDB module. Specifically, if $\mathcal{C}$ is a partial decomposition basis for G, then $\mathcal{C}_p = \{v_p(X) : X \in \mathcal{C}\}$ is a partial decomposition basis for G_p.*

Proof. It is clear that $\mathcal{C}_p$ is not empty and satisfies condition (1) of the definition by Lemma 3.4.1(2). To prove condition (2), let $v_p(X) \in \mathcal{C}_p$ for some $X \in \mathcal{C}$ and $x \otimes \frac{1}{n} \in G_p$, where $x \in G$, $n \in \mathbb{Z}$, and $p \nmid n$. Choose $Y \in \mathcal{C}$ such that $X \subseteq Y$ and $x \in \langle Y\rangle^0$. Then $x \otimes \frac{1}{n} \in \langle v_p(Y)\rangle^0$. □

As we hoped, the partial decomposition basis is preserved under partial isomorphism:

Theorem 5.1.2 (Jacoby [68]). *Let G and H be R-modules. Suppose $G \cong_p H$, and G has a partial decomposition basis. Then so does H. Specifically, if $\mathcal{C}$ is a partial decomposition basis for G, and $I : G \cong_p H$, then*

$$\mathcal{C}' = \{f(X) : X \in \mathcal{C}, f \in I, X \subseteq \mathrm{domain}(f)\}$$

is a partial decomposition basis for H.

Proof. First we prove that $\mathcal{C}'$ is not empty. We know that both $\mathcal{C}$ and I are nonempty, so we may take $X \in \mathcal{C}$ and $f \in I$. Since X is finite, there is some $g \in I$ such that $f \subseteq g$ and $X \subseteq \mathrm{domain}(g)$. Then $g(X)$ is in $\mathcal{C}'$.

Next, if $f(X) \in \mathcal{C}'$ for some $f \in I$ and $X \in \mathcal{C}$, $f(X)$ is a finite independent set since X is and f is an isomorphism. Furthermore, $f(X)$ is a decomposition set since f is height-preserving (Theorem 4.1.2).

To prove the extension property, suppose $f(X) \in \mathcal{C}'$ and $y \in H$. Choose $g \in I$ such that $f \subseteq g$ and $y \in \mathrm{range}(g)$, say $y = g(x)$. Then since $X \in \mathcal{C}$, there is some $Y \in \mathcal{C}$ such that $X \subseteq Y$ and $x \in \langle Y\rangle^0$. Next, find a $\tilde{g} \in I$ such that $g \subseteq \tilde{g}$ and $Y \subseteq \mathrm{domain}(\tilde{g})$. Then $y = \tilde{g}(x) \in \langle \tilde{g}(Y)\rangle^0$ and $\tilde{g}(Y) \in \mathcal{C}'$, as required. □

The theorem above shows that $\prod_\omega \mathbb{Z}$ has a partial decomposition basis, since it is partially isomorphic to $\bigoplus_\omega \mathbb{Z}$. However, later we will see that it has no decomposition basis (Corollary 5.7.2).

There are multiple partial decomposition bases for any group. In fact, we may extend any partial decomposition basis for closure under various operations. This will be useful several times later.

Lemma 5.1.3 (Jacoby [68], Jacoby–Loth [75]). *Let G be a module over $\mathbb{Z}$ or $\mathbb{Z}_p$ with partial decomposition basis $\mathcal{C}$. Then G has a partial decomposition basis $\mathcal{C}' \supseteq \mathcal{C}$ such that*

(i) *if* $Y \subseteq G$ *is a decomposition set and* $\langle Y\rangle = \langle X\rangle$ *for some* $X \in \mathcal{C}'$, *then* $Y \in \mathcal{C}'$;
(ii) *if* $x_1, \dots, x_k \in X$ *for some* $X \in \mathcal{C}'$, *and* $a_1, \dots, a_k \in \mathbb{Z} \setminus \{0\}$, *then* $X' \in \mathcal{C}'$ *with* $X' = \{a_1x_1, \dots, a_kx_k\}$.

Proof. We construct the expanded partial decomposition basis by successively adding elements to satisfy each of the two conditions in turn, and then we prove that the union is a partial decomposition basis that satisfies both conditions.

We define $\mathcal{C}_n$ by induction on n, beginning with $\mathcal{C}_0 = \mathcal{C}$. If n is odd, let

$$\mathcal{C}_n = \{Y \subseteq G : Y \text{ is a decomposition set, and } \langle Y\rangle = \langle X\rangle \text{ for some } X \in \mathcal{C}_{n-1}\}.$$

If $n > 0$ is even, let

$$\mathcal{C}_n = \{\{a_1x_1, \dots, a_mx_m\} : m \in \mathbb{N}, a_i \in \mathbb{Z} \setminus \{0\}, x_i \in X\ (i = 1, \dots, m), X \in \mathcal{C}_{n-1}\}.$$

It may be verified that each $\mathcal{C}_n$ is a partial decomposition basis (see Exercise 1). Now let $\mathcal{C}' = \bigcup_{n<\omega} \mathcal{C}_n$. This is a partial decomposition basis since each $\mathcal{C}_n$ is.

To prove $\mathcal{C}'$ satisfies (i), suppose $\langle Y\rangle = \langle X\rangle$ for some $X \in \mathcal{C}'$ and decomposition set $Y \subseteq G$. Choose n even such that $X \in \mathcal{C}_n$. Then, by construction, $Y \in \mathcal{C}_{n+1} \subseteq \mathcal{C}'$.

For item (ii), let $x_1, \dots, x_k \in X$ for some $X \in \mathcal{C}'$, and let $a_1, \dots, a_k \in \mathbb{Z} \setminus \{0\}$. Choose n odd such that $X \in \mathcal{C}_n$. Then $\{a_1x_1, \dots, a_kx_k\} \in \mathcal{C}_{n+1} \subseteq \mathcal{C}'$. □

We will call any partial decomposition basis satisfying (i) and (ii) of this lemma a **rich partial decomposition basis.** This proves that every PDB group has a rich partial decomposition basis. Many of the proofs that follow will begin by choosing partial decomposition bases that are rich for all groups of interest.

The following theorem allows us to relate the elements of two different partial decomposition bases for a group.

Theorem 5.1.4 **(Jacoby [68], Jacoby–Loth [74]).** *Let* G *be an* R*-module which has partial decomposition bases* $\mathcal{C}$ *and* $\mathcal{D}$. *Let* $X \in \mathcal{C}$, *and* $Y \in \mathcal{D}$. *Then there are decomposition sets* X' *and* Y' *with* $X \subseteq X'$ *and* $Y \subseteq Y'$ *such that* X' *and* Y' *are unions of ascending chains of elements of* $\mathcal{C}$ *and* $\mathcal{D}$, *respectively, and* $\langle X'\rangle^0 = \langle Y'\rangle^0$.

Proof. Define by induction on $i < \omega$ sets $X_i \in \mathcal{C}$ and $Y_i \in \mathcal{D}$, starting with $X_0 = X$ and $Y_0 = Y$. Suppose X_i and Y_i have been chosen. Choose X_{i+1} such that $X_i \subseteq X_{i+1}$ and $Y_i \subseteq \langle X_{i+1}\rangle^0$. Then choose Y_{i+1} containing Y_i such that $X_{i+1} \subseteq \langle Y_{i+1}\rangle^0$.

Now let $X' = \bigcup_{i<\omega} X_i$, and $Y' = \bigcup_{i<\omega} Y_i$. We claim that $\langle X'\rangle^0 \subseteq \langle Y'\rangle^0$. Suppose $z \in \langle X'\rangle^0$, say $az = a_1x_1 + \cdots + a_nx_n$ for some $n \in \mathbb{Z}$, $a, a_1, \dots, a_n \in R \setminus \{0\}$, and $x_1, \dots, x_n \in X'$. Choose $k < \omega$ such that $\{x_1, \dots, x_n\} \subseteq X_k \subseteq \langle Y_k\rangle^0$. Then for some $b \neq 0$, $\{bx_1, \dots, bx_n\} \subseteq \langle Y_k\rangle$. Then $baz \in \langle Y_k\rangle \subseteq \langle Y'\rangle$, and so $z \in \langle Y'\rangle^0$, as desired. Likewise, we see that $\langle Y'\rangle^0 \subseteq \langle X'\rangle^0$. □

One characteristic of Warfield groups that made their classification possible was the fact that the decomposition basis X is nice in the Warfield group G (that is, $\langle X\rangle$

is locally nice, thus nice in G). This facilitated extending an isomorphism on $\langle X\rangle$ to arbitrary elements of G. As we will see later in Example 5.7.4, a decomposition basis need not be nice. Hence, we have no reason to expect that the elements of a partial decomposition basis will be nice.

As we will see, not only does each element of a partial decomposition basis generate a locally nice subgroup, but any finitely generated subgroup of a PDB group is locally nice. This result allowed a streamlining and correction of early work in this area.

Lemma 5.1.5 (Jacoby–Loth [72]). *Let G be a group with partial decomposition basis $\mathcal{C}$, and let $Y \subseteq X \in \mathcal{C}$. Then $\langle Y\rangle$ is locally nice in G.*

Proof. Let p be a prime and $v_p : G \to G_p$ the localization map. For a decomposition set X in G, define $X_p = v_p(X)$. Now let $x \in G_p$, and $X \in \mathcal{C}$. Since $\{X_p : X \in \mathcal{C}\}$ is a partial decomposition basis for G_p (Lemma 5.1.1), there is an $X' \in \mathcal{C}$ such that $X_p \subseteq X'_p$ and $x \in \langle X'_p\rangle^0 \subseteq G_p$. Consider the group $H = \langle X'_p\rangle^0$ with decomposition basis X'_p. By Lemma 3.2.5, since $Y_p \subseteq X'_p$, $\langle Y\rangle_p = \langle Y_p\rangle$ is p-nice in H. In particular, $x + \langle Y\rangle_p$ has an element y of maximal p-height in H. Since H is isotype in G_p, y has maximal p-height in $x + \langle Y\rangle_p$ within G_p. Thus, $\langle Y\rangle$ is locally nice in G. □

It turns out that we can say even more, using the following result of Moore, rephrased in terms of valuated coproducts.

Theorem 5.1.6 (Moore [119]). *Let A, B, and K be submodules of a $\mathbb{Z}_p$-module G such that $A \oplus B$ is a valuated coproduct in G, and $K \subseteq A \oplus B$. Suppose the following conditions hold:*

1. *For every submodule S of A, each coset $a + S$ ($a \in A$) has an element of maximal p-height;*
2. *For every submodule T of B, each coset $b + T$ ($b \in B$) has an element of maximal p-height.*

Then each coset $z + K$ ($z \in A \oplus B$) has an element of maximal p-height. (All p-heights are computed in G.)

Proof. All p-heights in this proof are computed in G. Let K be a submodule of $A \oplus B$, and $C \subseteq A \oplus B$ a coset of K. We first claim that for every $(a, b) \in C$ with $a \in A, b \in B$, there is an (a^*, b^*) such that $|(a^*, b^*)| \geq |(a, b)|$ and if $(\tilde{a}, \tilde{b}) \in C$, then

1. $|\tilde{a}| \geq |a^*|$ implies $|\tilde{b}| \leq |b^*|$;
2. $|\tilde{b}| \geq |b^*|$ implies $|\tilde{a}| \leq |a^*|$.

Let (a, b) be an arbitrary element of C. Let

$$S = \{c \in A : (c, d) \in K \text{ and } |d| \geq |b| \text{ for some } d \in B\}.$$

Since S is a submodule of A, there is a $c' \in S$ with $|a+c'| \geq |a+c|$ for all $c \in S$. Since $c' \in S$, we have $(c', d') \in K$ and $|d'| \geq |b|$ for some d'. Let $a' = a + c'$ and $b' = b + d'$; hence, $(a', b') \in C$. Then by the choice of c', $|a'| \geq |a|$. Furthermore, $|b'| = |b + d'| \geq |b|$ since $|d'| \geq |b|$, so $|(a', b')| \geq |(a, b)|$. Now let

$$T = \{d \in B : (c, d) \in K, \text{ and } |c| \geq |a'| \text{ for some } c \in A\}.$$

Similarly, we may choose $(c^*, d^*) \in K$ such that $|b' + d^*| \geq |b' + d|$ for all $d \in T$, and $|c^*| \geq |a'|$. Let $a^* = a' + c^*$, and $b^* = b' + d^*$. Then $(a^*, b^*) \in C$, $|a^*| \geq |a'|$, and $|b^*| \geq |b'|$, so $|(a^*, b^*)| \geq |(a', b')| \geq |(a, b)|$.

Now let $(\tilde{a}, \tilde{b})$ be any element of C with $|\tilde{a}| \geq |a^*|$. Then $|\tilde{a} - a'| \geq |a'|$, and $(\tilde{a} - a', \tilde{b} - b') \in K$, so $\tilde{b} - b' \in T$ by definition of T. But then $|\tilde{b}| = |b' + (\tilde{b} - b')| \leq |b^*|$ since b^* is proper with respect to T.

Next suppose $(\tilde{a}, \tilde{b}) \in C$ with $|\tilde{b}| \geq |b^*|$. Then $|\tilde{b} - b| \geq |b|$, and so $\tilde{a} - a \in S$; hence, $|a^*| \geq |a'| = |a + c'| \geq |a + (\tilde{a} - a)| = |\tilde{a}|$. This completes the proof of the claim.

Now let $C' = \{(a^*, b^*) : (a, b) \in C\}$, $L = \{(a, b) \in C' : |a| \leq |b|\}$, and $U = \{(a, b) \in C' : |a| \geq |b|\}$. Choose $(g, h) \in L$ such that $|h| = \min\{|b| : (a, b) \in L\}$. Suppose $(a, b) \in L$. Then $|h| \leq |b|$ since $|h|$ is minimal. By the definition of C' and the claim, $|a| \leq |g|$. Then $|(a, b)| \leq |a| \leq |g|$. Also, $|(g, h)| = |g|$, since $(g, h) \in L$. This proves that $|(a, b)| \leq |(g, h)|$ for all $(a, b) \in L$.

Similarly, there is a $(g', h') \in U$ such that $|(a, b)| \leq |(g', h')|$ for all $(a, b) \in U$.

Finally, let $(a, b) \in C$. Since there is an associated (a^*, b^*) with $|(a, b)| \leq |(a^*, b^*)|$, and (a^*, b^*) is either in L or in U, either $|(a^*, b^*)| \leq |(g, h)|$, or $|(a^*, b^*)| \leq |(g', h')|$. Whichever of (g, h) or (g', h') has the greater height is the element of the coset of maximal height. □

Note that a finitely generated torsion $\mathbb{Z}_p$-module is finite since a sub*group* of a torsion $\mathbb{Z}_p$-module is a submodule. This basic observation will be used frequently.

Theorem 5.1.7. *Every finitely generated subgroup of a group with a decomposition basis is locally nice.*

Proof. Let G be a group with decomposition basis X. As before, we may localize by $G \to G_p$. So it suffices to prove the result for G a $\mathbb{Z}_p$-module. We claim that if $\{x_1, \ldots, x_n\} \subseteq X$, then for any submodule K of $\langle x_1, \ldots, x_n \rangle$, each coset $z + K$ ($z \in \langle x_1, \ldots, x_n \rangle$) has an element of maximal p-height in G. First, $\langle x_1 \rangle$ is p-nice in G by Lemma 3.2.5. In fact, any nonzero submodule of $\langle x_1 \rangle$ is p-nice in G, since it is of the form $\langle ax_1 \rangle$ for some $a \neq 0$, and $(X \setminus \{x_1\}) \cup \{ax_1\}$ is a decomposition basis. Next, let $\{x_1, \ldots, x_n\} \subseteq X$. Applying Theorem 5.1.6 and induction on n, we find that for any submodule K of $(\langle x_1 \rangle \oplus \cdots \oplus \langle x_{n-1} \rangle) \oplus \langle x_n \rangle = \langle x_1, \ldots, x_n \rangle$, each coset $z + K$ ($z \in \langle x_1, \ldots, x_n \rangle$) has an element of maximal p-height computed in G, proving the claim. Since $\langle x_1, \ldots, x_n \rangle$ is p-nice in G, Lemma 2.1.5 shows that K is p-nice in G.

Now let H be a finitely generated submodule of G. Then there is a finite $Y \subseteq X$ such that $H \subseteq \langle Y \rangle^0$, say $nH \subseteq \langle Y \rangle$ for some $n > 0$. Since Y is finite, nH is p-nice in G. Furthermore, H/nH is torsion and finitely generated; hence finite, and therefore, p-nice in G/nH. Finally, by Lemma 2.1.4(2), H is p-nice in G. □

Corollary 5.1.8. *If a group G has a partial decomposition basis $\mathcal{C}$, then every finitely generated subgroup of G is locally nice.*

Proof. Let S be any finitely generated subgroup of G, and suppose $x \in G \setminus S$. Choose an $X \in \mathcal{C}$ such that $\langle X\rangle^0$ contains x and all of the generators of S. Apply the theorem to the group $\langle X\rangle^0$ and S. Considered as a group, $\langle X\rangle^0$ has a decomposition basis X, so S is locally nice in $\langle X\rangle^0$. Then for any prime p and $n \in \mathbb{Z}$ with $p \nmid n$, the coset $(x \otimes \frac{1}{n}) + S_p \in (\langle X\rangle^0)_p/S_p$ has an element p-proper with respect to S_p. Since p-heights are the same in $(\langle X\rangle^0)_p$ as in G_p, and x and n were arbitrary, S is locally nice in G. □

Our ultimate goal in the next sections is to prove that equality of the modified Ulm and Warfield invariants guarantees that groups with partial decomposition bases are partially isomorphic. As a first step, we prove that equality of the Ulm invariants allows us to extend an isomorphism on subgroups to any element with a multiple in the subgroup. We will prove this key theorem in enough generality that we can apply it in both local and global cases and for classification in $L^{\lambda}_{\infty\omega}$.

Let S and T be subsets of R-modules G and H, respectively, $p \in R$ a prime and α an ordinal. Recall from Section 4.5 that a function $f : S \to T$ is said to preserve p-heights up to α if

$$|f(x)|_p^H \wedge \alpha = |x|_p^G \wedge \alpha$$

for all $x \in S$. From Section 1.13, we recall that, in defining the relative Ulm invariants, we defined a map

$$\phi : D_{\alpha,p}(S) = \frac{\{x \in S \cap p^\alpha G : px \in p^{\alpha+2}G\}}{S \cap p^{\alpha+1}G} \to \frac{(p^\alpha G)[p]}{(p^{\alpha+1}G)[p]}$$

by $\phi(x + S \cap p^{\alpha+1}G) = x - y + (p^{\alpha+1}G)[p]$, where $y \in p^{\alpha+1}G$ and $px = py$.

The following extension theorem is based on Lemma B of Barwise–Eklof [9], a similar result for torsion groups, which is based on Kaplansky's proof of Ulm's theorem.

Theorem 5.1.9 (Jacoby–Loth [74]). *Let G and H be R-modules, p a prime in R, and f an isomorphism between a finitely generated submodule S of G and a submodule T of H that preserves p-heights up to α for some ordinal α. Suppose that $\hat{u}_p(\sigma, G) \le \hat{u}_p(\sigma, H)$ for all $\sigma < \alpha$. If $x \in G$ is proper with respect to S, $px \in S$, and $|x|_p + 1 < \alpha$, then for a suitable y in H, f can be extended to an isomorphism $f^* : \langle S, x\rangle \to \langle T, y\rangle$ that preserves p-heights up to α.*

Proof. Let x be as given, and suppose $|x|_p = \sigma$. We may assume that x has been chosen such that $|px|_p > \sigma + 1$ if this is possible for any proper element of the coset, since all proper coset elements satisfy the stated conditions on x. We consider two cases based on the height of px. In each case, we will produce a $y \in H$ proper with respect to T, satisfying $|y|_p = \sigma$ and $py = f(px)$.

Case 1: $|px|_p = \sigma + 1 < \alpha$. Then $|f(px)|_p = \sigma + 1$, since $px \in S$ and $f : S \to T$ is α-height-preserving. There is then a $y \in H$ such that $f(px) = py$ and $|y|_p = \sigma$.

We claim that y is proper with respect to T. Suppose that for some $z \in T$, $|y+z|_p > |y|_p = \sigma$. The only way this can happen is if $|z|_p = \sigma$. This also says that $|p(y+z)|_p > |y+z|_p \geq \sigma+1$. Since f, and hence f^{-1}, preserves heights up to α, we get $|f^{-1}(z)|_p = \sigma$, and $\sigma+1 < |f^{-1}(p(y+z))|_p = |p(x+f^{-1}(z))|_p$. But since $z \in T, f^{-1}(z) \in S$, and so $|x+f^{-1}(z)| \leq \sigma$ since x is proper. At the same time, $|x+f^{-1}(z)|_p \geq |x|_p \wedge |f^{-1}(z)|_p = \sigma$. Then $x+f^{-1}(z)$ is proper with height σ and has a p-multiple of height $> \sigma + 1$, contradicting the choice of x. This proves y is proper with respect to T, completing Case 1.

Case 2: $|px|_p > \sigma + 1$. We then have $px = py'$ for some $y' \in p^{\sigma+1}G$. Then $x - y' \in G[p]$, and $|x - y'|_p = |x|_p = \sigma$ since $|y'|_p > \sigma$. Now let z be any element of S. Then $|x - y' + z|_p = |(x + z) - y'|_p = |x + z|_p \leq |x|_p = \sigma$. Thus, $x - y'$ is proper with respect to S. This gives us an element, $x - y' \in G[p]$ of height σ proper with respect to S. By Corollary 4.5.5, we have $\dim D_{\sigma,p}(S) < u_p(\sigma, G)$, and since S is finitely generated, $D_{\sigma,p}(S)$ is of finite dimension. Thus, $\dim D_{\sigma,p}(S) < \hat{u}_p(\sigma, G) \leq \hat{u}_p(\sigma, H)$. Therefore, by Lemma 4.5.3, we obtain $\dim D_{\sigma,p}(T) < \hat{u}_p(\sigma, H)$. By Corollary 4.5.5, there is a $w_1 \in H$, proper with respect to T, such that $|w_1|_p = \sigma$ and $pw_1 = 0$.

Since f is height-preserving up to $\alpha > \sigma + 1$, $|f(px)|_p > \sigma + 1$, so we may choose a $w_2 \in H$ such that $pw_2 = f(px)$ and $|w_2|_p > \sigma$. Let $y = w_1 + w_2$. Then $py = pw_2 = f(px)$, and $|y|_p = |w_1|_p \wedge |w_2|_p = \sigma$. We claim that y is proper with respect to T. Let $z \in T$. Then $|y + z|_p = |(w_1 + z) + w_2|_p = |w_1 + z|_p \leq \sigma$. This completes Case 2.

In either case, we have a y that is proper with respect to T, with $|y|_p = \sigma$, and $py = f(px)$. In particular, $y \notin T$ since $|y| \neq \infty$. Now we extend f by sending x to y, giving us the isomorphism $f^* : \langle S, x\rangle \to \langle T, y\rangle$ (Lemma 1.12.4(2)). We need only prove that f^* preserves p-heights up to α. Let $s + ax$ be an arbitrary element of $\langle S, x\rangle$, where $s \in S$ and $a \in R$. If $p|a$, then $s+ax \in S$. If not, we may write $1 = bp+ra$ for some $b, r \in R$. Then $r(s+ax) = rs-bpx+x \in S+x$. Thus, every $z \in \langle S, x\rangle$ is either in S, or else $rz \in S+x$ for some $r \in R$ not divisible by p. In the first case, we are done since $f^*(z) = f(z)$. So suppose $rz = s + x$ for some $s \in S$ and r not divisible by p. Then

$$|z|_p = |rz|_p = |s + x|_p = |s|_p \wedge |x|_p$$

since x is proper with respect to S, and similarly,

$$|f^*(z)|_p = |f(s)|_p \wedge |y|_p = |s|_p \wedge |x|_p$$

since y is proper with respect to T. □

The following version of this theorem is a key extension theorem for proving that two groups are partially isomorphic.

Theorem 5.1.10. *Let G and H be R-modules, p a prime in R, and f an isomorphism between a finitely generated submodule S of G and a submodule T of H that preserves p-heights. Suppose that $\hat{u}_p(\sigma, G) \leq \hat{u}_p(\sigma, H)$ for all ordinals σ. If $x \in G$ is proper with respect to S, $px \in S$, and $|x|_p \neq \infty$, then for a suitable y in H, f can be extended to an isomorphism $f^* : \langle S, x\rangle \to \langle T, y\rangle$ that preserves p-heights.*

Proof. Let $\alpha > \max\{p\text{-length}(G), p\text{-length}(H)\} + 1$, and apply Theorem 5.1.9. □

In the next sections, we will be proving classification theorems both up to partial isomorphism and up to $\equiv_\lambda$ for certain ordinals λ. We will prove in the various cases that, if the appropriate modified Ulm and Warfield invariants agree for PDB groups G and H, then we get extension properties that allow us to apply the following theorem:

Theorem 5.1.11. *Let G and H be R-modules with rich partial decomposition bases $\mathcal{C}$ and $\mathcal{D}$, respectively, where $R = \mathbb{Z}$ or $R = \mathbb{Z}_p$ for some prime p. Let λ be an ordinal, and suppose that the following conditions hold for any height-preserving (or height-preserving up to $\omega(\nu+1)$ for some $\nu < \lambda$) isomorphism $f : S \to T$, where $S \subseteq G$ and $T \subseteq H$ are finitely generated submodules:*

(a) *If $X \subseteq S \subseteq \langle X\rangle^0$ for some X in $\mathcal{C}$, where $f(X) \in \mathcal{D}$, and if $X \cup \{x\} \in \mathcal{C}$ for some $x \in G$, then there are a nonzero multiple x′ of x and a $y' \in H$ such that $f(X) \cup \{y'\} \in \mathcal{D}$ and $g : \langle S, x'\rangle \to \langle T, y'\rangle$, extending f by $x' \mapsto y'$, is an isomorphism that preserves heights (or preserves heights up to $\omega(\nu+1)$);*
(b) *If $a \in G$ has a nonzero multiple in S, then there is $b \in H$ such that $g : \langle S, a\rangle \to \langle T, b\rangle$, extending f by $a \mapsto b$, is an isomorphism that preserves heights (or preserves heights up to $\omega\nu$);*

and analogous conditions on partial isomorphisms from H to G.

Then $G \cong_p H$ (or $G \equiv_\lambda H$).

Specifically, suppose $X_0 \in \mathcal{C}$, $Y_0 \in \mathcal{D}$, possibly empty, and $f_0 : X_0 \to Y_0$ is a bijection such that $\|x\| = \|f_0(x)\|$ for all $x \in X_0$. Let I (or I_ν with $\nu \leq \lambda$) be the set of all maps $f : S \to T$ with associated $X \in \mathcal{C}$ and $Y \in \mathcal{D}$ such that
(i) *S and T are finitely generated submodules of G and H, respectively;*
(ii) *f is a height-preserving (or ων-height-preserving) isomorphism;*
(iii) *$X \subseteq S \subseteq \langle X\rangle^0$, and $Y \subseteq T \subseteq \langle Y\rangle^0$;*
(iv) *$f(X) = Y$;*
(v) *$X_0 \subseteq X$, and $f \restriction X_0 = f_0$.*

Then I (or $\{I_\nu : \nu \leq \lambda\}$) is a Karp system.

Proof. This system is not empty, since it contains f_0. Note that if $\mu \leq \nu$, then $I_\nu \subseteq I_\mu$. We will prove that this system has the back-and-forth property.

Let $f : S \to T$ be in I (or $I_{\nu+1}$) with associated X and Y, and let $a \in G$. Then, since $\mathcal{C}$ is a partial decomposition basis, there is some $X' \in \mathcal{C}$ such that $X \subseteq X'$ and $a \in \langle X'\rangle^0$, say $X' = X \cup \{x_1, \ldots, x_n\}$. We will define, by induction on $0 \leq i \leq n$, suitable maps $\tilde{f}_i : \tilde{S}_i \to \tilde{T}_i$ in I (or $I_{\nu+1}$) with associated $\tilde{X}_i = X \cup \{x'_1, \ldots, x'_i\}$ and $\tilde{Y}_i$, where $\tilde{S}_i = \langle S, x'_1, \ldots, x'_i \rangle$, and each x'_i is a nonzero multiple of x_i. Start the induction with $\tilde{f}_0 = f, \tilde{X}_0 = X, \tilde{S}_0 = S$ and $\tilde{T}_0 = T$. Now suppose we have defined $\tilde{f}_i, \tilde{S}_i, \tilde{T}_i, \tilde{X}_i$, and $\tilde{Y}_i$ for $0 \le i < n$. By (a), there is a height-preserving (or $\omega(\nu+1)$-height-preserving) isomorphism $\tilde{f}_{i+1} : \tilde{S}_{i+1} = \langle \tilde{S}_i, x'_{i+1} \rangle \to \langle \tilde{T}_i, y'_{i+1} \rangle$, for some x'_{i+1} a multiple of x_{i+1} and y'_{i+1} such that $\tilde{Y}_i \cup \{y'_{i+1}\} \in \mathcal{D}$. Let $\tilde{Y}_{i+1} = \tilde{Y}_i \cup \{y'_{i+1}\}, \tilde{X}_{i+1} = \tilde{X}_i \cup \{x'_{i+1}\}$ and $\tilde{T}_{i+1} = \langle \tilde{T}_i, y'_{i+1} \rangle$. It is then clear that $\tilde{f}_{i+1}$ satisfies all of (i) through (v) and so is in I (or $I_{\nu+1}$). This completes the induction and gives us $\tilde{f}_n \in I$ (or $I_{\nu+1}$) with associated $\tilde{S}_n, \tilde{T}_n, \tilde{X}_n$, and $\tilde{Y}_n$, where $\tilde{X}_n = X \cup \{x'_1, \ldots, x'_n\}$.

Since a has a nonzero multiple in $\langle X' \rangle$, it has one in $\langle \tilde{X}_n \rangle \subseteq \tilde{S}_n$. So by (b), there is a $b \in H$ such that $g : \langle \tilde{S}_n, a \rangle \to \langle \tilde{T}_n, b \rangle$, extending f_n, is an isomorphism that preserves heights (or preserves heights up to $\omega\nu$). Then the map g with associated $\tilde{X}_n \in \mathcal{C}$ and $\tilde{Y}_n \in \mathcal{C}$ satisfies all of (i) through (v) and so is in I (or I_ν). Also, $a \in \operatorname{domain}(g)$, as desired. The other direction follows by symmetry. □

Exercises.

1. Prove that C_n in the proof of Lemma 5.1.3 is a partial decomposition basis for every n.
2. Construct a partial decomposition basis for $\prod_\omega \mathbb{Z}$. (Hint: Use the proof of Theorem 4.1.4.)

5.2 Classification of local groups up to partial isomorphism

In this section, we focus on p-local PDB groups for some fixed prime p. This means we may use any of the previous results that apply to groups, $\mathbb{Z}_p$-modules, or arbitrary R-modules. We wish to construct an invariant analogous to the Warfield invariant for PDB modules, one that is maintained under $\cong_p$. Recall that Barwise and Eklof used the modified Ulm invariant $\hat{u}_p(\alpha, G) = u_p(\alpha, G) \wedge \omega$ to classify torsion groups up to partial isomorphism.

We may do the same thing with the invariant $s(e, G)$ and define a **modified Stanton invariant**: For e an equivalence class of Ulm sequences and G a $\mathbb{Z}_p$-module, we define

$$\hat{s}(e, G) = s(e, G) \wedge \omega.$$

We cannot do the same for $w(e, G)$ since the definition relies on the existence of a decomposition basis, but we can define the **modified Warfield invariant** as follows:

For e an equivalence class of Ulm sequences and G a $\mathbb{Z}_p$-module with partial decomposition basis $\mathcal{C}$, let

$$\hat{w}_{\mathcal{C}}(e, G) = \text{the maximum } n < \omega \text{ such that there are } X \in \mathcal{C} \text{ and } x_1, \ldots, x_n \in X \text{ with } \|x_i\|_p \in e \text{ for } 1 \le i \le n, \text{ if such a maximum exists, and } \omega \text{ otherwise.}$$

Theorem 5.2.1 (Jacoby [68], Jacoby–Loth [74]). *Let G be a $\mathbb{Z}_p$-module with partial decomposition bases C and C'. Suppose for some equivalence class e of Ulm sequences $\hat{w}_C(e,G) \geq n$. Then for any $Y \in C'$, there is a $\tilde{Y} \in C'$ such that $Y \subseteq \tilde{Y}$ and $\tilde{Y}$ contains at least n elements y such that $\|y\|_p \in e$.*

Proof. By the definition of $\hat{w}$, there is an $X \in C$ containing at least n elements with Ulm sequences in e. Choose X' and Y' as in Theorem 5.1.4, building on X and Y. Now consider the submodule $\langle X'\rangle^0 = \langle Y'\rangle^0$ of G as a $\mathbb{Z}_p$-module. Heights are the same in this module as they are in the whole of G; therefore, X' is a decomposition basis, as is Y'. Since X' contains at least n elements with Ulm sequence in e, we have $w(e, \langle X'\rangle^0) \geq n$. By Theorem 3.2.3, the Warfield invariant is independent of the decomposition basis, so the decomposition basis Y' must contain n such elements as well. Since Y' was built as an ascending chain of elements of C', we may choose some $\tilde{Y}$ in this chain that contains these n elements. □

Corollary 5.2.2. *Let G be a $\mathbb{Z}_p$-module with partial decomposition bases C and C'. Then for any equivalence class e of Ulm sequences, $\hat{w}_C(e,G) = \hat{w}_{C'}(e,G)$.*

Proof. Suppose $\hat{w}_C(e,G) \geq n$. Then, using any $Y \in C'$, the theorem says that there is some $\tilde{Y} \in C'$ with at least n elements with Ulm sequence in e. It follows that $\hat{w}_{C'}(e,G) \geq n$, and so $\hat{w}_C(e,G) \leq \hat{w}_{C'}(e,G)$. The other direction follows by symmetry. □

The corollary shows that $\hat{w}_C(e,G)$ is indeed an invariant, in the sense that it is independent of the choice of partial decomposition basis. Therefore, we will drop the subscript C in what follows. We may now restate Theorem 5.2.1.

Theorem 5.2.3. *Let G be a $\mathbb{Z}_p$-module with partial decomposition basis C. Suppose for some equivalence class e of Ulm sequences $\hat{w}(e,G) \geq n$. Then for any $X \in C$, there is an $\tilde{X} \in C$ such that $X \subseteq \tilde{X}$ and $\tilde{X}$ contains at least n elements x with $\|x\|_p \in e$.*

Furthermore, $\hat{w}(e,G)$ is invariant under partial isomorphism, as we can prove by an appropriate choice of partial decomposition basis:

Corollary 5.2.4. *Let G and H be $\mathbb{Z}_p$-modules such that G has a partial decomposition basis C and $I : G \cong_p H$. Then for any equivalence class e of Ulm sequences, $\hat{w}(e,H)$ is defined, and we have $\hat{w}(e,H) = \hat{w}(e,G)$.*

Proof. Let $C' = \{f(X) : X \in C, f \in I, X \subseteq \text{domain}(f)\}$. Theorem 5.1.2 tells us that C' is a partial decomposition basis for H. Since H has a partial decomposition basis, $\hat{w}(e,H)$ is defined for any equivalence class of Ulm sequences e. Now suppose for some e, $\hat{w}(e,G) \geq n$. Then by the theorem, there is some $X \in C$ and some $x_1, \ldots, x_n \in X$ with $\|x_i\|_p \in e$ for all $1 \leq i \leq n$. Since X is finite, we may choose an $f \in I$ with $X \subseteq \text{domain}(f)$. Since f is height-preserving, $\|f(x_i)\|_p \in e$ for $i = 1, \ldots, n$. Then $f(X) \in C'$ includes n elements with Ulm sequence equivalent to e, and so $\hat{w}(e,H) \geq n$. Conversely, suppose $\hat{w}(e,H) \geq n$. Then there are $f(x_1), \ldots, f(x_n) \in f(X)$ for some $f \in I$ and $X \in C$ with the property that for each $i = 1, \dots, n$, $\|f(x_i)\|_p \in e$. Then for each $1 \leq i \leq n$, $x_i \in X$ and $\|x_i\|_p \in e$. □

We will need two more lemmas before we can prove the local classification theorem.

Lemma 5.2.5 (Jacoby [68]). *Let G be a reduced $\mathbb{Z}_p$-module of finite torsion-free rank with a decomposition basis $\{x_1, \dots, x_n\}$. Then there is a torsion $\mathbb{Z}_p$-module T and submodules G_i of G ($i = 1, \dots, n$) such that*

$$G \oplus T \cong_p G_1 \oplus \cdots \oplus G_n,$$

and $x_i \in G_i$ for all $i = 1, \dots, n$. In fact, the set I of all height-preserving isomorphisms between finitely generated submodules of $G \oplus T$ and $G_1 \oplus \cdots \oplus G_n$, extending the canonical map $\varphi : \langle x_1, \dots, x_n\rangle \to \langle x_1, \dots, x_n\rangle$, satisfies the conditions of Karp's Theorem 4.3.2.

Proof. For each $1 \leq i \leq n$, let $G_i = \langle x_i\rangle^0 \subseteq G$, and let T be the direct sum of $n-1$ copies of the torsion part of G. Let S and L be the respective submodules of $G \oplus T$ and $G_1 \oplus \cdots \oplus G_n$ generated by $\{x_1, \dots, x_n\}$. Let $\varphi : S \to L$ be the isomorphism such that $\varphi(x_i) = x_i$ for $i = 1, \dots, n$. Note that $\{x_1, \dots, x_n\}$ is a decomposition basis for $G \oplus T$ and for $G_1 \oplus \cdots \oplus G_n$ and that the height of any multiple of any x_i is the same in G, $G \oplus T$, or $G_1 \oplus \cdots \oplus G_n$. It follows that φ is height-preserving.

Now we will construct a suitable Karp system.

Let I be the set of all $f : A \to B$ such that

(i) $A \subseteq G \oplus T$ and $B \subseteq G_1 \oplus \cdots \oplus G_n$ are finitely generated submodules,
(ii) $S \subseteq A$, $L \subseteq B$,
(iii) f is a height-preserving isomorphism and
(iv) $\varphi \subseteq f$.

Since $\varphi \in I$, I is not empty. We need only prove that I has the back-and-forth property. Let $f : A \to B$ be a map in I, and $z + w \in G \oplus T$, where $z \in G$ and $w \in T$. For some $m \geq 0$, $p^m(z + w) \in S$ since $G \oplus T$ is torsion over S.

We define by induction on k, $0 \leq k \leq m$, maps $f_k : A_k \to B_k$ in I such that $f \subseteq f_k$, $p^{m-k}(z + w) \in A_k$, and A_k/A is finite. Let $f_0 = f$. Suppose f_k has been defined. Then, since S is p-nice in $G \oplus T$ by Lemma 5.1.5 and A_k/S is finitely generated and torsion, A_k is p-nice as well (Lemma 2.1.4(2)). We may then choose x proper in $p^{m-(k+1)}(z+w)+A_k$. Note that $px \in A_k$.

If $x \in S$, let $f_{k+1} = f_k$. Suppose $x \notin S$. Notice that $G \oplus T$ and $G_1 \oplus \cdots \oplus G_n$ have the same Ulm invariants since their torsion parts agree. By Theorem 5.1.10, we may extend f_k to an isomorphism $f_{k+1} : \langle A_k, x\rangle \to \langle B_k, y\rangle$ that preserves heights. Let $A_{k+1} = \langle A_k, x\rangle$, and $B_{k+1} = \langle B_k, y\rangle$. Then it may be verified that $f_{k+1} : A_{k+1} \to B_{k+1}$ is in I, $p^{m-(k+1)}(z + w) \in x + A_k \subseteq A_{k+1}$, and A_{k+1}/A_k is finite. This completes the induction. Finally, let $g = f_m$. Then $z + w \in A_m = \text{domain}(g)$, $g \in I$, and $f \subseteq g$, as desired.

Similarly, if $z_1+\cdots+z_n \in G_1\oplus\cdots\oplus G_n$ with $z_i \in G_i$ $(i = 1,\ldots,n)$, then for some $m \geq 0$, $p^m z_i \in \langle x_i\rangle$ for all $1 \leq i \leq n$. Then $p^m(z_1+\cdots+z_n) \in \langle x_1,\ldots,x_n\rangle$, so we may extend f^{-1} to $\langle B, z_1+\cdots+z_n\rangle$ as in the other case. This proves that I has the back-and-forth property, and so the proof is complete. □

Lemma 5.2.6 (Jacoby–Loth [72]). *Let G be a $\mathbb{Z}_p$-module, X a decomposition basis for G, and S a finitely generated submodule of G such that $S \cap \langle X\rangle = \langle S \cap X\rangle$. If $y \in X$ such that $y \notin S$ then there is an $n < \omega$ such that*

$$|rp^n y + s| = |rp^n y| \wedge |s|$$

for all $r \in \mathbb{Z}_p$ and $s \in S$.

Proof. Since $S \cap \langle y\rangle \subseteq \langle S \cap X\rangle$ and $y \notin S$, $S \cap \langle y\rangle = 0$. We may write $S \cap X = \{x_1,\ldots,x_m\}$ since S is finitely generated. We may assume that the first k elements are exactly the ones with Ulm sequence not equivalent to $(\infty,\infty,\ldots)$, for some $0 \leq k \leq m$. We may assume $\|y\|_p \not\sim (\infty,\infty,\ldots)$, since otherwise $|p^n y| = \infty$ for some n, and the claim follows.

Now let $H = (S \oplus \langle y\rangle)^0$, and let D be the maximal divisible subgroup of H. Then $\{y, x_1,\ldots,x_m\}$ is a decomposition basis for H. Also, $\langle y, x_1,\ldots,x_k\rangle \cap D = 0$. Therefore, by Theorem 1.6.3, there is a subgroup R of H such that $\langle y, x_1,\ldots,x_k\rangle \subseteq R$ and $H = R \oplus D$. Since R is a reduced $\mathbb{Z}_p$-module with decomposition basis $\{y, x_1,\ldots,x_k\}$, we may apply Lemma 5.2.5 to get $I : R \oplus T \cong_p H_0 \oplus\cdots\oplus H_k$, where T is torsion, $y \in H_0$, and $x_i \in H_i$ for $i = 1,\ldots,k$. For any $f : A \to B$ in I, we define $f' : A \oplus D \to B \oplus D$ by $(a,x) \mapsto (f(a), x)$.

Then $I' = \{f' : f \in I\}$ is a Karp system between $R\oplus T\oplus D \cong H\oplus T$ and $H_0\oplus\cdots\oplus H_k\oplus D$. Note that $f'(x) = x$ for all $f' \in I'$ and $x \in \{y, x_1,\ldots,x_m\}$. Then the finitely generated subgroup $S \oplus \langle y\rangle$ is in the domain of some $f' \in I'$. Now let π be the projection of $H_0 \oplus\cdots\oplus H_k \oplus D$ onto H_0. We claim $\pi f'(S)$ is torsion. Let $z \in S$. Then for some $r \in \mathbb{Z}_p$, $rz \in S \cap \langle X\rangle = \langle S \cap X\rangle$. Then $r\pi f'(z) = \pi f'(rz) = \pi rz = 0$, proving the claim. Since S is finitely generated, $\pi f'(S)$ is as well. This says that $\pi f'(S)$ takes on only finitely many heights. But since $\|y\|_p \not\sim (\infty,\infty,\ldots)$ and f' is height-preserving, each $p^n f'(y)$ takes on a different height. So there is some $n < \omega$ such that $|p^n f'(y)| > |s_0|$ for all $s_0 \in \pi f'(S)$. In fact, $|rp^n f'(y)| > |s_0|$ for all $r \in \mathbb{Z}_p \setminus \{0\}$, and $s_0 \in \pi f'(S)$.

Now let $s \in S$ and $r \in \mathbb{Z}_p$. Write $f'(s) = s_0 + \cdots + s_k + d$, where $s_i \in H_i$ for $0 \leq i \leq k$ and $d \in D$. Then $|rp^n y + s| = |f'(rp^n y + s)| = |f'(rp^n y) + s_0| \wedge |s_1| \wedge\cdots\wedge |s_k| \wedge |d| = |f'(rp^n y)| \wedge |f'(s)| = |rp^n y| \wedge |s|$. □

Now we are ready for the local classification theorem for PDB groups.

Theorem 5.2.7 (Jacoby [68]). *Let G and H be PDB modules. Then $G \cong_p H$ if and only if for every $\alpha \in \mathrm{Ord}_\infty$ and equivalence class e of Ulm sequences, $\hat{u}(\alpha, G) = \hat{u}(\alpha, H)$ and $\hat{w}(e, G) = \hat{w}(e, H)$.*

In that case, if $\mathcal{C}$ and $\mathcal{C}'$ are rich partial decomposition bases of G and H, respectively, and if $X_0 \in \mathcal{C}$, $Y_0 \in \mathcal{C}'$, and $f_0 : X_0 \to Y_0$ is a bijection such that $\|x\|_p = \|f_0(x)\|_p$ for all $x \in X_0$, then $I : G \cong_p H$ may be chosen to be the set of all maps $f : S \to T$, for which there exist $X \in \mathcal{C}$ and $Y \in \mathcal{C}'$, satisfying the following properties:

(i) *S and T are finitely generated submodules of G and H, respectively;*
(ii) *f is a height-preserving isomorphism;*
(iii) *$X \subseteq S \subseteq \langle X \rangle^0$, and $Y \subseteq T \subseteq \langle Y \rangle^0$;*
(iv) *$f(X) = Y$;*
(v) *$X_0 \subseteq X$, and $f \restriction X_0 = f_0$.*

Proof. First, suppose $G \cong_p H$. Then all $\hat{u}$ and $\hat{w}$ invariants agree by Corollaries 4.4.6 and 5.2.4.

Now suppose that G and H have the same $\hat{u}$ and $\hat{w}$ invariants. Let $\mathcal{C}$ and $\mathcal{C}'$ be respective rich partial decomposition bases for G and H. By Theorem 5.1.11, it suffices to prove that the following conditions hold for any height-preserving isomorphism $f : S \to T$, where $S \subseteq G$ and $T \subseteq H$ are finitely generated submodules:

(a) If $X \subseteq S \subseteq \langle X \rangle^0$ for some X in $\mathcal{C}$, where $f(X) \in \mathcal{C}'$, and if $X \cup \{x\} \in \mathcal{C}$ for some $x \in G$, then there is a nonzero multiple x' of x and a $y' \in H$ such that $f(X) \cup \{y'\} \in \mathcal{C}'$ and $g : \langle S, x' \rangle \to \langle T, y' \rangle$, extending f by $x' \mapsto y'$, is an isomorphism that preserves heights;
(b) If $a \in G$ has a nonzero multiple in S, then there is a $b \in H$ such that $g : \langle S, a \rangle \to \langle T, b \rangle$, extending f by $a \mapsto b$, is an isomorphism that preserves heights;

and analogous conditions on functions from H to G.

Case (a): We may assume $x \notin S$, since otherwise the proof is trivial; hence, $x \notin X$. Suppose $X' = X \cup \{x\} \in \mathcal{C}$. Let $Y = f(X)$, and let e be the equivalence class of $\|x\|_p$. Since f is height-preserving, we have $\hat{w}(e, \langle X \rangle^0) = |\{x \in X : \|x\|_p \in e\}| = \hat{w}(e, \langle Y \rangle^0)$. Then $\hat{w}(e, \langle Y \rangle^0) = \hat{w}(e, \langle X \rangle^0) < \hat{w}(e, \langle X \cup \{x\} \rangle^0) \le \hat{w}(e, G) = \hat{w}(e, H)$. By Theorem 5.2.3, there is a $Y' \in \mathcal{C}'$ where $Y \subseteq Y'$, and a $y \in Y' \setminus Y$ with $\|y\|_p \sim \|x\|_p$. Since $\mathcal{C}'$ is rich, we may assume $Y' = Y \cup \{y\}$. Now apply Lemma 5.2.6 to $\langle S, x \rangle^0$, the submodule S, the decomposition basis X', and $x \in X'$. This applies since $S \cap \langle X' \rangle = \langle X \rangle = \langle S \cap X' \rangle$, and $x \notin S$. Apply it similarly to $\langle T, y \rangle^0$, T, Y', and y. This gives us an $n < \omega$ such that for all $s \in S$, $t \in T$, and $r \in \mathbb{Z}_p$, $|rp^n x + s| = |rp^n x| \wedge |s|$ and $|rp^n y + t| = |rp^n y| \wedge |t|$. Since $\|p^n x\|_p \sim \|p^n y\|_p$, there are x' and y', nonzero multiples of $p^n x$ and $p^n y$, such that $\|x'\|_p = \|y'\|_p$. Define $g : \langle S, x' \rangle \to \langle T, y' \rangle$, where $g \restriction S = f$ and $g(x') = y'$. By the choice of n, g is height-preserving. This completes Case (a).

Case (b): It suffices to consider the case of $a \in G \setminus S$ and $pa \in S$. By Lemma 5.1.5, $\langle X \rangle$ is p-nice in G. Since $\langle X \rangle \subseteq S \subseteq \langle X \rangle^0$, $S/\langle X \rangle$ is torsion and finitely generated, hence finite. It follows by Lemma 2.1.4(2) that S is p-nice, so we may assume a is proper with respect to S. If $|a| \neq \infty$, we may extend f to a by Theorem 5.1.10, specifically for some $b \in H$, there is a height-preserving $g : \langle S, a \rangle \to \langle T, b \rangle$ with $f \subseteq g$.

Suppose $|a| = \infty$. Then $|f(pa)| = \infty$, so $f(pa) = pb'$ for some $b' \in H$ with $|b'| = \infty$. If $b' \notin T$, we may extend f to an isomorphism by $a \mapsto b'$ by Lemma 1.6.5. So suppose $b' \in T$, say $b' = f(z)$ for some $z \in S$. Then $f(pa - pz) = 0$, and so $pa - pz = 0$, which means $a - z \in (p^\infty G)[p]$. Since $a - z \notin S$, $\dim(p^\infty H \cap T)[p] = \dim(p^\infty G \cap S)[p] < \dim(p^\infty G \cap \langle S, a - z \rangle)[p] \le \hat{u}(\infty, G) = \hat{u}(\infty, H)$. It follows that there exists an element

$\tilde{b} \in (p^\infty H)[p] \setminus (p^\infty H \cap T)[p]$. In particular, $\tilde{b} \notin T$, $|\tilde{b}| = \infty$, and $p\tilde{b} = 0$. Let $b = b' + \tilde{b}$. Then $pb = f(pa)$, $b \notin T$, and $|b| = \infty$, as desired. We may now extend f by $a \mapsto b$, and the extended function is a height-preserving isomorphism.

The specific structure of I comes from the construction in the proof of Theorem 5.1.11. □

Note that, in the theorem above, we can always choose $f_0 = \emptyset$, and $X_0 = Y_0 = \emptyset$, making (v) trivial.

5.3 Classification of general groups up to partial isomorphism

The classification of local groups serves as a stepping-stone to classifying general, global groups. A key piece to get us there is Warfield's local-global lemma (see Lemma 3.4.5):

Let G and H be groups and A and B subgroups of G and H, respectively, such that G/A and H/B are torsion. Suppose $f : A \to B$ is a homomorphism such that for every prime p, the induced homomorphism $f_p : A_p \to B_p$ extends to a homomorphism $g(p) : G_p \to H_p$. Then f extends to a homomorphism $g : G \to H$ such that $g_p = g(p)$ for all primes p. If every map $g(p)$ is an isomorphism, then g is an isomorphism.

We will use modified global Warfield and Stanton invariants based on those developed for the isomorphism case. Recall the global Warfield and Stanton invariants from Section 3.7:

For a compatibility class c of Ulm matrices, prime p, and equivalence class e of Ulm sequences, the (global) Stanton invariant of a group G is defined by

$$s(c, p, e, G) = \sup\{r_{p,0}(G(M)/G(M^*, p)) : M \in c \text{ and } M_p \in e\},$$

and if G has a decomposition basis X, the (global) Warfield invariant of G is given by

$$w_X(c, p, e, G) = \big|\{x \in X : \|x\| \in c \text{ and } \|x\|_p \in e\}\big|.$$

Recall that if G has a decomposition basis X, then the two invariants coincide for all c, p, e (Theorem 3.7.3), and so $w_X(c, p, e, G)$ is independent of the choice of decomposition basis. This allows us to write simply $w(c, p, e, G)$ without ambiguity.

We define the **modified Warfield invariant** for a group G with partial decomposition basis $\mathcal{C}$, compatibility class c, prime p, and equivalence class e as

> $\hat{w}_{\mathcal{C}}(c, p, e, G)$ = the largest $n < \omega$ if it exists, such that there are $X \in \mathcal{C}$ and $x_1, \ldots, x_n \in X$ with $\|x_i\| \in c$ and $\|x_i\|_p \in e$ for $1 \le i \le n$. If no such n exists, we let $\hat{w}_{\mathcal{C}}(c, p, e, G) = \omega$.

Note that if G has a decomposition basis X and $\mathcal{C}$ consists of all finite subsets of X, this cardinal is equal to the **modified Stanton invariant**, $\hat{s}(c, p, e, G) = s(c, p, e, G) \wedge \omega$

for all c, p, and e. Later we will see that this holds for any group with a partial decomposition basis.

As in the local case, this invariant is independent of the choice of $\mathcal{C}$, so we may drop the subscript $\mathcal{C}$:

Theorem 5.3.1. *If a group G has partial decomposition bases $\mathcal{C}$ and $\mathcal{C}'$, then for any compatibility class c, prime p, and equivalence class e of Ulm sequences, $\hat{w}_{\mathcal{C}}(c,p,e,G) = \hat{w}_{\mathcal{C}'}(c,p,e,G)$. In fact, if $\hat{w}_{\mathcal{C}}(c,p,e,G) \geq n$ and $Y \in \mathcal{C}'$, there is a $\tilde{Y} \in \mathcal{C}'$ such that $Y \subseteq \tilde{Y}$ and $\tilde{Y}$ contains distinct elements $y_1,\dots,y_n$ such that $\|y_i\| \in c$ and $\|y_i\|_p \in e$ for $i = 1,\dots,n$.*

Proof. Suppose $\hat{w}_{\mathcal{C}}(c,p,e,G) \geq n$ for some c, p, and e, say $x_1,\dots,x_n \in X$, where $X \in \mathcal{C}$, and for all $1 \leq i \leq n$, $\|x_i\| \in c$, and $\|x_i\|_p \in e$. Construct X' and Y' based on this X and the given Y as in Theorem 5.1.4. Then consider $\langle X'\rangle^0 = \langle Y'\rangle^0$ as a group with heights the same as in G. Both X' and Y' are decomposition bases, so by Theorem 3.7.3, $w(c,p,e,\langle Y'\rangle^0) = w(c,p,e,\langle X'\rangle^0) \geq n$. Thus, Y' contains the desired $y_1,\dots,y_n$. Since Y' is the union of an ascending chain of elements of $\mathcal{C}'$, we may choose $\tilde{Y} \subseteq Y'$ such that $\tilde{Y} \in \mathcal{C}'$, $Y \subseteq \tilde{Y}$, and $y_1,\dots,y_n \in \tilde{Y}$. □

As we would expect, the modified Warfield invariant is in fact invariant under $\cong_p$.

Lemma 5.3.2. *Let G and H be PDB groups. If $I : G \cong_p H$, then $\hat{w}(c,p,e,G) = \hat{w}(c,p,e,H)$ for each compatibility class c of Ulm matrices, prime p, and equivalence class e of Ulm sequences.*

Proof. Let $\mathcal{C}$ be a partial decomposition basis for G. Then $\{f(X) : X \in \mathcal{C}, f \in I\}$ is a partial decomposition basis for H (Theorem 5.1.2). Suppose $\hat{w}(c,p,e,G) \geq n$. Then there is some $X \in \mathcal{C}$ containing at least n elements with $\|x\| \in c$ and $\|x\|_p \in e$. Choose $f \in I$ such that $X \subseteq \operatorname{domain}(f)$. Then $f(X)$ contains at least n elements y with $\|y\| \in c$ and $\|y\|_p \in e$. This proves $\hat{w}(c,p,e,G) \leq \hat{w}(c,p,e,H)$ and the other direction follows by symmetry. □

Lemma 5.3.3. *Let G and H be groups with partial decomposition bases $\mathcal{C}$ and $\mathcal{C}'$, respectively. Let c be a compatibility class of Ulm matrices, p a prime, and e an equivalence class of Ulm sequences. Suppose $\hat{w}(c,p,e,G) = \hat{w}(c,p,e,H)$, $X \in \mathcal{C}$, $Y \in \mathcal{C}'$, and $\hat{w}(c,p,e,\langle X\rangle^0) > \hat{w}(c,p,e,\langle Y\rangle^0)$. Then there is a $Y' \in \mathcal{C}'$ such that $Y \subseteq Y'$, and a $y \in Y'\setminus Y$ such that $\|y\| \in c$ and $\|y\|_p \in e$.*

Proof. Letting $n = \hat{w}(c,p,e,\langle Y\rangle^0)$, we get $\hat{w}(c,p,e,H) = \hat{w}(c,p,e,G) \geq \hat{w}(c,p,e,\langle X\rangle^0) > \hat{w}(c,p,e,\langle Y\rangle^0) = n$. Thus, $\hat{w}(c,p,e,H) \geq n+1$. By Theorem 5.3.1, there is a $Y' \in \mathcal{C}'$ such that $Y \subseteq Y'$ and Y' contains at least $n+1$ elements y with $\|y\| \in c$ and $\|y\|_p \in e$. Since Y contains only n such elements, one must be in $Y' \setminus Y$. □

Lemma 5.3.4. *Suppose G is a PDB group, p is a prime, and e is an equivalence class of Ulm sequences. Then $\hat{w}(e,G_p) = (\sum_c \hat{w}(c,p,e,G)) \wedge \omega$.*

Proof. Let $\mathcal{C}$ be a partial decomposition basis for G. The image of $\mathcal{C}$ under the p-height-preserving natural map $G \to G_p$ is a partial decomposition basis for G_p; hence, $\hat{w}(e, G_p) \geq n$ implies $\sum_c \hat{w}(c, p, e, G) \geq n$. Suppose that $\sum_c \hat{w}(c, p, e, G) \geq n$, say $\sum_{i=1}^{k} \hat{w}(c_i, p, e, G) \geq n$ for some $1 \leq k \leq n$. By induction, define $X_1 \subseteq \cdots \subseteq X_k$, where each $X_i \in \mathcal{C}$ and contains at least $\hat{w}(c_i, p, e, G)$ elements x such that $\|x\| \in c_i$ and $\|x\|_p \in e$. This can be done by Theorem 5.3.1. The natural map $G \to G_p$ preserves heights, so the image of X_k contains at least n elements x with $\|x\|_p \in e$, completing the proof. □

Lemma 5.3.3 allows us to find an additional element in H associated with given c, p and e as long as the Warfield invariant is large enough. To prove the classification theorem, we will need to find such an element that matches the c, p, and e for a given $x \in G$ for every c, p, and e. The following variation on a result of Stanton will help us get there. It allows us to modify a decomposition basis one prime at a time, giving us the extension lemma that follows it.

Lemma 5.3.5 (Stanton [140]). *Let G be a group with a rich partial decomposition basis $\mathcal{C}$. Suppose $X \cup \{x_1, x_2\} \in \mathcal{C}$, where x_1 and x_2 have compatible Ulm matrices, and let p be a prime. Then there are elements y_1 and y_2 in G such that $\|y_1\|_p = \|x_2\|_p$, $\|y_2\|_p = \|x_1\|_p$, and $\|y_1\|_q = \|x_1\|_q$, $\|y_2\|_q = \|x_2\|_q$ for all primes $q \neq p$. Moreover, if $Y = X \cup \{y_1, y_2\}$, then $Y \in \mathcal{C}$ and $\langle X \cup \{x_1, x_2\}\rangle = \langle Y\rangle$.*

Proof. Since $X \cup \{x_1, x_2\}$ is a decomposition set, $\langle x_1\rangle \oplus \langle x_2\rangle$ is a valuated coproduct in G. By Proposition 3.7.4, we then get y_1 and y_2 with the desired Ulm sequences. Also, $\langle X \cup \{x_1, x_2\}\rangle = \langle X\rangle \oplus \langle x_1\rangle \oplus \langle x_2\rangle = \langle X\rangle \oplus \langle y_1\rangle \oplus \langle y_2\rangle = \langle Y\rangle$. In particular, $Y \in \mathcal{C}$ since $\mathcal{C}$ is rich. □

Lemma 5.3.6. *Let G and H be groups with rich partial decomposition bases $\mathcal{C}$ and $\mathcal{D}$, respectively. Suppose $\hat{w}(c, p, e, G) = \hat{w}(c, p, e, H)$ for every compatibility class c of Ulm matrices, prime p, and equivalence class e of Ulm sequences. Then if $X \cup \{x\} \in \mathcal{C}$, $Y \in \mathcal{D}$, and $\hat{w}(c, p, e, \langle X\rangle^0) = \hat{w}(c, p, e, \langle Y\rangle^0)$ for all c, p, and e, then there is a $y \in H$ such that $Y \cup \{y\} \in \mathcal{D}$ and $\hat{w}(c, p, e, \langle X \cup \{x\}\rangle^0) = \hat{w}(c, p, e, \langle Y \cup \{y\}\rangle^0)$ for all c, p, and e. In fact, $\|x\|$ and $\|y\|$ are equivalent Ulm matrices; that is, $k\|x\| = l\|y\|$ for some positive integers k, l.*

Proof. We start with a fixed prime p_0. Let c be the compatibility class of $\|x\|$ and e the equivalence class of $\|x\|_{p_0}$. Choose a $z \in H \setminus Y$ as in Lemma 5.3.3. Specifically, $\|z\| \in c$, $\|z\|_{p_0} \in e$, and $Y \cup \{z\} \in \mathcal{D}$. Then $\|z\|$ and $\|x\|$ are compatible. In particular, $\|z\|_p$ and $\|x\|_p$ are equivalent for all but finitely many primes $p_1, \ldots, p_n$. By induction on n, we will find our desired y, where $Y \cup \{y\} \in \mathcal{D}$, $\|y\| \in c$, and $\|y\|_p \sim \|x\|_p$ for all p. Since $\|y\|_p = \|x\|_p$ for almost all p, it will follow that $\|y\|$ and $\|x\|$ are equivalent.

If $n = 0$, z works as the desired y. Suppose the claim is true for $n - 1$. Let $e_n = [\|x\|_{p_n}]$. We know by assumption on p_n that $\|z\|_{p_n} \nsim \|x\|_{p_n}$, so $\|z\|_{p_n} \notin e_n$. It follows that $\hat{w}(c, p_n, e_n, \langle Y \cup \{z\}\rangle^0) = \hat{w}(c, p_n, e_n, \langle Y\rangle^0) < \hat{w}(c, p_n, e_n, \langle X \cup \{x\}\rangle^0)$. Choose a $z' \in H$ as in Lemma 5.3.3 applied to $Y \cup \{z\}$ and $X \cup \{x\}$. Then $Y \cup \{z, z'\} \in \mathcal{D}$, $\|z'\| \in c$, and $\|z'\|_{p_n} \in e_n$. Now we use Lemma 5.3.5 to swap $\{z, z'\}$ for some $\{y, y'\}$. Specifically,

$\langle Y \cup \{y, y'\}\rangle = \langle Y \cup \{z, z'\}\rangle$, $\|y\|_{p_n} = \|z'\|_{p_n}$, and $\|y\|_q = \|z\|_q$ for all $q \neq p_n$. Then $\|y\|_{p_n} = \|z'\|_{p_n} \sim \|x\|_{p_n}$, and $\|y\|_q = \|z\|_q \sim \|x\|_q$ for all $q \notin \{p_1, \ldots, p_n\}$. In particular, $\|y\|_p \sim \|x\|_p$ for all $p \notin \{p_1, \ldots, p_{n-1}\}$. The result follows by the induction hypothesis. □

Lemma 5.3.7. *Let G be a group, X a decomposition basis for G, and S a finitely generated subgroup of G such that $S \cap \langle X\rangle = \langle S \cap X\rangle$. Then if $y \in X$ and $y \notin S$, there is a positive integer n such that $|mny + s|_p = |mny|_p \wedge |s|_p$ for all $m \in \mathbb{Z}$, $s \in S$, and primes p.*

Proof. This lemma was proved in the local case, Lemma 5.2.6. We will localize to use that result. Since S is finitely generated and $S \subseteq \langle X\rangle^0$, we may find an integer $k \neq 0$ such that $ks \in \langle X\rangle$ for all $s \in S$. Let p be a prime and $v_p : G \to G_p$ the natural map. We claim $v_p(y) \notin S \otimes \mathbb{Z}_p$. Suppose to the contrary that $m(y \otimes 1) = s' \otimes 1$ for some $m \in \mathbb{Z} \setminus \{0\}$ and $s' \in S$. Then $my - s'$ is torsion (Lemma 1.17.8(3)); hence, $m'y \in S$ for some $m' \in \mathbb{Z} \setminus \{0\}$. Then $m'y \in S \cap \langle X\rangle = \langle S \cap X\rangle$, which means that $y \in S \cap X$, contradicting the original assumption that $y \notin S$, completing the claim.

Now apply Lemma 5.2.6 to G_p with decomposition basis $v_p(X)$, and $v_p(y) \notin S \otimes \mathbb{Z}_p$. There is an $n_p \in \mathbb{Z}$ such that $|rp^{n_p}v_p(y) + v_p(s)|_p = |rp^{n_p}v_p(y)|_p \wedge |v_p(s)|_p$ for all $r \in \mathbb{Z}_p$ and $s \in S$.

Now let $n = \Pi_{p|k} p^{n_p}$. We will prove that this is our desired n. First suppose p divides k, and hence, p^{n_p} divides n. Then for any $m \in \mathbb{Z}$ and $s \in S$, we have $|mny + s|_p = |mnv_p(y) + v_p(s)|_p = |mnv_p(y)|_p \wedge |v_p(s)|_p$.

Next suppose p does not divide k, and let $s \in S$. Since $ks \in S \cap \langle X\rangle = \langle S \cap X\rangle$, we may write $ks = a_1x_1 + \cdots + a_lx_l$ for some $x_1, \ldots, x_l \in S \cap X$. In particular, none of $x_1, \ldots, x_l$ is y. Then since p does not divide k, we may write $v_p(s) = \frac{a_1}{k}v_p(x_1) + \cdots + \frac{a_n}{k}v_p(x_n)$, and so $v_p(s) \in \langle v_p(\{x_1, \ldots, x_n\})\rangle$. Finally, since $v_p(\{x_1, \ldots, x_n, y\})$ is a decomposition set, $|mny + s|_p = |v_p(mny) + v_p(s)|_p = |v_p(mny)|_p \wedge |v_p(s)|_p = |mny|_p \wedge |s|_p$, as required. □

Now we are ready for the global classification theorem.

Theorem 5.3.8. *Let G and H be PDB groups. Then $G \cong_p H$ if and only if for every $\alpha \in$ Ord_∞, prime p, compatibility class c of Ulm matrices, and equivalence class e of Ulm sequences, $\hat{u}_p(\alpha, G) = \hat{u}_p(\alpha, H)$ and $\hat{w}(c, p, e, G) = \hat{w}(c, p, e, H)$.*

Proof. First suppose that $G \cong_p H$. Then by Corollary 4.4.6 and Lemma 5.3.2, the modified Ulm and Warfield invariants agree.

Now suppose the modified Ulm and Warfield invariants agree. We may choose rich partial decomposition bases $\mathcal{C}$ and $\mathcal{C}'$ for G and H, respectively. By Theorem 5.1.11, we need only prove that the following conditions hold for any height-preserving isomorphism $f : S \to T$, where $S \subseteq G$ and $T \subseteq H$ are finitely generated subgroups:

(a) If $X \subseteq S \subseteq \langle X\rangle^0$ for some X in $\mathcal{C}$, where $f(X) \in \mathcal{C}'$, and if $X \cup \{x\} \in \mathcal{C}$ for some $x \in G$, then there is a nonzero multiple x' of x and a $y' \in H$ such that $f(X) \cup \{y'\} \in \mathcal{C}'$ and $g : \langle S, x'\rangle \to \langle T, y'\rangle$, extending f by $x' \mapsto y'$, is an isomorphism that preserves heights;

(b) If $a \in G$ has a nonzero multiple in S, then there is a $b \in H$ such that $g : \langle S, a\rangle \to \langle T, b\rangle$, extending f, is an isomorphism that preserves heights;

and analogous conditions on functions from H to G.

Case (a): Suppose $X \cup \{x\} \in \mathcal{C}$. We will extend f to some multiple of x. Let $Y = f(X)$. By Lemma 5.3.6, there is a $y \in H \setminus Y$ such that $Y \cup \{y\} \in \mathcal{C}'$ and $\|x\|$ and $\|y\|$ are equivalent. Then there are multiples x' and y' of x and y, respectively, such that $\|x'\| = \|y'\|$. Consider the group $\langle S, x'\rangle^0$ with decomposition basis $X' = X \cup \{x'\}$. Since $S \cap \langle X'\rangle = \langle X\rangle = \langle S \cap X'\rangle$, we may apply Lemma 5.3.7 to X' and S, and similarly to $Y' = Y \cup \{y'\}$ and T. That gives us a positive integer n such that $|mnx' + s|_p = |mnx'|_p \wedge |s|_p$ and $|mny' + t|_p = |mny'|_p \wedge |t|_p$ for all $m \in \mathbb{Z}$, p prime, $s \in S$, and $t \in T$. Now define $g : \langle S, nx'\rangle \to \langle T, ny'\rangle$, extending f by $nx' \mapsto ny'$.

Case (b): Suppose a has a nonzero multiple in S. Let p be a prime. Then the localization map v_p maps $\mathcal{C}$ and $\mathcal{C}'$ to partial decomposition bases of G_p and H_p, respectively. By Lemma 5.3.4, $\hat{w}(e, G_p) = (\sum_{\mathcal{C}} \hat{w}(c, p, e, G)) \wedge \omega = (\sum_{\mathcal{C}} \hat{w}(c, p, e, H)) \wedge \omega = \hat{w}(e, H_p)$. Also $\hat{u}(\alpha, G_p) = \hat{u}(\alpha, H_p)$ for all $\alpha \in \mathrm{Ord}_\infty$ by Lemma 1.17.11.

Now we may apply the local classification theorem 5.2.7 to get $I_p : G_p \cong_p H_p$. Since I_p is a Karp system, there is a function in I_p extending f_p with $v_p(a)$ in its domain. Let $G' = \langle S, a\rangle$, $H' = T^0$. Let $g(p)$ be that function restricted to G'_p. Now apply Lemma 3.4.5, the local-global lemma, to the groups G' and H', torsion over their respective subgroups S and T, and $f : S \to T$. By that lemma, f extends to an injective homomorphism $g : \langle S, a\rangle \to T^0$ such that $g_p = g(p)$ for all primes p. Thus, g is an isomorphism onto its range. Finally, g is height-preserving since each $g(p)$ is p-height-preserving. □

Exercises.

1. (Göbel–Leistner–Loth–Strüngmann [37]) Give a counterexample to disprove the following statement: Let $f : S \to T$ be a height-preserving isomorphism, where S and T are submodules of some $\mathbb{Z}_p$-modules G and H. Then if $x \in G$ and $y \in H$ such that for some $n > 0$, $p^n x \in S$, $p^n y \in T$, $p^{n-1}x \notin S$, $p^{n-1}y \notin T$, and $f(p^n x) = p^n y$, then f can be extended to a height-preserving isomorphism $\langle S, x\rangle \to \langle T, y\rangle$ by $x \mapsto y$.

5.4 Classification up to an ordinal

Now that we have completed the classification up to partial isomorphism, we return to the question of equivalence up to some ordinal. Recall from the previous chapter that one version of Karp's Theorem 4.3.1 gives us two different ways of looking at such an equivalence:

Let G and H be groups and $\lambda \in \mathrm{Ord}_\infty$. Then the following are equivalent:

1. $G \equiv_\lambda H$;
2. *For each ordinal $\nu \le \lambda$, there is a nonempty set I_ν of isomorphisms on finitely generated subgroups of G into H such that*

(a) *if $\nu \leq \mu$, then $I_\mu \subseteq I_\nu$;*
(b) *if $\nu < \lambda$, $f \in I_{\nu+1}$, and $x \in G$ (or $y \in H$), then f extends to a map $f' \in I_\nu$ such that $x \in \mathrm{domain}(f')$ (or $y \in \mathrm{range}(f')$).*

Recall that a system of sets I_ν ($\nu \leq \lambda$) as in item (2) is called a **Karp system**. We will prove classification theorems in $L^\lambda_{\infty\omega}$ by constructing appropriate Karp systems. These will depend highly on the idea of "up to λ" for various ordinals. This viewpoint focuses on what happens below some threshold ordinal, and ignores differences, for example in heights, above the threshold. The following definitions will give the flavor of this approach and will be used throughout this section.

Let α be an ordinal. Recall from Section 4.4 the following definitions: Two ordinals β and γ are called **equal up to α**, written

$$\beta =_\alpha \gamma$$

if $\beta \wedge \alpha = \gamma \wedge \alpha$. Then $\beta \geq_\alpha \gamma$ is defined similarly. Given groups G and H, a map $f : S \to T$ with $S \subseteq G$ and $T \subseteq H$ is **α-height-preserving** if and only if $|f(x)|^H_p =_\alpha |x|^G_p$ for all $x \in S$ and primes p. For Ulm sequences and Ulm matrices, $=_\alpha$ and $\geq_\alpha$ are defined analogously. Two Ulm sequences $u = (\beta_i)_{i<\omega}$ and $u' = (\beta'_i)_{i<\omega}$ are **α-equivalent**, written $u \sim_\alpha u'$, exactly if $(\beta_{i+m})_{i<\omega} =_\alpha (\beta'_{i+n})_{i<\omega}$ for some $m, n > 0$. Then we write $e \sim_\alpha e'$ for equivalence classes e, e' of Ulm sequences whenever $u \sim_\alpha u'$ for some $u \in e$ and $u' \in e'$. We also write $u \sim_\alpha e$ in order to simplify notation. For Ulm matrices, $\sim_\alpha$ is defined analogously. Specifically, two Ulm matrices M and N are called **α-compatible**, written $M \sim_\alpha N$, whenever $mM \geq_\alpha N$ and $nN \geq_\alpha M$ for some $m, n > 0$. Then we write $c \sim_\alpha c'$ for compatibility classes c and c' of Ulm matrices whenever $M \sim_\alpha N$ for some $M \in c$ and $N \in c'$. Again, we often make use of the notation $M \sim_\alpha c$. For more details and properties, see Section 4.4.

Lemma 5.4.1. *Let c and c' be compatibility classes of Ulm matrices and α an ordinal. Then $c \sim_\alpha c'$ if and only if there are $M \in c$ and $N \in c'$, such that $M =_\alpha N$.*

Proof. Suppose $c \sim_\alpha c'$, $M = [m_{p,i}] \in c$, and $\tilde{N} = [n_{p,i}] \in c'$, where $M \sim_\alpha \tilde{N}$, say $mM \geq_\alpha \tilde{N}$ and $n\tilde{N} \geq_\alpha M$ for $m, n > 0$. Note that if $p \nmid mn$, then $M_p =_\alpha \tilde{N}_p$. By taking multiples, we may assume that for all $p \mid mn$, if $n_{p,i} \geq \alpha$ for some i, then $n_{p,0} \geq \alpha$, and similarly for $m_{p,i}$. Now suppose $m_{p,0} \geq \alpha$ for some $p \mid mn$. Then since $M \sim_\alpha \tilde{N}$, $n_{p,k} \geq m_{p,0} \geq \alpha$ for some k, and so $n_{p,0} \geq \alpha$. By symmetry we see that $m_{p,0} \geq \alpha$ if and only if $n_{p,0} \geq \alpha$.

Define N such that $N_p = \tilde{N}_p$ if $m_{p,0} \geq \alpha$ and $N_p = M_p$ otherwise. Then $M =_\alpha N$. Then $mN_p = m\tilde{N}_p \geq \tilde{N}_p$ if $m_{p,0} \geq \alpha$, and $mN_p = mM_p \geq_\alpha \tilde{N}_p$, if $m_{p,0} < \alpha$. Combining these cases, we get $mN \geq_\alpha \tilde{N}$. Since $n\tilde{N} \geq_\alpha M =_\alpha N$, it follows that $N \sim_\alpha \tilde{N}$, and so $N \in c'$. The converse is obvious. □

The following are easily verified:
(a) If $c \sim_\alpha c'$, then $M \sim_\alpha N$ for all $M \in c$ and $N \in c'$.
(b) $M \sim_\alpha N$ if and only if $M' =_\alpha N'$ for some $M' \sim M$ and $N' \sim N$.

(c) If $M \sim_\alpha N$, then $M_p =_\alpha N_p$ for almost all primes p.
(d) α-compatibility is an equivalence relation, and so partitions the compatibility classes.

We may now define a variation of the Warfield invariant to be used for $\equiv_\lambda$. Let α be an ordinal. First, we consider the p-local case. Assuming that G is a PDB module, we define

$$\hat{w}_\alpha(e,G) = \Big(\sum_{e' \sim_\alpha e} \hat{w}(e',G)\Big) \wedge \omega.$$

Turning to the global case, we assume now that G is a PDB group and define

$$\hat{w}_\alpha(c,p,e,G) = \Big(\sum_{c' \sim_\alpha c, e' \sim_\alpha e} \hat{w}(c',p,e',G)\Big) \wedge \omega.$$

In the previous chapter, we saw how Barwise and Eklof classified p-groups up to $L^\lambda_{\infty\omega}$-equivalence (Theorem 4.5.7) using modified Ulm and Warfield invariants. We wish to generalize this result to mixed groups.

As in the previous sections, we will first prove the classification theorem in the local case. So here we will consider $\mathbb{Z}_p$-modules for some fixed prime p, and all heights are p-heights.

As before, we will prove the classification theorem by proving that suitable isomorphisms may be extended to elements of a partial decomposition basis and to elements with multiples in the domain. Recall Theorem 5.1.9, which allows us to extend isomorphisms to elements with multiples in the domain, while maintaining the height-preserving characteristics, restated here for $\mathbb{Z}_p$-modules.

Theorem 5.4.2 (Jacoby [68]). *Let G and H be $\mathbb{Z}_p$-modules, α an ordinal, and $f : S \to T$ an α-height-preserving isomorphism, where S and T are finitely generated submodules of G and H, respectively. Suppose that $\hat{u}(\sigma,G) \le \hat{u}(\sigma,H)$ for all $\sigma < \alpha$. If $x \in G$ is proper with respect to S, $px \in S$, and $|x| + 1 < \alpha$, then for a suitable y in H, f can be extended to an α-height-preserving isomorphism $g : \langle S,x\rangle \to \langle T,y\rangle$ by mapping x onto y.*

The following lemma allows us to extend isomorphisms to elements with heights of α or greater.

Lemma 5.4.3 (Göbel–Leistner–Loth–Strüngmann [37], Jacoby [69]). *Let G and H be $\mathbb{Z}_p$-modules, α an ordinal, and $f : S \to T$ an α-height-preserving isomorphism, where S and T are submodules of G and H, respectively. Suppose that $x \in G$, $y \in H$, $|x| \ge \alpha$, and $|y| \ge \alpha$. Suppose further that $x + S$ and $y + T$ have the same finite order p^n, and $f(p^n x) = p^n y$. Then f can be extended to an α-height-preserving isomorphism $f' : \langle S,x\rangle \to \langle T,y\rangle$ by mapping x onto y.*

Proof. First, we note that by Lemma 1.12.4(2), f extends to an isomorphism f'. We claim that f' is α-height-preserving. First, suppose that $|s+mx| < \alpha$ for some $s \in S$ and $m \in \mathbb{Z}$. Then since $|mx| \geq \alpha$, we must have $|s| < \alpha$, and so $|f(s)| < \alpha$. Then $|s+mx| = |s| \wedge |mx| = |s| = |f(s)| = |f(s)| \wedge |my| = |f(s+mx)|$.

Now suppose $|s+mx| \geq \alpha$. This can happen only if $|s| \geq \alpha$, and so $|f(s)| \geq \alpha$. Then $|f(s)+my| \geq |f(s)| \wedge |my| \geq \alpha$. □

In the "up to ∞" case of the preceding section, we were able to use Theorem 5.1.9 to extend an isomorphism to an element with an arbitrary multiple in the domain. Now, however, the extension may not preserve heights to the original level; therefore, we must keep track of it each step of the way. The following lemma makes this explicit.

Lemma 5.4.4. *Let G and H be $\mathbb{Z}_p$-modules, and let ν be an ordinal such that $\hat{u}(\alpha, G) = \hat{u}(\alpha, H)$ for all $\alpha < \omega(\nu+1)$ and if* $\text{length}(tG) < \omega(\nu+1)$, *then $\hat{u}(\infty, G) = \hat{u}(\infty, H)$. Suppose $f : S \to T$ is an isomorphism that preserves heights up to $\omega\nu+k+n+1$ for some $k, n < \omega$, where S and T are p-nice, finitely generated submodules of G and H, respectively. Let $a \in G$, and suppose*
1. *$p^{n+1}a \in S$, and $p^n a \notin S$;*
2. *for all $0 \leq m \leq n$ the following is true: if for all $x \in p^m a + \langle S, p^{m+1}a\rangle$ we have $|x| < \omega(\nu+1)$, then $|x|+1 < \omega\nu+k$ for all such x;*
3. *if* $\text{length}(tG) < \omega(\nu+1)$, *then* $\text{length}(tG) < \omega\nu+k$.

Then f extends to an $\omega\nu+k$-height-preserving isomorphism $g : \langle S, a\rangle \to T'$ for some submodule T' of H.

Proof. Case 1: First, suppose that for all $0 \leq m \leq n$, and $x \in p^m a + \langle S, p^{m+1}a\rangle$, $|x| < \omega(\nu+1)$. Then by (2), $|x|+1 < \omega\nu+k$ for all such x.

By induction on m, $0 \leq m \leq n+1$, we will prove that there exists an isomorphism $f_m \supseteq f$ with domain $S_m = \langle S, p^{n+1-m}a\rangle$ that preserves heights up to $\omega\nu+k$. Let $f_0 = f$. Then domain$(f_0) = S = \langle S, p^{n+1}a\rangle = S_0$, as desired.

Suppose f_m has been constructed for some $0 \leq m \leq n$. We wish to extend f_m to $y = p^{n-m}a$. If $y \in S_m$, let $f_{m+1} = f_m$. Suppose $y \notin S_m$. By the assumption for this case, every $x \in y+S_m = p^{n-m}a+\langle S, p^{n+1-m}a\rangle$ satisfies $|x| < \omega(\nu+1)$, and so by (2), $|x|+1 < \omega\nu+k$. We may choose x proper with respect to S_m, since S_m is a finite extension of S, and hence, p-nice. Now we apply Theorem 5.1.9 to extend f_m to f_{m+1} with domain $\langle S_m, x\rangle = \langle S_m, y\rangle = \langle S, p^{n-m}a\rangle$ such that f_{m+1} preserves heights up to $\omega\nu+k$. This completes the induction. Let $g = f_{n+1}$ with domain $\langle S, a\rangle$, as desired.

Case 2: Now suppose that there is an $x \in p^m a + \langle S, p^{m+1}a\rangle$ such that $|x| \geq \omega(\nu+1)$ for some $0 \leq m \leq n$. We may assume that x has been chosen so that m is minimal. Choose $j \leq n-m$ least so that $p^{j+1}x \in S$. We seek a $b \in H$ as the image of x such that $|b| \geq \omega\nu+m+k$, $p^j b \notin T$ and $p^{j+1}b = f(p^{j+1}x)$. We have $|p^{j+1}x| \geq |x|+j+1 \geq \omega(\nu+1)+j+1 \geq \omega\nu+m+k+j+1$. Since f preserves heights up to $\omega\nu+k+n+1 \geq \omega\nu+m+k+j+1$, we have $|f(p^{j+1}x)| \geq \omega v + m + k + j + 1$. Thus, there is a $b_0 \in H$ such that $|b_0| \geq \omega v + m + k$ and $p^{j+1}b_0 = f(p^{j+1}x)$. If $p^j b_0 \notin T$, we may take b_0 as our desired b.

So suppose $p^j b_0 \in T$, say $p^j b_0 = f(a_0)$ for some $a_0 \in S$. Now we seek a $b_1 \in H$ such that $|b_1| \geq \omega v + m + k$, $p^j b_1 \notin T$ and $p^{j+1}b_1 = 0$. Then $b_0 + b_1$ will be our desired b. There are two cases depending on length(tG).

Case 2a: Suppose length$(tG) < \omega(v+1)$. Then length$(tG) < \omega v + k$ by (3). Let $\alpha = \omega v + k + n + 1$. Consider $z = p^j x - a_0$. Then $f(pz) = p^{j+1}b_0 - p^{j+1}b_0 = 0$, so $pz = 0$. Also $|z| \geq |p^j x| \wedge |a_0| \geq \omega v + m + k$ since $|p^j b_0| \geq \omega v + m + k$. But length$(tG) < \omega v + k$, so we must have $|z| = \infty$.

Then we have $\dim(p^\infty H)[p] \wedge \omega = \hat{u}(\infty, H) = \hat{u}(\infty, G) = \dim(p^\infty G)[p] \wedge \omega \geq \dim(p^\infty G \cap \langle S, z\rangle[p]) = \dim(p^\infty G \cap (S \oplus \langle z\rangle)[p]) = \dim((p^\infty G \cap S[p]) \oplus \langle z\rangle)$, where the latter equality follows from the modular law, noting that $\langle z\rangle + S[p] = (S + \langle z\rangle)[p]$. This gives us $\dim(p^\infty H)[p] > \dim(S[p] \cap p^\infty G) = \dim(S[p] \cap p^\infty(tG)) = \dim(S[p] \cap p^\alpha(tG))$, since length$(tG) < \omega v + k < \alpha$. Since f is α-height-preserving, this is $\dim(T[p] \cap p^\alpha H)$. This proves that $\dim(p^\infty H)[p] > \dim(T[p] \cap p^\alpha H)$. Therefore, there is an element in $(p^\infty H)[p]$ that is not in T. We may write it as $p^j b_1$, where $|b_1| = \infty$. This is our desired b_1. We may now take $b_0 + b_1$ as our desired b.

Case 2b: Now suppose length$(tG) \geq \omega(v+1)$. Then length$(tH) \geq \omega(v+1)$ as well, since otherwise the Ulm invariants would disagree between length(tH) and $\omega(v+1)$. Again, we seek a b as the image of x. By Lemma 4.5.1, applied to $tT \subseteq tH$, the limit ordinal $\omega(v+1) \leq$ length(tH), and $\beta = \omega v + m + k$, we get $b_1 \in p^{\omega v+m+k}H$ such that $p^j b_1 \notin T$ and $p^{j+1}b_1 = 0$. As before, take $b = b_0 + b_1$. This completes Case 2b.

In either case, we have produced a $b \in H$ such that $p^j b \notin T$, $p^{j+1}b = f(p^{j+1}x)$, and $|b| \geq \omega v + m + k$. By Lemma 5.4.3, we may define $f' : \langle S, x\rangle \to \langle T, b\rangle$, extending f by $x \mapsto b$, an $\omega v + m + k$-height-preserving isomorphism. Now we will extend f' to include a.

First, we claim that $\langle S, x\rangle = \langle S, p^m a\rangle$. Recall that $x \in p^m a + \langle S, p^{m+1}a\rangle$, say $x = p^m a + s + rp^{m+1}a$ for some $r \in \mathbb{Z}_p$ and $s \in S$. Let $y = p^m a + rp^{m+1}a = (1 + rp)p^m a$. Then $\langle S, x\rangle = \langle S, y\rangle$, since $x - y \in S$. Since $p \nmid (1 + rp)$, $\langle y\rangle = \langle p^m a\rangle$, and so domain$(f') = S' = \langle S, x\rangle = \langle S, p^m a\rangle$.

Now let $l \leq m - 1$, and $z \in p^l a + \langle S', p^{l+1}a\rangle = p^l a + \langle S, p^{l+1}a\rangle$. By the minimality of m, $|z| < \omega(v+1)$. Since z and l were arbitrary, we may apply Case 1 to $f' : S' \to T'$, and $m - 1$. Thus, we may extend f' to g, an $\omega v + k$-height-preserving isomorphism with domain $\langle S', a\rangle$. □

Now we look at extending isomorphisms to elements of a partial decomposition basis. This theorem provides the foundation.

Theorem 5.4.5. *Let G be a $\mathbb{Z}_p$-module with partial decomposition basis $\mathcal{C}$, α an ordinal, and n a positive integer. Suppose $\hat{w}_\alpha(e, G) \geq n$, and $X \in \mathcal{C}$. Then there is an $X' \in \mathcal{C}$ such that $X \subseteq X'$ and X' has $\geq n$ elements x with $\|x\|_p \sim_\alpha e$.*

Proof. Choose a finite set $\{e_1, \dots, e_m\}$ of distinct equivalence classes with $e_i \sim_\alpha e$ for $i = 1, \dots, m$, such that $\sum_{i=1}^m w(e_i, G) \geq n$. If $w(e_i, G) \geq \omega$ for some i, the result follows from Theorem 5.2.1. So suppose for each $1 \leq i \leq m$, $w(e_i, G)$ is finite, say $w(e_i, G) = n_i$.

Now we will define by induction on $0 \leq i \leq m$, $X_i \in C$, having at least n_i elements x with $\|x\|_p \in e_i$, and such that $X = X_0 \subseteq X_1 \subseteq \cdots \subseteq X_m$. Given X_i, Theorem 5.2.1 gives us $X_{i+1} \supseteq X_i$ with $\geq n_{i+1}$ elements x with $\|x\|_p \in e_{i+1}$. This completes the induction. Note that these elements are distinct since the e_i are distinct. Let $X' = X_m$. □

Now we are ready to prove the classification theorem. If the modified Ulm and Warfield invariants agree up to a certain level, depending on λ, then the groups are λ-equivalent.

Theorem 5.4.6 (Jacoby–Loth [72]). *Let G and H be PDB modules. Let λ be an ordinal such that*
1. $\hat{u}(\alpha, G) = \hat{u}(\alpha, H)$ *for all* $\alpha < \omega\lambda$;
2. $\hat{w}_{\omega(\nu+1)}(e, G) = \hat{w}_{\omega(\nu+1)}(e, H)$ *for all* $\nu < \lambda$ *and equivalence classes e of Ulm sequences;*
3. *if* $\text{length}(tG) < \omega\lambda$, *then* $\hat{u}(\infty, G) = \hat{u}(\infty, H)$.

Then $G \equiv_\lambda H$.

Specifically, for any (possibly empty) X_0 and Y_0 in the respective partial decomposition bases C and C', and for any bijection $f_0 : X_0 \to Y_0$ such that $\|x\|_p = \|f_0(x)\|_p$ for all $x \in X_0$, there is a Karp system $\{I_\nu : \nu \leq \lambda\}$ with each set I_ν consisting of all maps $f : S \to T$ with associated $X \in C$ and $Y \in C'$ such that
(i) *S and T are finitely generated submodules of G and H, respectively;*
(ii) *f is an $\omega\nu$-height-preserving isomorphism;*
(iii) $X \subseteq S \subseteq \langle X \rangle^0$, *and* $Y \subseteq T \subseteq \langle Y \rangle^0$;
(iv) $f(X) = Y$;
(v) $X_0 \subseteq X$ *and* $f \restriction X_0 = f_0$.

Proof. We may choose rich partial decomposition bases C and C'. By Theorem 5.1.11, it suffices to prove that the following conditions hold for any ordinal $\nu < \lambda$ and any $\omega(\nu+1)$-height-preserving isomorphism $f : S \to T$, where $S \subseteq G$ and $T \subseteq H$ are finitely generated submodules:
(a) If $X \subseteq S \subseteq \langle X \rangle^0$ for some X in C, where $f(X) \in C'$, and if $X \cup \{x\} \in C$ for some x, then there is a nonzero multiple x' of x and a $y' \in H$ such that $f(X) \cup \{y'\} \in C'$ and $g : \langle S, x' \rangle \to \langle T, y' \rangle$, extending f by $x' \mapsto y'$, is an isomorphism that preserves heights up to $\omega(\nu+1)$;
(b) If $a \in G$ has a nonzero multiple in S, then there is a $b \in H$ such that $g : \langle S, a \rangle \to \langle T, b \rangle$, extending f by $a \mapsto b$, is an isomorphism that preserves heights up to $\omega\nu$;

and analogous conditions on partial isomorphisms from H to G.

Now let $f : S \to T$ be such a mapping.

Case (a): Let X and x be as in (a), and $e = [\|x\|]$. Then $\hat{w}_{\omega(v+1)}(e, H) = \hat{w}_{\omega(v+1)}(e, G) \geq \hat{w}_{\omega(v+1)}(e, \langle X \cup \{x\}\rangle^0) > \hat{w}_{\omega(v+1)}(e, \langle X\rangle^0) = \hat{w}_{\omega(v+1)}(e, \langle Y\rangle^0)$, where $Y = f(X)$. By Theorem 5.4.5, there is a $Y' \in C'$ such that $Y \subseteq Y'$ and

$$\hat{w}_{\omega(v+1)}(e, \langle Y'\rangle^0) \geq \hat{w}_{\omega(v+1)}(e, \langle Y\rangle^0) + 1.$$

Therefore, there must be a $y \in Y' \backslash Y$ such that $\|y\|_p \sim_{\omega(v+1)} e$, and hence, $\|y\|_p \sim_{\omega(v+1)} \|x\|_p$. Then there are $n', m' < \omega$ such that $\|p^{n'} x\|_p =_{\omega(v+1)} \|p^{m'} y\|_p$.

Assume $x \notin S$, since the other case is trivial. Now apply Lemma 5.2.6 to the $\mathbb{Z}_p$-module $\langle S, x\rangle^0$ with decomposition basis $X \cup \{x\}$ and submodule S. This applies since $X \cup \{x\} \subseteq \langle S, x\rangle$, and so, since $S \subseteq \langle X\rangle^0$,

$$S \cap \langle X \cup \{x\}\rangle = \langle X\rangle = \langle S \cap (X \cup \{x\})\rangle.$$

We may similarly apply Lemma 5.2.6 to $\langle T, y\rangle^0$, $Y \cup \{y\}$, and T, since $T \subseteq \langle Y\rangle^0$ says $y \notin T$. This gives us $k < \omega$ such that for all $s \in S$, $t \in T$, and $r \in \mathbb{Z}_p$,

$$|rp^k x + s| = |rp^k x| \wedge |s| \quad \text{and}$$
$$|rp^k y + t| = |rp^k y| \wedge |t|.$$

Now let $n = k + n'$, $m = k + m'$, and define g extending f by sending $p^n x$ to $p^m y$. It is straightforward to verify that g is an isomorphism that preserves heights to $\omega(v + 1)$.

Case (b): Choose $n \geq 0$ least such that $p^{n+1} a \in S$. Choose $k < \omega$ sufficiently large so that for any given $j \leq n$, if every $x \in p^j a + \langle S, p^{j+1} a\rangle$ has height $< \omega(v + 1)$, then every such x has height $< \omega v + k$, and if $\operatorname{length}(tG) < \omega(v+1)$ then $\operatorname{length}(tG) < \omega v + k$. This is possible since the submodules involved are p-nice by Corollary 5.1.8, and so each coset has an element of maximal height. The map f preserves heights up to $\omega(v + 1)$; hence, it preserves heights up to $\omega v + k + n + 1$. By Lemma 5.4.4, f extends to an $\omega v + k$-height-preserving isomorphism g with $\operatorname{domain}(g) = \langle S, a\rangle$. Thus, g is an ωv-height-preserving isomorphism. □

This completes the classification of PDB modules. Now we consider arbitrary PDB groups. Recall that

$$\hat{w}_\alpha(c, p, e, G) = \Bigg(\sum_{c' \sim_\alpha c, e' \sim_\alpha e} \hat{w}(c', p, e', G)\Bigg) \wedge \omega.$$

Recall also that if G has a partial decomposition basis C, then $\hat{w}(c, p, e, G)$ is the largest n, if it exists, such that for some $X \in C$ and $x_1, \ldots, x_n \in X$, $\|x_i\| \in c$, and $\|x_i\|_p \in e$. In particular, for any X, $\hat{w}(c, p, e, \langle X\rangle^0) = |\{x \in X : \|x\| \in c, \text{ and } \|x\|_p \in e\}|$. Hence, $\hat{w}(c, p, e, G) \geq \hat{w}(c, p, e, \langle X\rangle^0)$. It then follows that $\hat{w}_\alpha(c, p, e, G) \geq \hat{w}_\alpha(c, p, e, \langle X\rangle^0)$.

As before, the local classification gives us the foundation for the global classification theorem. We previously used Lemma 5.3.5 to build the required system of partial isomorphisms. Here we prove the analogous lemma up to α.

Lemma 5.4.7. *Let G be a group with decomposition basis X, p a prime, and $x_1, x_2 \in X$ with α-compatible Ulm matrices. Then there are elements $y_1, y_2 \in \langle X \rangle$ such that $\|y_1\|_p =_\alpha \|x_2\|_p$, $\|y_2\|_p =_\alpha \|x_1\|_p$, and $\|y_1\|_q =_\alpha \|x_1\|_q$, $\|y_2\|_q =_\alpha \|x_2\|_q$ for all primes $q \neq p$. The set $Y = (X \setminus \{x_1, x_2\}) \cup \{y_1, y_2\}$ is a decomposition basis for G, and $\langle X \rangle = \langle Y \rangle$.*

Proof. Since $\|x_1\| \sim_\alpha \|x_2\|$, we have some positive integer m such that $m\|x_1\| \geq_\alpha \|x_2\|$ and $m\|x_2\| \geq_\alpha \|x_1\|$. Write $m = p^j n$, where $p \nmid n$. Now we follow the proof of Proposition 3.7.4. This gives us integers s and t such that $sp^{2j} - tn = 1$ and proves that, if we define

$$\begin{pmatrix} y_1 \\ y_2 \end{pmatrix} = \begin{pmatrix} sp^j & tn \\ n & p^j \end{pmatrix} \begin{pmatrix} x_1 \\ x_2 \end{pmatrix},$$

then $\langle x_1 \rangle \oplus \langle x_2 \rangle = \langle y_1 \rangle \oplus \langle y_2 \rangle$ and $\langle y_1 \rangle \oplus \langle y_2 \rangle$ is a valuated coproduct. Then since $\langle X \rangle = \langle X \setminus \{x_1, x_2\} \rangle \oplus \langle x_1 \rangle \oplus \langle x_2 \rangle = \langle X \setminus \{x_1, x_2\} \rangle \oplus \langle y_1 \rangle \oplus \langle y_2 \rangle = \langle Y \rangle$ is a valuated coproduct, Y is a decomposition set.

Following the proof further, we get $\|y_1\|_p = \|sp^j x_1\|_p \wedge \|tnx_2\|_p =_\alpha \|x_2\|_p$, since $\|sp^j x_1\|_p = \|smx_1\|_p \geq_\alpha \|sx_2\|_p$ and $\|tnx_2\|_p = \|x_2\|_p$. Similarly, $\|y_2\|_p =_\alpha \|x_1\|_p$. Now suppose $q \neq p$. Note that q does not divide both n and s, and so at least one of them is relatively prime to q. It may be verified that in either case $\|y_1\|_q =_\alpha \|x_1\|_q$, and $\|y_2\|_q =_\alpha \|x_2\|_q$. □

Theorem 5.4.8. *Let G be a group with partial decomposition basis $\mathcal{C}$, α an ordinal, and n a positive integer. If $\hat{w}_\alpha(c, p, e, G) \geq n$ and $Y \in \mathcal{C}$, then there exists $Y' \in \mathcal{C}$ such that $Y \subseteq Y'$ and Y' contains elements $y_1, \ldots, y_n$ such that $\|y_i\| \sim_\alpha c$ and $\|y_i\|_p \sim_\alpha e$ for all $i = 1, \ldots, n$.*

Proof. Let $E = \{(c', e') : \hat{w}(c', p, e', G) \neq 0, c' \sim_\alpha c, \text{ and } e' \sim_\alpha e\}$. Choose distinct elements $(c_1, e_1), \ldots, (c_k, e_k)$ of E such that $k \leq n$ and $\sum_{1 \leq i \leq k} m_i \geq n$, where $m_i = \hat{w}(c_i, p, e_i, G)$ for each $1 \leq i \leq k$. This is possible since $\hat{w}(c_i, p, e_i, G) \geq 1$ for each $1 \leq i \leq k$.

We define Y_i inductively for each $0 \leq i \leq k$. Let $Y_0 = Y$. Given Y_i, by Theorem 5.3.1, there is a $Y_{i+1} \in \mathcal{C}$ such that $Y_i \subseteq Y_{i+1}$ and Y_{i+1} contains at least m_{i+1} elements x such that $\|x\| \in c_{i+1} \sim_\alpha c$ and $\|x\|_p \in e_{i+1} \sim_\alpha e$. Let $Y' = \bigcup_{i=0}^k Y_i$. Then $Y' \supseteq Y$ contains at least m_i elements x such that $\|x\| \in c_i$ and $\|x\|_p \in e_i$ for each i, and such elements for m_i and m_j are distinct for $i \neq j$. Hence, there are at least $m_1 + \cdots + m_k \geq n$ elements in Y' with $\|x\| \sim_\alpha c$ and $\|x\|_p \sim_\alpha e$. □

Corollary 5.4.9. *Let G be a group with partial decomposition basis $\mathcal{C}$, α an ordinal, c a compatibility class of Ulm matrices, p a prime, and e an equivalence class of Ulm sequences. Then $\hat{w}_\alpha(c, p, e, G)$ is the largest integer n, if it exists, such that there are $X \in \mathcal{C}$ and $x_1, \ldots, x_n \in X$ with $\|x_i\| \sim_\alpha c$ and $\|x_i\|_p \sim_\alpha e$ for all $i = 1, \ldots, n$. If no such n exists, $\hat{w}_\alpha(c, p, e, G) = \omega$.*

Proof. First, suppose there is a largest n for which there is an $X \in \mathcal{C}$ and $x_1, \ldots, x_n \in X$, such that $\|x_i\| \sim_\alpha c$ and $\|x_i\|_p \sim_\alpha e$ for all $1 \leq i \leq n$. We know that $\hat{w}_\alpha(c, p, e, G) \geq$

$\hat{w}_\alpha(c,p,e,\langle X\rangle^0) = n$. By the theorem, if $\hat{w}_\alpha(c,p,e,G) > n$, then there is some $Y \in \mathcal{C}$ and $y_1,\ldots,y_{n+1} \in Y$ such that $\|y_i\| \sim_\alpha c$ and $\|y_i\|_p \sim_\alpha e$ for all $1 \le i \le n+1$, contradicting maximality. If there is no largest n, then for any n, there is an $X \in \mathcal{C}$ and $x_1,\ldots,x_n \in X$ with $\|x_i\| \sim_\alpha c$ and $\|x_i\|_p \sim_\alpha e$ for all $1 \le i \le n$. Then $\hat{w}_\alpha(c,p,e,G) \ge \hat{w}_\alpha(c,p,e,\langle X\rangle^0) \ge n$. Since n was arbitrary, $\hat{w}_\alpha(c,p,e,G) = \omega$. □

Corollary 5.4.10. *Let G and H be groups with partial decomposition bases $\mathcal{C}$ and $\mathcal{D}$, respectively, α an ordinal, c a compatibility class of Ulm matrices, p a prime, and e an equivalence class of Ulm sequences. Suppose $\hat{w}_\alpha(c,p,e,G) = \hat{w}_\alpha(c,p,e,H)$, $X \in \mathcal{C}$, $Y \in \mathcal{D}$, and*

$$\hat{w}_\alpha(c,p,e,\langle X\rangle^0) > \hat{w}_\alpha(c,p,e,\langle Y\rangle^0).$$

Then there exists a $Y' \in \mathcal{D}$ such that $Y \subseteq Y'$ and there is an element $y \in Y'\backslash Y$ such that $\|y\| \sim_\alpha c$ and $\|y\|_p \sim_\alpha e$.

Proof. Suppose $\hat{w}_\alpha(c,p,e,\langle Y\rangle^0) = n$. Then $\hat{w}_\alpha(c,p,e,G) \ge \hat{w}_\alpha(c,p,e,\langle X\rangle^0) \ge n+1$. By the theorem, there is a $Y' \in \mathcal{C}$ such that $Y \subseteq Y'$ and Y' contains elements $y_1,\ldots,y_{n+1}$ such that $\|y_i\| \sim_\alpha c$ and $\|y_i\|_p \sim_\alpha e$ for all $1 \le i \le n+1$. Since Y contains only n such elements, one of them must be in $Y' \setminus Y$. □

To prove the classification theorem, we first need an extension theorem for partial decomposition bases.

Lemma 5.4.11 (Jacoby–Leistner–Loth–Strüngmann [76]). *Let G and H be groups with rich partial decomposition bases $\mathcal{C}$ and $\mathcal{D}$, respectively. Suppose α is an ordinal such that $\hat{w}_\alpha(c,p,e,G) = \hat{w}_\alpha(c,p,e,H)$ for every compatibility class c of Ulm matrices, prime p, and equivalence class e of Ulm sequences. Assume $X \cup \{x\} \in \mathcal{C}$ and $Y \in \mathcal{D}$ such that*

$$\hat{w}_\alpha(c,p,e,\langle X\rangle^0) = \hat{w}_\alpha(c,p,e,\langle Y\rangle^0)$$

for all c, p, and e. Then there exists an element $y \in H$ such that $Y \cup \{y\} \in \mathcal{D}$ and

$$\hat{w}_\alpha(c,p,e,\langle X \cup \{x\}\rangle^0) = \hat{w}_\alpha(c,p,e,\langle Y \cup \{y\}\rangle^0)$$

for all c, p, and e. In fact, $\|x\| \sim_\alpha \|y\|$ and $\|x\|_p \sim_\alpha \|y\|_p$ for all primes p.

Proof. The case of $x \in X$ is trivial, so assume $x \in G \setminus X$. Let $c_0 = [\|x\|]$, p_0 an arbitrary prime, and $e_0 = [\|x\|_{p_0}]$. Then

$$\hat{w}_\alpha(c_0,p_0,e_0,\langle X \cup \{x\}\rangle^0) = \hat{w}_\alpha(c_0,p_0,e_0,\langle X\rangle^0) + 1 > \hat{w}_\alpha(c_0,p_0,e_0,\langle Y\rangle^0).$$

By Corollary 5.4.10, there is a $Y' \in \mathcal{D}$ such that $Y \subseteq Y'$, and a $z \in Y' \setminus Y$ with $\|z\| \sim_\alpha c_0$ and $\|z\|_{p_0} \sim_\alpha e_0$. Then $\|x\| \sim_\alpha \|z\|$, and that implies that $\|x\|_p =_\alpha \|z\|_p$ for almost all primes p, so we may assume that $\|x\|_p \sim_\alpha \|z\|_p$ for all primes p, except $\{p_1,\ldots,p_n\}$.

We will prove, by induction on n, that if there is a $\tilde{z}$ such that $Y \cup \{\tilde{z}\} \in \mathcal{D}$, $\|x\| \sim_\alpha \|\tilde{z}\|$, and $\|x\|_p \sim_\alpha \|\tilde{z}\|_p$ for all but n values of p, then there is a $y \in H \setminus Y$ such that

$$Y \cup \{y\} \in \mathcal{D}, \quad \|y\| \sim_\alpha c_0, \quad \text{and} \quad \|x\|_p \sim_\alpha \|y\|_p \quad \text{for all primes } p.$$

Suppose $n = 0$. Since $\mathcal{D}$ is rich, $Y \cup \{z\} \in \mathcal{D}$, so we simply take $y = z$. Now assume the assertion is true for $n - 1$. Let $e_n = [\|x\|_{p_n}]$. By assumption, $\|z\|_{p_n} \nsim_\alpha \|x\|_{p_n}$; that is, $\|z\|_{p_n} \nsim_\alpha e_n$, and so

$$\hat{w}_\alpha(c_0, p_n, e_n, \langle Y \cup \{z\}\rangle^0) = \hat{w}_\alpha(c_0, p_n, e_n, \langle Y\rangle^0) < \hat{w}_\alpha(c_0, p_n, e_n, \langle X \cup \{x\}\rangle^0).$$

By Corollary 5.4.10, there is an element $z' \in H$ such that $Y \cup \{z, z'\} \in \mathcal{D}$, $\|z'\| \sim_\alpha c_0$, and $\|z'\|_{p_n} \sim_\alpha e_n$. Now apply Lemma 5.4.7 to the group $\langle Y \cup \{z, z'\}\rangle^0$ with decomposition basis $Y \cup \{z, z'\}$, the elements z and z', and the prime p_n. This then gives us $y, y' \in \langle Y \cup \{z, z'\}\rangle$ such that $\langle Y \cup \{z, z'\}\rangle = \langle Y \cup \{y, y'\}\rangle$, $\|y\|_{p_n} =_\alpha \|z'\|_{p_n}$, $\|y'\|_{p_n} =_\alpha \|z\|_{p_n}$, and for all $p \neq p_n$, $\|y\|_p =_\alpha \|z\|_p$, $\|y'\|_p =_\alpha \|z'\|_p$. In particular, $\|y\|_{p_n} \sim_\alpha \|z'\|_{p_n} \sim_\alpha e_n = [\|x\|_{p_n}]$, and for all $p \notin \{p_0, \ldots, p_n\}$, $\|y\|_p \sim_\alpha \|z\|_p \sim_\alpha \|x\|_p$. Furthermore, $\|y\|_{p_0} =_\alpha \|z\|_{p_0} \sim_\alpha \|x\|_{p_0}$, so $\|y\|_p \sim_\alpha \|x\|_p$ for all $p \notin \{p_1, \ldots, p_{n-1}\}$.

By the induction hypothesis, there is a $\tilde{y}$ such that $Y \cup \{\tilde{y}\} \in \mathcal{D}$, $\|\tilde{y}\| \sim_\alpha c_0 \sim_\alpha \|x\|$, and $\|x\|_p \sim_\alpha \|\tilde{y}\|_p$ for all primes p. This $\tilde{y}$ is the desired y. □

We will use the local classification theorem to prove the global one. First, we relate the local and global Warfield invariants.

Lemma 5.4.12. *Let G and H be PDB groups, p a prime, α an ordinal, and e an equivalence class of Ulm sequences. If $\hat{w}_\alpha(c, p, e, G) = \hat{w}_\alpha(c, p, e, H)$ for all compatibility classes c of Ulm matrices, then $\hat{w}_\alpha(e, G_p) = \hat{w}_\alpha(e, H_p)$.*

Proof. Let C be a collection of representatives from the α-compatibility classes, with exactly one from each class. Using Lemma 5.3.4, we obtain

$$\begin{aligned}
\hat{w}_\alpha(e, G_p) &= \Big(\sum_{e' \sim_\alpha e} \hat{w}(e', G_p)\Big) \wedge \omega = \Big(\sum_{e' \sim_\alpha e} \sum_{c} \hat{w}(c, p, e', G)\Big) \wedge \omega \\
&= \Big(\sum_{e' \sim_\alpha e} \sum_{c \in C} \sum_{c' \sim_\alpha c} \hat{w}(c', p, e', G)\Big) \wedge \omega \\
&= \Big(\sum_{c \in C} \sum_{e' \sim_\alpha e} \sum_{c' \sim_\alpha c} \hat{w}(c', p, e', G)\Big) \wedge \omega \\
&= \Big(\sum_{c \in C} \hat{w}_\alpha(c, p, e, G)\Big) \wedge \omega.
\end{aligned}$$

By the same argument, we get

$$\hat{w}_\alpha(e, H_p) = \Big(\sum_{c \in C} \hat{w}_\alpha(c, p, e, H)\Big) \wedge \omega = \Big(\sum_{c \in C} \hat{w}_\alpha(c, p, e, G)\Big) \wedge \omega = \hat{w}_\alpha(e, G_p). \qquad \square$$

Now we are ready for the global classification theorem up to λ.

Theorem 5.4.13. *Let G and H be PDB groups, and let λ be an ordinal. Suppose the following:*
1. $\hat{u}_p(\alpha, G) = \hat{u}_p(\alpha, H)$ *for all primes p and* $\alpha < \omega\lambda$;
2. $\hat{w}_{\omega(\nu+1)}(c, p, e, G) = \hat{w}_{\omega(\nu+1)}(c, p, e, H)$ *for every compatibility class c of Ulm matrices, prime p, equivalence class e of Ulm sequences, and* $\nu < \lambda$;
3. *if* $\text{length}(t(G_p)) < \omega\lambda$, *then* $\hat{u}_p(\infty, G) = \hat{u}_p(\infty, H)$.

Then $G \equiv_\lambda H$.

Proof. Let $\mathcal{C}$ and $\mathcal{D}$ be rich partial decomposition bases for G and H, respectively. By Theorem 5.1.11, it suffices to prove that the following conditions hold for any $\nu < \lambda$ and any $\omega(\nu+1)$-height-preserving isomorphism $f : S \to T$, where $S \subseteq G$ and $T \subseteq H$ are finitely generated:
(a) If $X \subseteq S \subseteq \langle X\rangle^0$ for some X in $\mathcal{C}$, where $f(X) \in \mathcal{D}$, and if $X \cup \{x\} \in \mathcal{C}$ for some $x \in G$, then there is a nonzero multiple x' of x and a $y' \in H$ such that $f(X) \cup \{y'\} \in \mathcal{D}$ and $g : \langle S, x'\rangle \to \langle T, y'\rangle$, extending f by $x' \mapsto y'$, is an isomorphism that preserves heights up to $\omega(\nu+1)$;
(b) If $a \in G$ has a nonzero multiple in S, then there is a $b \in H$ such that $g : \langle S, a\rangle \to \langle T, b\rangle$, extending f by $a \mapsto b$, is an isomorphism that preserves heights up to $\omega\nu$;

and analogous conditions on partial isomorphisms from H to G.

Therefore, suppose $f : S \to T$ is an $\omega(\nu+1)$-height-preserving isomorphism, where $\nu < \lambda$, and $S \subseteq G$ and $T \subseteq H$ are finitely generated.

Case (a): Suppose for some $X \in \mathcal{C}$, $X \subseteq S \subseteq \langle X\rangle^0$, $Y = f(X) \in \mathcal{D}$, and $X \cup \{x\} \in \mathcal{C}$. Since the case of $x \in S$ is trivial, we may assume $x \notin S$. Then $\hat{w}_{\omega(\nu+1)}(c, p, e, \langle X\rangle^0) = \hat{w}_{\omega(\nu+1)}(c, p, e, \langle Y\rangle^0)$ for all c, p, and e since f is $\omega(\nu+1)$-height-preserving. Then by Lemma 5.4.11, there is a $y \in H$ such that $Y \cup \{y\} \in \mathcal{D}$ and $\hat{w}_{\omega(\nu+1)}(c, p, e, \langle X \cup \{x\}\rangle^0) = \hat{w}_{\omega(\nu+1)}(c, p, e, \langle Y \cup \{y\}\rangle^0)$ for all c, p, e. In fact, $\|x\| \sim_{\omega(\nu+1)} \|y\|$ and $\|x\|_p \sim_{\omega(\nu+1)} \|y\|_p$ for all p. Since $\|x\| \sim_{\omega(\nu+1)} \|y\|$, $\|x\|_p =_{\omega(\nu+1)} \|y\|_p$ for all but finitely many primes p. This means that there are positive integers k and l such that $\|kx\|_p =_{\omega(\nu+1)} \|ly\|_p$ for all primes p. Let $x' = kx$, and $y' = ly$. As in the proof of the global classification in $L_{\infty\omega}$, we let $\tilde{X} = X \cup \{x'\}$. Then $S \cap \langle\tilde{X}\rangle = \langle S \cap \tilde{X}\rangle$, so by Lemma 5.3.7, there is a positive integer n such that for all $m \in \mathbb{Z}, s \in S$ and p prime, $|mnx' + s|_p = |mnx'|_p \wedge |s|_p$, and similarly for T, $|mny' + t|_p = |mny'|_p \wedge |t|_p$, for all $m \in \mathbb{Z}, t \in T$, and prime p. Now let $S' = \langle S, nx'\rangle$, and $T' = \langle T, ny'\rangle$. Then extend f to $g' : S' \to T'$ by sending nx' to ny'. By the choice of n, g' is $\omega(\nu+1)$-height-preserving. Since $\mathcal{C}$ and $\mathcal{D}$ are rich, $X' = X \cup \{nx'\} \in \mathcal{C}$, and $Y' = f(X) \cup \{ny'\} \in \mathcal{D}$.

Case (b): Suppose $x \in G$ and $rx \in S$ for some positive integer r. As in the $L_{\infty\omega}$ case, we will localize and then use Warfield's local-global lemma. Let $\alpha < \omega\lambda, \nu < \lambda$, p a prime, and e an equivalence class of Ulm sequences. Then $\hat{u}(\alpha, G_p) = \hat{u}(\alpha, H_p)$ and similarly, $\hat{u}(\infty, G_p) = \hat{u}(\infty, H_p)$ if $\text{length}(t(G_p)) < \omega\lambda$ (Lemma 1.17.11). By Lemma 5.4.12, we have $\hat{w}_{\omega(\nu+1)}(e, G_p) = \hat{w}_{\omega(\nu+1)}(e, H_p)$.

Recall that G_p and H_p have induced partial decomposition bases, say $\mathcal{C}_{G_p}$ and $\mathcal{C}_{H_p}$, respectively. Hence, G_p and H_p satisfy the conditions of the local Theorem 5.4.6. As part of that proof, we proved an analogous assertion (b), which states, in this case, that if $\nu < \lambda$, f is an $\omega(\nu+1)$-height-preserving isomorphism, $f : S \to T$, $S \subseteq G_p$, $T \subseteq H_p$ are finitely generated submodules, and $a \in G_p$ has a nonzero multiple in S, then there is a $b \in H$ such that $g : \langle S, a\rangle \to \langle T, b\rangle$, extending f by $a \mapsto b$, is an isomorphism that preserves heights up to $\omega\nu$. Apply this to the induced map $f_p : S_p \to T_p$ with associated $\{x \otimes 1 : x \in X\} \in \mathcal{C}_{G_p}$ and $\{y \otimes 1 : y \in Y\} \in \mathcal{C}_{H_p}$. Then f_p can be extended to an $\omega\nu$-height-preserving isomorphism $g(p)$ with $x_p = x \otimes 1 \in \text{domain}(g(p))$. By the local-global Lemma 3.4.5, there is a homomorphism $g : \langle S, x\rangle \to T^0$, with $g(x) = y$ for some $y \in T^0$, and $g_p = g(p)$ for all p. Since each $g(p)$ is $\omega\nu$-height-preserving and injective for each p, so is g; thus, $g : \langle S, x\rangle \to \langle T, y\rangle$ is an $\omega\nu$-height-preserving isomorphism. □

We have not quite finished the classifications in $L^{\lambda}_{\infty\omega}$. We still need to prove the converse: that λ-equivalence implies that the invariants agree. Recall that Karp's theorem allows us to take a model-theoretic view of the groups. If $G \equiv_\lambda H$, then any statement of quantifier rank no more than λ that is true in one group will be true in the other. So all we need is such a statement that says, in essence, "$\hat{w}_\alpha(c, p, e, G) \geq n$". The problem is that, as we will see, "decomposition set" is not expressible in $L_{\infty\omega}$, so neither is "$\hat{w}_\alpha(c, p, e, G) \geq n$". The solution is to find an equivalent statement that does not refer to decomposition sets. We have already seen that the modified Stanton invariant $\hat{s}(c, p, e, G)$ agrees with $\hat{w}(c, p, e, G)$ whenever G has a decomposition basis. We seek an analogous invariant that agrees with $\hat{w}_\alpha(c, p, e, G)$ and is expressible in $L_{\infty\omega}$. Recall the following definitions from Section 4.4. Let G be a group, p a prime, M an Ulm matrix, and α an ordinal:

$$G_\alpha(M) = \{x \in G : \|x\| \geq_\alpha M\};$$
$$G_\alpha(M^*)^\diamond = \{x \in G_\alpha(M) : \|x\| \nsim_\alpha M \text{ or } x \text{ is torsion}\};$$
$$G_\alpha(M_p^*, p)^\diamond = \{x \in G : \|x\|_p \geq_\alpha M_p \text{ and } |p^i x|_p \neq_\alpha m_{p,i} \text{ for infinitely many } i\};$$
$$G_\alpha(M^*, p)^\diamond = G_\alpha(M^*)^\diamond \cup G_\alpha(M_p^*, p)^\diamond;$$
$$G_\alpha(M^*, p) = G_\alpha(M) \cap \langle G_\alpha(M^*, p)^\diamond\rangle.$$

Then for c a compatibility class and e an equivalence class, we modified the Stanton invariant defining

$$\hat{s}_\alpha(c, p, e, G) = \sup\{\hat{r}_{p,0}(G_\alpha(M)/G_\alpha(M^*, p))\},$$

where the supremum is taken over all Ulm matrices M with $M \sim_\alpha c$, $M_p \sim_\alpha e$.

We proved in Section 4.4 that $\hat{s}_\alpha$ is expressible in $L_{\infty\omega}$ (Lemma 4.4.7). Now we will prove that it agrees with $\hat{w}_\alpha$.

Lemma 5.4.14. *Let G be a group, α an ordinal, p a prime, and $M = [m_{q,i}]$ an Ulm matrix. Then we have the following:*

1. *The map $x \mapsto px$ induces an isomorphism*
$$f\colon G_\alpha(M)/G_\alpha(M^*,p) \to G_\alpha(pM)/G_\alpha((pM)^*,p);$$
2. *If $m_{p,0} \geq \alpha$, then $G_\alpha(M) = G_\alpha(pM)$ and $G_\alpha(M^*,p) = G_\alpha((pM)^*,p)$.*

Proof. The proof of (1) is the same as that for Lemma 3.7.2, with the obvious adjustments for $\sim_\alpha, =_\alpha$, and $\geq_\alpha$.

To prove (2), suppose $m_{p,0} \geq \alpha$. Then, for all i, $m_{p,i+1} \geq m_{p,i} \geq \alpha$, so $m_{p,i+1} =_\alpha m_{p,i}$, and $M_p =_\alpha (pM)_p$. Furthermore, $M_q = (pM)_q$ for all $q \neq p$, so $M =_\alpha pM$. Then $\|x\| \geq_\alpha M$ if and only if $\|x\| \geq_\alpha pM$, and so it follows from the definitions that $G_\alpha(M) = G_\alpha(pM)$, $G_\alpha(M^*)^\diamond = G_\alpha((pM)^*)^\diamond$, and $G_\alpha(M_p^*,p)^\diamond = G_\alpha((pM)_p^*,p)^\diamond$. □

Applying the lemma repeatedly gives the following:

Corollary 5.4.15. $G_\alpha(M)/G_\alpha(M^*,p) \cong G_\alpha(p^kM)/G_\alpha((p^kM)^*,p)$ *for every group G, prime p, ordinal α, $k \geq 0$, and Ulm matrix M.*

In Section 3.7, we saw that Warfield and Stanton invariants agree for groups with decomposition bases. Here we wish to prove the analogous result for PDB groups up to a given ordinal α. Note that then the result for PDB groups up to partial isomorphism is a simple corollary. We cannot use that previous result (Theorem 3.7.3) here, since there is no easy relationship between $\hat{s}_\alpha$ and $\hat{s}$. In fact, we cannot even use its proof, since it depends on results about $*$-decomposition sets. However, the following lemma, which is analogous to some of the $*$-decomposition results, will be useful.

Lemma 5.4.16. *Suppose G is a group with partial decomposition basis $\mathcal{C}$, α an ordinal, p a prime, M an Ulm matrix, $x_1,\dots,x_n \in X$ for some $X \in \mathcal{C}$, where for each i, $x_i \in G_\alpha(M)$, $\|x_i\| \sim_\alpha M$, and $\|x_i\|_p =_\alpha M_p$. If $a_1x_1 + \dots + a_nx_n \in G_\alpha(M^*,p)$ with $a_1,\dots,a_n \in \mathbb{Z}$, then $a_ix_i \in G_\alpha(M^*,p)^\diamond$ for all $i = 1,\dots,n$.*

Proof. Suppose $a_1x_1 + \dots + a_nx_n = z_1 + \dots + z_m$ for some $z_1,\dots,z_m \in G_\alpha(M^*,p)^\diamond$. Let $z \in \{z_1,\dots,z_m\}$. Since $\mathcal{C}$ is a partial decomposition basis for G, we may write $p^kaz = b_1x_1 + \dots + b_nx_n + y$, for some integers $a, b_1,\dots,b_n$, with $p \nmid a$, and $y \in \langle Y \setminus \{x_1,\dots,x_n\}\rangle$ for some $Y \in \mathcal{C}$ with $X \subseteq Y$. In fact, we may choose p^k, a, and Y such that this holds for all $z \in \{z_1,\dots,z_m\}$.

We claim that $p^kaz = p^{k+1}x + y$ for some $x \in \langle x_1,\dots,x_n\rangle$ and $y \in \langle Y \setminus \{x_1,\dots,x_n\}\rangle$. First, suppose $z \in G_\alpha(M_p^*,p)^\diamond$. Then $\|z\|_p \geq_\alpha M_p$ and $|p^iz|_p \neq_\alpha m_{p,i}$ for infinitely many i. In particular, $m_{p,i} < \alpha$ for all i. For any $j_0 < \omega$, we may find j such that $j + k \geq j_0$ and $|p^{j+k}z|_p \neq_\alpha m_{p,j+k}$. Then for all i, $|p^{j+k}x_i|_p = m_{p,j+k} < |p^jp^kaz|_p \leq |p^jb_ix_i|_p$, since X is a decomposition set. It then follows that $p^{k+1} \mid b_i$ for each i, proving the claim in this case.

Next, suppose $z \in G_\alpha(M^*)^\diamond$. If z is torsion, then $p^k az = 0$, proving the claim. Suppose $\|z\| \nsim_\alpha M$. Suppose further that $b_i \neq 0$ for some i. Then since $\|x_i\| \sim_\alpha M$, for some $r > 0$, $rM \geq_\alpha \|b_i x_i\| \geq \|p^k az\| \geq \|z\| \geq_\alpha M$, contradicting $\|z\| \nsim_\alpha M$. Hence, $b_i = 0$ for all i, and $p^k az \in \langle Y \setminus \{x_1, \dots, x_n\}\rangle$. Again the claim is satisfied.

We then get $p^k a(a_1x_1 + \cdots + a_nx_n) = p^{k+1}x + y$ for some $x \in \langle x_1, \dots, x_n\rangle$ and $y \in \langle Y \setminus \{x_1, \dots, x_n\}\rangle$, since this is true for each z. Equating x_i terms, we get $p \mid a_i$ for each i.

First, suppose $m_{p,j} < \alpha$ for all j. Then $|p^{j+k}a_ix_i|_p = |p^j p^k aa_ix_i|_p > |p^{j+k}x_i|_p = m_{p,j+k}$ for all j, and so $a_ix_i \in G_\alpha(M) \cap G_\alpha(M_p^*, p)^\diamond \subseteq G_\alpha(M^*, p)^\diamond$ for each i. Now consider the case in which $m_{p,i} \geq \alpha$ for some i. Let $z \in \{z_1, \dots, z_m\}$. Then $|p^k z|_p \geq_\alpha m_{p,k} \geq \alpha$ for each $k \geq i$, so $|p^k z|_p =_\alpha m_{p,k}$, and $z \notin G_\alpha(M_p^*, p)^\diamond$. Therefore, we have $z \in G_\alpha(M^*)^\diamond$. If z is torsion, $p^k az = 0$. If $\|z\| \nsim_\alpha M$, then it follows that $\|b_ix_i\| \nsim_\alpha M$ for all $1 \leq i \leq n$ with $b_i \neq 0$; hence, $b_i = 0$ for all i and $p^k az \in \langle Y \setminus \{x_1, \dots, x_n\}\rangle$. In any case, equating x_i terms gives $p^k aa_ix_i = 0$, which means a_ix_i is torsion, and hence, in $G_\alpha(M^*, p)^\diamond$. □

Theorem 5.4.17 (Jacoby–Loth [70]). *Suppose that G is a PDB group, and α is an ordinal. Then $r_{p,0}(G_\alpha(M)/G_\alpha(M^*, p)) =_\omega \hat{w}_\alpha(c, p, e, G)$ for every compatibility class c of Ulm matrices, prime p, equivalence class e of Ulm sequences, and $M \in c$ with $M_p \in e$.*

Proof. We may assume that G has a rich partial decomposition basis $\mathcal{C}$. Suppose $M = [m_{q,i}] \in c$ and $M_p \in e$. Note that by taking p-multiples, which we are allowed to do by Corollary 5.4.15, we may assume that if $m_{p,i} \geq \alpha$ for some i, then $m_{p,i} \geq \alpha$ for all i.

First, suppose that $\hat{w}_\alpha(c, p, e, G) \geq n$ for some $n < \omega$, say there are $x_1, \dots, x_n \in X$ for some $X \in \mathcal{C}$ and for all $1 \leq i \leq n$, $\|x_i\| \sim_\alpha c$ and $\|x_i\|_p \sim_\alpha e$. For each $1 \leq i \leq n$, $\|x_i\| \sim_\alpha M$, so for some $m_i > 0$, $\|m_ix_i\| \geq_\alpha M$. Then if we let $m = \prod_{i=1}^n m_i$, we get $\|mx_i\| \geq_\alpha M$ for all $1 \leq i \leq n$. Also, $\|mx_i\|_p \sim \|x_i\|_p \sim_\alpha M_p$, so there are k_i and k_i' such that $\|p^{k_i}mx_i\|_p =_\alpha p^{k_i'}M_p$. Let $k' = \sum_{i=1}^n k_i'$ and $l_i = k_i + \sum_{j \neq i} k_j'$ for each i. Then it is easily verified that $\|p^{l_i}mx_i\|_p =_\alpha (p^{k'}M)_p$, and for $q \neq p$, $\|p^{l_i}mx_i\|_q \geq_\alpha (p^{k'}M)_q$. Since $\mathcal{C}$ is rich, we may use $\{p^{l_i}mx_i : 1 \leq i \leq n\}$ for the decomposition set and write x_i in place of $p^{l_i}mx_i$. Furthermore, by Corollary 5.4.15, it suffices to prove the assertion for $p^{k'}M$, so we may write M in place of $p^{k'}M$. Hence, we may assume $x_i \in G_\alpha(M)$ and $\|x_i\|_p =_\alpha M_p$ for all $1 \leq i \leq n$.

Now we will prove that $x_1, \dots, x_n$ are independent representatives of $G_\alpha(M)$ over $G_\alpha(M^*, p)$, and of either p-power or infinite order. Suppose $a_1x_1 + \cdots + a_nx_n \in G_\alpha(M^*, p)$. Then $a_ix_i \in G_\alpha(M^*, p)$ by Lemma 5.4.16. It remains to prove that the order of each $x_i + G_\alpha(M^*, p)$ is a p-power or infinite. First, suppose $m_{p,k} < \alpha$ for all k. For each k, since $m_{p,k} < \alpha$ and $|p^k px_i|_p > |p^k x_i|_p = m_{p,k}$, we get $|p^k px_i|_p >_\alpha m_{p,k}$, and thus, $px_i \in G_\alpha(M_p^*, p)^\diamond \subseteq G_\alpha(M^*, p)$. Suppose $x_i \in G_\alpha(M^*, p)$. Then by the lemma, $x_i \in G_\alpha(M^*, p)^\diamond$, a contradiction since $\|x_i\|_p =_\alpha M_p$, $\|x_i\| \sim_\alpha M$, and x_i is not torsion. This proves that each $x_i + G_\alpha(M^*, p)$ has order p; thus, $r_p(G_\alpha(M)/G_\alpha(M^*, p)) \geq n$ in that case. Now suppose $m_{p,k} \geq \alpha$ for all k. We claim the order of each $x_i + G_\alpha(M^*, p)$ is infinite. Suppose $mx_i \in G_\alpha(M^*, p)$ for some $m \neq 0$. Then $mx_i \in G_\alpha(M^*, p)^\diamond$, a contradiction since x_i is not torsion, $\|mx_i\| \sim_\alpha M$, and for all j, $|p^j mx_i| \geq_\alpha m_{p,j} \geq \alpha$; hence, $|p^j mx_i|_p =_\alpha m_{p,j}$. Thus, we have $r_0(G_\alpha(M)/G_\alpha(M^*,p)) \geq n$. This proves that $r_{p,0}(G_\alpha(M)/G_\alpha(M^*,p)) \geq \hat{w}_\alpha(c,p,e,G)$.

Now suppose $\hat{w}_\alpha(c,p,e,G) = n < \omega$, and $\{x_1,\dots,x_n\} \in C$ is a maximal set with $\|x_i\| \sim_\alpha M$ and $\|x_i\|_p \sim_\alpha M_p$ for all $i = 1,\dots,n$. As before, we may assume $x_i \in G_\alpha(M)$ and $\|x_i\|_p =_\alpha M_p$ for all i. By the paragraph above, $x_1,\dots,x_n$ are independent representatives of $G_\alpha(M)/G_\alpha(M^*,p)$. We will show that $\{x_1 + G_\alpha(M^*,p),\dots,x_n + G_\alpha(M^*,p)\}$ is a maximal independent set of elements of p-power or infinite order in $G_\alpha(M)/G_\alpha(M^*,p)$.

Suppose to the contrary that $x_1,\dots,x_n,y$ are independent representatives, where $y + G_\alpha(M^*,p)$ has either p-power or infinite order. Choose $Y \in C$ such that $X \subseteq Y$ and $y \in \langle Y\rangle^0$. Then write $ay = a_1x_1 + \cdots + a_nx_n + b_1y_1 + \cdots + b_my_m$, where $a > 0$ and $y_1,\dots,y_m \in Y \setminus \{x_1,\dots,x_n\}$. We may write $a = p^k a'$, where $p \nmid a'$. Then since Y is a decomposition set and $\|y\| \geq_\alpha M$, we get $p^k M \leq_\alpha \|ay\| \leq \|a_ix_i\|$ for all i, and similarly for b_jy_j, so a_ix_i and b_jy_j are in $G_\alpha(p^kM)$ for all $1 \leq i \leq n$ and $1 \leq j \leq m$. Now for each $1 \leq i \leq n$ with $a_i \neq 0$, write $a_i = p^{k(a_i)}a_i'$, where $p \nmid a_i'$ if $a_i' \neq 0$. If $a_i' = 0$, let $k(a_i) = k$. Let $k' = \min\{k, k(a_1),\dots,k(a_n)\}$. By Corollary 5.4.15, $p^{k'}y, p^{k'}x_1,\dots,p^{k'}x_n$ are independent representatives of $G_\alpha(p^{k'}M)/G_\alpha((p^{k'}M)^*,p) \cong G_\alpha(M)/G_\alpha(M^*,p)$.

Now we will prove that $b_1y_1 + \cdots + b_my_m \in G_\alpha((p^{k'}M)^*,p)$. Consider y_i for some $1 \leq i \leq m$. Since $\{x_1,\dots,x_n,b_iy_i\} \in C$, since C is rich, and $\{x_1,\dots,x_n\}$ is maximal, we have either $\|b_iy_i\| \nsim_\alpha c$ or $\|b_iy_i\|_p \nsim_\alpha e$. First suppose $\|b_iy_i\| \nsim_\alpha c$. Then $\|b_iy_i\| \nsim_\alpha M \sim p^{k'}M$, so $b_iy_i \in G_\alpha((p^{k'}M)^*)^\diamond$.

Now suppose $\|b_iy_i\|_p \nsim_\alpha e$. Then $\|b_iy_i\|_p \nsim_\alpha M_p \sim p^{k'}M_p$. Suppose there is some j_0 such that $|p^jb_iy_i|_p =_\alpha m_{p,j+k'}$ for all $j \geq j_0$. Then $\|p^{j_0}b_iy_i\|_p =_\alpha p^{j_0+k'}M_p$, and so, $p^{k'}M_p \sim_\alpha \|b_iy_i\|_p \nsim_\alpha e \sim M_p \sim p^{k'}M_p$, a contradiction. This gives us $|p^jb_iy_i|_p \neq_\alpha m_{p,j+k'}$ for infinitely many j and $b_iy_i \in G_\alpha((p^{k'}M)^*_p,p)^\diamond$. In either case, $b_iy_i \in G_\alpha((p^{k'}M)^*,p)$.

Then $ay + G_\alpha((p^{k'}M)^*,p) = a_1x_1 + \cdots + a_nx_n + G_\alpha((p^{k'}M)^*,p)$. By the independence assumption, we then have $ay, a_1x_1,\dots,a_nx_n \in G_\alpha((p^{k'}M)^*,p)$. First, suppose $k' = k(a_i)$ for some i with $a_i \neq 0$. Then $p^{k'}a_i'x_i \in G_\alpha((p^{k'}M)^*,p)$. By Corollary 5.4.15, $a_i'x_i \in G_\alpha(M^*,p)$; thus, $a_i'x_i \in G_\alpha(M^*,p)^\diamond$ by Lemma 5.4.16 since $\|a_i'x_i\| \sim \|x_i\| \sim_\alpha M$. But then we get a contradiction since $a_i'x_i$ is not torsion, $\|a_i'x_i\|_p = \|x_i\|_p =_\alpha M_p$, and $\|a_i'x_i\| \sim_\alpha M$. Thus, $r_{p,0}(G_\alpha(M)/G_\alpha(M^*,p)) = n$, and the assertion follows. Similarly, if $k' = k$, then $y \in G_\alpha(M^*,p)$, also a contradiction. □

Corollary 5.4.18. *Suppose G is a PDB group. Then $\hat{s}_\alpha(c,p,e,G) = \hat{w}_\alpha(c,p,e,G)$ for all α, c, p, and e.*

Corollary 5.4.19. *Suppose G is a PDB group. Then $\hat{s}(c,p,e,G) = \hat{w}(c,p,e,G)$ for all c, p, and e.*

Proof. Choose $\alpha = \sup\{p\text{-length}(G) : p \text{ prime}\}$. □

Now we see that for PBD groups, the modified Stanton invariant is independent of the choice of M.

Corollary 5.4.20. *Let G be a PDB group and α an ordinal. Then $\hat{s}_\alpha(c,p,e,G) = \hat{r}_{p,0}(G_\alpha(M)/G_\alpha(M^*,p))$ for all $M \in c$ with $M_p \in e$.*

Theorem 5.4.21 (Jacoby–Loth [70]). *Suppose G and H are PDB groups. If $G \equiv_\lambda H$, where $\lambda = \omega\gamma$ for some γ a limit ordinal, then for all $\alpha < \omega\lambda$, $\hat{w}_\alpha(c,p,e,G) = \hat{w}_\alpha(c,p,e,H)$ for all c,p,e.*

Proof. Suppose for some c,p,e, and $\alpha < \omega\lambda$, $\hat{w}_\alpha(c,p,e,G) \geq m$. Then by Corollary 5.4.18, $\hat{s}_\alpha(c,p,e,G) \geq m$. Write α as $\omega\delta + n$ and δ as $\omega\delta' + n'$ $(n,n' < \omega)$. By Lemma 4.4.7, there is a formula φ of quantifier rank $\leq \delta + \omega + m$ such that $G \models \varphi$ if and only if $\hat{s}_\alpha(c,p,e,G) \geq m$. Note that $\delta' < \gamma$, so $\delta' + 3 < \gamma$, since γ is a limit ordinal. Then $\lambda = \omega\gamma > \omega(\delta' + 3) > \omega\delta' + n' + \omega + m = \delta + \omega + m$. Since $\hat{s}_\alpha(c,p,e,G) \geq m$, $G \models \varphi$. Since the quantifier rank of φ is $< \lambda$ and $G \equiv_\lambda H$, we get $H \models \varphi$ and $\hat{w}_\alpha(c,p,e,H) \geq m$. By symmetry $\hat{w}_\alpha(c,p,e,G) = \hat{w}_\alpha(c,p,e,H)$. □

Corollary 5.4.22. *Let G and H be PDB groups such that $G \equiv_\lambda H$, where $\lambda = \omega\gamma$ and γ is a limit ordinal. Then the following hold:*

1. *$\hat{u}_p(\alpha, G) = \hat{u}_p(\alpha, H)$ for all primes p and $\alpha < \omega\lambda$;*
2. *$\hat{w}_{\omega(\nu+1)}(c,p,e,G) = \hat{w}_{\omega(\nu+1)}(c,p,e,H)$ for every compatibility class c of Ulm matrices, prime p, equivalence class e of Ulm sequences, and $\nu < \lambda$;*
3. *If $\mathrm{length}(t(G_p)) < \omega\lambda$, then $\hat{u}_p(\infty, G) = \hat{u}_p(\infty, H)$.*

Proof. (1) and (3) follow from Theorem 4.5.7, and (2) follows from Theorem 5.4.21 by noting that $\nu + 1 < \lambda$ since λ is a limit ordinal. □

Note that this attempt at a converse of Theorem 5.4.13 has been stated and proved only for certain λ. This is to be expected since the converse part of Theorem 4.5.7, on which this is based, applies only to limit ordinals. It is an open question whether this corollary can be broadened. (See Problem 2.)

Note that one direction of the global classification theorem 5.3.8 follows easily:

Corollary 5.4.23. *Let G and H be PDB groups. If $G \equiv_\infty H$, then $\hat{w}(c,p,e,G) = \hat{w}(c,p,e,H)$ for all c,p,e.*

5.5 PDB groups and k-groups

Now we examine the class of PDB groups that we have just classified. First, we recall the following definitions and facts from Chapter 3: A direct sum $\bigoplus_{i\in I} A_i \subseteq G$ is said to be a **valuated coproduct in** G if for all primes p and ordinals α, we have $(\bigoplus_{i\in I} A_i) \cap p^\alpha G = \bigoplus_{i\in I}(A_i \cap p^\alpha G)$. We call the valuated coproduct $A = \bigoplus_{i\in I} A_i$ in G a **$*$-valuated coproduct in** G if for each Ulm matrix M, Ulm sequence $\overline{\alpha}$, prime p, and group $F \in \{G(\overline{\alpha}^*,p), G(M^*), G(M^*,p)\}$, $A \cap F = \bigoplus_{i\in I}(A_i \cap F)$. An element x in a group G is called **primitive** if for any Ulm matrix M, prime p, and positive integer

n, $nx \in G(M^*, p)$ implies that $\|x\| \nleftarrow M$ or $nx \in G(M_p^*, p)^\diamond$. A decomposition set X in G is called a **$*$-decomposition set** if $\langle X\rangle = \bigoplus_{x\in X}\langle x\rangle$ is a $*$-valuated coproduct in G and each element $x \in X$ is primitive. A subgroup A of a group G is called **knice** if for every finite subset S of G, there is a finite (possibly empty) set of primitive elements $x_1, \ldots, x_n \in G$ such that $B = A \oplus \langle x_1\rangle \oplus \cdots \oplus \langle x_n\rangle$ is a $*$-valuated coproduct in G with $\langle S, B\rangle/B$ finite. If 0 is a knice subgroup of G, then G is called a **k-group**. By Corollary 3.10.1, if G has a decomposition basis X, then X is a $*$-decomposition set. Recall also that, by Theorem 3.12.3, if $B = A \oplus \langle x_1\rangle \oplus \cdots \oplus \langle x_n\rangle$ is a $*$-valuated coproduct in a group G such that if A is knice in G and each element x_i is primitive, then B is knice in G.

The relationship between k-groups and PDB groups was first explored in the local case by Jacoby and Loth [71] and then extended to the global case in [73], from which these results were taken. For any group G, we define

$$\mathcal{C}_G = \{Y : Y \text{ is a finite } *\text{-decomposition set in } G\}.$$

Lemma 5.5.1. *If G is a k-group, then $\mathcal{C}_G$ is a partial decomposition basis for G.*

Proof. First, $\mathcal{C}_G \neq \emptyset$ since $\emptyset \in \mathcal{C}_G$. Each $*$-decomposition set is a decomposition set, so we only need to verify the extension property. Suppose $X = \{x_1, \ldots, x_n\} \in \mathcal{C}_G$ and $x \in G$. By Theorem 3.12.3 with $A = 0$, $\langle x_1\rangle \oplus \cdots \oplus \langle x_n\rangle$ is knice in G. By the definition of knice, there is a subset $Y = \{y_1, \ldots, y_m\}$ of G such that each y_i is primitive and $\langle x_1\rangle \oplus \cdots \oplus \langle x_n\rangle \oplus \langle y_1\rangle \oplus \cdots \oplus \langle y_m\rangle$ is a $*$-valuated coproduct with $x \in \langle X \cup Y\rangle^0$. Thus, $X \cup Y \in \mathcal{C}_G$. □

Lemma 5.5.2 (Jacoby–Loth [71]). *Let G be a group with partial decomposition basis $\mathcal{C}$, $X \in \mathcal{C}$, and $\{x_1, \ldots, x_n\} \subseteq X$. Suppose $y = a_1x_1 + \cdots + a_nx_n$, where $a_1, \ldots, a_n$ are nonzero integers, and $y = z_1 + \cdots + z_m$ for some $z_1, \ldots, z_m \in G$, and let p be a prime. Then for every $1 \le j \le n$, there is $1 \le i_0 \le m$ such that z_{i_0} is not torsion, and*

1. *for any Ulm sequence $\overline{\alpha}$, if $y \in G(\overline{\alpha}, p)$ and $z_{i_0} \in G(\overline{\alpha}^*, p)^\diamond$, then $a_jx_j \in G(\overline{\alpha}^*, p)^\diamond$;*
2. *for every Ulm matrix M, if $y \in G(M)$ and $z_{i_0} \in G(M^*)^\diamond$, then $a_jx_j \in G(M^*)^\diamond$.*

Proof. Let $Y \in \mathcal{C}$ such that $X \subseteq Y$ and $z_1, \ldots, z_m \in \langle Y\rangle^0$. We apply the analogous lemma for decomposition bases, Lemma 3.6.3 to $H = \langle Y\rangle^0$, giving an $1 \le i_0 \le m$ for each $1 \le j \le n$ as in that lemma. Then if $\overline{\alpha}$ is an Ulm sequence, $y \in G(\overline{\alpha}, p)$, and $z_{i_0} \in G(\overline{\alpha}^*, p)^\diamond$, then since H is isotype in G, we have $y \in H(\overline{\alpha}, p)$, $z_{i_0} \in H(\overline{\alpha}^*, p)^\diamond$; and thus, $a_jx_j \in H(\overline{\alpha}^*, p)^\diamond \subseteq G(\overline{\alpha}^*, p)^\diamond$. The second part is similar. □

Lemma 5.5.3 (Jacoby–Loth [71]). *Suppose G is a group with partial decomposition basis $\mathcal{C}$. Let $X \in \mathcal{C}$, $x_1, \ldots, x_n \in X$, $a_1, \ldots, a_n$ nonzero integers, M an Ulm matrix, $\overline{\alpha}$ an Ulm sequence, p a prime, and $F \in \{G(\overline{\alpha}^*, p), G(M^*), G(M^*, p)\}$. If $a_1x_1 + \cdots + a_nx_n \in F$, then $a_ix_i \in F^\diamond$ for all $1 \le i \le n$.*

Proof. The result follows from Lemma 5.5.2 exactly as Lemma 3.6.4 followed from Lemma 3.6.3. □

Lemma 5.5.4. *Suppose G is a group with partial decomposition basis $\mathcal{C}$. If $X \in \mathcal{C}$, then X is a $*$-decomposition set in G.*

Proof. Let $X \in \mathcal{C}$, $x \in X$, p prime, $\overline{\alpha}$ an Ulm sequence, and M an Ulm matrix. Since X is a decomposition set, the elements are independent. To prove they are primitive, suppose $kx \in G(M^*, p)$ for some $x \in X$ and positive integer k. By Lemma 5.5.3, $kx \in G(M^*, p)^\diamond$. If $kx \in G(M^*)^\diamond$, then $\|x\| \sim \|kx\| \not\to M$. If not, $kx \in (M_p^*, p)^\diamond$.

Now we wish to show that $\langle X\rangle = \bigoplus_{x\in X}\langle x\rangle$ is a $*$-valuated coproduct. It is a valuated coproduct since X is a decomposition set. Suppose that F is any group in $\{G(\overline{\alpha}^*, p), G(M^*), G(M^*, p)\}$, and $y = a_1x_1 + \cdots + a_nx_n \in F$ for some $x_1, \ldots, x_n \in X$ and $a_1, \ldots, a_n$ nonzero integers. For any $1 \le i \le n$, $a_ix_i \in F^\diamond$ by Lemma 5.5.3. This proves $(\bigoplus_{x\in X}\langle x\rangle) \cap F \subseteq \bigoplus_{x\in X}(\langle x\rangle \cap F)$, as desired. □

Theorem 5.5.5 (Jacoby–Loth [71]). *A group has a partial decomposition basis if and only if it is a k-group.*

Proof. Suppose G has a partial decomposition basis $\mathcal{C}$, and let S be a finite subset of G. Then there is an $X \in \mathcal{C}$ such that $S \subseteq \langle X\rangle^0$, say $X = \{x_1, \ldots, x_n\}$. By Lemma 5.5.4, X is a $*$-decomposition set, specifically, $x_1, \ldots, x_n$ are primitive and $\langle x_1\rangle \oplus \cdots \oplus \langle x_n\rangle$ is a $*$-valuated coproduct. This means that 0 is knice, and so G is a k-group. Conversely, if G is a k-group, then it has a partial decomposition basis $\mathcal{C}_G$ by Lemma 5.5.1. □

Now we may use what we know about k-groups to answer a long-standing question about PDB groups.

Corollary 5.5.6. *The class of PDB groups is closed under direct summands.*

Proof. By Theorem 3.12.11, a direct summand of a k-group is a k-group. □

Likewise, we may use what we know about PDB groups to shed light on k-groups.

Corollary 5.5.7. *Every finitely generated subgroup of a k-group is locally nice.*

Proof. By Corollary 5.1.8, this is true for PDB groups. □

We have often expanded a given partial decomposition basis to give us what we need to complete a proof or construction. Now we see that there is a largest partial decomposition basis, and, furthermore, we know what it looks like.

Proposition 5.5.8. *Suppose that G is a group with partial decomposition basis $\mathcal{C}$. Then the set $\mathcal{C}_G$ of all finite $*$-decomposition sets in G is a partial decomposition basis for G, and we have $\mathcal{C} \subseteq \mathcal{C}_G$.*

Proof. Suppose G has a partial decomposition basis $\mathcal{C}$. Then G is a k-group by Theorem 5.5.5, and it has a partial decomposition basis $\mathcal{C}_G = \{Y : Y \text{ is a finite}$ $*$-decomposition set in $G\}$ by Lemma 5.5.1. Let $X \in \mathcal{C}$. Then by Lemma 5.5.4, X is a $*$-decomposition set in G, and so $X \in \mathcal{C}_G$. □

Corollary 5.5.9. *Let G be a k-group and X a finite $*$-decomposition set in G. If Y is a finite decomposition set in G such that $\langle Y\rangle = \langle X\rangle$, then Y is a $*$-decomposition set in G.*

Proof. The k-group G has a partial decomposition basis $\mathcal{C}_G$ (Lemma 5.5.1). By Lemma 5.1.3, G has a rich partial decomposition basis $\mathcal{C} \supseteq \mathcal{C}_G$. But by Proposition 5.5.8, $\mathcal{C} \subseteq \mathcal{C}_G$; hence, $\mathcal{C}_G = \mathcal{C}$ is rich. In particular, if $X \in \mathcal{C}_G$ and $\langle X\rangle = \langle Y\rangle$, then $Y \in \mathcal{C}_G$. □

5.6 Partial subbases and properties of $*$-decomposition sets

We often take advantage of isotype subgroups to transfer characteristics of the larger group to the subgroup. For example, if $\mathcal{C}$ is a partial decomposition basis and $X \in \mathcal{C}$, then X is a decomposition basis for $\langle X\rangle^0$, which allows us to apply what we know about decomposition bases to partial decomposition bases. We would have liked to do something similar to prove Lemma 5.5.4, using the analogous Lemma 3.10.1 for decomposition bases, but we cannot argue that X a $*$-decomposition set in $\langle X\rangle^0$ implies that it is one in all of G. The following example shows a set that is a $*$-decomposition set in H, an isotype subgroup of G, but not a $*$-decomposition set in G. We will be exploring which properties do and do not transfer between a group and an isotype subgroup.

Example 5.6.1 (Hill–Megibben [49]). There is a group G, an isotype subgroup H of G, and an $X \subseteq H$ such that X is a finite $*$-decomposition set in H, but not in G.

Proof. Define H_a, H_b, and H_c as in Corollary 2.7.5, with $\|a\|_p = (0, 3, 4, 7, 8, 11, 12, \dots)$ and $\|b\|_p = (1, 2, 5, 6, 9, 10, \dots)$ for some prime p, and $\|c\|_q = (|q^i a|_q \wedge |q^i b|_q)_{i<\omega}$ for all primes q. In particular, $\|c\|_p = (0, 2, 4, 6, 8, \dots)$. Then let $G = H_a \oplus H_b \oplus H_c$, $x = a + b + c$, and $H = \langle x, c\rangle^0$. We claim that $\{x, c\}$ is a decomposition set. If $r, s \in \mathbb{Z}$, $|rx + sc|_q = |ra + rb + rc + sc|_q = |ra|_q \wedge |rb|_q \wedge |(r + s)c|_q = |rc|_q \wedge |sc|_q = |rx|_q \wedge |sc|_q$ for all primes q. Hence, $\{x, c\}$ is a decomposition basis for H, and thus, a $*$-decomposition set in H.

Now suppose $\{x, c\}$ is a $*$-decomposition set in G. Then $\{x, c\} \in \mathcal{C}_G$. Let $\overline{\alpha} = (0, 2, 4, 6, \dots)$. Note that $a, b \in G(\overline{\alpha}^*, p)^\diamond$, and so $x - c = a + b \in G(\overline{\alpha}^*, p)$. Then by Lemma 5.5.3, $c \in G(\overline{\alpha}^*, p)^\diamond$, a contradiction since $\|c\|_p = \overline{\alpha}$. Thus, $\{x, c\}$ is a $*$-decomposition set in H, but not in G. □

Let H be a subgroup of a group G. We define

$$\mathcal{C}_H^G = \{Y \cap H : Y \text{ is a finite } * \text{-decomposition set in } G\}.$$

Note the following:

1. For any H, $\mathcal{C}_H^G \neq \emptyset$ since $\emptyset \in \mathcal{C}_H^G$.
2. If $H \subseteq H'$, then $\mathcal{C}_H^G \subseteq \mathcal{C}_{H'}^G$.
3. $\mathcal{C}_G^G = \mathcal{C}_G$.

Any nonempty $C \subseteq C_H^G$ is called a **partial subbasis** for H if for every $X \in C$ and $x \in H$, there is a $Y \in C$ such that $X \subseteq Y$ and $x \in \langle Y \rangle^0$.

Remark 5.6.2. Let H be a subgroup of a group G. Then C_H^G and C_H may not be the same, even if H is isotype in G.

In the example above, we note that $\{x, c\}$ is in C_H, but not in C_H^G, supporting the remark. Note that if C is a partial subbasis of H, and H is isotype in G, then C is a partial decomposition basis for H, since any subset of H is a decomposition set in H whenever it is one in G. Therefore we have the following:

Remark 5.6.3. If H is an isotype subgroup of G and has a partial subbasis, then C_H^G is a subset of C_H. By the example above, the inclusion may be proper.

Remark 5.6.4. If X is a $*$-decomposition set in G and H is a subgroup of G such that $H/\langle X \cap H \rangle$ is torsion, then $C = \{Y \cap H : Y \subseteq X, Y \text{ finite}\}$ is a partial subbasis for H.

Proof. Since any subset of a $*$-decomposition set is a $*$-decomposition set, we have $C \subseteq C_H^G$. Let $Y \cap H \in C$ with $Y \subseteq X$ and Y finite. If $x \in H$, then $x \in \langle H \cap X \rangle^0$ by assumption, so there is a finite $X' \subseteq X$ with $x \in \langle H \cap X' \rangle^0$. Letting $Y' = X' \cup Y$, we obtain $Y' \cap H \in C$ and $x \in \langle Y' \cap H \rangle^0$, as desired. □

Lemma 5.6.5. *Let H be an isotype subgroup of G. If H has a partial subbasis, then H has a partial decomposition basis. In fact, any partial subbasis for H is a partial decomposition basis.*

Proof. Let $C \subseteq C_H^G$ be a partial subbasis for H, $X = \{x_1, \dots, x_n\} \in C$, and $x = a_1x_1 + \cdots + a_nx_n$. Then for any prime p, $|x|_p^H = |x|_p^G = \min_{1 \le i \le n}\{|a_ix_i|_p^G\} = \min_{1 \le i \le n}\{|a_ix_i|_p^H\}$, and so X is a decomposition set in H. The extension property is from the definition. □

We will see that subgroups with partial subbases are related to knice subgroups. First, we need some results on valuated and $*$-valuated coproducts.

Theorem 5.6.6 (Hill–Megibben [50]). *Let G be a group and A a subgroup of G.*

1. *If A is nice in G and $H = A \oplus \langle x \rangle$ is a valuated coproduct in G, then H is nice in G.*
2. *Suppose that $A \oplus \langle y_1 \rangle \oplus \cdots \oplus \langle y_n \rangle$ is a valuated coproduct in G such that every element y_i has infinite order. If B is a subgroup of G such that B/A is finite, then there are nonzero multiples y_i' of y_i $(i = 1, \dots, n)$ such that $B \oplus \langle y_1' \rangle \oplus \cdots \oplus \langle y_n' \rangle$ is a valuated coproduct in G.*

Proof. (1) follows from Lemma 3.8.12, and (2) is the Seesaw lemma, Proposition 3.6.5. □

We wish to extend many of our notions to allow us to view them within subgroups, just as a partial subbasis extends the definition of partial decomposition basis to subgroups. Suppose $A \subseteq H$ are subgroups of G. Then we say A is **$*$-extendable in** H, if for every finite $S \subseteq H$ there is a (possibly empty) subset $\{x_1, \dots, x_n\}$ of H such that each

$x_1, \ldots, x_n$ is primitive in G and $B = A \oplus \langle x_1 \rangle \oplus \cdots \oplus \langle x_n \rangle$ is a $*$-valuated coproduct in G with $S \subseteq B^0$. If, in addition, A is nice in G, we say A is an **H-knice subgroup of** G. Note that a subgroup of G is knice if and only if it is G-knice. Note also that 0 is H-knice if and only if 0 is $*$-extendable in H.

The next theorem slightly extends Theorem 3.12.3.

Theorem 5.6.7. *Let A and H be subgroups of G, and suppose that $B = A \oplus \langle x_1 \rangle \oplus \cdots \oplus \langle x_n \rangle$ is a $*$-valuated coproduct in G such that every element x_i is primitive in G and $B \subseteq H \subseteq G$. If A is $*$-extendable in H, then B is $*$-extendable in H. If A is H-knice in G, then B is H-knice in G.*

Proof. The proof follows that of Theorem 3.12.3. Clearly, it suffices to consider the case of $B = A \oplus \langle x_1 \rangle$, a $*$-valuated coproduct in G, where x_1 is primitive in G, $B \subseteq H$, and A is $*$-extendable in H. Let S be a finite subset of H. Then since A is $*$-extendable in H, there are primitive $y_1, \ldots, y_m \in G$ such that $\{y_1, \ldots, y_m\} \subseteq H$, $A' = A \oplus \langle y_1 \rangle \oplus \cdots \oplus \langle y_m \rangle$ is a $*$-valuated coproduct in G, and $S \cup \{x_1\} \subseteq (A')^0$. In particular, $rx_1 \in A'$ for some positive r. Let $x_1' = rx_1$. Now we may apply Theorem 3.10.14 to $A \oplus \langle x_1' \rangle$ and $A \oplus \langle y_1 \rangle \oplus \cdots \oplus \langle y_m \rangle$, giving primitive elements $x_2, \ldots, x_m \in G$ such that $C = A \oplus \langle x_1' \rangle \oplus \langle x_2 \rangle \oplus \cdots \oplus \langle x_m \rangle$ is a $*$-valuated coproduct in G with $C \subseteq A'$ and A'/C finite.

Then $x_2, \ldots, x_m \in C \subseteq A' \subseteq H$. Since $B/(A \oplus \langle x_1' \rangle)$ is finite and $A \oplus \langle x_1' \rangle \oplus \langle x_2 \rangle \oplus \cdots \oplus \langle x_m \rangle$ is a valuated coproduct in G, we may apply Theorem 5.6.6(2) to $(A \oplus \langle x_1' \rangle) \oplus \langle x_2 \rangle \oplus \cdots \oplus \langle x_m \rangle$ and $B/(A \oplus \langle x_1' \rangle)$. This gives us nonzero multiples x_i' for each x_i, $2 \leq i \leq n$, such that $B \oplus \langle x_2' \rangle \oplus \cdots \oplus \langle x_m' \rangle$ is a valuated coproduct in G. These elements are primitive in G. Since B is torsion over $A \oplus \langle x_1' \rangle$ and $A \oplus \langle x_1' \rangle \oplus \cdots \oplus \langle x_m' \rangle$ is a $*$-valuated coproduct in G, it follows that $B' = B \oplus \langle x_2' \rangle \oplus \cdots \oplus \langle x_m' \rangle$ is a $*$-valuated coproduct in G as well. Also, $S \subseteq (A')^0 \subseteq (B')^0$. Therefore, B is $*$-extendable in H. If A is H-knice, then A is nice, and by Theorem 5.6.6(1), B is nice, and hence, H-knice. □

Recall that a subgroup H of G is a k-subgroup exactly if for every finite $S \subseteq H$ there is a finite (possibly empty) $*$-decomposition set X in G such that $X \subseteq H$ and $S \subseteq \langle X \rangle^0$ (see Section 3.11). Now we may characterize k-subgroups in terms of partial subbases.

Theorem 5.6.8. *Let G be a group and H a subgroup of G. Then the following conditions are equivalent:*

1. *H is a k-subgroup of G;*
2. *0 is H-knice in G;*
3. *$\mathcal{C}_H^G$ is a partial subbasis for H;*
4. *H has a partial subbasis.*

Proof. It follows from the definitions that (1) and (2) are equivalent. To show that (2) implies (3), suppose 0 is H-knice in G. Let $X = \{x_1, \ldots, x_n\} \in \mathcal{C}_H^G$ and $x \in H$. By Theorem 5.6.7, since 0 is H-knice in G, $\langle x_1 \rangle \oplus \cdots \oplus \langle x_n \rangle$ is H-knice in G. This means that for $S = \{x\}$, there are $y_1, \ldots, y_m \in H$, each primitive in G, such that $\langle x_1 \rangle \oplus \cdots \oplus \langle x_n \rangle \oplus \langle y_1 \rangle \oplus \cdots \oplus \langle y_m \rangle$ is a $*$-valuated coproduct in G containing a nonzero multiple of x. Now let $Y = \{x_1, \ldots, x_n, y_1, \ldots, y_m\}$. Then Y is a ∗-decomposition set in G; hence, $Y \in \mathcal{C}_H^G$. Furthermore, $X \subseteq Y$, and $x \in \langle Y \rangle^0$. This proves that $\mathcal{C}_H^G$ is a partial subbasis for H.

Clearly, (3) implies (4). Finally, to prove (4) implies (2), suppose $\mathcal{C}$ is a partial subbasis for H, and let S be a finite subset of H. Then the extension property of $\mathcal{C}$, starting with any $X \in \mathcal{C}$, applied successively to each element of S, gives a $Y \in \mathcal{C}$ such that $S \subseteq \langle Y \rangle^0$. Since $\mathcal{C} \subseteq \mathcal{C}_H^G$, Y is a ∗-decomposition set in G; hence, 0 is H-knice. □

Corollary 5.6.9. *Let H be an isotype subgroup of a group G. If H is a k-subgroup of G, then H is a k-group.*

Proof. By the theorem, H has a partial subbasis, which is a partial decomposition basis by Lemma 5.6.5. Then by Theorem 5.5.5, H is a k-group. □

By Remark 5.6.3, we have the following:

Corollary 5.6.10. *If H is an isotype k-subgroup of a group G, then* $\mathcal{C}_H^G$ *is a subset of* $\mathcal{C}_H$.

Corollary 5.6.11 (Hill–Megibben [53]). *An isotype knice subgroup of a k-group is a k-group.*

Proof. Suppose H is an isotype knice subgroup of a k-group G. As a knice subgroup of a k-group, H is a k-subgroup by Theorem 3.13.1(1). By Corollary 5.6.9, H is a k-group. □

Next we examine properties of ∗-decomposition sets, and whether they transfer from a group G to an isotype subgroup H, and conversely. Because we are looking at subgroups, we will adopt notation to make it clear whether we are talking about properties relative to all of G or to a subgroup H. Hence, we will sometimes refer to $F_G \in \{G(\overline{\alpha}^*, p), G(M^*), G(M^*, p)\}$ and the corresponding $F_G^\diamond$. The notation for isotype subgroup H is the same, and refers to H as a group in its own right.

Lemma 5.6.12. *Let G be a group, H an isotype subgroup, p a prime,* $\overline{\alpha}$ *an Ulm sequence, M an Ulm matrix, and* $F_G \in \{G(\overline{\alpha}^*, p), G(M^*), G(M^*, p)\}$. *Then* $F_H^\diamond = F_G^\diamond \cap H$ *and* $F_H \subseteq F_G \cap H$.

Proof. Since H is isotype, for any $x \in G$, $\|x\|$ is the same in H and G, so $F_H^\diamond = F_G^\diamond \cap H$. To verify the second claim, let $x \in F_H$. Then $x = b_1 + \cdots + b_n$ for some $b_1, \ldots, b_n \in F_H^\diamond \subseteq F_G^\diamond$, and hence, $x \in F_G$. □

Theorem 5.6.13. *If H is an isotype subgroup of a group G and* $x \in H$ *is primitive in G, then it is primitive in H.*

Proof. Let M be an Ulm matrix and p a prime. Suppose nx is an element of $H(M^*, p) \subseteq G(M^*, p)$ for some positive integer n. Then since x is primitive in G, $\|x\| \nrightarrow M$ or $nx \in G(M_p^*, p)^\diamond$. In the latter case, $nx \in G(M_p^*, p)^\diamond \cap H = H(M_p^*, p)^\diamond$. □

Theorem 5.6.14. *Let G be a group, H an isotype subgroup, and* $A = \bigoplus_{i \in I} A_i \subseteq H$. *Then A is a valuated coproduct in H if and only if it is a valuated coproduct in G.*

Proof. Let $a = a_1 + \cdots + a_n$ with $a_k \in A_{i_k}$ for $k = 1, \ldots, n$, and let M be an Ulm matrix. Suppose that $A = \bigoplus_{i \in I} A_i$ is a valuated coproduct in H. If $a \in A \cap G(M)$, then $a \in G(M) \cap H = H(M)$, and so $a_i \in H(M) \subseteq G(M)$ for all $i = 1, \ldots, n$. Conversely, suppose A is a valuated coproduct in G, and $a \in A \cap H(M) \subseteq G(M)$. Then each a_i is an element of $G(M) \cap H = H(M)$, as required. □

Though many properties transfer from G to the isotype subgroup H, or from H to G, they may not go the other way. To show this, we construct counterexamples.

Theorem 5.6.15. *Let H be an isotype subgroup of a group G, p a prime, $\overline{\alpha}$ an Ulm sequence, M an Ulm matrix, and $F_G \in \{G(\overline{\alpha}^*, p), G(M^*), G(M^*, p)\}$. Then the following hold:*
1. *F_H need not equal $F_G \cap H$.*
2. *If x is primitive in H, it need not be primitive in G.*
3. *If X is a $*$-decomposition set in H, it need not be one in G.*
4. *If $A = \bigoplus_{i \in I} A_i$ is a $*$-valuated coproduct in H, it need not be one in G.*

Proof. For the following examples (Hill–Meggiben [49]), define H_a, H_b, and H_c as in Example 5.6.1, with $\|a\|_p = (0, 3, 4, 7, 8, 11, \ldots)$, $\|b\|_p = (1, 2, 5, 6, 9, 10, \ldots)$, and $\|c\|_p = (0, 2, 4, 6, 8, 10, \ldots)$ for every prime p.

To prove (1), let $G = H_a \oplus H_b$, $x = a + b$, and $H = \langle x \rangle^0$ as a subgroup of G. H is isotype in G, and $\{x\}$ is a decomposition basis for H. Let $M = \|x\|$; that is, $M_p = (0, 2, 4, 6, \ldots)$ for all p. Then a and b are in both $G(M_p^*, p)^\diamond$ and $G(M^*)^\diamond$, and $x \in G(M)$, so x is in all of $G(M_p^*, p)$, $G(M^*)$, and $G(M^*, p)$. Now suppose that x is in $H(M_p^*, p)$, $H(M^*)$, or $H(M^*, p)$. Then since x is primitive by Lemma 5.5.4, $x \in H(M_p^*, p)^\diamond$ or $\|x\| \not\sim M$, a contradiction. This proves for all F that $x \in F_G$, but $x \notin F_H$.

We may use the same example to prove (2). Since $\{x\}$ is a decomposition basis for H, x is primitive in H. On the other hand, $x \in G(M^*, p)$, though $\|x\| \sim M$, and $x \notin G(M_p^*, p)^\diamond$, so x is not primitive in G.

For (3), Example 5.6.1 demonstrated this with $G = H_a \oplus H_b \oplus H_c$, using $x = a + b + c$, and $H = \langle x, c \rangle^0$. Then $\{x, c\}$ is a $*$-decomposition set in H, but not in G.

The example of (3) serves to prove (4). Let $A = \langle x \rangle \oplus \langle c \rangle$. Then A is a $*$-valuated coproduct in H since $\{x, c\}$ is a $*$-decomposition set in H. But it is not a $*$-valuated coproduct in G, as we saw in (3). □

Let H be a subgroup of a group G. Following Megibben and Ullery [115], we say H is **$*$-isotype** in G if H is isotype in G and for every p a prime, $\overline{\alpha}$ an Ulm sequence, M an Ulm matrix, and $F_G \in \{G(\overline{\alpha}^*, p), G(M^*), G(M^*, p)\}$, we have $F_H = F_G \cap H$.

Theorem 5.6.16 (Megibben–Ullery [115]). *An isotype knice subgroup of a group G is $*$-isotype in G.*

Proof. Suppose H is an isotype knice subgroup of G and $h \in F_G \cap H$ for some $F_G \in \{G(\overline{\alpha}^*, p), G(M^*), G(M^*, p)\}$, some Ulm sequence $\overline{\alpha}$, Ulm matrix M, and prime p. Suppose $h = z_1 + \cdots + z_n$, where each $z_i \in F_G^\diamond$. Since H is knice, there are $x_1, \ldots, x_m \in G$

such that $B = H \oplus \langle x_1 \rangle \oplus \cdots \oplus \langle x_m \rangle$ is a $*$-valuated coproduct and $z_1, \ldots, z_n \in B^0$. Let $A = \langle x_1 \rangle \oplus \cdots \oplus \langle x_m \rangle$. Then for some $r > 0$, $rz_i \in B = H \oplus A$ for all $1 \le i \le n$. Let $1 \le i \le n$. Then $rz_i = y_i + a_i$, for some $y_i \in H$ and $a_i \in A$. Now we define $rF_G^\diamond$ as follows: $rF_G^\diamond = G((r\overline{\alpha})^*, p)^\diamond$ if $F_G = G(\overline{\alpha}^*, p)$; $rF_G^\diamond = G((rM)^*)^\diamond$ if $F_G = G(M^*)$; and $rF_G^\diamond = G((rM)^*, p)^\diamond$ if $F_G = G(M^*, p)$. Since $z_i \in F_G^\diamond$, it may be verified that $rz_i \in rF_G^\diamond$. Furthermore, since $\|y_i\| \ge \|rz_i\|$, $y_i \in rF_G^\diamond$. In fact, $y_i \in rF_G^\diamond \cap H = rF_H^\diamond$, since H is isotype. Returning to h, we have $rh = rz_1 + \cdots + rz_n = y_1 + \cdots + y_n + a_1 + \cdots + a_n = y_1 + \cdots + y_n \in rF_H$. If $F_H = H(M^*, p)$, then Section 3.6(c), (d) say that $rh \in H((rM)^*, p) = rH(M^*, p)$ implies $h \in H(M^*, p)$. Similarly, we may prove the analogous statements for the other possibilities of F. Thus, $h \in F_H$, proving $F_G \cap H \subseteq F_H$. The other inclusion is obvious. □

Theorem 5.6.17. *Let G be a group, H a knice isotype subgroup of G, p a prime, $\overline{\alpha}$ an Ulm sequence, M an Ulm matrix, and $F_G \in \{G(\overline{\alpha}^*, p), G(M^*), G(M^*, p)\}$. Then the following hold:*
1. $F_H = F_G \cap H$.
2. *If x is primitive in H, it is primitive in G.*
3. *If X is a $*$-decomposition set in H, it is one in G.*
4. *If $A = \bigoplus_{i \in I} A_i$ is a $*$-valuated coproduct in H, it is one in G.*

Proof. Since H is knice and isotype, it is $*$-isotype by Theorem 5.6.16, which gives (1). To prove (2), assume x is primitive in H and $nx \in G(M^*, p)$ for some $n > 0$. Then $nx \in G(M^*, p) \cap H = H(M^*, p)$, and so, since x is primitive in H, $\|x\| \nsim M$ or $|p^i nx|_p \ne m_{p,i}$ for infinitely many $i < \omega$. Next, to prove (4), suppose A is a $*$-valuated coproduct in H. Then, since $A \subseteq H$, $A \cap F_G = A \cap F_G \cap H = A \cap F_H = \bigoplus_{i \in I}(A_i \cap F_H) = \bigoplus_{i \in I}(A_i \cap F_G)$. Finally, (3) follows from (2) and (4). □

Theorem 5.6.18. *Let H be an isotype knice subgroup of a group G. Then G is a k-group if and only if H is a k-group.*

Proof. If G is a k-group, then so is H by Corollary 5.6.11. Now suppose H is a k-group. Let $S \subseteq G$ be finite. Since H is knice in G, there are primitive $y_1, \ldots, y_m \in G$ such that $B = H \oplus \langle y_1 \rangle \oplus \cdots \oplus \langle y_m \rangle$ is a $*$-valuated coproduct in G and $S \subseteq B^0$, so $r\langle S \rangle \subseteq B$ for some $r > 0$. Let $S' = \{\pi(rs) : s \in S\}$, where $\pi : B \to H$ is the projection map, which is a finite subset of H. Since H is a k-group, there is a $*$-decomposition set $\{x_1, \ldots, x_n\}$ in H, and hence in G, such that $S' \subseteq \langle x_1, \ldots, x_n \rangle^0$. It follows that $\{x_1, \ldots, x_n, y_1, \ldots, y_m\}$ is a $*$-decomposition set in G such that $S \subseteq \langle x_1, \ldots, x_n, y_1, \ldots, y_m \rangle^0$. □

At this point, we have not resolved the following questions for G a group and H an isotype subgroup:

Problem 1. If $X \subseteq H$ is a $*$-decomposition set in G, is it necessarily one in H?

Problem 2. If $A = \bigoplus_{i \in I} A_i \subseteq H$ is a $*$-valuated coproduct in G, is it necessarily one in H?

5.7 PDB groups and Warfield groups

Recall that Warfield groups may be characterized as those groups G such that for some $X \subseteq G$,
1. X is a decomposition basis for G;
2. $\langle X \rangle$ is nice in G;
3. $G/\langle X \rangle$ is simply presented.

The class of PDB groups is strictly larger than the class of Warfield groups. In fact, for each of these three conditions, we will produce an example of a group that fails the condition and yet is a PDB group. We will give examples that are neither torsion nor torsion-free, and in fact do not split into torsion and torsion-free summands; recall that such mixed groups are said to be nonsplitting.

In the following, for a direct summand A of a group G, we let $\pi_A : G \to A$ denote the projection map.

Theorem 5.7.1 (Jacoby [68]). *Let H be a reduced group of torsion-free rank 1, say $H/\langle x \rangle$ is torsion. Let $H_i = H$ for all $i < \omega$. Define a subgroup G of $\prod_{i<\omega} H_i$ as $G = (\prod_{i<\omega} \langle x \rangle)^0$. Then G has a partial decomposition basis, but no decomposition basis.*

Proof. Suppose that G has a decomposition basis X. By taking multiples, we may assume that $X \subseteq \tilde{G} = \prod_{i<\omega} \langle x \rangle$. We will use a countability argument. Let p be an arbitrary prime and $S = \prod_{i<\omega} \langle p^i x \rangle$. Clearly, S is uncountable. Consider the set $\mathcal{X}$ of all $A \subseteq X$, which can be written as $A = \dot{\bigcup}_{i \in I_A} A(i)$ with $A(i)$ finite and $\langle A(i) \rangle \cap S \neq 0$ for all $i \in I_A$. Then $\mathcal{X}$ is partially ordered by declaring $A \leq A'$ whenever $I_A \subseteq I_{A'}$ and $A(i) = A'(i)$ for all $i \in I_A$. By Zorn's lemma, $\mathcal{X}$ has a maximal element B.

We will prove that B is uncountable. Suppose B is countable, say $B = \{x_1, x_2, \dots\}$. Since X is a decomposition basis, for any $s \in S$, there are $x_{i_1}, \dots, x_{i_k} \in B$, $y_1, \dots, y_n \in X \backslash B$ and nonzero integers $m, n_1, \dots, n_k, m_1, \dots, m_n$ such that

$$ms = n_1 x_{i_1} + \cdots + n_k x_{i_k} + m_1 y_1 + \cdots + m_n y_n.$$

Considering that there are only countably many choices for $(m, n_1, i_1, \dots, n_k, i_k)$, there must be distinct s_1 and s_2 in S, for which they agree. Then $ms_1 - ms_2$ is a nonzero element of S which is in $\langle y_1, \dots, y_n \rangle$. But $\{y_1, \dots, y_n\}$ is disjoint from B, and hence, from all subsets $B(i)$, which contradicts the maximality of B. This proves B is uncountable.

For each $i \in I_B$, choose a nonzero $y(i) \in S \cap \langle B(i) \rangle$, and let $Y = \{y(i) : i \in I_B\}$. The elements of Y are distinct since the subsets $B(i)$ of B are pairwise disjoint, and Y is a decomposition set since the $B(i)$ are disjoint subsets of a decomposition set. Let $y \in Y$ with p-height α, say $y = (y_i x)_{i<\omega}$ and $|y_k x|_p^H = \alpha$ for some $k < \omega$. Since $y \in S$, p^i divides y_i for all $i < \omega$; thus, there is some $m > k$ such that $|y_i x|_p^H > \alpha$ for all $i \geq m$. So for each $y \in Y$, we have an associated finite sequence of integers $(y_1, \dots, y_m)$. Since there are only countably many such sequences and Y is uncountable, there must be distinct y

and y', for which they agree. But then $\alpha = |y|_p^G = |y_k x|_p^H = |y_k' x|_p^H = |y'|_p^G$, and y and y' are members of the decomposition set Y. Yet $y - y'$ consists of terms beyond m, all with height $> \alpha$, giving us $\alpha = |y| \wedge |y'| \neq |y - y'|$, a contradiction, proving G has no decomposition basis.

Now we will construct a partial decomposition basis for G using techniques similar to those used in Lemma 4.1.3 and Theorem 4.1.4. For any set $\{i_1, \dots, i_k\}$ of nonnegative integers, let $\tilde{G}_{i_1,\dots,i_k}$ denote the subgroup of $\tilde{G}$ consisting of all elements whose projection on $H_{i_1} \oplus \dots \oplus H_{i_k}$ is zero. In the case of $k = 0$, we define $\tilde{G}_\emptyset = \tilde{G}$.

Let $\mathcal{C}$ be the collection of all $\{c_1, \dots, c_k\} \subseteq G$ with the following properties:

1. $\{c_1, \dots, c_k\}$ is a decomposition set;
2. For any $k > 0$, there are distinct nonnegative integers $i_1, \dots, i_k$ such that $\tilde{G} = \langle c_1 \rangle \oplus \dots \oplus \langle c_k \rangle \oplus \tilde{G}_{i_1,\dots,i_k}$;
3. For every $1 \le j \le k$ and prime p, the height of any multiple of c_j equals the p-height of its projection on H_{i_j};
4. For every $1 \le j \le k$, $c_j \in \tilde{G}_{i_1,\dots,i_{j-1}}$.

Since $\emptyset \in \mathcal{C}$, $\mathcal{C}$ is not empty. Therefore, we need only to prove the extension property. Suppose $\{c_1, \dots, c_k\} \in \mathcal{C}$ and $b \in G$. By taking multiples, we may assume $b \in \tilde{G}$. By (2), we may write $b = c + d$ for some $c \in \langle c_1 \rangle \oplus \dots \oplus \langle c_k \rangle$ and $d \in \tilde{G}_{i_1,\dots,i_k}$. Since $d \in \tilde{G}$, we may write $d = (d_i x)_{i<\omega}$, where each $d_i \in \mathbb{Z}$. We will prove by induction on $m = \min_{i<\omega}\{|d_i| : d_i \neq 0\}$ that for some $\{c_1, \dots, c_n\} \in \mathcal{C}$, with $n \ge k$, $b \in \langle c_1, \dots, c_n \rangle^0$. Here we use the notation $|d_i|$ for the absolute value. As usual, we will use the subscript p when discussing heights.

Suppose $m = 1$. We will show that $\{c_1, \dots, c_k, d\} \in \mathcal{C}$. There is a j such that $|d_j| = 1$. Then for any prime p, the p-height of any multiple of d must be the p-height of the j-th term, which is the same as the p-height of its projection on H_j. Taking $c_{k+1} = d$, and $i_{k+1} = j$, this proves (3). We already have (4) by the choice of d. Now we wish to prove (2). Let $(a_i x)_{i<\omega}$ be an arbitrary element of $\tilde{G}_{i_1,\dots,i_k}$. Then, depending on the sign of the integer d_j, either $(a_i x)_{i<\omega} - a_j d$ or $(a_i x)_{i<\omega} + a_j d \in \tilde{G}_{i_1,\dots,i_k,j}$, so $\tilde{G}_{i_1,\dots,i_k} = \langle d \rangle + \tilde{G}_{i_1,\dots,i_k,j}$. Furthermore, the sum is direct since x is not torsion. This gives us $\tilde{G} = \langle c_1 \rangle \oplus \dots \oplus \langle c_k \rangle \oplus \langle d \rangle \oplus \tilde{G}_{i_1,\dots,i_k,j}$, property (2).

We claim that $\{c_1, \dots, c_k, d\}$ is a decomposition set. Let $n_1, \dots, n_{k+1} \in \mathbb{Z}$, and let p be a prime. We may assume $|n_1 c_1 + \dots + n_k c_k|_p = |n_{k+1} d|_p$ since otherwise the result is trivial. Since $\{c_1, \dots, c_k\}$ is a decomposition set, $|n_1 c_1 + \dots + n_k c_k|_p = |n_l c_l|_p$ for some $1 \le l \le k$. By (4), $\pi_{H_{i_l}}(n_r c_r) = 0$ for all $k \ge r > l$. Then

$$\begin{aligned} |n_1 c_1 + \dots + n_k c_k + n_{k+1} d|_p &\le \left|\pi_{H_{i_l}}(n_1 c_1) + \dots + \pi_{H_{i_l}}(n_k c_k) + \pi_{H_{i_l}}(n_{k+1} d)\right|_p \\ &= \left|\pi_{H_{i_l}}(n_1 c_1 + \dots + n_k c_k)\right|_p \\ &= \left|\pi_{H_{i_l}}(n_1 c_1) + \dots + \pi_{H_{i_l}}(n_l c_l)\right|_p \\ &\le \left|\pi_{H_{i_l}}(n_l c_l)\right|_p = |n_l c_l|_p = |n_1 c_1 + \dots + n_k c_k|_p \end{aligned}$$

$$= |n_1c_1 + \cdots + n_kc_k|_p \wedge |n_{k+1}d|_p$$
$$= \min\{|n_1c_1|_p, \ldots, |n_kc_k|_p, |n_{k+1}d|_p\},$$

proving that $\{c_1, \ldots, c_k, d\}$ is a decomposition set, and completing the verification that $\{c_1, \ldots, c_k, d\} \in \mathcal{C}$.

Now suppose $m > 1$. For each $i < \omega$, write $d_i = q_im+r_i$, where $q_i \in \mathbb{Z}$ and $0 \le r_i < m$, and let $q = (q_ix)_{i<\omega}$ and $r = (r_ix)_{i<\omega}$. Then $d = mq + r$ and, since $d_i = 0$ for $i = i_1, \ldots, i_k$, q and r are in $\tilde{G}_{i_1,\ldots,i_k}$. Since $|q_j| = 1$ for some j, we may apply the previous case to get $\{c_1, \ldots, c_k, q\} \in \mathcal{C}$. Since $r_j = 0$, $r \in \tilde{G}_{i_1,\ldots,i_k,j}$. Furthermore, $|r_i| < m$ for all i, so $\min\{|r_i| : |r_i| \neq 0\} < m$, so we may apply the induction hypothesis to $\{c_1, \ldots, c_k, q\}$ and r, proving that $\{c_1, \ldots, c_k, q, r\} \in \mathcal{C}$. Finally, $d \in \langle c_1, \ldots, c_k, q, r\rangle^0$, and so is b. □

Corollary 5.7.2. *$\prod_\omega \mathbb{Z}$ has a partial decomposition basis, but no decomposition basis.*

Proof. Let $H = \mathbb{Z}$, and $x = 1$. □

Corollary 5.7.3. *For any prime p, there is a nonsplitting mixed $\mathbb{Z}_p$-module with a partial decomposition basis, but no decomposition basis.*

Proof. Let H be a nonsplitting, reduced $\mathbb{Z}_p$-module of torsion-free rank 1, for example, H_x of Corollary 2.7.5, where $\|x\|_p$ has gaps. Construct G as in the theorem. Suppose $G = T \oplus B$ for some torsion T and torsion-free B. We will get a contradiction by looking at the first summand H_0 of $\prod_{i<\omega} H_i$. Let $H' = \{(z, 0, 0, \ldots) : z \in H_0\} \subseteq G$. We claim that $H_0 = \pi_{H_0}\pi_T(H') + \pi_{H_0}\pi_B(H')$, and that the summands are, respectively, torsion and torsion-free. Let $z \in H_0$. Then $z' = (z, 0, 0, \ldots) \in G = T \oplus B$, so there are $t \in T$ and $b \in B$ such that $z' = t + b$; say $t = (t_i)_{i<\omega}$, and $b = (b_i)_{i<\omega}$. Then $z = t_0 + b_0 \in \pi_{H_0}\pi_T(H') + \pi_{H_0}\pi_B(H')$, so $H_0 \subseteq \pi_{H_0}\pi_T(H') + \pi_{H_0}\pi_B(H')$, and the other inclusion is clear.

Now let $0 \neq z \in \pi_{H_0}\pi_B(H')$, say $z = \pi_{H_0}\pi_B(h)$ for some $h \in H' \subseteq G = T \oplus B$. Write $h = t + b$ with $b = (b_i)_{i<\omega} \in B$. Then $z = b_0$ and $nt = 0$ for some $n > 0$. It follows that $nb = nh \in H'$, so $nb_i = 0$ for all $i \geq 1$. If z is torsion, then so is $b = (z, b_1, b_2, \ldots)$, which is impossible since B is torsion-free. Since $z \neq 0$ was arbitrary, $\pi_{H_0}\pi_B(H')$ is torsion-free. Next suppose $y \in \pi_{H_0}\pi_T(H')$, say $(y, c_1, c_2, \ldots) \in T$. Then y is torsion.

Finally, $\pi_{H_0}\pi_T(H') \cap \pi_{H_0}\pi_B(H') = 0$ since they are, respectively, torsion and torsion-free, giving $H_0 = \pi_{H_0}\pi_T(H') \oplus \pi_{H_0}\pi_B(H')$, contradicting the assumption that H_0 does not split. □

Next we will give examples to show that having a decomposition basis is not sufficient to make a group a Warfield group. Specifically, properties (2) or (3) may fail. In those cases, the group will have a partial decomposition basis. The examples that follow construct the groups of interest. The proofs that the examples have the stated characteristics are sketched, with reference to results in Hunter–Richman–Walker [65] and Hunter–Richman [61]. The interested reader is referred to those papers for more details. The examples are $\mathbb{Z}_p$-modules for an arbitrary prime p, so they hold as counterexamples, even if we restrict ourselves to local groups.

Example 5.7.4 (Hunter–Richman–Walker [65]). There is a nonsplitting mixed $\mathbb{Z}_p$-module with a decomposition basis, but no nice decomposition basis.

Proof. Let S be the set of all sequences $s = (s_n)_{n<\omega}$ of ordinals of the form $s_n = \omega n + m_n(s)$, where $m_n(s) < \omega$. For each $s \in S$, construct H_{x_s} as in Corollary 2.7.5, with $\|x_s\| = s$. Let A be the direct sum of the $\langle x_s \rangle$ and let $X = \{x_s : s \in S\}$. By [65, Corollary 35], applied to the height valuation relative to $\bigoplus_s H_{x_s}$, there is a $\mathbb{Z}_p$-module B such that A is an isotype submodule of B, B/A is torsion, and whenever (x_j) is a sequence in a decomposition basis for A such that $|x_j| < |x_{j+1}|$ and $|px_j| \geq \sup_{i<\omega}|x_i|$ for all j, then there is a $y \in B$ such that $|y + x_1 + \cdots + x_n| = |x_{n+1}|$ for $n = 0, 1, 2, \ldots$. Now let X' be an arbitrary subordinate of X. We may write $X' = \{p^{f(s)}x_s : s \in S\}$. We will show that X' is not nice in B.

We claim $\{s_n : f(s) = n\}$ is infinite for some n. Suppose not. Then for any n, there must be a maximum $m(s)$ in the finite set $\{s_n = \omega n + m(s) : f(s) = n\}$; we denote it by $k(n)$. Then take $s' = (\omega n + k(n) + 1)_{n<\omega} \in S$. Then $s'_n > s_n$ for all n for which $f(s) = n$. But $f(s') = n$ for some n, a contradiction. This proves that there is n such that for any $k < \omega$, there is $s \in S$ such that $s_n > \omega n + k$ and $f(s) = n$. Then if we take $x' = p^{f(s)}x_s$, $|x'| > \omega n + k$. Since k was arbitrary, this gives us $x_1, x_2, \ldots \in X'$ with $|x_1| < |x_2| < \cdots$. Also, for each such x, $|px| \geq \omega(n+1) > \omega n + l$ for all l, and so $|px_j| \geq \sup_{i<\omega}|x_i|$ for all $j < \omega$. Then by the construction of B, there is $y \in B$ such that $|y + x_1 + \cdots + x_n| = |x_{n+1}|$ for $n = 0, 1, 2, \ldots$. Now suppose $y + A$ has an element $y + g$ of maximal height. In particular, $|y + g| \geq \sup_{i<\omega}|x_i|$. Then for all n, $|x_1 + \cdots + x_n - g| \geq |y + x_1 + \cdots + x_n| \wedge |y + g| \geq |x_{n+1}|$. Since $g \in A$, it is a linear combination of at most finitely many elements of $\{x_s\}_{s\in S}$, which implies there is some n such that the x_n term of g is 0. We then get $|x_n| \geq |x_1 + \cdots + x_n - g| \geq |y + x_1 + \cdots + x_n| \wedge |y + g| \geq |x_{n+1}|$, which is a contradiction.

By a lemma of Hunter and Richman [61, Corollary 9.5], if G is a group with a nice decomposition basis, then every decomposition basis of G has a nice subordinate. Since B has a decomposition basis with no nice subordinate, we conclude that B has a decomposition basis, but no nice decomposition basis. □

It should be mentioned that the original proof in [65] cites a lemma of Warfield that states that if X and Y are decomposition bases for a group, then there are respective subordinates X' and Y' such that $\langle X' \rangle = \langle Y' \rangle$. They then conclude that B has no nice decomposition basis. This lemma was subsequently found to be false. In fact, [61, Example 7.1] gives a countable rank, local counterexample of this lemma.

Example 5.7.5. There is a nonsplitting mixed module G with a decomposition basis such that $G/\langle X \rangle$ is not simply presented for any decomposition basis X.

Proof. Let T be a p-group that is not simply presented. Let $H = H_x$, where $\|x\|$ has at least one gap. Then H is nonsplitting and of torsion-free rank 1. Now let $G = H \oplus T$.

Suppose G splits, say $G = B \oplus T'$, where B is torsion-free, and T' is torsion. Then it may be verified that $H = \pi_H \pi_B H \oplus \pi_H \pi_{T'} H$. But then H splits into these summands, which are, respectively, torsion-free and torsion. This is a contradiction, so G does not split. Clearly, G has a decomposition basis, namely $\{x\}$.

We claim that $G/\langle p^n x\rangle$ is not simply presented for any n. Suppose to the contrary that $G/\langle p^n x\rangle$ is simply presented. Since $G/\langle p^n x\rangle \cong H/\langle p^n x\rangle \oplus T$, we have that T is simply presented by Theorem 2.8.20, a contradiction. Now suppose Y is a decomposition basis such that $G/\langle Y\rangle$ is simply presented. By the invariance of the Warfield invariants (see Theorem 3.2.3), all decomposition bases must have the same cardinality, so $Y = \{y\}$ for some y. Furthermore, $\langle y\rangle$ is nice by Lemma 5.1.5. This means G is a Warfield module by the characterization given at the beginning of this section. By [61, Corollary 9.4], every decomposition basis of a Warfield group has a subordinate which generates a nice subgroup with simply presented cokernel. But this fails for G and $\{x\}$, giving a contradiction. □

Note that $\prod_\omega \mathbb{Z}$, which has a partial decomposition basis, but no decomposition basis, is partially isomorphic to $\bigoplus_\omega \mathbb{Z}$, a Warfield group. This suggests the possibility that the class of PDB groups may consist simply of those groups partially isomorphic to Warfield groups. We will see that this is not the case. We will use some results based on the characteristics of the functions that can occur as Ulm invariants or modified Ulm invariants of simply presented p-groups. Recall that, by Theorem 2.8.19, the Ulm function of a simply presented p-group is admissible.

Barwise and Eklof [9] examined functions that can arise as $\hat{u}$. Let f be a function from an ordinal τ, considered as the set of smaller ordinals, into $\omega \cup \{\omega\}$. Let κ be the first ordinal such that $|\kappa| > 2^{\aleph_0}$. Write $\tau = \tau_1 + \tau_2$, where $\tau_1 = \kappa\lambda_1$ and $\tau_2 < \kappa$, and write $\tau_2 = \omega\lambda_2 + m$ for $m < \omega$. We call f **permissible** if for any α and β, limit ordinals or 0,

(i) if $\tau = \mu + 1$, then $f(\mu) \neq 0$;

(ii) if $\alpha < \beta < \tau$, then there are infinitely many γ with $\alpha < \gamma < \beta$ and $f(\gamma) \neq 0$;

(iii) if $\alpha < \beta < \tau_1$, then there are infinitely many γ with $\alpha < \gamma < \beta$ and $f(\gamma) = \omega$.

We give the following result without proof:

Theorem 5.7.6 (Barwise–Eklof [9]). *Let τ be an infinite ordinal. If G is a p-group of length τ, then $\hat{u}_G \restriction \tau$ is a permissible function. Conversely, suppose $f : \tau \to \omega \cup \{\omega\}$ is permissible and $\mathfrak{m}$ is a cardinal $\geq |\tau|$.*

(i) *If $f(n) = \omega$ for some $n < \omega$, then there is a reduced p-group G of cardinality $\mathfrak{m}$ and length τ such that $f = \hat{u}_G \restriction \tau$.*

(ii) *If for all $n < \omega$, $f(n) \neq \omega$ (in which case $|\tau| \leq 2^{\aleph_0}$), then there is a reduced p-group G of cardinality $\mathfrak{m}$ and length τ such that $f = \hat{u}_G \restriction \tau$ if and only if $\mathfrak{m} \leq 2^{\aleph_0}$.* □

Theorem 5.7.7. *There is a p-group G such that $G \not\equiv_p W$ for all Warfield modules W.*

Proof. We begin by constructing a simple permissible function f. Let τ be a limit ordinal with $|\tau| = 2^{\aleph_0}$. Considering τ as a set of ordinals, let $f : \tau \to \{1\}$ be identically 1. We claim that f is permissible by reference to the definition. We note that $\tau < \kappa$, so we get $\tau_1 = 0$ and $\tau_2 = \tau$, so (iii) is vacuous. Since τ is a limit ordinal, (i) is vacuous. Now suppose α and β are limit ordinals or 0 such that $\alpha < \beta < \tau$. Then there are infinitely many γ with $\alpha < \gamma < \beta$, and for all such γ, $f(\gamma) = 1 \neq 0$, satisfying (ii).

Since f is permissible, by Theorem 5.7.6(ii), with $\mathfrak{m} = 2^{\aleph_0}$, there is a reduced p-group G of length τ such that $f = \hat{u}_G \restriction \tau$. Suppose $G \cong_p H$ for some Warfield module H. Then H is torsion since G is, and so H is simply presented since its decomposition basis is empty. Furthermore, H is reduced of length τ since partial isomorphisms are height-preserving. By Theorem 2.8.19, u_H is τ-admissible. Let $g = u_H$.

By the classification theorem 5.2.7, $\hat{g} = \hat{u}_H = \hat{u}_G$, which equals f on τ. Then $g(\beta) = 1$ for all $\beta \in \tau$, and so, taking $\alpha = 0$, $\sum_{\alpha\le\beta<\alpha+\omega} g(\beta) = \sum_{\beta<\omega} 1 = |\omega| = \aleph_0$, whereas $\sum_{\beta\ge\omega} g(\beta) = \sum_{\tau\ge\beta\ge\omega} g(\beta) = \sum_{|\tau|} 1 = |\tau| > \aleph_0$, contradicting the fact that g is admissible. □

Since every p-group is a PDB module, we have the following:

Corollary 5.7.8. *There is a PDB module that is not partially isomorphic to any Warfield module.*

Hunter, Richman, and Walker [66] provide a theorem that lets us extend the result beyond torsion modules. We cite without proof their theorem on valuated groups, applied to the height valuation:

Theorem 5.7.9 (Hunter–Richman–Walker [66]). *Let C be a direct sum of infinite cyclic $\mathbb{Z}_p$-modules and f a function from the ordinals to the cardinals that vanishes beyond some ordinal. Then there exists a Warfield module A with $u_A = f$ and $w_A = w_C$ if and only if f is admissible and $f \geq u_C$.* □

Corollary 5.7.10. *There is a nontorsion PDB module that is not partially isomorphic to any Warfield module.*

Proof. Let f and G be the function and p-group of length τ defined in the proof of Theorem 5.7.7. Recall that f is not admissible. We may assume that f is defined for all ordinals by setting $f(\alpha) = 0$ for all $\alpha \geq \tau$, so we have $f = u_G$. Consider $G \oplus \mathbb{Z}_p$, and suppose $G \oplus \mathbb{Z}_p \cong_p A$ for some Warfield module A. Then $\hat{u}_A = \hat{u}_{G\oplus\mathbb{Z}_p} = \hat{u}_G$; hence, $u_A = f$. Next, consider the Warfield invariant $w_A = w_{G\oplus\mathbb{Z}_p} = w_{\mathbb{Z}_p}$. Taking $C = \mathbb{Z}_p$ in the theorem, we conclude that f is admissible, a contradiction. □

Corollary 5.7.11. *There is a PDB group that is not partially isomorphic to any Warfield group.*

Proof. Let G be the $\mathbb{Z}_p$-module in the previous corollary, and suppose $G \cong_p W$ for some Warfield group W. By the global classification theorem 5.3.8, G and W have the same $\hat{u}$ and $\hat{w}$ invariants. Then G_p and W_p have the same $\hat{u}$ and $\hat{w}$ invariants by Lemmas 1.17.11

and 5.3.4. By the local classification theorem 5.2.7, $G = G_p \cong_p W_p$, contradicting the previous corollary since W_p is a Warfield module. □

Corollary 5.7.12. *The class of k-groups is strictly larger than the class of Axiom 3 groups. In fact, there are k-groups not even partially isomorphic to any Axiom 3 group.*

Proof. By Theorem 5.5.5, the k-groups are exactly the PDB groups. The Axiom 3 groups are exactly the Warfield groups, and we saw above how PDB groups and Warfield groups differ. □

5.8 The class of PDB groups

We now examine the class of PDB groups. We will see that it is a natural generalization of the class of Warfield groups relative to $L_{\infty\omega}$. Recall Karp's theorem, which equates equivalence in $L_{\infty\omega}$ and $\cong_p$. We will use the two concepts interchangeably. We summarize the key characteristics of PDB groups in both the global and local cases. As usual, we view the local groups as modules over $\mathbb{Z}_p$.

Lemma 5.8.1. *The class of PDB groups (or modules) has the following characteristics:*
1. *It is closed under taking finite direct sums.*
2. *It is closed under taking direct summands.*
3. *It contains the Warfield groups (or modules).*
4. *It is closed under $\cong_p$, or equivalently, $\equiv_\infty$.*
5. *It is classified up to $\cong_p$ by $\hat{u}$ and $\hat{w}$.*

Proof. Suppose G and H are PDB groups (or modules), say with partial decomposition bases $\mathcal{C}$ and $\mathcal{C}'$, respectively. Then it is easy to verify that $\{X \cup Y : X \in \mathcal{C}, Y \in \mathcal{C}'\}$ is a partial decomposition basis for $G \oplus H$, proving (1). The class is closed under direct summands by Corollary 5.5.6. The class includes the Warfield groups (or modules), since they have decomposition bases, and thus, partial decomposition bases. Theorem 5.1.2 proves (4), where the equivalence follows from Karp's theorem 4.3.2. Finally, (5) is Theorem 5.3.8, or Theorem 5.2.7 in the local case. □

Recall that for any group G, there is a simply presented group H such that $G \oplus H$ and H have the same Ulm and Warfield invariants (Theorem 3.16.2).

Corollary 5.8.2. *The class of PDB groups (or modules) is the smallest class of groups (or $\mathbb{Z}_p$-modules) that contains all Warfield groups (or modules) and is closed under $L_{\infty\omega}$-equivalence and the formation of direct summands.*

Proof. Let C be the class of PDB groups (or modules). First, C satisfies the three conditions by Lemma 5.8.1. Now suppose C' is a class of groups (or $\mathbb{Z}_p$-modules) satisfying the conditions, and let G be an arbitrary PDB group (or module). Then by Theorem 3.16.2, there is a simply presented group (or $\mathbb{Z}_p$-module) H such that $G \oplus H$ and H have the same Ulm and Warfield invariants. In particular, $\hat{u}$ and $\hat{w}$ agree. Since H is simply presented, it is a PDB group (or module). By Lemma 5.8.1(5), $G \oplus H \cong_p H$. Since H is a Warfield group (or module), it is in C'. Then $G \oplus H \in C'$ by closure under $\cong_p$. Finally, $G \in C'$ by closure under summands, proving $C \subseteq C'$. □

Theorem 5.8.3. *The following are equivalent for a group G:*
1. *G has a partial decomposition basis;*
2. *There is a simply presented group H such that* $G \oplus H \cong_p H$*;*
3. *There is a Warfield group H such that* $G \oplus H \cong_p H$*;*
4. *There is a group H with partial decomposition basis such that* $G \oplus H \cong_p H$.

Proof. Suppose (1). Then $G \oplus H \cong_p H$ for some simply presented H, by Theorem 3.16.2 and Lemma 5.8.1(5), proving (2). Since a simply presented group is a Warfield group, which has a partial decomposition basis, we get (2)⇒(3)⇒(4). Now suppose $G \oplus H \cong_p H$ for some PDB group H. Then by Lemma 5.8.1(4), $G \oplus H$ is a PDB group, and so is G by Lemma 5.8.1(2), proving (1). □

Theorem 5.8.4. *The class of groups with partial decomposition bases is the smallest class of groups that* (1) *contains all simply presented groups,* (2) *is closed under* $\cong_p$*, and* (3) *is closed under direct summands.*

Proof. The class of PDB groups satisfies (1) through (3) by Lemma 5.8.1. Suppose C is a class of groups satisfying (1) through (3). Let G be a PDB group. Then $G \oplus H \cong_p H$ for some simply presented H. Then $H \in C$ by (1), $G \oplus H \in C$ by (2), and $G \in C$ by (3). □

Recall that the class of Warfield groups is the unique class of groups, which contains all simply presented groups, is closed under finite direct sums and direct summands, and in which the Ulm and Warfield invariants distinguish between nonisomorphic groups (Theorem 3.16.3). We have an analogous result for PDB groups:

Theorem 5.8.5. *The class of groups with partial decomposition bases is the unique class of groups that* (1) *contains all simply presented groups, is closed under* (2) $\cong_p$ *and under* (3) *finite direct sums, and* (4) *direct summands, and in which* (5) *the invariants* $\hat{u}$ *and* $\hat{w}$ *distinguish between nonpartially isomorphic groups.*

Proof. The class of PDB groups satisfies (1) through (5) by Lemma 5.8.1. Let C be a class satisfying (1) through (5). Then C contains all PDB groups by Theorem 5.8.4. Let $G \in C$. Then by Theorem 3.16.2, $G \oplus H$ and H have the same Ulm and Warfield invariants for some simply presented H. Then $H \in C$ by (1), $G \oplus H \in C$ by (3), and $G \oplus H \cong_p H$ by (5). Since H is a PDB group, so is $G \oplus H$ as well by Lemma 5.8.1(4), and so is G by Lemma 5.8.1(2). □

Corollary 5.8.6. *The class of groups with partial decomposition bases is the unique class of groups that contains all Warfield groups, is closed under partial isomorphism, finite direct sums and direct summands, and in which the invariants* $\hat{u}$ *and* $\hat{w}$ *distinguish between nonpartially isomorphic groups.*

In the local case, we may take advantage of Theorem 3.3.5, which gives, for any $\mathbb{Z}_p$-module G, simply presented H and T, where T is torsion and $G \oplus T$ and H share the same Ulm and Warfield invariants. This will give us an alternative characterization of PDB modules.

Theorem 5.8.7. *If G is a $\mathbb{Z}_p$-module, there are simply presented $\mathbb{Z}_p$-modules H and T, where T is torsion, such that $G \oplus T$ and H have the same Ulm and Warfield invariants. In particular, if G has a partial decomposition basis, $G \oplus T \cong_p H$. If G is torsion-free, we may take H to be completely decomposable and $T = 0$.*

Proof. The first and third statements follow directly from Theorem 3.3.5. The second follows from the first and the classification theorem 5.2.7. □

Corollary 5.8.8. *The following conditions are equivalent for a $\mathbb{Z}_p$-module G:*
(i) *G is a PDB module;*
(ii) *There is a simply presented p-group T such that $G \oplus T \cong_p H$ for some simply presented $\mathbb{Z}_p$-module H;*
(iii) *There is a p-group T such that $G \oplus T \cong_p H$ for some Warfield module H.*

Proof. By Theorem 5.8.7, (i) implies (ii) immediately. Any simply presented $\mathbb{Z}_p$-module is a Warfield module; hence, (ii) implies (iii). Suppose (iii) holds for G. Then H is a PDB module, and hence, so is $G \oplus T$. By Lemma 5.8.1(2), G is a PDB module. □

Corollary 5.8.9. *A torsion-free $\mathbb{Z}_p$-module is a PDB module if and only if it is partially isomorphic to a direct sum of $\mathbb{Z}_p$-modules of rank 1.*

Proof. Suppose G is a torsion-free PDB module. Then by Theorem 5.8.7, $G \cong_p H$ for some completely decomposable H. Conversely, suppose $G \cong_p H$, where H is completely decomposable. Then H has a decomposition basis, and hence, a partial decomposition basis. Thus, G has a partial decomposition basis as well. □

Theorem 5.8.10. *The class of $\mathbb{Z}_p$-modules with partial decomposition bases is the smallest class C of $\mathbb{Z}_p$-modules closed under $L_{\infty\omega}$-equivalence that contains the Warfield modules, such that $G \oplus T \in C$ for some p-group T implies $G \in C$.*

Proof. The class C of PDB modules satisfies the conditions by Lemma 5.8.1, noting that every p-group is a PDB module. Now suppose C' is a class of $\mathbb{Z}_p$-modules satisfying the conditions, and let G be an arbitrary PDB module. Then $G \oplus T \cong_p H$ for some simply presented T and H, with T torsion, by Theorem 5.8.7. Then H is in C', $G \oplus T \in C'$, and finally, $G \in C'$, proving $C \subseteq C'$. □

Theorem 5.8.11. *The class C of PDB modules is the unique class such that*
(i) *C contains the Warfield modules;*
(ii) *the members G of C are classified in $L_{\infty\omega}$ by the invariants $\hat{u}(\infty, G), \hat{u}(\alpha, G)$ and $\hat{w}(e, G)$;*
(iii) *if $G \in C$ and $H \in C$, then $G \oplus H \in C$;*

(iv) *C is closed under* $\equiv_\infty$;
(v) *if T is torsion and* $G \oplus T \in C$, *then* $G \in C$.

Proof. We have already seen that the class of PDB modules satisfies these conditions. Suppose C' satisfies the conditions. Then $C \subseteq C'$ by Theorem 5.8.10. Now let $G \in C'$. By Theorem 5.8.7, there are simply presented H and T, where T is torsion and $G \oplus T$ and H share the same Ulm and Warfield invariants. Then $T \in C'$ by (i), $G \oplus T \in C'$ by (iii), and $G \oplus T \cong_p H$ by (ii). Since H is simply presented, it is a PDB module, and then so is $G \oplus T$, and finally, so is G, proving that $C' \subseteq C$. □

5.9 Infinitary logic perspective

Barwise and Eklof [9] generalized Ulm's theorem in the context of $L_{\infty\omega}$ (see Section 4.5). In fact, they identified a class of groups, the torsion groups, and defined an invariant, $\hat{u}$, that are analogues relative to $\equiv_\infty$ of Ulm's results relative to isomorphism. Ideally the class can be described in $L_{\infty\omega}$. Specifically, we say a class of groups is **definable** if there is a formula φ of $L_{\infty\omega}$ such that for any group G, $G \models \varphi$ if and only if G is a member of the class. In this case, the class of torsion groups is definable, as we saw earlier. In summary, the class of torsion groups has the following properties:
(i) The class includes the groups classified by Ulm's theorem.
(ii) The class is closed under $\equiv_\infty$.
(iii) The invariants agree up to ω with the Ulm invariants for groups covered by Ulm's theorem.
(iv) The invariants classify members of the class up to $L_{\infty\omega}$-equivalence.
(v) The invariants are expressible in $L_{\infty\omega}$.
(vi) The class is definable in $L_{\infty\omega}$.

PDB groups were developed to similarly take the classification of Warfield groups into $L_{\infty\omega}$. Here we examine how well they have succeeded. The class of PDB groups satisfies the following properties:
(i) The class includes the groups classified by Warfield's theorem.
(ii) The class is closed under $\equiv_\infty$.
(iii) The $\hat{w}$ invariants agree up to ω with the Warfield invariants for Warfield groups, and the $\hat{u}$ invariants agree up to ω with the Ulm invariants.
(iv) The invariants classify members of the class up to $L_{\infty\omega}$-equivalence.
(v) The invariants are expressible in $L_{\infty\omega}$.

In fact, items (i) through (v) have all been proved earlier, and it was shown that PDB groups are a suitable generalization of Warfield groups in $L_{\infty\omega}$.

The remaining question is whether (vi) is satisfied for the class of PDB groups, that is, whether this class is definable in $L_{\infty\omega}$. Now, in order to answer that, we look at PDB groups and what may be said about them in $L_{\infty\omega}$. We will see that though the modified Warfield invariant is expressible in $L_{\infty\omega}$, "decomposition set" is not. We will also see that our class is not definable in our language; that is, there is no sentence of $L_{\infty\omega}$ that says, "This is a PDB group." This raises the question of whether there might be some other class that better generalizes both Warfield's and Barwise and Eklof's work. However, we will see that no class that generalizes the Warfield modules in any reasonable way is definable in the language of $L_{\infty\omega}$.

Theorem 5.9.1 (Jacoby [68]). *"X is a decomposition set" is not expressible in $L_{\infty\omega}$.*

Proof. We will prove that it is not expressible even if we restrict our attention to $\mathbb{Z}_p$-modules for any prime p, and in fact "$\{x_1,\dots,x_n\}$ is a decomposition set" is not expressible for any $n \geq 2$. Suppose to the contrary that for some $n \geq 2$, there is a formula $\psi_n(x_1,\dots,x_n)$ such that for any $\mathbb{Z}_p$-module G and $x_1,\dots,x_n \in G$, $G \models \psi_n[a_1,\dots,a_n]$ if and only if $\{a_1,\dots,a_n\}$ is a decomposition set in G. Let δ be the quantifier rank of ψ_n.

Now construct the group $G = \bigoplus_{i=1}^{n} \mathbb{Z}_p$. We will represent this group as $G = \mathbb{Z}_p(1,\dots,1) \oplus \bigoplus_{i=1}^{n-1} \mathbb{Z}_p$. We define a p-valuation on each of these two summands, specifically,

$$v(m,\dots,m) = \omega\delta + |m|_p + 2,$$

where $|m|_p$ represents the ordinary height in $\mathbb{Z}_p$, and

$$v(m_1,\dots,m_{n-1},0) = \begin{cases} \omega\delta + \min\{|m_i|_p\} + 2 & \text{if } |m_i|_p > 0 \text{ for all } i \leq n-1, \\ \omega\delta + 1 & \text{otherwise.} \end{cases}$$

Then define v on G as the minimum of the valuations on the summands. Note that v is a valuation on $\mathbb{Z}_p(1,\dots,1)$ since height is. To prove that v is a valuation on $\bigoplus_{i=1}^{n-1} \mathbb{Z}_p$, recall that for v to be a valuation, the following conditions must hold: (a) $v(x)$ is an ordinal or ∞; (b) $v(x+y) \geq v(x) \wedge v(y)$ for all x and y in G; (c) $v(px) > v(x)$; (d) $v(kx) = v(x)$ if $p \nmid k$. In this case, (a) is clear. To prove (b), consider $(m_1,\dots,m_{n-1},0)+(l_1,\dots,l_{n-1},0)$. First, suppose $|m_i + l_i|_p > 0$ for all $1 \leq i \leq n-1$. Then

$$v(m_1 + l_1,\dots,m_{n-1} + l_{n-1},0) = \omega\delta + \min_{1\leq i\leq n-1}\{|m_i + l_i|_p\} + 2,$$

which is $\geq v(m_1,\dots,m_{n-1},0) \wedge v(l_1,\dots,l_{n-1},0)$. Now suppose $|m_i + l_i|_p = 0$ for some i, in which case $|m_i|_p = 0$ or $|l_i|_p = 0$. Then

$$v(m_1 + l_1,\dots,m_{n-1} + l_{n-1},0) = \omega\delta + 1 = v(m_1,\dots,m_{n-1},0) \wedge v(l_1,\dots,l_{n-1},0).$$

To prove (c), note that

$$v(pm_1,\dots,pm_{n-1},0) = \omega\delta + \min_{1\leq i\leq n-1}\{|pm_i|_p\} + 2 > \omega\delta + \min_{1\leq i\leq n-1}\{|m_i|_p\} + 2,$$

which is $\geq v(m_1, \ldots, m_{n-1}, 0)$. Finally, (d) follows since $|km_i|_p = |m_i|_p$ for all $1 \leq i \leq n-1$ and $p \nmid k$.

By Theorem 2.7.3, there is a $\mathbb{Z}_p$-module H with submodule $G' \cong G$, such that H/G' is torsion and simply presented, and for any $x \in G$ and its image $x' \in G'$, $|x'|_p = v(x)$. Now consider $a_i = (0, \ldots, 0, 1, 0, \ldots, 0) \in G$, where the 1 is in the i-th position. If $i < n$, and identifying a_i with its image and considering p-heights in H, we have $|a_i| = v(a_i) = \omega\delta + 1$ for all $1 \leq i \leq n-1$, and $|a_n| = |(1, \ldots, 1) - (1, \ldots, 1, 0)| = |(1, \ldots, 1)| \wedge |(1, \ldots, 1, 0)| = (\omega\delta + 2) \wedge (\omega\delta + 1) = \omega\delta + 1$. But $|a_1 + \cdots + a_n| = |(1, 1, \ldots, 1)| = \omega\delta + 2$, and so $\{a_1, \ldots, a_n\}$ is not a decomposition set in H.

Now consider $b_i = pa_i$ for $i = 1, \ldots, n$. We claim that $\{b_i\}$ is a decomposition set. Let $m_1, \ldots, m_n \in \mathbb{Z}$. Then

$$m_1 b_1 + \cdots + m_n b_n = (m_n p, \ldots, m_n p) + ((m_1 - m_n)p, \ldots, (m_{n-1} - m_n)p, 0).$$

The height of the first term is $\omega\delta + |m_n p| + 2$, whereas that of the second is $\omega\delta + \min_{1 \leq i \leq n-1}\{|m_i p - m_n p|\} + 2$. Thus, $|m_1 b_1 + \cdots + m_n b_n|_p = \omega\delta + \min_{1 \leq i \leq n}\{|m_i p|\} + 2 = \min_{1 \leq i \leq n}\{|m_i b_i|\}$.

Now define $f : \langle a_1, \ldots, a_n \rangle \to \langle b_1, \ldots, b_n \rangle$ by $f(a_i) = b_i$, let $S = \langle a_1, \ldots, a_n \rangle$, $T = \langle b_1, \ldots, b_n \rangle$, $X = \{b_1, \ldots, b_n\}$, $Y = \{pb_1, \ldots, pb_n\}$, and $\varphi = f \restriction X$. Note that since all heights are above $\omega\delta$, $f \in I_\delta$, a member of the Karp system as defined in Theorem 5.4.6, with both modules being H, and f_0 being $\emptyset$. Then by Karp's Theorem 4.3.1, $H \models \psi_n(a_1, \ldots, a_n)$ if and only if $H \models \psi_n(b_1, \ldots, b_n)$. But this says $\{a_1, \ldots, a_n\}$ is a decomposition set if and only if $\{b_1, \ldots, b_n\}$ is, a contradiction. □

This theorem does not automatically rule out the possibility that the class of PDB groups is definable in $L_{\infty\omega}$. In fact, we saw that the class of k-groups is the same class, and does not use decomposition bases in its definition. However, the following theorem says that, no matter how it may be described, neither the class of Warfield groups nor the class of PDB groups is definable in $L_{\infty\omega}$. In fact, this is true, even if we restrict our attention to the local case. First, we need a lemma constructing a local group to be used in the counterexample.

Lemma 5.9.2 (Jacoby [69]). *For any ordinal δ and prime p, there is a $\mathbb{Z}_p$-module G containing elements x and y with the following properties:*

(i) *G is an isotype submodule of $\prod_{i<\omega} H_i$, where each H_i is a Warfield module of torsion-free rank 1.*

(ii) *$\bigoplus_{i<\omega} H_i \subseteq G$.*

(iii) *The only torsion elements of G are in $\bigoplus_{i<\omega} H_i$.*

(iv) *Every element of G has a nonzero multiple in the form $ax+by+z$, where $z \in \bigoplus_{i<\omega} H_i$, $|z|_p^G = \infty$ and $a, b \in \mathbb{Z}_p$.*

(v) *For every $m \geq 0$, $|p^m x|_p^G = \omega\delta + m + \lfloor m/2 \rfloor + 1$, where $\lfloor m/2 \rfloor$ indicates the largest integer $\leq m/2$, occurring at $i = \lfloor m/2 \rfloor + 1$.*

(vi) *For every $m \geq 0$, $|p^m y|_p^G = \omega\delta + 3m + 2$, and this occurs at $i = m + 1$.*

(vii) *There is no decomposition set $X \subseteq G$ such that $x, y \in \langle X \rangle^0$.*

Proof. For each $i < \omega$, define an Ulm sequence $(\overline{\alpha}(i)_n)_{n<\omega}$ by

$$\overline{\alpha}(i)_n = \begin{cases} \omega\delta + n & \text{if } n < 3i \\ \infty & \text{if } n \geq 3i. \end{cases}$$

Let H_i be the Warfield module H_{x_i} built on this valuation per Corollary 2.7.7, containing x_i of infinite order with $\|x_i\|_p = (\overline{\alpha}(i)_n)_{n<\omega}$ and $H_{x_i}/\langle x_i\rangle$ torsion.

Now, considering $\prod_{i<\omega} H_i$, let $x = (p^i x_i)_{i<\omega}$ and $y = (p^{2i} x_i)_{i<\omega}$, and let

$$G = \Big\{z = (z_i) \in \prod H_i : \text{ for some } a, b \in \mathbb{Z}_p \text{ and } 0 \leq k \leq l < \omega, \text{ we have } z_i = ap^{i-k} x_i + bp^{2i-k} x_i \text{ for all } i \geq l\Big\}.$$

It may be verified that G is a $\mathbb{Z}_p$-module. Note that it includes $\bigoplus H_i$, by taking $a = b = 0$, as well as x and y. This proves (ii).

Now we consider heights in G, specifically p-heights. First, consider those $z \in G$ with associated l, for which $z_i = 0$ for all $i < l$. We will prove by transfinite induction on $\alpha \leq \omega\delta$ that $|z|^G \geq \alpha$ for each such z. If α is a limit ordinal, it is immediate from the definition of height. Suppose $\alpha - 1$ exists. Define z' by

$$z_i' = \begin{cases} 0 & \text{if } i < l+1 \\ ap^{i-k-1} x_i + bp^{2i-k-1} x_i & \text{if } i \geq l+1. \end{cases}$$

Then $z' \in G$ with leading zeroes, so by the induction hypothesis, $|z'|^G \geq \alpha - 1$. Note that $pz_i' = z_i$ for all $i \neq l$. Hence, $z = pz' + \overline{z}_l$ with $\overline{z}_l = (0, \dots, 0, z_l, 0, \dots)$, and so, $|z|^G \geq |pz'|^G \wedge |\overline{z}_l|^G$. Since $|\overline{z}_l|^G \geq |z_l|^{H_l} \geq |x_l|^{H_l} \geq \omega\delta \geq \alpha$, we have $|z|^G \geq (\alpha - 1) + 1 = \alpha$. This completes the induction, and proves $|z|^G \geq \omega\delta$ for those $z \in G$ such that $z_i = 0$ for all $i < l$.

Now for an arbitrary element z of G with associated l, if we subtract off components 1 through $l - 1$, we get an element with height $\geq \omega\delta$. This proves that:

Every element in G may be written in the form $w + z$ for some $z \in \bigoplus H_i$ and $|w|^G \geq \omega\delta$.

Since every element of H_i has a multiple in $\langle x_i\rangle$, z has a multiple in $\bigoplus\langle x_i\rangle$. By taking further multiples, we see that z has a multiple of height ∞ in $\bigoplus H_i$, thus in G. Also $p^k w$ and $ax + by$ agree for all components $\geq l$. This proves (iv).

Next, we will prove that G is isotype in $\prod H_i$, in other words, that $|z|^G = \min\{|z_i|^{H_i}\}$ for all $z \in G$. Let $z \in G$ with associated a, b, k, and l. The claim is immediate if $a = b = 0$ since it holds for $\bigoplus H_i$. So we may assume a and b are not both 0. Note that for any n, we may choose i large enough that $|p^n x_i|^{H_i} \neq \infty$, and hence, $|z_i|^{H_i} \neq \infty$ for sufficiently large i. In particular, $\min\{|z_i|^{H_i}\} = \omega\delta + m$ for some $m < \omega$ and occurs at some z_j. Choose n larger than the maximum of j and $l + m + 1$. Define z' by $z_i' = 0$ for $i < n$, and $z_i' = ap^{i-m-k-1}x_i + bp^{2i-m-k-1}x_i$ for $i \geq n$. Then $z = \overline{z}_1 + \cdots + \overline{z}_{n-1} + p^{m+1}z'$ with $\overline{z}_k = (0,\dots,0,z_k,0,\dots)$. Since $z' \in G$, $|z'|^G \geq \omega\delta$, so $|p^{m+1}z'|^G \geq \omega\delta + m + 1 > \min_{i<n}\{|z_i|^{H_i}\}$. It follows that $|z|^G = \min_{i<n}\{|z_i|^{H_i}\} = \min_{i<\omega}\{|z_i|^{H_i}\}$, proving (i).

Now let us examine heights of multiples of x and y. As we just proved, $|p^m x|^G = \min\{|p^m p^i x_i|^{H_i}\}$. But $|p^{m+i}x_i|^{H_i} \neq \infty$ if and only if $m + i < 3i$. Note that noninfinite heights increase with i, so the minimum $\min\{|p^m x_i|^{H_i}\}$ occurs uniquely at the least i such that $m + i < 3i$; thus, i is the least integer $> m/2$. In other words, $i - 1 = \lfloor m/2 \rfloor$, where the brackets indicate the greatest integer less than or equal to $m/2$. This gives us $|p^m x|^G = |p^{m+i}x_i|^{H_i}$ at $i = \lfloor m/2 \rfloor + 1$, which is $\omega\delta + m + \lfloor m/2 \rfloor + 1$. Similarly, we may show that $|p^m y|^G = \omega\delta + 3m + 2$; this occurs uniquely at $i = m + 1$. This proves (v) and (vi).

We claim that the only torsion elements of G are in $\bigoplus H_i$. Suppose $z \in G$ with associated a, b, l, and k, and $cz = 0$ for some $c \neq 0$. For any $i \geq l$, we have $z_i = ap^{i-k}x_i + bp^{2i-k}x_i$, so $0 = cz_i = (a + bp^i)cp^{i-k}x_i$. Since x_i has infinite order and $cp^{i-k} \neq 0$, we get $a + bp^i = 0$. Since $i \geq l$ was arbitrary, this says that $a = b = 0$, and hence, $z_i = 0$ for all $i \geq l$; thus, $z \in \bigoplus H_i$, proving (iii).

Suppose there is a decomposition set $\{y_1,\dots,y_n\}$ in G such that $x, y \in \langle y_1,\dots,y_n\rangle^0$. By (iv), each y_j has a multiple of the form $c_j x + d_j y + z_j$, where $z_j \in \bigoplus\langle x_i\rangle$, $|z_j|^G = \infty$, and $c_j, d_j \in \mathbb{Z}_p$, so by taking multiples, we may assume each y_j is in this form. From this point on, the subscript j will identify an element of G, rather than a component of one.

Now choose an integer $l > |c_j|$ for all $1 \leq j \leq n$, for which $c_j \neq 0$, where $|c_j|$ represents the p-height of c_j in $\mathbb{Z}_p$. Then for every $m \geq l$ and $c_j \neq 0$, $|p^m c_j x| \leq |p^{2m}x| = \omega\delta + m + 1 < \omega\delta + 3m + 2 \leq |p^m d_j y|$. In fact, if we replace each y_j by its multiple by p^l, and similarly for the c_j and d_j, we get $|p^m c_j x| < |p^m d_j y|$ for all $m < \omega$, whenever $c_j \neq 0$. In other words, if $c_j \neq 0$, $|ac_j x| < |ad_j y|$ for all $a \neq 0$.

There must be at least one y_j with c_j and d_j, not both 0, since x has no nonzero multiple in $\bigoplus H_i$. Suppose there is only one, which we may assume is y_1. Then for some $a, b, a_1, b_1 \in \mathbb{Z}_p \setminus \{0\}$, $z, z' \in \bigoplus H_i$, $ax = a_1 y_1 + z$, and $by = b_1 y_1 + z'$, since $y_2,\dots,y_n \in \bigoplus H_i$. Then $b_1 ax - a_1 by \in \bigoplus H_i$, which means the j-th components of $b_1 ax$ and $a_1 by$ are equal for sufficiently large j, a contradiction. Thus, there are at least two y_j with c_j and d_j not both 0. Without loss of generality, we will assume they are y_1 and y_2.

First, suppose $c_1 \neq 0$, and $c_2 \neq 0$. Consider the sum $(p-1)c_2 y_1 + c_1 y_2$. Substituting $y_1 = c_1 x + d_1 y + z_1$ and $y_2 = c_2 x + d_2 y + z_2$, and recalling that $|z_1|^G = |z_2|^G = \infty$, we get

$$|(p-1)c_2 y_1 + c_1 y_2| = |pc_1 c_2 x + (p-1)c_2 d_1 y + c_1 d_2 y| > |c_1 c_2 x|,$$

since all terms have height greater than $|c_1 c_2 x|$. Likewise, we find that $|(p-1)c_2 y_1| = |c_2 c_1 x + c_2 d_1 y| = |c_1 c_2 x|$, since $|c_2 d_1 y| > |c_2 c_1 x|$. This contradicts the assumption that $\{y_1, y_2\}$ is a decomposition set.

Next, suppose $c_1 = c_2 = 0$. Then $d_1, d_2 \neq 0$. By a similar argument,

$$|(p-1)d_2y_1 + d_1y_2| = |pd_1d_2y| > |d_1d_2y| = |(p-1)d_2y_1| \wedge |d_1y_2|,$$

again contradicting the assumption that $\{y_1, y_2\}$ is a decomposition set.

Finally, suppose, without loss of generality, $c_1 = 0$, and $c_2 \neq 0$. Note that $d_1 \neq 0$ since $c_1 = 0$. Choose $m \geq l$ such that $m \geq |d_1|$ and $2m+1 > |c_2|$. By taking multiples, we may assume that $y_1 = p^m y + z_1$ and $y_2 = p^{2m+1}x + d_2y + z_2$. Then $|y_1| = |p^m y| = \omega\delta + 3m + 2$, and the unique minimum occurs at $i = m+1$, the component $p^m p^{2(m+1)} x_{m+1} = p^{3m+2} x_{m+1}$. Now consider $|y_2|$. By our choice of l, $|p^{2m+1}x| < |d_2y|$, so $|y_2| = |p^{2m+1}x| = \omega\delta + 2m + 1 + \lfloor (2m+1)/2 \rfloor + 1 = \omega\delta + 3m + 2$, and this minimum occurs at $i = \lfloor (2m+1)/2 \rfloor + 1 = m+1$. So, in both cases, the unique minimum height occurs at the $m+1$ component. Next consider $y_1 + (p-1)y_2$. By assumption, $\{y_1, y_2\}$ is a decomposition set, so we must have $|y_1 + (p-1)y_2| = |y_1| \wedge |y_2| = \omega\delta + 3m + 2$ with the height occurring uniquely at the $m+1$ component. But

$$|y_1 + (p-1)y_2| = |p^m y + z_1 + (p-1)p^{2m+1}x + (p-1)d_2y + z_2| = |p^m y + (p-1)p^{2m+1}x|$$

since $|z_1| = |z_2| = \infty$ and $|d_2y| > |p^{2m+1}x|$. The $m+1$ component, which is $p^m p^{2(m+1)} x_{m+1} + (p-1)p^{m+1}p^{2m+1}x_{m+1} = p^{3m+3}x_{m+1}$, has height ∞, a contradiction. This proves that $\{y_1, \ldots, y_n\}$ is not a decomposition set, proving (vii). □

Theorem 5.9.3 (Jacoby [69]). *Let C be some class of $\mathbb{Z}_p$-modules that includes all Warfield modules and such that if a $\mathbb{Z}_p$-module G is in C, then any two elements of G have multiples in a submodule generated by a decomposition set. Then C is not definable in $L_{\infty\omega}$.*

Proof. Let C be a class of $\mathbb{Z}_p$-modules with the stated properties, and suppose φ is a sentence of $L_{\infty\omega}$ such that for any $\mathbb{Z}_p$-module G, $G \models \varphi$ if and only if $G \in C$. Let $\delta = qr(\varphi)$. Let G, x, y, and all H_i be as in Lemma 5.9.2.

Now let $H = \bigoplus_{i<\omega} H_i \oplus \mathbb{Q} \oplus \mathbb{Q}$. Then H is a Warfield module since each summand is. We will prove $G \equiv_\delta H$ by exhibiting a Karp system.

For $\nu \leq \delta$, let I_ν be the set of all isomorphisms $f : S \to T$ for S and T submodules of G and H, respectively, where

(i) f preserves heights up to $\omega\nu$;
(ii) $x, y \in S$;
(iii) f maps x and y, respectively, to the unit elements of the first and second copies of $\mathbb{Q}$;
(iv) $\bigoplus H_i \subseteq S$, and $\bigoplus H_i \subseteq T$;
(v) $f \restriction \bigoplus H_i = 1_{\bigoplus H_i}$.

We first note that for any ν, I_ν is not empty, since we may take $S = \bigoplus H_i \oplus \langle x, y\rangle$ and $T = \bigoplus H_i \oplus \mathbb{Z}_p \oplus \mathbb{Z}_p$, define f following (iii) and (v), and then note that f preserves heights up to $\omega\nu$, since $|x|$ and $|y|$ are both $\geq \omega\delta \geq \omega\nu$.

Now suppose $f \in I_{\nu+1}$, $z \in G \setminus S$. We wish to extend f to z. We may assume $|z| \geq \omega\delta$ since we may subtract off an element of $\bigoplus H_i \subseteq S$ to remove components of smaller height. Since z has a multiple of the form $ax + by + w$, where $w \in \bigoplus H_i$, by (ii) and (iv) of Lemma 5.9.2, $p^k z \in S$ for some $k < \omega$. We may assume k is minimal. Suppose $f(p^k z) = z_0 + z_1 + z_2$, where $z_0 \in \bigoplus H_i$, and z_1 and z_2 are in the first and second copies of $\mathbb{Q}$, respectively. We may write $z_1 = p^k z_1{}'$ and $z_2 = p^k z_2{}'$ for unique $z_1{}'$ and $z_2{}'$ in $\mathbb{Q}$. Since $|z| \geq \omega\delta \geq \omega(\nu+1)$, we have $|z_0| = |f(p^k z)| \geq \omega(\nu+1)$. Thus, we may choose $z_0{}'$ such that $|z_0{}'| \geq \omega\nu$ and $p^k z_0{}' = z_0$. This then gives us $p^k(z_0{}' + z_1{}' + z_2{}') = f(p^k z)$. We claim that the order of $z_0{}' + z_1{}' + z_2{}' + T$ is k. Suppose for some $l < k$, $p^l(z_0{}' + z_1{}' + z_2{}') \in T$, say $p^l(z_0{}' + z_1{}' + z_2{}') = f(z')$, where $z' \in S$. Then $f(p^{k-l} z') = p^k(z_0{}' + z_1{}' + z_2{}') = f(p^k z)$. Since f is injective, we get $0 = p^k z - p^{k-l} z' = p^{k-l}(p^l z - z')$. Since the only torsion elements of G are in $\bigoplus H_i$, $p^l z - z' \in \bigoplus H_i \subseteq S$, which puts $p^l z \in S$, contradicting minimality of k. Now by Lemma 5.4.3, f can be extended to an $\omega\nu$-height-preserving isomorphism $g : \langle S, z\rangle \to \langle T, z_0{}' + z_1{}' + z_2{}'\rangle$. Then $g \in I_\nu$, as desired.

Next suppose $f \in I_{\nu+1}$ and $z \in H \setminus T$, say $z = z_0 + z_1 + z_2$. Then $z + T = z_1 + z_2 + T$, so we may assume that $z_0 = 0$. For some $k < \omega$, $p^k z \in \mathbb{Z}_p \oplus \mathbb{Z}_p \subseteq T$. Choose k minimal such that $p^k z \in T$, say $f(w) = p^k z$ for $w \in S$. Then since $|p^k z| = \infty$ and f preserves heights up to $\omega(\nu+1)$, $|w| \geq \omega(\nu+1)$. Then $w = p^k w'$ for some $w' \in G$ with $|w'| > \omega\nu$. We may apply the same argument as above to show that $w' + S$ has order p^k, since all torsion elements of H are in $\bigoplus H_i \subseteq T$. Then by Lemma 5.4.3 again, we may extend f to a $g \in I_\nu$ such that $g(w') = z$.

This proves that we have a Karp system, so by Theorem 4.3.1, $G \equiv_\delta H$. This means, in particular, that $G \models \varphi$ if and only if $H \models \varphi$, in other words, $G \in C$ if and only if $H \in C$, a contradiction since by Lemma 5.9.2(vii), $G \notin C$. □

Corollary 5.9.4. *The class of PDB groups is not definable in $L_{\infty\omega}$.*

Proof. Suppose φ is a sentence of $L_{\infty\omega}$ such that for any group G, $G \models \varphi$ if and only if G is a PDB group. Then $G \models \varphi \wedge \forall x \bigwedge_{q \in \mathbb{P}, q \neq p} \exists y(x = qy)$ if and only if G is a p-local PDB group, that is, a PDB module. □

We now summarize the characteristics of the class of PDB modules analogous to the characteristics of the class classified by Barwise and Eklof, all of which we have now proven.

Theorem 5.9.5. *The following hold for the class of all PDB groups and the class of all PDB modules:*

(i) *The class includes all Warfield groups (or modules).*
(ii) *The class is closed under $\equiv_\infty$.*
(iii) *The $\hat{w}$ invariants agree up to ω with the Warfield invariants for Warfield modules, and the $\hat{u}$ invariants agree up to ω with the Ulm invariants.*
(iv) *The invariants classify members of the class up to $L_{\infty\omega}$-equivalence.*
(v) *The invariants are expressible in $L_{\infty\omega}$.*
(vi) *The class is not definable in $L_{\infty\omega}$.*

5.10 Going further

As discussed in Chapter 3, there are limits to the expansion of the classes of groups classified. This is true even for countable groups (Downey–Montalbán [20]). Since countable partially isomorphic groups are isomorphic, the same is true for partial isomorphisms.

We may be able to find some larger class of groups classifiable up to partial isomorphism using the modified Ulm and Stanton invariants, although by Theorem 5.8.11, we will need to give up some nice closure properties for the class to do so. One such candidate might be the class of isotype subgroups of PDB groups. Alternatively, we may be able to extend the classification to a larger class of groups with added invariants.

These techniques open the possibility of classifying other algebraic classes up to partial isomorphism. Whereas we have considered modules only over $\mathbb{Z}$ and $\mathbb{Z}_p$, Leistner [94], in her doctoral dissertation, considers modules over a larger class of rings and constructs a class of partially isomorphic modules, characterized by their Ulm invariants. Other classification results may be extended to a larger class by weakening the isomorphism requirement to partial isomorphism.

Now that we have some classifications, we look at how they might be used. Ideally, a set of invariants will help us answer questions about the groups under consideration. Kaplansky [78] formulated three "test questions" that are open questions for general abelian groups. A useful classification scheme will allow us to answer questions such as these for the classes of groups studied. See the Problems below.

In addition to classification, we touched on some instances of a broad inquiry, into which properties of isotype subgroups hold in the larger group and, conversely, which properties of isotype subgroups, when considered relative to the larger group, continue to hold in the isotype subgroup. See Example 5.6.1, Problems 1 and 2 at the end of Section 5.6 and related discussion. This only hints at the breadth of this question.

In Theorem 3.16.1, we gave numerous alternative characterizations of Warfield groups. We saw that the class of PDB groups is the same as the class of k-groups, but there may be many more ways to look at this class.

PDB groups include some very complex groups, but each may be approached from a finite point of view, as built on a set of finite subsets that can be extended finitely. This viewpoint may afford us access to group complexities that have not been addressed previously. In any case, PDB groups form a large class of groups that have not been studied extensively.

In the previous chapter, we raised the question of whether extending the language to $L_{\infty\alpha}$ for $\alpha > \omega$ would add useful and meaningful expressive richness. In particular, this may give new insights into PDB groups. Alternatively, this might suggest a new class of groups, just as partial isomorphism suggested PDB groups.

Problems.

1. Characterize the groups which can be embedded in PDB groups.
2. Barwise and Eklof [9, p. 64] extended Theorem 4.5.7 to a larger set of ordinals. Can Theorem 5.4.21 be extended to other ordinals?
3. Kaplansky test problem I [78]: Let G and H be PDB groups. If G is partially isomorphic to a direct summand of H and H is partially isomorphic to a direct summand of G, are G and H necessarily partially isomorphic?
4. Kaplansky test problem II [78]: Let G and H be PDB groups. If $G \oplus G$ is partially isomorphic to $H \oplus H$, are G and H necessarily partially isomorphic?
5. Kaplansky test problem III [78]: Let F, G, and H be PDB groups. If F is finitely generated and $F \oplus G$ is partially isomorphic to $F \oplus H$, are G and H necessarily partially isomorphic?
6. Characterize the fully invariant subgroups of PDB groups.
7. Are there characterizations of PDB groups similar to those for Warfield groups in Theorem 3.16.1?
8. Which characteristics transfer between PDB groups and isotype subgroups? (See Problems 1 and 2 in Section 5.6.)
9. If "finite" is replaced by "countable" in the definition of partial decomposition basis, how does the theory change?

6 Characters and Pontrjagin duality of locally compact abelian groups

We take a topological view of abelian groups and examine LCA (locally compact abelian) groups, a class of topological abelian groups, which contains all (discrete) abelian groups. We define counterparts of the key constructs on abelian groups, including isomorphism and direct summand, that speak to the topological characteristics. Every LCA group G has a dual group called the character group $\hat{G}$ with a compact-open topology, also an LCA group. By the Pontrjagin duality theorem, the LCA group G can be identified with the dual group of $\hat{G}$. We then explore the relations between properties of a group and its dual. For example, a group is discrete if and only if its dual is compact. Thus, every discrete group is the dual of a compact group, and every compact group is the dual of a discrete group. We discuss the annihilator mechanism, an indispensable tool for the study of LCA groups. Then we define counterparts of exact sequences and see that $\mathrm{Ext}(C, A) \cong \mathrm{Ext}(\hat{A}, \hat{C})$. The chapter is concluded with a study of topological p-groups, the analogues of (discrete) p-groups.

6.1 Preliminaries

In this section, we recall some definitions and basic facts on general topology and topological groups without proving them. We assume that the reader is familiar with both areas. For details and information on topological groups, particularly on locally compact abelian groups, we refer to the book by Hewitt and Ross, "Abstract Harmonic Analysis I" [41].

Let G be a set that is an abelian group and also a topological space. Then G is called a **topological group** if

1. the map $G \times G \to G$ defined by $(x, y) \mapsto x + y$ is continuous, where $G \times G$ is endowed with the product topology;
2. the map $G \to G$ given by $x \mapsto -x$ is continuous.

For example, the additive group $\mathbb{R}$ of real numbers with its usual topology is a topological group. It is clear that every group can be considered as a topological group, equipped with the discrete topology (that is, every subset is open). A topological group endowed with the discrete topology is simply called a **discrete group**.

Let G and H be topological groups. Then the following hold:

(a) For any $a \in G$, the map $x \mapsto a + x$ is a homeomorphism of G.
(b) A homomorphism $\phi : G \to H$ is continuous if and only if ϕ is continuous at $0 \in G$.

If A is a subgroup of G, then A is a topological group with the relative topology as a subspace of G, and the factor group G/A is a topological group with the quotient

https://doi.org/10.1515/9783110427684-006

topology. A subgroup of G is open if and only if it has nonempty interior, and every open subgroup of G is closed (see [41, Theorem 5.5]).

For subsets S and T of G, we let $S + T = \{s + t : s \in S, t \in T\}$. Then we have (see [41, Theorem 4.4]):

(c) $S + T$ is open if S is open and T is arbitrary.
(d) $S + T$ is closed whenever S is closed and T is compact.
(e) If both S and T are compact, then $S + T$ is compact.

The **closure of** S in G is denoted by $\overline{S}$. Given an element $x \in G$, we call an open set containing x a **neighborhood of** x. A Hausdorff topological abelian group G (see Exercise 3) is said to be **locally compact** (and we call G an **LCA group**) if every element $x \in G$ has a neighborhood U such that $\overline{U}$ is compact. For instance, the group $\mathbb{R}$ is locally compact. Given an LCA group G, we let G_d denote the group G with the discrete topology.

If G is an LCA group, then one-element subsets of G are closed (see Exercise 3). A subgroup A of an LCA group G is an LCA group if and only if A is closed in G (see [41, Theorem 5.11]). If A is a closed subgroup of an LCA group G, then G/A is an LCA group [41, Theorems 5.21, 5.22]. By the Baire category theorem, an LCA group cannot be written as a countable union of closed subsets with empty interior (see [41, Lemma 5.28]):

Theorem 6.1.1. *Let G be an LCA group, and let $A_n \subseteq G$ ($n \in \mathbb{N}$) such that $G = \bigcup_{n\in\mathbb{N}} A_n$. Then there is an $m \in \mathbb{N}$ such that the set A_m has nonempty interior.*

By the above theorem, a countable LCA group is automatically discrete. Thus, the groups $\mathbb{Z}$, $\mathbb{Z}/n\mathbb{Z}$, and $\mathbb{Q}$, considered as LCA groups, are discrete.

Let G be an LCA group. An element $x \in G$ is called **compact** if $\overline{\langle x\rangle}$ is a compact subgroup of G. The set of all compact elements of G, written bG, is a closed subgroup of G [41, Theorem 9.10]. A topological space is called **connected** if it is not the union of two disjoint nonempty open subsets. The **identity component** (also called **connected component**) of G is the largest connected subset of G containing 0 and is denoted by G_0. It is a closed subgroup of G [41, Theorem 7.1]. Clearly, G is connected if and only if $G = G_0$. If $G_0 = 0$, then G is called **totally disconnected.** Note that G/G_0 is totally disconnected [41, Theorem 7.3] and that G_0 is the intersection of all open subgroups of G [41, Theorem 7.8]. If G is totally disconnected, then G has **arbitrarily small compact open subgroups**; that is, every neighborhood of 0 in G contains a compact open subgroup of G [41, Theorem 7.7].

By a **compact group**, we mean a compact LCA group. An important example of a compact group is the **circle group** $\mathbb{T}$, which is defined to be the multiplicative group of complex numbers z with absolute value $|z| = 1$ (with the topology induced by the complex plane $\mathbb{C}$). For $\varepsilon > 0$, we define

$$\mathbb{T}_\varepsilon = \{z \in \mathbb{T} : |z - 1| < \varepsilon\}.$$

Clearly, we have:

(f) $\mathbb{T}$ does not have any proper connected nontrivial subgroups.

(g) The set $\mathbb{T}_{\sqrt{3}}$ does not contain any nontrivial subgroups of $\mathbb{T}$.

A topological space is called **σ-compact** if it is a countable union of compact subspaces. The **weight** of a topological space X is the least cardinal number $w(X)$ of an open basis of X. A topological space X is said to be **metrizable** if its topology is induced by some metric on the set X (see Engelking [21]). A Hausdorff topological group is metrizable if and only if it has a countable open basis at 0 [41, Theorem 8.3]. Since the weight of a compact metric space is countable [21, Chapter 4, Theorem 6], it follows that a compact group G is metrizable if and only if $w(G) \leq \aleph_0$.

For LCA groups G and H, we let $\mathrm{Hom}(G, H)$ denote the group of continuous homomorphisms from G to H. A map $\phi \in \mathrm{Hom}(G, H)$ is called **open** if $\phi(U)$ is open in H for every open subset U of G. Similarly, ϕ is called **closed** if $\phi(U)$ is closed in H whenever U is a closed subset of G. We say that G and H are **topologically isomorphic**, written

$$G \cong H$$

if there is a continuous, open isomorphism $\phi : G \to H$. In this case, ϕ is called a **topological isomorphism.** For instance, we have $\mathbb{R}/\mathbb{Z} \cong \mathbb{T}$ via $x + \mathbb{Z} \mapsto \exp(2\pi i x)$. Pontrjagin's open mapping theorem (see [41, Theorem 5.29]) will be needed:

Theorem 6.1.2 (Pontrjagin). *Suppose that G and H are LCA groups such that G is σ-compact. If $\phi \in \mathrm{Hom}(G, H)$ is surjective, then ϕ is an open map.*

If A is any subgroup of G, then the natural map $\varphi : G \to G/A$ is continuous and open [41, Theorems 5.16, 5.17]. Thus, G/A is discrete if and only if A is open in G [41, Theorem 5.21]. If A is compact, then φ is a closed map [41, Theorem 5.18]. If F is a compact subset of G, then $\varphi(F)$ is compact. On the other hand, if F is a compact subset of G/A, then we have $F = \varphi(F')$ for some compact subset F' of G [41, 5.24(b)]. We have the following **isomorphism theorems** for LCA groups (see [41, Theorems 5.27, 5.33, 5.35]):

(h) If $\alpha : G \to H$ is a continuous, open epimorphism, then $G/\ker\alpha \cong H$.

(i) Let A and B be closed subgroups of G such that $A + B$ is locally compact. If A is σ-compact, then $(A + B)/B \cong A/(A \cap B)$.

(j) If A and B are subgroups of G with $A \subseteq B$, then $G/B \cong (G/A)/(B/A)$.

Let $\{G_i : i \in I\}$ be a nonempty collection of LCA groups. Then the group $G = \prod_{i \in I} G_i$, equipped with the product topology, is called the **(topological) direct product** of the groups G_i. The group G is compact exactly if each group G_i is compact, and G is locally compact if and only if all but finitely many groups G_i are compact (see [41, Theorem 6.4]). If there is a group H such that $G_i = H$ for all $i \in I$, we write $G = H^I$ or $G = H^{\mathfrak{m}}$ with $\mathfrak{m} = |I|$. In case I is finite, say $I = \{1, \ldots, n\}$, we sometimes write $G_1 \times \cdots \times G_n$ instead of $\prod_{i \in I} G_i$.

If A is a closed subgroup of an LCA group G, then we say that A **splits from** G if there exists a closed subgroup B of G with $G = A \oplus B$ such that the map

$$\begin{array}{rccc} \phi: & A \times B & \to & G \\ & (a,b) & \mapsto & a+b \end{array}$$

is a topological isomorphism. In this case, A is called a **direct factor** or a **topological direct summand of** G, and we write $G = A \bigoplus B$.

Notice that the map ϕ above is automatically a continuous, algebraic isomorphism whenever $G = A \oplus B$. If the closed subgroups A and B of G are σ-compact, then $A \times B$ is σ-compact, in which case ϕ is a topological isomorphism by Theorem 6.1.2. We keep track of this fact:

Proposition 6.1.3. *Let A and B be closed, σ-compact subgroups of an LCA group G. If $G = A \oplus B$, then $G = A \bigoplus B$.*

Exercises.

1. Prove (a) and (b).
2. If G is a topological group, then the map $G \to G$ given by $x \mapsto -x$ is a homeomorphism of G.
3. Let X be a topological space. Then X is called a T_0**-space** if for any distinct elements $x, y \in X$ at least one element has a neighborhood not containing the other element. The topological space X is called a T_1**-space** if for any distinct elements $x, y \in X$ each element has a neighborhood not containing the other element, and X is called a **Hausdorff space** or a T_2**-space** if any distinct elements $x, y \in X$ have disjoint neighborhoods. Prove:
   (a) X is a T_1-space exactly if every one-element subset of X is closed.
   (b) For a topological group G, the following conditions are equivalent:
   – (i) G is a T_0-space;
   – (ii) G is a T_1-space;
   – (iii) G is a T_2-space.
4. Let A be a subgroup of a σ-compact group G. Then G/A is σ-compact. If A is closed in G, then A is σ-compact.
5. (Fulp–Griffith [35]) If A and B are closed subgroups of an LCA group such that $A+B$ is locally compact and B is σ-compact, then $(A+B)/B \cong A/(A \cap B)$.
6. (Fulp–Griffith [35]) If A and B are closed subgroups of an LCA group G such that $G = A \oplus B$ and A is σ-compact, then $G = A \bigoplus B$.
7. (Ahern–Jewett [1]) Let $G = \mathbb{Z}(p^\infty) \times \mathbb{R}/\mathbb{Z}$ and $A = \{(px, x) : x \in \mathbb{Z}(p^\infty)\}$. Then A is a discrete (and thus, closed) subgroup of G and $A \cong \mathbb{Z}(p^\infty)$, but A does not split from G.
8. If A and B are closed subgroups of an LCA group G, then $A + B$ need not be closed. [Hint: Consider $G = \mathbb{R}$.]

6.2 Character groups and Pontrjagin duality

In this section, we introduce the character group of an LCA group and state (without proof) the celebrated Pontrjagin–van Kampen duality theorem and related results, which will be needed. By a **character** of an LCA group G, we mean a map $\chi \in \mathrm{Hom}(G, \mathbb{T})$. The sum of characters χ_1 and χ_2 of G is defined by

$$(\chi_1 + \chi_2)(x) = \chi_1(x) \cdot \chi_2(x) \quad (x \in G).$$

With this operation "+", the set of all characters of G is an abelian group; it is denoted by $\widehat{G}$ and is called the **character group** or **dual (group)** of G. We equip $\widehat{G}$ with the **compact-open topology**: For every compact subset F of G and $\varepsilon > 0$ let

$$P(F, \varepsilon) = \{\chi \in \widehat{G} : \chi(F) \subseteq \mathbb{T}_\varepsilon\},$$

and take the set $\{P(F, \varepsilon) : F \text{ compact } \subseteq G \text{ and } \varepsilon > 0\}$ to be an open basis at the identity 0 in $\widehat{G}$. The topology of $\widehat{G}$ is designed to turn $\widehat{G}$ into an LCA group (see for instance [121, Theorem 11]):

Theorem 6.2.1. *Suppose that G is an LCA group. If U is any neighborhood of $0 \in G$ with $\overline{U}$ compact and $0 < \varepsilon < \sqrt{2}$, then $P(\overline{U}, \varepsilon)$ is compact. Thus, $\widehat{G}$ is an LCA group.*

Next we describe some examples of dual groups:

(a) The map $(\mathbb{Z}/n\mathbb{Z})\widehat{\ } \to \mathbb{T}$ with $\chi \mapsto \chi(1)$ yields $(\mathbb{Z}/n\mathbb{Z})\widehat{\ } \cong \mathbb{Z}/n\mathbb{Z}$.

(b) The map $\varphi : \mathbb{T} \to \widehat{\mathbb{Z}}$ defined by $t \mapsto \chi_t$ with $\chi_t : z \mapsto t^z$ is an algebraic isomorphism. To show that φ is continuous, it suffices to prove continuity at 0. Let $\varepsilon > 0$ and F a compact (hence finite) subset of $\mathbb{Z}$. If $F \neq \{0\}$, then for $m = \max\{|z| : z \in F\}$, we have $\varphi(\mathbb{T}_{\varepsilon/m}) \subseteq P(F, \varepsilon)$. Thus, φ is continuous at 0, so by Theorem 6.1.2, we obtain

$$\widehat{\mathbb{Z}} \cong \mathbb{T}.$$

(c) The map $\psi : \mathbb{R} \to \widehat{\mathbb{R}}$ given by $r \mapsto \eta_r$ with $\eta_r : x \mapsto \exp(irx)$ is an algebraic isomorphism (see [41, 23.27(e)]). To prove that ψ is continuous at 0, let $\varepsilon > 0$ and F a compact subset of $\mathbb{R}$. Then $F \subseteq [-c, c]$ for some nonzero $c \in \mathbb{R}$, and we have $\psi((-\frac{\varepsilon}{c}, \frac{\varepsilon}{c})) \subseteq P(F, \varepsilon)$. Again, we apply Theorem 6.1.2 and conclude that

$$\widehat{\mathbb{R}} \cong \mathbb{R}.$$

(d) The dual of the discrete group $\mathbb{Q}$ is simply denoted by $\widehat{\mathbb{Q}}$.

Pontrjagin [126] proved that G and $\widehat{G}$ have the same weight whenever G is an LCA group (see also [41, Theorem 24.14]):

Theorem 6.2.2. *If G is an LCA group, then $w(G) = w(\widehat{G})$.*

We state the famous duality theorem for LCA groups [41, Theorem 24.8], which allows us to view an LCA group G as the dual group of $\widehat{G}$:

Theorem 6.2.3 (**Pontrjagin–van Kampen duality theorem**). *Let G be an LCA group. Then the map*

$$\begin{array}{rcccl} \Phi: & G & \longrightarrow & (\widehat{G})\widehat{} & \\ & x & \longmapsto & x': & \widehat{G} \to \mathbb{T} \\ & & & & \chi \mapsto \chi(x) \end{array}$$

is a topological isomorphism.

A proof of the duality theorem for discrete torsion groups and their duals can be found in Fuchs [26]. Van Kampen [77] proved that LCA groups have enough characters to separate points, a fact which can be easily recovered from the duality theorem:

Theorem 6.2.4. *Let G be an LCA group and $x, y \in G$ with $x \neq y$. Then there exists a $\chi \in \widehat{G}$ such that $\chi(x) \neq \chi(y)$.*

Let G, H be LCA groups and $\phi \in \mathrm{Hom}(G, H)$. Then ϕ induces its **adjoint map**

$$\begin{array}{rccl} \phi^*: & \widehat{H} & \to & \widehat{G} \\ & \chi & \mapsto & \chi\phi, \end{array}$$

which is a continuous homomorphism, that is, $\phi^* \in \mathrm{Hom}(\widehat{H}, \widehat{G})$. The following properties are easily verified:

(i) $\phi^{**} = \phi$ [here we identify $(\widehat{G})\widehat{}$ and $(\widehat{H})\widehat{}$ with G and H, respectively].

(ii) $(\psi\phi)^* = \phi^*\psi^*$ whenever $\psi \in \mathrm{Hom}(H, K)$.

(iii) $(n\phi)^* = n\phi^*$ for each $n \in \mathbb{N}$.

(iv) $(1_G)^* = 1_{\widehat{G}}$.

(v) ϕ is a topological isomorphism if and only if ϕ^* is a topological isomorphism.

Exercises.

1. Let $\phi \in \mathrm{Hom}(G, H)$. Then:
   (a) If ϕ is surjective, then ϕ^* is injective.
   (b) If ϕ is an open map and injective, then ϕ^* is surjective.
2. (Dikranjan [18]) Let G be a discrete group, and consider the group $\widetilde{G} = \mathrm{Hom}(G, \mathbb{T})$ equipped with the relative topology as a subspace of the compact group $\mathbb{T}^G$. Prove:
   (a) $\widetilde{G}$ is closed in $\mathbb{T}^G$.
   (b) The topology of $\widetilde{G}$ coincides with the compact-open topology of $\widehat{G}$.
3. Let G be a totally disconnected LCA group. Prove that $\ker\chi$ is open in G for every $\chi \in \widehat{G}$.

6.3 Discrete and compact groups

Theorem 6.3.1. *The dual of a discrete group is compact, and the dual of a compact group is discrete.*

Proof. Suppose that G is discrete. For any $0 < \varepsilon < \sqrt{2}$, the set $\overline{P(\overline{U}, \varepsilon)}$ is compact whenever U is a neighborhood of $0 \in G$ with $\overline{U}$ compact (Theorem 6.2.1); thus, $\widehat{G} = \overline{P(\{0\}, \varepsilon)}$ is compact.

Now assume that G is compact. Then $P(G, \sqrt{3})$ is an open subset of $\widehat{G}$. Since $\mathbb{T}_{\sqrt{3}}$ does not contain any nontrivial subgroups of $\mathbb{T}$, we have $P(G, \sqrt{3}) = \{0\}$; therefore, $\widehat{G}$ is discrete. □

By the duality theorem we have the following:

Corollary 6.3.2. *A group is discrete (compact) if and only if its dual group is compact (discrete).*

Since a compact group is metrizable if and only if its weight is countable, Corollary 6.3.2 and Theorem 6.2.2 yield:

Corollary 6.3.3. *A discrete group G is countable if and only if $\widehat{G}$ is metrizable.*

Theorem 6.3.4. *Let $\{G_i : i \in I\}$ be a collection of discrete groups. For every $(\chi_i) \in \prod_{i\in I} \widehat{G_i}$ define a map $[\chi_i] : \bigoplus_{i\in I} G_i \to \mathbb{T}$ by $\sum_{k=1}^{n} x_{i_k} \mapsto \prod_{k=1}^{n} \chi_{i_k}(x_{i_k})$ $(x_{i_k} \in G_{i_k}, k = 1, \ldots, n)$. Then $[\chi_i] \in (\bigoplus_{i\in I} G_i)\widehat{}$, and*

$$\begin{array}{rccc} f: & \prod_{i\in I} \widehat{G_i} & \to & (\bigoplus_{i\in I} G_i)\widehat{} \\ & (\chi_i) & \mapsto & [\chi_i] \end{array}$$

is a topological isomorphism.

Proof. It is easily verified that f is an algebraic isomorphism between the two compact groups. To show that f is continuous at 0, let $\varepsilon > 0$ and F a compact (hence finite) subset of $G = \bigoplus_{i\in I} G_i$. Then there is a finite subset J of I such that $F \subseteq \bigoplus_{i\in I} G_i'$ with $G_j' = G_j$ $(j \in J)$ and $G_i' = 0$ $(i \in I \setminus J)$. Letting $\pi_i : G \to G_i$ be the projection map and $n = |J|$, we define $U_i = P(\pi_i(F), \frac{\varepsilon}{n}) \subseteq \widehat{G_i}$. Notice that U_i coincides with $\widehat{G_i}$ for all $i \in I \setminus J$. For any $\chi = (\chi_i) \in U = \prod_{i\in I} U_i$ and $x = (x_i) \in F$, we have $|\chi_j(x_j) - 1| < \frac{\varepsilon}{n}$ if $j \in J$ and $\chi_i(x_i) = 1$ whenever $i \in I \setminus J$. Then

$$|[\chi_i](x) - 1| = \left|\prod_{j\in J} \chi_j(x_j) - 1\right| \leq \sum_{j\in J} |\chi_j(x_j) - 1| < \varepsilon;$$

hence, $f(\chi) \in P(F, \varepsilon)$. Therefore, f is continuous at 0, so by Theorem 6.1.2, f is a topological isomorphism. □

Theorem 6.3.5. *A compact torsion group is bounded. Thus, a discrete group G is bounded if and only if $\widehat{G}$ is torsion.*

Proof. Let C be a compact torsion group, and write $C = \bigcup_{n\in\mathbb{N}} C[n]$. Since each subgroup $C[n]$ of C is closed, the interior of $C[m]$ is nonempty for some $m \in \mathbb{N}$ by Theorem 6.1.1; thus, $C[m]$ is an open subgroup of C. But then $C/C[m]$ is both compact and discrete, hence finite. Therefore, C is bounded. □

Exercises.

1. Let G be a discrete group and p a prime. Then G is a $\mathbb{Z}_p$-module if and only if $\widehat{G}$ is a $\mathbb{Z}_p$-module.
2. Let G be a compact group. Then $\widehat{G}$ is torsion if and only if $\chi(G)$ is finite for all $\chi \in \widehat{G}$.
3. The group $\widehat{\mathbb{Q}}$ is divisible and torsion-free.

6.4 Annihilators

In this section, we introduce and explore annihilators, which are indispensable in the study of LCA groups. Most of the results are taken from Hewitt–Ross [41] and Kiefer [83]. For a subset S of an LCA group G, the **annihilator of** S **in** $\widehat{G}$ is defined to be the set

$$(\widehat{G}, S) = \{\chi \in \widehat{G} : \chi(x) = 1 \text{ for all } x \in S\}.$$

We have:

Lemma 6.4.1. *Let S be a subset of an LCA group G. Then:*

1. *If $S \neq \{0\}$, then $(\widehat{G}, S) \neq \widehat{G}$.*
2. *$(\widehat{G}, S)$ is a closed subgroup of $\widehat{G}$.*
3. *If $S \subseteq T \subseteq G$, then $(\widehat{G}, T) \subseteq (\widehat{G}, S)$.*
4. *$(\widehat{G}, S) = (\widehat{G}, \langle S\rangle) = (\widehat{G}, \overline{\langle S\rangle})$.*

Proof. (1) follows from Theorem 6.2.4. Clearly, $(\widehat{G}, S)$ is a subgroup of $\widehat{G}$. To show that the set $\widehat{G} \setminus (\widehat{G}, S)$ is open, let $\chi \in \widehat{G} \setminus (\widehat{G}, S)$. Then $\chi(s) \neq 1$ for some $s \in S$, so for any $\psi \in U = P(\{s\}, |\chi(s) - 1|)$, we obtain $\chi(s)\psi(s) \neq 1$ since $|\psi(s) - 1| < |\chi(s) - 1|$; thus, $\chi + U \subseteq \widehat{G} \setminus (\widehat{G}, S)$. Therefore, (2) is true.

(3) is obvious. It remains to verify (4). Clearly, $(\widehat{G}, S)$ coincides with $(\widehat{G}, \langle S\rangle)$, so in view of (3), it suffices to show that $(\widehat{G}, \langle S\rangle) \subseteq (\widehat{G}, \overline{\langle S\rangle})$. Let $\chi \in (\widehat{G}, \langle S\rangle)$ and $x \in \overline{\langle S\rangle}$. Then there exists a net $(x_i)_{i\in\Lambda}$ in $\langle S\rangle$ with $x_i \to x$. Since χ is continuous, we have $1 = \chi(x_i) \to \chi(x)$, and therefore, $\chi(x) = 1$. □

Given a closed subgroup A of G, the dual group of G/A can be identified with the annihilator of A in $\widehat{G}$:

Lemma 6.4.2. *Let A be a closed subgroup of an LCA group G. Then for each $\chi \in (G/A)\widehat{}$, we have $\chi\varphi \in (\widehat{G}, A)$, where $\varphi : G \to G/A$ denotes the natural map. The map*

$$\begin{array}{cccc} \psi : & (G/A)\widehat{} & \to & (\widehat{G}, A) \\ & \chi & \mapsto & \chi\varphi \end{array}$$

is a topological isomorphism. If B is a closed subgroup of G containing A, then ψ maps $((G/A)\widehat{\ }, B/A)$ onto $(\widehat{G}, B)$.

Proof. The first claim is obvious. Clearly, the map ψ is a homomorphism, and $\chi \in (G/A)\widehat{\ }$ is the zero map whenever $\chi\varphi$ is; thus, ψ is injective. To show that ψ is surjective, let $\eta \in (\widehat{G}, A)$, and define $\chi : G/A \to \mathbb{T}$ by $\chi(x + A) = \eta(x)$. Then χ is a well-defined map and is a homomorphism. For any $\varepsilon > 0$, the set $U = \eta^{-1}(\mathbb{T}_\varepsilon)$ is open in G. Thus, $\chi^{-1}(\mathbb{T}_\varepsilon) = \{x + A : x \in U\}$ is open in G/A. Therefore, χ is a continuous map, and since $\eta = \chi\varphi$, we conclude that ψ is surjective.

For every compact subset F of G/A, there is a compact subset F' of G such that $\varphi(F') = F$. Thus, ψ maps the set $\{\chi \in (G/A)\widehat{\ } : \chi(F) \subseteq \mathbb{T}_\varepsilon\}$ onto $\{\eta \in (\widehat{G}, A) : \eta(F') \subseteq \mathbb{T}_\varepsilon\}$. Consequently, ψ maps an open basis at 0 in $(G/A)\widehat{\ }$ onto an open basis at 0 in $(\widehat{G}, A)$; thus, ψ is a homeomorphism.

Finally, we need to verify that ψ maps the annihilator $X = ((G/A)\widehat{\ }, B/A)$ onto $(\widehat{G}, B)$. Let $b \in B$. If $\chi \in X$, then $\chi\varphi(b) = \chi(b + A) = 1$; thus, $\psi(X) \subseteq (\widehat{G}, B)$. To show that the reverse inclusion holds as well, assume that $\eta \in (\widehat{G}, B)$. Since ψ is surjective, we have $\eta = \chi\varphi$ for some $\chi \in (G/A)\widehat{\ }$. Then $\chi(b + A) = \eta(b) = 1$ shows that $\chi \in X$; thus, $\eta = \psi(\chi) \in \psi(X)$ as desired. □

Lemma 6.4.3. *Let A be a closed subgroup of an LCA group G. If $x \in G \setminus A$, then there exists an $\eta \in (\widehat{G}, A)$ such that $\eta(x) \neq 1$.*

Proof. By Theorem 6.2.4, there is a $\chi \in (G/A)\widehat{\ }$ such that $\chi(x + A) \neq 1$. Letting $\varphi : G \to G/A$ be the natural map, we have $\eta = \chi\varphi \in (\widehat{G}, A)$ and $\eta(x) = \chi(x+A) \neq 1$, as desired. □

By the duality theorem, we can identify the dual group of $\widehat{G}$ with G. Then any closed subgroup A of G can be viewed as the annihilator of $(\widehat{G}, A)$ in G:

Lemma 6.4.4. *Let A be a closed subgroup of an LCA group G. Then $A = (G, (\widehat{G}, A))$.*

Proof. We claim that the topological isomorphism $\Phi : G \to (\widehat{G})\widehat{\ }$ in Theorem 6.2.3 maps A onto $((\widehat{G})\widehat{\ }, (\widehat{G}, A))$. Thus, we need to show that the sets

$$A' = \bigcup_{a \in A} \{f_a \in (\widehat{G})\widehat{\ } : f_a(\chi) = \chi(a) \text{ for all } \chi \in \widehat{G}\}$$

and

$$G' = \bigcup_{x \in G} \{f_x \in ((\widehat{G})\widehat{\ }, (\widehat{G}, A)) : f_x(\chi) = \chi(x) \text{ for all } \chi \in \widehat{G}\}$$

coincide. If $f_a \in A'$ and $\chi \in (\widehat{G}, A)$, then $f_a(\chi) = \chi(a) = 1$; thus, $f_a \in G'$. Now assume that there is an $f_x \in G' \setminus A'$. Then $x \in G \setminus A$, so by Lemma 6.4.3, there is an $\eta \in (\widehat{G}, A)$ with $\eta(x) \neq 1$. It follows that $f_x(\eta) = \eta(x) \neq 1$; thus, $f_x \notin ((\widehat{G})\widehat{\ }, (\widehat{G}, A))$, which is impossible. Thus, $A' = G'$. □

In view of Lemma 6.4.1(3), we obtain:

Corollary 6.4.5. *Let A and B be closed subgroups of an LCA group G. Then $A \subseteq B$ if and only if $(\widehat{G}, B) \subseteq (\widehat{G}, A)$. Consequently, $A = B$ if and only if $(\widehat{G}, A) = (\widehat{G}, B)$.*

Since $\widehat{\mathbb{T}} \cong \mathbb{Z}$ and factor groups of $\mathbb{Z}$ are cyclic, we have:

Corollary 6.4.6. *Let A be a proper closed subgroup of $\mathbb{T}$. Then A is finite and cyclic.*

The following result can be found in Armacost [3]:

Proposition 6.4.7. *Let G be an LCA group and $\chi \in \widehat{G}$. Then $\ker\chi = (G, \{\chi\})$ and $\overline{\langle\chi\rangle} = (\widehat{G}, \ker\chi)$.*

Proof. Let $\Phi : G \to (\widehat{G})\widehat{}$ be the topological isomorphism given by $x \mapsto x'$ as in Theorem 6.2.3. Then $\Phi(\ker\chi) = \{x' \in (\widehat{G})\widehat{} : \chi(x) = 1\} = \{x' \in (\widehat{G})\widehat{} : x'(\chi) = 1\} = ((\widehat{G})\widehat{}, \{\chi\})$, and so, the first assertion follows. Using this result and Lemma 6.4.4, we obtain $\overline{\langle\chi\rangle} = (\widehat{G}, (G, \overline{\langle\chi\rangle})) = (\widehat{G}, \ker\chi)$, as claimed. □

Next we describe the annihilators of image and kernel of a continuous homomorphism between LCA groups:

Proposition 6.4.8. *Let $\phi \in \mathrm{Hom}(G, H)$, where G and H are LCA groups. Then $\ker\phi^* = (\widehat{H}, \mathrm{im}\,\phi)$ and $\overline{\mathrm{im}\,\phi^*} = (\widehat{G}, \ker\phi)$.*

Proof. Letting $\chi \in \widehat{H}$, we have $\phi^*(\chi) = 0$ exactly if $\chi(\phi(x)) = 1$ for all $x \in G$; thus, the first claim is evident. Then $\ker\phi = (G, \mathrm{im}\,\phi^*)$, so by Lemma 6.4.4, we obtain $\overline{\mathrm{im}\,\phi^*} = (\widehat{G}, (G, \mathrm{im}\,\phi^*)) = (\widehat{G}, \ker\phi)$, so the second assertion holds. □

To prove that a closed subgroup A of G splits from G exactly if its annihilator $(\widehat{G}, A)$ splits from $\widehat{G}$, the following fact will be needed:

Theorem 6.4.9 (Weil [156]). *Let A be a closed subgroup of an LCA group G. Then A splits from G if and only if there is a continuous homomorphism $\alpha : G \to A$ such that $\alpha \restriction A = 1_A$. In this case, we have $G = A \bigoplus \ker\alpha$.*

Proof. Suppose that $G = A \bigoplus B$ for some closed subgroup B of G. Let $\phi : A \times B \to G$ be the natural topological isomorphism and $\pi : A \times B \to A$ the projection map. Then $\alpha = \pi\phi^{-1} \in \mathrm{Hom}(G, A)$ with $\alpha \restriction A = 1_A$ and $\ker\alpha = B$.

Conversely, assume that there is a map $\alpha \in \mathrm{Hom}(G, A)$ with $\alpha \restriction A = 1_A$, and let $B = \ker\alpha$. Clearly, we have $G = A \oplus B$ since $A \cap B = 0$ and each element $x \in G$ can be written as $x = \alpha(x) + (x - \alpha(x)) \in A + B$. Then the natural map $\phi : A \times B \to G$ is a continuous algebraic isomorphism. We need to prove that ϕ^{-1} is continuous. It suffices to show continuity at 0. Therefore, assume that $(x_i)_{i\in\Lambda}$ is a net in G with $x_i \to 0$. Then $\alpha(x_i) \to 0$ in A, and $x_i - \alpha(x_i) \to 0$ in B; thus, $\phi^{-1}(x_i) = (\alpha(x_i), x_i - \alpha(x_i)) \to (0, 0)$ in $A \times B$. Consequently, ϕ^{-1} is continuous at 0, and we obtain $G = A \bigoplus B$. □

Corollary 6.4.10. *Let A be a closed subgroup of an LCA group G. Then A splits from G if and only if $(\widehat{G}, A)$ splits from $\widehat{G}$.*

Proof. If A splits from G, then by Theorem 6.4.9, there is an $\alpha \in \mathrm{Hom}(G, A)$ with $\alpha \restriction A = 1_A$. Define a map $\beta : G \to G$ by $x \mapsto x - \alpha(x)$. Then $\beta \in \mathrm{Hom}(G, G)$, and the adjoint map β^* satisfies $\mathrm{im}\, \beta^* \subseteq (\widehat{G}, A)$. For any $x \in G$ and $\chi \in (\widehat{G}, A)$, we have $\chi\beta(x) = \chi(x)$; thus, $\beta^* \restriction (\widehat{G}, A) = 1_{(\widehat{G},A)}$. Therefore, $(\widehat{G}, A)$ splits from $\widehat{G}$. For the converse use Lemma 6.4.4. □

Corollary 6.4.11. *If $\mathcal{C}$ is any class of LCA groups that is closed under direct factors, then so is the dual class $\mathcal{C}^* = \{\widehat{C} : C \in \mathcal{C}\}$.*

The dual of a closed subgroup A of G can be expressed in terms of its annihilator in $\widehat{G}$:

Lemma 6.4.12. *Let A be a closed subgroup of an LCA group G. Then every character of A can be written as $\chi\sigma$ for some $\chi \in \widehat{G}$ with $\sigma : A \to G$ the inclusion map. The map*

$$\begin{array}{rccc} \phi : & \widehat{G}/(\widehat{G}, A) & \to & \hat{A} \\ & \chi + (\widehat{G}, A) & \mapsto & \chi\sigma \end{array}$$

is a topological isomorphism. If S is a subset of A, then ϕ maps $(\widehat{G}, S)/(\widehat{G}, A)$ onto $(\hat{A}, S)$.

Proof. Again, we identify the dual group of $\widehat{G}$ with G and have $A = (G, (\widehat{G}, A))$ by Lemma 6.4.4. Let $\varphi : \widehat{G} \to \widehat{G}/(\widehat{G}, A)$ be the natural map. Then by Lemma 6.4.2,

$$\begin{array}{rccc} \tau : & (G, (\widehat{G}, A)) & \to & (\widehat{G}/(\widehat{G}, A))\widehat{\ } \\ & \chi\varphi & \mapsto & \chi \end{array}$$

is a topological isomorphism. Since the adjoint map φ^* sends χ to $\chi\varphi$, the composite $\sigma = \varphi^*\tau : A \to G$ is the inclusion map. Its adjoint map $\sigma^* = \tau^*\varphi : \widehat{G} \to \hat{A}$ is a continuous, open epimorphism with $\ker \sigma^* = (\widehat{G}, A)$; thus, we have a topological isomorphism

$$\begin{array}{rccc} \phi : & \widehat{G}/(\widehat{G}, A) & \to & \hat{A} \\ & \chi + (\widehat{G}, A) & \mapsto & \sigma^*(\chi) \end{array}$$

as claimed. For any subset S of A, we have $(\widehat{G}, A) \subseteq (\widehat{G}, S)$ by Lemma 6.4.1(3). Using the fact that every character of A has the form $\chi\sigma$ for some $\chi \in \widehat{G}$, we obtain $\phi((\widehat{G}, S)/(\widehat{G}, A)) = \sigma^*((\widehat{G}, S)) = \{\chi\sigma : \chi \in (\widehat{G}, S)\} = (\hat{A}, S)$. □

Annihilators of sums and intersections of closed subgroups can be described as follows:

Lemma 6.4.13. *If $\{A_i : i \in I\}$ is a nonempty collection of closed subgroups of an LCA group G, then $(\widehat{G}, \overline{\sum_{i\in I} A_i}) = \bigcap_{i\in I}(\widehat{G}, A_i)$ and $(\widehat{G}, \bigcap_{i\in I} A_i) = \overline{\sum_{i\in I}(\widehat{G}, A_i)}$.*

Proof. If $\chi \in \bigcap_{i\in I}(\widehat{G}, A_i)$, then $\chi(a) = 1$ for all $a \in \sum_{i\in I} A_i$. By Lemma 6.4.1(3), the first equality is evident. Using this result and Lemma 6.4.1(4), we obtain

$$\Big(\widehat{G}, \bigcap_{i\in I}(G, (\widehat{G}, A_i))\Big) = \Big(\widehat{G}, \Big(G, \sum_{i\in I}(\widehat{G}, A_i)\Big)\Big) = \Big(\widehat{G}, \Big(G, \overline{\sum_{i\in I}(\widehat{G}, A_i)}\Big)\Big).$$

Then Lemma 6.4.4 shows that $(\widehat{G}, \bigcap_{i\in I} A_i) = \overline{\sum_{i\in I}(\widehat{G}, A_i)}$. □

For a subgroup A of G and $n \in \mathbb{N}$, we define $n^{-1}A = \{x \in G : nx \in A\}$. Notice that $n^{-1}A$ is closed in G whenever A is since the map $n1_G$ is continuous. The following useful fact can be found in [96]:

Lemma 6.4.14. *If A is a closed subgroup of an LCA group G and $n \in \mathbb{N}$, then $(\widehat{G}, n^{-1}A) = n(\widehat{G}, A)$ and $(\widehat{G}, nA) = n^{-1}(\widehat{G}, A)$.*

Proof. Let $\varphi : G \to G/A$ be the natural map. By Lemma 6.4.2, its adjoint map φ^* satisfies $\operatorname{im}\varphi^* = (\widehat{G}, A)$; thus, $\operatorname{im}(n\varphi)^* = \operatorname{im}(n\varphi^*) = n(\widehat{G}, A)$. Using Proposition 6.4.8 and Lemma 6.4.4, we obtain $(\widehat{G}, \ker(n\varphi)) = (\widehat{G}, (G, \operatorname{im}(n\varphi)^*)) = (\widehat{G}, (G, n(\widehat{G}, A))) = n(\widehat{G}, A)$. Since $n^{-1}A = \ker(n\varphi)$, the first assertion follows.

To show the second claim, we use our first result and apply Lemma 6.4.4 once more. Then $(\widehat{G}, nA) = (\widehat{G}, n(G, (\widehat{G}, A))) = (\widehat{G}, (G, n^{-1}(\widehat{G}, A))) = n^{-1}(\widehat{G}, A)$. □

Since $G[n] = n^{-1}\{0\}$, we obtain:

Lemma 6.4.15. *For any LCA group G and $n \in \mathbb{N}$, we have $(\widehat{G}, G[n]) = \overline{n\widehat{G}}$ and $(\widehat{G}, nG) = \widehat{G}[n]$.*

The annihilators of the subgroup bG of compact elements of G and of the identity component G_0 are described as follows:

Theorem 6.4.16. *If G is an LCA group, then $\widehat{G}_0 = (\widehat{G}, bG)$ and $b\widehat{G} = (\widehat{G}, G_0)$.*

Proof. By Lemma 6.4.4, it suffices to verify the first equation. Let $\{C_i : i \in I\}$ be the collection of all compact subgroups of G. If $i \in I$, then $(\widehat{G}, C_i)$ is open in $\widehat{G}$ since $\widehat{G}/(\widehat{G}, C_i) \cong (C_i)\widehat{}$ is discrete (Lemma 6.4.12 and Theorem 6.3.1). Conversely, if A is an open subgroup of $\widehat{G}$, then $A = (\widehat{G}, (G, \widehat{A}))$ and $(G, \widehat{A}) \cong (\widehat{G}/A)\widehat{}$ is compact (Lemma 6.4.2 and Theorem 6.3.1). Since $\widehat{G}_0$ is the intersection of all open subgroups of $\widehat{G}$, it follows that $\widehat{G}_0 = \bigcap_{i\in I}(\widehat{G}, C_i) = (\widehat{G}, \sum_{i\in I} C_i) = (\widehat{G}, bG)$. □

Since $bG = tG$ whenever G is a discrete group, we obtain:

Theorem 6.4.17. *If G is discrete, then $\widehat{G}_0 = (\widehat{G}, tG)$ and $tG = (G, \widehat{G}_0)$. Thus, a discrete group G is torsion if and only if $\widehat{G}$ is totally disconnected.*

Proposition 6.4.18. *Let G be an LCA group. If $\chi \in (\widehat{G}, G_0)$, then $\ker\chi$ is open.*

Proof. Letting $\varphi : G \to G/G_0$ be the natural map, we can write $\chi = \chi'\varphi$ for some $\chi' \in (G/G_0)\widehat{}$ by Lemma 6.4.2. Let U be a neighborhood of 0 in G/G_0 such that $\chi'(U) \subseteq \mathbb{T}_{\sqrt{3}}$. Since G/G_0 is totally disconnected, U contains a compact open subgroup K/G_0 of G/G_0. On the other hand, $\mathbb{T}_{\sqrt{3}}$ does not contain any nontrivial subgroups. Hence, $K/G_0 \subseteq \ker\chi'$, and therefore, $\ker\chi'$ is open. But then $\ker\chi$ is open since it contains $\varphi^{-1}(\ker\chi')$. □

Theorem 6.4.19. *A discrete or compact group G is torsion-free if and only if $\widehat{G}$ is divisible. A compact group is divisible if and only if it is connected.*

Proof. Let $n \in \mathbb{N}$. Then $G[n] = (G, n\widehat{G})$ and $\overline{n\widehat{G}} = (\widehat{G}, G[n])$ by Lemma 6.4.15. Since $\widehat{G}$ is compact or discrete (Theorem 6.3.1), and the map $n1_{\widehat{G}}$ is continuous, we have $\overline{n\widehat{G}} = n\widehat{G}$. It follows that $G[n] = 0$ if and only if $n\widehat{G} = \widehat{G}$. Thus, G is torsion-free if and only if $\widehat{G}$ is divisible.

Now suppose G is compact. Using Theorem 6.4.17, we obtain $G_0 = (G, t\widehat{G}) = (G, \sum_{n\in\mathbb{N}} \widehat{G}[n]) = \bigcap_{n\in\mathbb{N}}(G, \widehat{G}[n]) = \bigcap_{n\in\mathbb{N}} nG$; hence, $G = G_0$ exactly if G is divisible. □

Corollary 6.4.20. *If G is discrete and torsion-free and A is a subgroup of G, then A is pure in G if and only if $(\widehat{G}, A)$ is connected.*

Proof. A subgroup A of a torsion-free group G is pure exactly if G/A is torsion-free. Now use Theorem 6.4.19 and Lemma 6.4.2. □

A compact group C is called **coreduced** if it has no proper closed subgroup A such that C/A is torsion-free. Then:

Corollary 6.4.21. *A discrete group G is reduced if and only if $\widehat{G}$ is coreduced.*

Proof. A subgroup A of G is divisible if and only if $\widehat{A} \cong \widehat{G}/(\widehat{G}, A)$ is torsion-free (Theorem 6.4.19 and Lemma 6.4.12). □

The following characterizations of direct sums will be useful:

Lemma 6.4.22. *Let A be a subgroup of a discrete group G, and let I be a finite set. Assume that A_i are subgroups of A ($i \in I$), and define $H_i = \sum_{j\neq i} A_j$ for each $i \in I$. Then the following statements are equivalent:*

1. $A = \bigoplus_{i\in I} A_i$;
2. $A = H_i \oplus \bigcap_{j\neq i} H_j$ *for all* $i \in I$;
3. $\widehat{G}/(\widehat{G}, A) = \bigoplus_{i\in I}(\widehat{G}, H_i)/(\widehat{G}, A)$.

If (3) holds, then for each $i \in I$, the annihilator $(\widehat{G}, A_i)$ is the kernel of the natural map $\phi_i : \widehat{G} \to (\widehat{G}, H_i)/(\widehat{G}, A)$.

Proof. It is clear that (1) ⇔ (2). By Lemma 6.4.4, (2) is equivalent to $\widehat{G} = (\widehat{G}, \bigcap_{i\in I} H_i) = \sum_{i\in I}(\widehat{G}, H_i)$ and $(\widehat{G}, H_i) \cap \sum_{j\neq i}(\widehat{G}, H_j) = (\widehat{G}, A)$ for all $i \in I$. Thus, (2) ⇔ (3). To verify the last assertion, let $i \in I$. Then $\ker \phi_i = \sum_{j\neq i}(\widehat{G}, H_j) = (\widehat{G}, \bigcap_{j\neq i} H_j) = (\widehat{G}, A_i)$. □

Exercises.

1. Let A be a closed subgroup of an LCA group G. Then A is open (compact) if and only if $(\widehat{G}, A)$ is compact (open).
2. (Weil [156]) Let A and B be closed subgroups of an LCA group G such that $G = A \oplus B$. If A is open in G, then $G = A \bigoplus B$.
3. (Armacost [3]) Let D be a closed divisible subgroup of an LCA group G, and let U be an open subgroup of G with $D \subseteq U$. Show: If D splits from U, then D splits from G. In particular, any open divisible subgroup of G splits from G.

4. (Robertson [133]) Let $G = \mathbb{R}\times(\mathbb{Q}/\mathbb{Z})$. Then $A = \{(x, x+\mathbb{Z}) : x \in \mathbb{Q}\}$ is a discrete (and thus, closed) subgroup of G, and $A \cong \mathbb{Q}$, but A does not split from G.
5. (Preston [128]) Let A be a closed subgroup of a totally disconnected LCA group G. If A is discrete divisible, then A splits from G.
6. Let A be a closed subgroup of an LCA group G and $\chi \in \widehat{A}$. Then χ can be extended to a character $\chi' \in \widehat{G}$. If $g \in G \setminus A$, then χ' can be chosen such that $\chi'(g) \neq 1$.
7. Let G be a discrete $\mathbb{Z}_p$-module and A a subgroup of G. Then A is a submodule of G if and only if $(\widehat{G}, A)$ is a submodule of the $\mathbb{Z}_p$-module $\widehat{G}$ (see Section 6.3, Exercise 1).
8. Let G be a compact group and p a prime. Then $\bigcap_{n\in\mathbb{N}} p^n G = 0$ if and only if $\widehat{G}$ is p-torsion.
9. Recall that for a given subgroup A of a discrete group G, A^0 denotes the group $\{x \in G : nx \in A \text{ for some } n \in \mathbb{N}\}$. Show that $(\widehat{G}, A^0) = (\widehat{G}, A)_0$.

6.5 Proper homomorphisms and extensions of LCA groups

A continuous homomorphism $\phi : G \to H$ is called **proper** if $\phi(U)$ is open in $\operatorname{im}\phi$ for every open subset U of G. If $\phi \in \operatorname{Hom}(G, H)$ is proper, then $\operatorname{im}\phi \cong G/\ker\phi$; thus, $\operatorname{im}\phi$ is an LCA group and is, therefore, a closed subgroup of H.

Lemma 6.5.1. *Let $\phi \in \operatorname{Hom}(G, H)$ be proper, where G and H are LCA groups. Then we have the following:*
1. *ϕ^* is proper.*
2. *$\operatorname{im}\phi^* = (\widehat{G}, \ker\phi)$.*
3. *ϕ is injective if and only if ϕ^* is surjective.*
4. *ϕ is surjective if and only if ϕ^* is injective.*

Proof. Let $\nu : G/\ker\phi \to \operatorname{im}\phi = \overline{\operatorname{im}\phi}$ be the natural topological isomorphism, $\sigma : \operatorname{im}\phi \to H$ the inclusion map, and $\varphi : G \to G/\ker\phi$ the natural map. By Lemmas 6.4.2 and 6.4.12, we have topological isomorphisms

$$\widehat{H}/(\widehat{H}, \operatorname{im}\phi) \xrightarrow{\alpha} (\operatorname{im}\phi)\widehat{\ } \xrightarrow{\nu^*} (G/\ker\phi)\widehat{\ } \xrightarrow{\psi} (\widehat{G}, \ker\phi)$$

with $\alpha : \chi + (\widehat{H}, \operatorname{im}\phi) \mapsto \chi\sigma$, $\nu^* : \xi \mapsto \xi\nu$ and $\psi : \gamma \mapsto \gamma\varphi$. Then $\tau = \psi\nu^*\alpha$ is a topological isomorphism with $\tau : \chi + (\widehat{H}, \operatorname{im}\phi) \mapsto \chi\sigma\nu\varphi = \chi\phi$. Since the natural map $\varphi' : \widehat{H} \to \widehat{H}/(\widehat{H}, \operatorname{im}\phi)$ is open, the homomorphism $\tau\varphi'$ is open. Finally, we let $\sigma' : (\widehat{G}, \ker\phi) \to \widehat{G}$ be the inclusion map and obtain the proper homomorphism $\phi^* = \sigma'\tau\varphi' : \widehat{H} \to \widehat{G}$, as desired.

Next we show (2). Since ϕ^* is proper by (1), $\operatorname{im}\phi^*$ is closed in $\widehat{G}$. By Proposition 6.4.8, the proof is complete.

For (3) and (4), use (2) and Proposition 6.4.8. □

A short exact sequence of LCA groups

$$E : 0 \to A \xrightarrow{\mu} B \xrightarrow{\nu} C \to 0$$

with proper homomorphisms μ and ν is called a **proper exact sequence.** In this case, we call E an **extension** of A by C and have $A \cong \mu(A)$ and $C \cong B/\mu(A)$. For example, given any LCA group G and a closed subgroup A, we obtain the proper exact sequence $0 \to A \to G \to G/A \to 0$. By Proposition 6.4.8 and Lemma 6.5.1, the following result is evident:

Proposition 6.5.2 (Moskowitz [122]). *Let* $E : 0 \to A \xrightarrow{\mu} B \xrightarrow{\nu} C \to 0$ *be an exact sequence of LCA groups. Then E is proper exact if and only if its dual sequence* $\widehat{E} : 0 \to \widehat{C} \xrightarrow{\nu^*} \widehat{B} \xrightarrow{\mu^*} \widehat{A} \to 0$ *is proper exact.*

Fulp and Griffith [35] defined two proper exact sequences $E : 0 \to A \xrightarrow{\mu} B \xrightarrow{\nu} C \to 0$ and $E' : 0 \to A \xrightarrow{\mu'} B' \xrightarrow{\nu'} C \to 0$ to be **equivalent**, written $E \equiv E'$, if there is a continuous homomorphism $\beta : B \to B'$ such that the diagram

$$\begin{array}{ccccccccc} 0 & \to & A & \xrightarrow{\mu} & B & \xrightarrow{\nu} & C & \to & 0 \\ & & \| & & \downarrow\beta & & \| & & \\ 0 & \to & A & \xrightarrow{\mu'} & B' & \xrightarrow{\nu'} & C & \to & 0 \end{array}$$

is commutative. By a version of the five Lemma 1.2.3 in terms of proper homomorphisms (for details see [35, Proposition 2.8]), the map β is automatically a topological isomorphism, so $\equiv$ is an equivalence relation. As in the discrete case, the set $\mathrm{Ext}(C, A)$ of all equivalence classes of extensions of A by C can be turned into an abelian group:

Theorem 6.5.3 (Fulp–Griffith [35]). *Let A and C be LCA groups. Then the set* $\mathrm{Ext}(C, A)$ *is an abelian group under the operation defined by* $[E_1] + [E_2] = [E_1 + E_2]$, *where*

$$E_1 + E_2 = \nabla_A(E_1 \oplus E_2)\Delta_C$$

with extensions E_1 *and* E_2 *of A by C. The identity of* $\mathrm{Ext}(C, A)$ *is the equivalence class of the split extension* $E_0 : 0 \to A \to A \times C \to C \to 0$.

By Proposition 6.5.2, we have the following:

Theorem 6.5.4 (Fulp–Griffith [35]). *Let A and C be LCA groups. Then we have an isomorphism* $\mathrm{Ext}(C, A) \xrightarrow{\sim} \mathrm{Ext}(\widehat{A}, \widehat{C})$ *given by* $E \mapsto \widehat{E}$.

Exercises.

1. Show that a proper exact sequence $0 \to A \xrightarrow{\mu} B \to C \to 0$ of LCA groups is equivalent to the split extension $0 \to A \to A \times C \to C \to 0$ if and only if $\mu(A)$ splits from B.

2. (Armacost [3]) The exact sequence $0 \to \mathbb{Z} \xrightarrow{\mu} \mathbb{R}_d \xrightarrow{\nu} \mathbb{T} \to 0$ with $\mu = 1_{\mathbb{Z}}$ and $\nu : x \mapsto \exp(2\pi i x)$ gives rise to the dual sequence $0 \to \widehat{\mathbb{T}} \xrightarrow{\nu^*} (\mathbb{R}_d)\widehat{} \xrightarrow{\mu^*} \widehat{\mathbb{Z}} \to 0$, which is not exact.
3. (Loth [108]) An extension $0 \to A \to B \to C \to 0$ of LCA groups is called **almost pure** if A is almost pure in B; that is, if $A \cap \overline{nB} \subseteq \overline{nA}$ for all $n \in \mathbb{N}$. Let Apext(C, A) denote the set of all elements of Ext(C, A) represented by almost pure extensions.
(a) Show that an extension equivalent to an almost pure extension is again almost pure and that Apext(C, A) is a subgroup of Ext(C, A).
(b) Show that the annihilator of a closed almost pure subgroup of an LCA group is almost pure in the dual group. Conclude that Apext$(C, A) \cong$ Apext$(\widehat{A}, \widehat{C})$.
4. Let A and C be LCA groups. If C is discrete, then Ext$(C, A) \cong$ Ext(C, A_d).
5. Let C be a compact group. Prove the following:
(a) Ext$(C, A) = 0$ for all compact groups A if and only if C is torsion-free.
(b) Ext$(A, C) = 0$ for all compact groups A if and only if $C \cong \mathbb{T}^{\mathfrak{m}}$ for some cardinal $\mathfrak{m}$.
[Hint: Section 1.16, Exercises 6 and 7.]

6.6 Topological *p*-groups

Let G be an LCA group and p a prime. Then the ***p*-component of** G is defined to be the set

$$G_{(p)} = \left\{x \in G : \lim_{n\to\infty} p^n x = 0\right\}.$$

It is clear that $G_{(p)}$ is a subgroup of G. Following Vilenkin [147] and Robertson [133], we call an LCA group G a **topological *p*-group** if $G = G_{(p)}$. Independently from Vilenkin [147], Braconnier [10] investigated the same class of groups, calling them *p-primary groups* (see Exercise 2). The following facts are easily verified:

(a) The class of topological p-groups is closed under taking closed subgroups and factor groups by closed subgroups.
(b) A discrete group is a topological p-group if and only if it is a p-group.
(c) The compact group $\prod_{n=1}^{\infty} \mathbb{Z}/p^n\mathbb{Z}$ is a topological p-group whose torsion part is a proper dense subgroup.

The p-component of an LCA group need not be closed. For instance, the p-component of the circle group $\mathbb{T}$ is a proper dense subgroup since it coincides with the p-torsion part of $\mathbb{T}$:

Lemma 6.6.1 (Armacost [3]). *If p is a prime, then $\mathbb{T}_{(p)} = t_p(\mathbb{T})$.*

Proof. Identifying $\mathbb{T}$ with the additive group $\mathbb{R}/\mathbb{Z}$, let $x \in [0, 1)$, and assume that $\lim_{n\to\infty}(p^n x + \mathbb{Z}) = 0 + \mathbb{Z}$. We need to show that $p^{n_0} x \in \mathbb{Z}$ for some $n_0 < \omega$. Let

$\varepsilon = \frac{1}{3p}$, and write $p^n x = z_n + r_n$ with $z_n \in \mathbb{Z}$, $r_n \in \mathbb{R}$ $(n < \omega)$. Then there is an $n_0 < \omega$ such that $|r_n| < \varepsilon$ for all $n \geq n_0$. Since $pz_n + pr_n = p^{n+1}x = z_{n+1} + r_{n+1}$, we obtain

$$|pz_n - z_{n+1}| = |pr_n - r_{n+1}| \leq |pr_n| + |r_{n+1}| < (p+1)\varepsilon < 2/3$$

for all $n \geq n_0$. For such n, we have $pz_n - z_{n+1} = 0$ since $pz_n - z_{n+1} \in \mathbb{Z}$; thus, $pr_n = r_{n+1}$. It follows that $|p^k r_{n_0}| = |r_{n_0+k}| < \varepsilon$ for all $k \in \mathbb{N}$. But this implies that $r_{n_0} = 0$, hence $p^{n_0}x \in \mathbb{Z}$. □

Proposition 6.6.2. *For an LCA group G and a prime p, consider the following properties:*
1. *G is a topological p-group.*
2. *If $\chi \in \widehat{G}$, then $\operatorname{im}\chi \subseteq t_p(\mathbb{T})$.*
3. *G is totally disconnected.*

Then we have: (1) ⇒ (2) ⇒ (3).

Proof. Let $\chi \in \widehat{G}$ and $x \in G$. If (1) holds, then $\chi(x) \in \mathbb{T}_{(p)} = t_p(\mathbb{T})$ by Lemma 6.6.1, so we obtain $\operatorname{im}\chi \subseteq t_p(\mathbb{T})$ as desired. Suppose that (2) holds. If $G_0 \neq 0$, then $\chi(G_0) \neq 0$ for some $\chi \in \widehat{G}$ (Theorem 6.2.4). By (2), this is impossible since the continuous image of a connected space is connected and $\mathbb{T}$ has no proper connected nontrivial subgroups. Thus, $G_0 = 0$. □

Our next result shows that the converse of (1) ⇒ (2) in the above proposition holds as well. We follow the proof given by Armacost [3].

Theorem 6.6.3 (Robertson [133]). *Let G be an LCA group and p a prime. Then the following conditions are equivalent:*
1. *G is a topological p-group;*
2. *For each $\chi \in \widehat{G}$, we have $\operatorname{im}\chi \subseteq t_p(\mathbb{T})$;*
3. *G has arbitrarily small compact open subgroups such that the corresponding factor groups are discrete p-groups.*

Proof. (1) ⇒ (2) was shown in Proposition 6.6.2.

Now suppose that (2) holds. Then G is totally disconnected (Proposition 6.6.2), so G has arbitrarily small compact open subgroups. Letting C be any compact open subgroup of G, we will show that G/C is a p-group. Assume that G/C contains an element of infinite order or of order q^n with $q \neq p$ a prime and $n \in \mathbb{N}$. By injectivity of the divisible groups $\mathbb{Q}$ and $\mathbb{Z}(q^\infty)$ (Theorem 1.6.2), we obtain a nonzero $\eta \in (G/C)\widehat{}$ with $\operatorname{im}\eta$ torsion-free or q-torsion. Letting $\varphi : G \to G/C$ be the natural map, we obtain $\chi = \eta\varphi \in \widehat{G}$ with $\operatorname{im}\chi \nsubseteq t_p(\mathbb{T})$, contradicting (2). Thus, G/C is a p-group, and we have (2) ⇒ (3).

To prove that (3) implies (1), let $x \in G$ and U a neighborhood of 0 in G. By assumption, G has a compact open subgroup C such that $C \subseteq U$ and G/C is a p-group. Then

$p^n(x + C) = 0 + C$ for almost all n; that is, $p^n x \in C \subseteq U$ for almost all n. Therefore, $\lim_{n\to\infty} p^n x = 0$, as desired. □

Corollary 6.6.4 (Braconnier [10], Robertson [133], Vilenkin [147]). *Let G be an LCA group. Then G is a topological p-group if and only if $\widehat{G}$ is a topological p-group.*

Proof. By duality, only necessity needs to be proved. Let $\Phi : G \to (\widehat{G})\widehat{}$ be the topological isomorphism as in Theorem 6.2.3. Then for $\chi \in \widehat{G}$ and $x \in G$, we have $\Phi : x \mapsto x'$, where $x'(\chi) = \chi(x) \in t_p(\mathbb{T})$ (Proposition 6.6.2), so $\operatorname{im} x' \subseteq t_p(\mathbb{T})$. By Theorem 6.6.3(2) ⇒ (1), $\widehat{G}$ is a topological p-group. □

Corollary 6.6.5. *A discrete group G is a p-group if and only if $\widehat{G}$ is a topological p-group.*

The (compact) dual group of $\mathbb{Z}(p^\infty)$ is denoted by $\mathbb{J}_p$. Algebraically, it is the group of all endomorphisms of $\mathbb{Z}(p^\infty)$. The group $\mathbb{J}_p$ is called the additive group of ***p*-adic integers**.

Theorem 6.6.6. *Let p be a prime. Then we have the following:*
1. *$\mathbb{J}_p$ is a torsion-free topological p-group.*
2. *If $n \in \mathbb{N}$, then $\mathbb{J}_p/p^n\mathbb{J}_p \cong \mathbb{Z}/p^n\mathbb{Z}$.*
3. *Each subgroup $p^n\mathbb{J}_p$ is open in $\mathbb{J}_p$, and $\bigcap_{n\in\mathbb{N}} p^n\mathbb{J}_p = 0$. Thus, $\mathbb{J}_p$ is totally disconnected.*
4. *The only proper closed subgroups of $\mathbb{J}_p$ are 0 and $p^n\mathbb{J}_p$ ($n \in \mathbb{N}$).*

Proof. We let $n \in \mathbb{N}$ and write $D = \mathbb{Z}(p^\infty)$. By Lemma 6.4.15, we have $\mathbb{J}_p[n] = (\mathbb{J}_p, nD) = (\mathbb{J}_p, D) = 0$; thus, $\mathbb{J}_p$ is torsion-free. Now let $\chi \in \mathbb{J}_p$, $\varepsilon > 0$ and F a finite subset of D. There is an $m_0 \in \mathbb{N}$ such that $p^{m_0}x = 0$ for all $x \in F$. Thus, $p^m\chi \in P(F, \varepsilon)$ for all $m \geq m_0$, and we obtain $\lim_{m\to\infty} p^m\chi = 0$.

To prove (2), consider the short exact sequence $0 \to \ker \nu \to D \xrightarrow{\nu} D \to 0$ with $\nu = p^n 1_D$. Then $\ker \nu \cong \mathbb{Z}/p^n\mathbb{Z}$, and the induced sequence $0 \to \mathbb{J}_p \xrightarrow{\nu^*} \mathbb{J}_p \to (\ker \nu)\widehat{} \to 0$ is proper exact (Proposition 6.5.2) with $\nu^* = p^n 1_{\mathbb{J}_p}$; thus, $\mathbb{J}_p/p^n\mathbb{J}_p \cong (\mathbb{Z}/p^n\mathbb{Z})\widehat{} \cong \mathbb{Z}/p^n\mathbb{Z}$.

Since finite groups are discrete, (2) shows that $p^n\mathbb{J}_p$ is open in $\mathbb{J}_p$ for all n. By Lemma 6.4.13, we have $\bigcap_{n\in\mathbb{N}} p^n\mathbb{J}_p = \bigcap_{n\in\mathbb{N}}(\mathbb{J}_p, D[p^n]) = (\mathbb{J}_p, \sum_{n\in\mathbb{N}} D[p^n]) = (\mathbb{J}_p, D) = 0$. Thus, (3) is evident.

For (4), use the fact that the only nontrivial subgroups of D are D and $D[p^n]$ with $n \in \mathbb{N}$ (Proposition 1.6.1). Then the only proper closed subgroups of $\mathbb{J}_p$ are $(\mathbb{J}_p, D) = 0$ and $(\mathbb{J}_p, D[p^n]) = p^n\mathbb{J}_p$. □

The next result is taken from Armacost [3] and is slightly reworded. It shows that the group $\mathbb{J}_p$ can be used to describe the p-component of an LCA group:

Proposition 6.6.7 (Armacost [3]). *Let G be an LCA group and p a prime. Then $G_{(p)} = \{x \in G : \overline{\langle x\rangle} \cong \mathbb{Z}/p^n\mathbb{Z}$ for some $n < \omega$ or $\overline{\langle x\rangle} \cong \mathbb{J}_p\}$.*

Proof. It suffices to prove the left to right inclusion of the set equation. By duality, we may view G as the dual of $H = \widehat{G}$, and let $\chi \in G_{(p)}$. For any $z \in H$, we have $\chi(z) \in \mathbb{T}_{(p)} = t_p(\mathbb{T})$ (Lemma 6.6.1); thus, $\operatorname{im}\chi \subseteq t_p(\mathbb{T})$. Since $\chi(H_0)$ is a connected subgroup of $\mathbb{T}$, it follows that $\chi(H_0) = 0$. By Proposition 6.4.18, the kernel of χ is open in H; thus, $H/\ker\chi$ is discrete. Since $H/\ker\chi$ is algebraically isomorphic to $\operatorname{im}\chi$, we have either $H/\ker\chi \cong \mathbb{Z}/p^n\mathbb{Z}$ for some $n < \omega$, or else $H/\ker\chi \cong \mathbb{Z}(p^\infty)$. Using Proposition 6.4.7, we obtain $\overline{\langle\chi\rangle} = (\widehat{H}, \ker\chi) \cong (H/\ker\chi)\widehat{\ }$; therefore, $\overline{\langle\chi\rangle} \cong \mathbb{Z}/p^n\mathbb{Z}$ or $\overline{\langle\chi\rangle} \cong \mathbb{J}_p$. □

Proposition 6.6.8 (Braconnier [10], Vilenkin [147]). *Let G be a compact torsion-free group. Then G is a topological p-group if and only if $G \cong \mathbb{J}_p^{\mathfrak{m}}$ for some cardinal $\mathfrak{m}$.*

Proof. A group is a compact torsion-free topological p-group if and only if its dual is a discrete divisible p-group. Now use Theorems 1.6.7 and 6.3.4. □

Recall that a discrete torsion group is the direct sum of its p-torsion parts (see Theorem 1.3.2). Dually, a compact totally disconnected group is topologically isomorphic to the direct product of its p-components:

Proposition 6.6.9. *Let C be a compact totally disconnected group. Then each p-component $C_{(p)}$ is a closed subgroup of C and we have $C \cong \prod_{p\in\mathbb{P}} C_{(p)}$.*

Proof. First, we show that each subgroup $C_{(p)}$ is closed in C, so let $x \in \overline{C_{(p)}}$. If U is a neighborhood of 0 in C, then it contains a compact open subgroup A of C since C is totally disconnected. Since $x + A$ is a neighborhood of x, we have $C_{(p)} \cap (x + A) \neq \emptyset$. Letting $c \in C_{(p)} \cap (x + A)$, we write it as $c = x + a$ with $a \in A$. Then there is an $n_0 < \omega$ such that $p^n x + p^n a = p^n c \in A \subseteq U$ for all $n \geq n_0$; thus, $\lim_{n\to\infty} p^n x = 0$. Therefore, $x \in C_{(p)}$; hence, $C_{(p)}$ is closed in C.

Now view C as the dual group $\hat{B}$ with B a discrete torsion group (Theorem 6.4.17), and write $B = \bigoplus_{p\in\mathbb{P}} t_p B$. For every prime p, we define $B(p) = \sum_{q\neq p} t_q B$ and $C_{[p]} = (\widehat{B}, B(p))$. Then $C_{[p]}$ is a compact topological p-group since $(C_{[p]})\widehat{\ } \cong B/B(p) \cong t_p B$ is a discrete p-group (see Lemma 6.4.2 and Corollary 6.6.5), and we have $C \cong \prod_{p\in\mathbb{P}} C_{[p]}$ by Theorem 6.3.4.

It remains to show that $C_{[p]}$ coincides with $C_{(p)}$ for all p. Obviously, we have $C_{[p]} \subseteq C_{(p)}$ since $C_{[p]}$ is a topological p-group. Assume that $C_{[p]}$ is a proper subgroup of $C_{(p)}$. Then Corollary 6.4.5 shows that $(B, C_{(p)})$ is a proper subgroup of $(B, C_{[p]}) = B(p)$; thus, $B(p)/(B, C_{(p)})$ is a nontrivial subgroup of the p-group $B/(B, C_{(p)}) \cong (C_{(p)})\widehat{\ }$, which is impossible. Consequently, $C_{[p]} = C_{(p)}$ for all primes p. □

Exercises.

1. Let G be an LCA group and p a prime. Then $G_{(p)} = \{x \in G :$ there exists a continuous epimorphism $\phi : \mathbb{J}_p \to \overline{\langle x\rangle}\}$.
2. (Armacost [3]) Let p be a prime. Braconnier [10] defined an LCA group G to be a ***p*-primary group**, if for every $x \in G$ there is a $\phi \in \operatorname{Hom}(\mathbb{J}_p, G)$ such that $\phi(\mathbf{1}) = x$. Here $\mathbf{1}$ denotes the multiplicative identity of $\mathbb{J}_p$ (considered as the endomorphism ring of $\mathbb{Z}(p^\infty)$). Show that the following are equivalent:

(i) G is a topological p-group;
(ii) For each $x \in G$ there is a $\phi \in \mathrm{Hom}(\mathbb{J}_p, G)$ such that $x \in \mathrm{im}\,\phi$;
(iii) G is a p-primary group.

3. (Dikranjan–Prodanov–Stoyanov [19]) Let p be a prime. Show that $\widehat{\mathbb{Q}}_{(p)}$ is a dense subgroup of $\widehat{\mathbb{Q}}$.
4. An LCA group is called **monothetic** if it contains a dense cyclic subgroup. If G is a compact monothetic group, then the p-components of G ($p \in \mathbb{P}$) generate a dense subgroup of G.
5. (Loth [107]) Let G be an LCA group. Prove that G_0 is the intersection of all open subgroups A of G for which there is a prime p such that G/A is a cocyclic p-group (that is, G/A is a cyclic or quasicyclic p-group). [Hint: Exercise 4 and Theorem 6.4.16.]
6. (Robertson [133]) For an LCA group G, let $G! = \{x \in G : \lim_{n\to\infty} n!x = 0\}$, and call G a **topological torsion group** if $G! = G$. Show the following:
(a) If G is a topological p-group, then G is a topological torsion group.
(b) $G_{(p)} \subseteq G!$ whenever p is a prime.
(c) (Armacost [3]) $t(\mathbb{T}) \subsetneq \mathbb{T}! \subsetneq \mathbb{T}$.
7. (Robertson [133]) Prove that the following conditions are equivalent for an LCA group G (cf. Theorem 6.6.3):
(1) G is a topological torsion group (see Exercise 6);
(2) neither G nor $\widehat{G}$ has a surjective character;
(3) both G and $\widehat{G}$ are totally disconnected;
(4) $\chi(x)$ has finite order for all $\chi \in \widehat{G}$ and $x \in G$;
(5) G has arbitrarily small compact open subgroups such that the corresponding factor groups are discrete torsion groups.
Conclude that G is a topological torsion group if and only if $\widehat{G}$ is a topological torsion group.

6.7 Going further

Moskowitz [122] developed a homological theory in the category $\mathcal{L}$ of LCA groups (with continuous homomorphisms as morphisms) and studied the functors Hom, ⊗, Tor, and Ext on certain subcategories of $\mathcal{L}$. Fulp and Griffith [35, 36] extended Moskowitz's construction of the functor Ext to the category $\mathcal{L}$, obtaining various properties on Ext, including exactness of the two Hom–Ext sequences (see Theorem 1.16.10). Gutiérrez [39] studied properties of Ext in the category of topological abelian groups.

The papers [35, 36] investigate splitting problems in the category $\mathcal{L}$ and subcategories, involving connected and totally disconnected LCA groups. Continuing this investigation, Fulp [33] studied splitting problems, involving torsion and torsion-free LCA groups. Splitting problems, involving the identity component of LCA groups, were studied by Loth [99, 104] and Sahleh and Alijani [138]. Kryuchkov [87] introduced and investigated extensions of LCA groups, involving a generalization of the concept of cotorsion group. For a detailed discussion on splitting problems involving different classes of LCA groups, see the book by Armacost [3] and the references therein.

The study of purity in LCA groups goes back to Braconnier [10] and Vilenkin [147]. Later on, several authors investigated this topic further, for instance, Hartman and Hulanicki [40], M. A. Khan [82], Armacost [2], Fulp [34], J. A. Khan [81], Kryuchkov [86], Loth [106, 109, 110], and Sahleh and Alijani [135, 136, 137].

Extensions of compact groups by discrete groups were investigated by Fuchs and Hofmann [31, 32] and Kryuchkov [85]. Tensor products in LCA groups were introduced and studied by Hofmann [58].

For additional information on the theory of locally compact groups, we refer to the books by Dikranjan, Prodanov and Stoyanov [19], Hofmann and Morris [59] and Stroppel [143].

7 Classifications of compact abelian groups

We show that Pontrjagin duality can be used to obtain classifications of compact abelian groups using the group classifications from the previous chapters. To do this, we define counterparts of the key constructs used in those classifications and prove that they hold for a group if and only if the corresponding property holds for the dual. We define coinvariants corresponding to the Ulm, Baer, and Stanton invariants and use them to classify various classes of compact groups. Then we establish numerous characterizations of the considered classes of compact groups.

7.1 Dimension, *p*-dimension and *p*-order

As we saw earlier, large classes of (discrete) groups are classified up to isomorphism in terms of cardinal invariants. We will now show that Pontrjagin duality can be used to obtain corresponding classifications of compact groups.

First, we determine the dual counterparts of the torsion-free rank and p-rank of a discrete group. Let C be a compact group and consider the cardinal

$$\mathfrak{m} = \sup\{\kappa : C/A \cong \mathbb{T}^{\kappa}, \text{ and } A \text{ is a closed subgroup of } C\}.$$

Then $\mathfrak{m}$ coincides with the torsion-free rank of $\widehat{C}$ (see Lemmas 6.4.2 and 6.4.4); thus, dualization of Theorem 1.4.4 shows that $\mathfrak{m}$ is the uniquely determined cardinal satisfying

$$C/D \cong \mathbb{T}^{\mathfrak{m}}$$

for some closed totally disconnected subgroup D of C (see Theorem 6.4.17). We call it the **dimension of** C and write $\dim C = \mathfrak{m}$. Notice that in case $\mathfrak{m}$ is finite, $\dim C = \mathfrak{m}$ if and only if C has dimension $\mathfrak{m}$ in the sense of Hewitt and Ross [41]. If p is a prime, then

$$C/pC \cong (\mathbb{Z}/p\mathbb{Z})^{\mathfrak{n}}$$

for some cardinal $\mathfrak{n}$, which we call the ***p*-dimension of** C, written $\dim_p C = \mathfrak{n}$.

Lemma 7.1.1. *If C is a compact group, then* $\dim C = r_0(\widehat{C})$ *and* $\dim_p C = r_p(\widehat{C})$ *for all primes p.*

Proof. The first claim was already verified. Since $(C/pC)\widehat{\ } \cong (\widehat{C}, pC) = \widehat{C}[p]$ by Lemmas 6.4.2 and 6.4.15, $\dim_p C$ coincides with the p-rank of $\widehat{C}$. □

Lemma 7.1.2. *Let C be a compact group. Then we have the following:*
1. *C is totally disconnected if and only if* $\dim C = 0$.
2. *C is connected if and only if* $\dim_p C = 0$ *for all* $p \in \mathbb{P}$.

https://doi.org/10.1515/9783110427684-007

3. *If* $C \cong \mathbb{T}$ *or* $\widehat{\mathbb{Q}}$, *then* $\dim C = 1$.
4. *If* $C \cong \mathbb{J}_p$ *or* $\mathbb{Z}/p^n\mathbb{Z}$ *for some* $n \in \mathbb{N}$, *then* $\dim_p C = 1$ *and* $\dim_q C = 0$ *whenever* $q \neq p$.

Proof. The first assertion is immediate. (2) is true since C is connected if and only if $pC = C$ for all primes p (Theorem 6.4.19). It is obvious that $\dim \mathbb{T} = 1$, and we have $\dim \widehat{\mathbb{Q}} = r_0(\mathbb{Q}) = 1$ by Lemma 7.1.1; thus, (3) holds. (4) is clear. □

The dual counterparts of the groups $p^\alpha G$ and of p-heights were introduced by Kiefer [84]. If C is a compact group and p is a prime, then closed subgroups $C[p^\alpha]$ of C are defined for all ordinals α by

$$C[p^0] = 0,$$
$$C[p^{\alpha+1}] = \{x \in C : px \in C[p^\alpha]\}, \quad \text{and}$$
$$C[p^\alpha] = \overline{\bigcup_{\beta<\alpha} C[p^\beta]} \quad \text{if } \alpha \text{ is a limit ordinal.}$$

This yields a well-ordered ascending chain $0 = C[p^0] \subseteq C[p] \subseteq \cdots \subseteq C[p^\alpha] \subseteq \cdots$ of closed subgroups of C that becomes stationary. The ***p*-order of an element** $x \in C$ is defined to be $o_p(x) = \alpha$ whenever

$$x \in C[p^\alpha] \setminus \bigcup_{\beta<\alpha} C[p^\beta],$$

and we set $o_p(x) = \infty$ if $x \notin C[p^\infty] = \bigcup_\alpha C[p^\alpha]$. We put $o_p(0) = 0$. Then:
(a) If $x \in C$ and α is an ordinal, then $o_p(x) \leq \alpha$ exactly if $x \in C[p^\alpha]$.
(b) If $\phi : C \to D$ is a continuous homomorphism, then $\phi(C[p^\alpha]) \subseteq D([p^\alpha])$ for all ordinals α. Thus, $o_p(\phi(x)) \leq o_p(x)$ for all $x \in C$.

The following fact was shown for p-groups by Kiefer [84]; his proof applies to arbitrary groups as well:

Lemma 7.1.3 (Kiefer [84]). *Suppose that G is a discrete group. If p is a prime and* $\alpha \in \mathrm{Ord}_\infty$, *then* $\widehat{G}[p^\alpha] = (\widehat{G}, p^\alpha G)$.

Proof. For ordinals α, we prove the assertion by induction. Let $\alpha > 0$, and assume that the claim holds for all $\beta < \alpha$. If α is a limit ordinal, the claim follows from Lemma 6.4.13. Suppose that $\alpha - 1$ exists. To show that $\widehat{G}[p^\alpha] \subseteq (\widehat{G}, p^\alpha G)$, let $\chi \in \widehat{G}[p^\alpha]$ and $x \in p^\alpha G$. Then $p\chi \in \widehat{G}[p^{\alpha-1}]$, which coincides with $(\widehat{G}, p^{\alpha-1}G)$ by the induction hypothesis. There is a $y \in p^{\alpha-1}G$ such that $x = py$; thus, $\chi(x) = (p\chi)(y) = 1$ as desired. For the reverse inclusion, assume that $\chi \in (\widehat{G}, p^\alpha G)$. Then for any $x \in p^{\alpha-1}G$, we have $px \in p^\alpha G$; thus, $(p\chi)(x) = \chi(px) = 1$. It follows that $p\chi \in (\widehat{G}, p^{\alpha-1}G) = \widehat{G}[p^{\alpha-1}]$; therefore, $\chi \in \widehat{G}[p^\alpha]$, and the induction is complete.

The claim holds for $\alpha = \infty$ as well since it holds for the p-length of G. □

Since a compact group is coreduced exactly if its dual is reduced, we obtain:

Corollary 7.1.4. *A compact topological p-group C is coreduced if and only if $C = C[p^\alpha]$ for some ordinal α.*

Exercises.

1. Prove (a) and (b).
2. Let x, y be elements of a compact group and p a prime. If $p^n x = p^n y \neq 0$ for some $n \in \mathbb{N}$, then $o_p(x) = o_p(y)$.
3. (Loth [105]) Let $G = \{(x_i) \in \prod_{i=1}^{\infty} \mathbb{R}/\mathbb{Z} : px_1 = \cdots = p^n x_n = \cdots\}$, where p is a fixed prime. [This is the dual of the nonsplitting discrete group $\langle z_1, \ldots, z_n, \ldots : pz_1 = \cdots = p^n z_n = \cdots \rangle$ (see Section 2.5, Exercise 1).] Prove the following:
   (a) If $k < \omega$, then $G[p^{\omega+k}] = \{(x_i) \in G : p^{k+1}x_1 = 0\}$. Thus, $G[p^{\omega \cdot 2}] = G$.
   (b) Show directly that the identity component G_0 does not split from G. [Hint: Use (a).]
   (c) $G[p^\omega]$ is totally disconnected, and $G = G_0 + G[p^\omega]$.

7.2 Compact torsion and torsion-free groups

In this section, we use Pontrjagin duality to determine the structure of compact torsion and compact torsion-free groups, and to construct their topological isomorphism invariants.

Theorem 7.2.1. *A compact group C is torsion if and only if $C \cong \prod_{i \in I} C_i$, where each C_i is a cyclic group of prime power order such that only finitely many distinct prime powers occur.*

Proof. The statement follows from Theorems 6.3.5, 1.8.1 and duality. □

Next we need to dualize the Ulm invariants. The **co-Ulm invariants** of a compact group C are defined for p a prime and α an ordinal by

$$u_p^*(\alpha, C) = \mathfrak{m} \quad \text{if } (pC + C[p^{\alpha+1}])/(pC + C[p^\alpha]) \cong (\mathbb{Z}/p\mathbb{Z})^{\mathfrak{m}}$$

and

$$u_p^*(\infty, C) = \mathfrak{n} \quad \text{if } C/(pC + C[p^\infty]) \cong (\mathbb{Z}/p\mathbb{Z})^{\mathfrak{n}}.$$

Co-Ulm invariants are the dual counterparts of Ulm invariants:

Lemma 7.2.2. *Let G be a discrete group and p a prime. Then $u_p(\alpha, G) = u_p^*(\alpha, \widehat{G})$ for all $\alpha \in \mathrm{Ord}_\infty$.*

Proof. First, we assume that α is an ordinal. Using Lemmas 6.4.2, 6.4.12, 6.4.13, 6.4.15, and 7.1.3, we obtain

$$\begin{aligned}((p^\alpha G)[p]/(p^{\alpha+1}G)[p])\widehat{} &\cong (\widehat{G}, G[p] \cap p^{\alpha+1}G)/(\widehat{G}, G[p] \cap p^\alpha G)\\ &= ((\widehat{G}, G[p]) + (\widehat{G}, p^{\alpha+1}G))/((\widehat{G}, G[p]) + (\widehat{G}, p^\alpha G))\\ &= (p\widehat{G} + \widehat{G}[p^{\alpha+1}])/(p\widehat{G} + \widehat{G}[p^\alpha]).\end{aligned}$$

Since $(\mathbb{Z}/p\mathbb{Z})\widehat{\ } \cong \mathbb{Z}/p\mathbb{Z}$, it follows from Theorem 6.3.4 that $u_p(\alpha, G) = u_p^*(\alpha, \widehat{G})$. This equation holds for $\alpha = \infty$ as well since

$$((p^\infty G)[p])\widehat{\ } \cong \widehat{G}/(\widehat{G}, G[p] \cap p^\infty G) = \widehat{G}/(p\widehat{G} + \widehat{G}[p^\infty]). \qquad \square$$

Recall that if a discrete group G is a direct sum of cyclic groups of prime power order, then for $n < \omega$ the Ulm invariant $u_p(n, G)$ counts the number of direct summands of order p^{n+1}. Therefore, we have the following:

Corollary 7.2.3. *Two compact torsion groups are topologically isomorphic if and only if they have the same co-Ulm invariants u_p^* ($p \in \mathbb{P}$).*

To determine both structure and topological isomorphism invariants of compact torsion-free groups, we dualize Theorem 1.6.7 using the first part of Theorem 6.4.19 and Lemma 7.1.1:

Corollary 7.2.4. *Suppose that C is a compact group. Then C is torsion-free if and only if $C \cong \prod_{\mathfrak{m}} \widehat{\mathbb{Q}} \times \prod_{p \in \mathbb{P}} \prod_{\mathfrak{n}_p} \mathbb{J}_p$. The cardinals $\mathfrak{m} = \dim C$ and $\mathfrak{n}_p = \dim_p C$ ($p \in \mathbb{P}$) form a complete set of topological isomorphism invariants for C.*

Exercises.

1. If A is a closed subgroup of a compact group C such that C/A is torsion-free, then A splits from C.
2. A compact group is a topological direct product of a compact torsion-free group and a compact coreduced group.
3. (Pontrjagin [127]) Let G be a discrete torsion-free group of rank $\mathfrak{m}$. Then $\widehat{G}$ contains a subset homeomorphic to the topological space $(0, 1)^{\mathfrak{m}}$.

7.3 Simply given groups

Recall that a discrete group is simply presented exactly if it can be defined in terms of generators and relations of the form $px = 0$ or $px = y$, where x and y are generators, and $p \in \mathbb{P}$. In other words, the group can be written as F/H, where F is the free group with free generators x_i ($i \in I$), and H is the subgroup of F generated by elements

$$px_j \ (j \in J_p) \quad \text{and} \quad px_k - x_{f_p(k)} \ (k \in K_p)$$

with $p \in \mathbb{P}$ such that

(i) for each $p \in \mathbb{P}$, J_p and K_p are disjoint subsets of I and $f_p : K_p \to I$ is a map;
(ii) if $p \neq q$, then $J_p \cap J_q = \emptyset$ and $f_p(k) \neq f_q(k)$ whenever $k \in K_p \cap K_q$.

A compact group is called **simply given** if it is topologically isomorphic to

$$C = \left\{ (x_i) \in \prod_{i \in I} \mathbb{R}/\mathbb{Z} : px_j = 0 \text{ if } j \in J_p \text{ and } px_k = x_{f_p(k)} \text{ if } k \in K_p \text{ for all } p \in \mathbb{P} \right\},$$

where J_p, K_p, and f_p are defined as in (i), (ii) for all $p \in \mathbb{P}$.

For example, the cyclic group $\mathbb{Z}/p^n\mathbb{Z}$ is simply given since it can be identified with

$$\{(x_1,\ldots,x_n) \in (\mathbb{R}/\mathbb{Z})^n : px_n = 0 \text{ and } px_i = x_{i+1} \text{ for } i = 1,\ldots,n-1\}.$$

It is clear that the class of simply given groups is closed under direct products. Using Theorem 7.2.1, we obtain:

Corollary 7.3.1. *Compact torsion groups are simply given.*

The next result was shown for p-groups by Kiefer [84]; the proof applies to the general case as well.

Proposition 7.3.2 (Kiefer [84]). *Let G be a discrete group. Then G is simply presented if and only if $\widehat{G}$ is simply given.*

Proof. Suppose that G is simply presented. Then $G = F/H$, where F is the free group with free generators x_i $(i \in I)$, and H is the subgroup of F generated by elements px_j $(j \in J_p)$ and $px_k - x_{f_p(k)}$ $(k \in K_p)$ with $p \in \mathbb{P}$ satisfying conditions (i), (ii). By Theorem 6.3.4 and Lemma 6.4.2, $\widehat{F}$ can be identified with $\prod_{i\in I} \mathbb{R}/\mathbb{Z}$ and $\widehat{G}$ with $(\widehat{F},H)$. Thus, it suffices to show that $(\widehat{F},H)$ coincides with the simply given group C as constructed above.

Let $\chi \in \widehat{F}$ and $y = (y_i)_{i\in I} \in F$. We write $y_i = r_i x_i$ for some $r_i \in \mathbb{Z}$ $(i \in I)$ and $\chi = [\chi_i]$ as in Theorem 6.3.4. If χ annihilates H, then for all $p \in \mathbb{P}$ and $j \in J_p$, $k \in K_p$, we have $(p\chi_j)(y_j) = \chi(r_j p x_j) = 1$ and

$$(p\chi_k)(r_k x_k)\left[\chi_{f(k)}(r_k x_{f(k)})\right]^{-1} = \chi(r_k(px_k - x_{f(k)})) = 1;$$

thus, $p\chi_j = 0$ and $p\chi_k = \chi_{f(k)}$. Conversely, if $p\chi_j = 0$ and $p\chi_k = \chi_{f(k)}$ for all $j \in J_p$, $k \in K_p$ $(p \in \mathbb{P})$, then χ annihilates H. It follows that $(\widehat{F},H) = C$.

Conversely, if $\widehat{G} = C$ as above, then $G \cong \widehat{(\widehat{G})}$ is simply presented. □

7.4 Simply given totally disconnected groups

Compact totally disconnected groups are the duals of discrete torsion groups and can be written as direct products of compact topological p-groups (Theorem 6.4.17 and Proposition 6.6.9). In this section, we classify those compact topological p-groups which are simply given and establish a variety of characterizations. The theory was developed by Kiefer [84], who answered the following question given as Problem 65 in Fuchs' book [27]: "Which are the compact abelian groups whose duals are totally projective p-groups?"

We start with an application of Pontrjagin duality to Ulm's Theorem 1.14.2:

Corollary 7.4.1. *Two compact metrizable topological p-groups are topologically isomorphic if and only if they have the same co-Ulm invariants u_p^*.*

More generally, in view of Corollary 1.14.3, we have:

Corollary 7.4.2. *Two compact metrizable totally disconnected groups are topologically isomorphic if and only if they have the same co-Ulm invariants u_p^* ($p \in \mathbb{P}$).*

Since discrete countable torsion groups are simply presented (see Theorem 2.8.17), we obtain:

Corollary 7.4.3. *Compact metrizable totally disconnected groups are simply given.*

The first part of the next result extends Corollary 7.4.1:

Theorem 7.4.4 (Kiefer [84]). *Two simply given topological p-groups are topologically isomorphic if and only if they have the same co-Ulm invariants u_p^*. The class of simply given topological p-groups is closed under direct factors.*

Proof. A compact group C is a simply given topological p-group if and only if $\widehat{C}$ is a simply presented p-group (Proposition 7.3.2). The class of simply presented p-groups is closed under direct summands (see Corollary 2.8.21), and their isomorphism invariants are the Ulm invariants (see Theorem 2.3.3 and Proposition 2.8.7). Therefore, in view of Lemma 7.2.2, the proof is complete. □

Corollary 7.4.5. *Two simply given totally disconnected groups are topologically isomorphic if and only if they have the same co-Ulm invariants u_p^* ($p \in \mathbb{P}$). The class of simply given totally disconnected groups is closed under direct factors.*

Let F be a closed subgroup of a compact topological p-group C. Then F is called **smart (in C)** if

$$F[p^\alpha] = F \cap C[p^\alpha]$$

for all ordinals α. Thus, F is smart in C exactly if each element of F has the same p-order in F as in C.

Lemma 7.4.6. *Suppose that A is a subgroup of a discrete group G. Let $\psi : (G/A)\widehat{\ } \to (\widehat{G}, A)$ be the topological isomorphism as in Lemma 6.4.2, let p be a prime and α an ordinal. Then $((G/A)\widehat{\ }, p^\alpha(G/A))$ is mapped by ψ onto $(\widehat{G}, A)[p^\alpha]$, and $((G/A)\widehat{\ }, (p^\alpha G + A)/A)$ is mapped by ψ onto $\widehat{G}[p^\alpha] \cap (\widehat{G}, A)$.*

Proof. Since $((G/A)\widehat{\ }, p^\alpha(G/A))$ coincides with $(G/A)\widehat{\ }[p^\alpha]$ (Lemma 7.1.3), it is mapped by ψ onto $(\widehat{G}, A)[p^\alpha]$. The annihilator $((G/A)\widehat{\ }, (p^\alpha G + A)/A)$ is mapped by ψ onto $(\widehat{G}, p^\alpha G + A) = (\widehat{G}, p^\alpha G) \cap (\widehat{G}, A) = \widehat{G}[p^\alpha] \cap (\widehat{G}, A)$ (see Lemma 6.4.2). □

A subgroup of a p-group G is nice exactly if its annihilator is smart:

Proposition 7.4.7 (Kiefer [84]). *Let A be a subgroup of a discrete p-group G. Then A is nice in G if and only if $(\widehat{G}, A)$ is smart in $\widehat{G}$.*

Proof. Let α be an ordinal. Then by Lemma 7.4.6, we have $(\widehat{G},A)[p^\alpha] = \widehat{G}[p^\alpha] \cap (\widehat{G},A)$ if and only if $((G/A)\widehat{}\,, p^\alpha(G/A)) = ((G/A)\widehat{}\,, (p^\alpha G + A)/A)$. The latter identity holds exactly if $p^\alpha(G/A) = (p^\alpha G + A)/A$ (Corollary 6.4.5), so the proof is complete. □

Kiefer [84] dualized Hill's concept of a nice system (see Section 2.2) introducing smart systems for compact topological p-groups. For any given compact group C, we call a system $\mathcal{F}$ of closed subgroups of C an H^***-family** in C if the following conditions are satisfied:

(H1*) $C \in \mathcal{F}$;
(H2*) if $F_i \in \mathcal{F}$ $(i \in I)$, then $\bigcap_{i \in I} F_i \in \mathcal{F}$;
(H3*) if $F \in \mathcal{F}$, and K is a closed subgroup of F with F/K metrizable, then there is a $D \in \mathcal{F}$ such that $D \subseteq K$ and F/D is metrizable.

In the case where C is a topological p-group, and $\mathcal{F}$ consists entirely of smart subgroups of C, $\mathcal{F}$ is called a **smart system**. Similarly, the concepts of $G(\aleph_0)$-family and $F(\aleph_0)$-family (see Section 2.2) can be dualized. Letting C be any compact group, we call a system $\mathcal{F}$ of closed subgroups of C a G^***-family** in C if it satisfies (H1*), (H3*), and

(G2*) $\mathcal{F}$ is closed under descending intersections.

A well-ordered descending chain of closed subgroups

$$C = F_0 \supseteq F_1 \supseteq \cdots \supseteq F_\alpha \supseteq \cdots \supseteq F_\lambda = 0$$

of C is called an F^***-family** in C if

(i) $F_\alpha = \bigcap_{\beta<\alpha} F_\beta$ whenever α is a limit ordinal;
(ii) $F_\alpha/F_{\alpha+1}$ is metrizable for all $\alpha < \lambda$.

If in addition, each quotient $F_{\alpha+1}/F_\alpha$ is either cyclic of prime order or topologically isomorphic to $\mathbb{T}$, then this chain is called a **descending composition series for** C. A descending composition series of smart subgroups for a compact topological p-group C is called a **smart composition series for** C.

Proposition 7.4.8 (Kiefer [84]). *Let G be a discrete group, $\mathcal{C}$ a system of subgroups of G and $\mathcal{C}^* = \{(\widehat{G},N) : N \in \mathcal{C}\}$. Then we have the following:*

1. *$\mathcal{C}$ is an $H(\aleph_0)$-family in G if and only if $\mathcal{C}^*$ is an H^*-family in $\widehat{G}$.*
2. *$\mathcal{C}$ is a $G(\aleph_0)$-family in G if and only if $\mathcal{C}^*$ is a G^*-family in $\widehat{G}$.*
3. *$\mathcal{C}$ is an $F(\aleph_0)$-family in G if and only if $\mathcal{C}^*$ is an F^*-family in $\widehat{G}$.*
4. *$\mathcal{C}$ is a composition series for G if and only if $\mathcal{C}^*$ is a descending composition series for $\widehat{G}$.*

Proof. Assume that $\mathcal{C}$ is an $H(\aleph_0)$-family in G. Clearly, $0 \in \mathcal{C}$ implies that $\widehat{G} \in \mathcal{C}^*$, and since $\mathcal{C}$ is closed under arbitrary group unions, $\mathcal{C}^*$ is closed under arbitrary intersections (Lemma 6.4.13). Let $F = (\widehat{G}, N)$ with $N \in \mathcal{C}$, and let K be a closed subgroup of F such that F/K is metrizable. Identifying the dual of $\widehat{G}$ with G and using Lemmas 6.4.2, 6.4.12 and 6.4.4, we obtain

$$(G,K)/N = (G,K)/(G,F) \cong (\widehat{F},K) \cong (F/K)\widehat{}\,.$$

Then $(G,K) = \langle S, N\rangle$ for some countable subset S of G (Corollary 6.3.3), so there is an $M \in \mathcal{C}$ such that $\langle S, N\rangle \subseteq M$ and M/N is countable. Letting $D = (\widehat{G}, M)$, it follows that $D \subseteq K$ and $F/D = (\widehat{G},N)/(\widehat{G},M) \cong (\widehat{M},N) \cong (M/N)\widehat{}$ is metrizable, as desired.

Conversely, suppose that $\mathcal{C}^*$ is an H^*-family in $\widehat{G}$. Identifying the dual of $\widehat{G}$ with G again, we can write $\mathcal{C} = \{(G,F) : F \in \mathcal{C}^*\}$. Since $\widehat{G} \in \mathcal{C}^*$, we have $0 = (G,\widehat{G}) \in \mathcal{C}$ and if $(G,F_i) \in \mathcal{C}$ $(i \in I)$, then $A = \bigcap_{i\in I} F_i \in \mathcal{C}^*$; thus, $\sum_{i\in I}(G,F_i) = (G,A) \in \mathcal{C}$. Finally, let $N = (G,F) \in \mathcal{C}$, and assume that S is a countable subset of G. We define $K = (\widehat{G},S) \cap F$, and obtain

$$(F/K)\widehat{} \cong (\widehat{F},K) \cong (G,K)/(G,F) = (G,(\widehat{G},S)\cap F)/(G,F) = \langle S,N\rangle/N.$$

Then F/K is metrizable, so there is a $D \in \mathcal{C}^*$ such that $D \subseteq K$ and F/D is metrizable. It follows that $M = (G,D) \in \mathcal{C}$ and $\langle S,N\rangle = (G,K) \subseteq M$, where $M/N \cong (F/D)\widehat{}$ is countable. This shows that (1) is true.

Since $\mathcal{C}$ is closed under ascending unions exactly if $\mathcal{C}^*$ is closed under descending intersections, (2) is immediate. (3) is clear; for (4) we use the fact that finite groups are self-dual and that $\widehat{\mathbb{Z}} \cong \mathbb{T}$. □

Let G be a discrete group and p a prime. Recall that the p-nice subgroups of G are those subgroups N such that each nonzero coset $x+N$ has an element having the same p-height in G as $x+N$ has in G/N (Proposition 2.1.1(2)). We saw that in the case where G is a p-group, their annihilators are exactly the smart subgroups of $\widehat{G}$; that is, those closed subgroups F of $\widehat{G}$ such that every element of F has the same p-order in F as in $\widehat{G}$.

To determine the annihilators of p-isotype subgroups of G, we make the following definition: A closed subgroup D of a compact group C is called **p-duotype (in C)** if every nonzero coset $x+D$ has an element having the same p-order in C as $x+D$ has in C/D. We characterize p-duotype subgroups in the following lemma:

Lemma 7.4.9. *Let D be a closed subgroup of a compact group C and p a prime. Then the following conditions are equivalent:*
1. *D is p-duotype;*
2. *Each nonzero coset $x + D$ satisfies $o_p(x + D) = \min\{o_p(x + d) : d \in D\}$;*
3. *$(C/D)[p^\alpha] = (C[p^\alpha] + D)/D$ for all ordinals α.*

Proof. (1) implies (2). Let $x + D$ be a nonzero coset. By Section 7.1(b), the natural map $C \to C/D$ does not increase p-orders; thus, $o_p(x + D) \le o_p(x + d)$ for all $d \in D$ and the assertion follows.

(2) implies (3). Let α be an ordinal. All we need to verify is the inclusion $(C/D)[p^\alpha] \subseteq (C[p^\alpha]+D)/D$, so let $x+D \in (C/D)[p^\alpha]$. Then $o_p(x+D) \le \alpha$, so by our assumption, there is a $d \in D$ with $o_p(x+d) \le \alpha$; thus, $x+D = x+d+D \in (C[p^\alpha]+D)/D$ as desired.

(3) implies (1). Suppose $x+D$ is a nonzero coset, and let $\alpha = o_p(x+D)$. If $\alpha = \infty$, then $o_p(x) = \infty$ since the map $C \to C/D$ does not increase p-orders. If $\alpha \ne \infty$, then by assumption, we have $x+D \in (C/D)[p^\alpha] = (C[p^\alpha]+D)/D$. Therefore, there is a $d \in D$ such that $x+d \in C[p^\alpha]$; hence, $o_p(x+d) \le \alpha$. But then $o_p(x+d) = \alpha$ since $o_p(x+D) \le o_p(x+d)$. In any case, the coset $x+D$ has an element of p-order α; thus, D is p-duotype. □

Kiefer [84] showed that a subgroup of a p-group is isotype exactly if its annihilator is p-duotype. We isolate the main part of his proof; it applies to arbitrary discrete groups as well:

Lemma 7.4.10. *Let A be a subgroup of a discrete group G, p a prime, and α an ordinal. Then $p^\alpha A = p^\alpha G \cap A$ if and only if $(\widehat{G}/(\widehat{G},A))[p^\alpha] = (\widehat{G}[p^\alpha]+(\widehat{G},A))/(\widehat{G},A)$.*

Proof. Let $\phi : \widehat{G}/(\widehat{G},A) \to \widehat{A}$ be the topological isomorphism as in Lemma 6.4.12. Then

$$(\widehat{G},p^\alpha A)/(\widehat{G},A) \xrightarrow{\phi} (\widehat{A},p^\alpha A) = \widehat{A}[p^\alpha] \xrightarrow{\phi^{-1}} (\widehat{G}/(\widehat{G},A))[p^\alpha],$$

and $(\widehat{G},p^\alpha G\cap A)/(\widehat{G},A) = (\widehat{G}[p^\alpha]+(\widehat{G},A))/(\widehat{G},A)$. Thus, Corollary 6.4.5 shows that $p^\alpha A = p^\alpha G \cap A$ if and only if $(\widehat{G}/(\widehat{G},A))[p^\alpha] = (\widehat{G}[p^\alpha]+(\widehat{G},A))/(\widehat{G},A)$. □

A subgroup of C is called **duotype** if it is p-duotype for all primes p.

Corollary 7.4.11. *Let A be a subgroup of a discrete group G. Then we have the following:*
1. *A is isotype in G if and only if $(\widehat{G},A)$ is duotype in $\widehat{G}$.*
2. *A is pure in G if and only if $(\widehat{G},A)$ is pure in $\widehat{G}$.*

Proof. Use Lemma 7.4.10 and Section 1.7(f). □

Clearly the corollary remains valid if "isotype", "duotype", and "pure" are replaced by "p-isotype", "p-duotype" and "p-pure". Then (1) shows that a subgroup of a compact topological p-group is duotype exactly if it is p-duotype.

We call an exact sequence of compact topological p-groups $0 \to A \to B \to K \to 0$ **cobalanced-exact** if A is both smart and duotype in B. Then a compact topological p-group C is called **cobalanced-injective** if it has the injective property relative to all cobalanced-exact sequences; that is, if for every cobalanced-exact sequence $0 \to A \xrightarrow{\alpha} B \to K \to 0$ and continuous homomorphism $\phi : A \to C$, there is a continuous homomorphism $\beta : B \to C$ such that the diagram

$$\begin{array}{ccccccccc} 0 & \to & A & \xrightarrow{\alpha} & B & \to & K & \to & 0 \\ & & \phi\downarrow & \swarrow\beta & & & & & \\ & & C & & & & & & \end{array}$$

is commutative. By duality we have:

Proposition 7.4.12 (Kiefer [84]). *A discrete p-group G is balanced-projective if and only if $\widehat{G}$ is cobalanced-injective.*

Recall that the generalized Prüfer groups are those discrete reduced p-groups H_α which are uniquely determined (up to isomorphism) by properties (i–iii), Section 2.4. Therefore, for all cardinals α, their duals are compact coreduced topological p-groups K_α, which are uniquely determined (up to topological isomorphism) by the following properties:
(i) $K_\alpha[p^\alpha] = K_\alpha$ and $K_\alpha[p^\beta] \neq K_\alpha$ for all $\beta < \alpha$;
(ii) $K_{\alpha+1}/K_{\alpha+1}[p^\alpha]$ is cyclic of order p, and $K_\alpha \cong K_{\alpha+1}[p^\alpha]$;
(iii) $K_\alpha = \prod_{\beta<\alpha} K_\beta$ if α is a limit ordinal.

For instance, we have $K_0 = 0$, $K_n \cong \mathbb{Z}/n\mathbb{Z}$ for all $n < \omega$, and $K_\omega \cong \prod_{n=1}^{\infty} \mathbb{Z}/p^n\mathbb{Z}$. The dual of the Prüfer group $H_{\omega+1}$ is

$$K_{\omega+1} \cong \left\{ (x_i) \in \prod_{i=0}^{\infty} \mathbb{R}/\mathbb{Z} : px_0 = 0 \text{ and } p^i x_i = x_0 \text{ if } i = 1, 2, \dots \right\}.$$

Kiefer [84] obtained a variety of characterizations of compact coreduced topological p-groups, describing the duals of reduced totally projective p-groups. We rephrase his result and state it in a slightly extended form to include the non-coreduced case as well. Notice that the dual analogues of the various characterizations can be found in Theorem 2.10.1.

Theorem 7.4.13. *The following conditions are equivalent for a compact topological p-group C:*
1. *C is the dual of a simply presented p-group;*
2. *C is simply given;*
3. *C has a smart system;*
4. *C has a G^*-family of smart subgroups;*
5. *C has a smart composition series;*
6. *C is cobalanced-injective;*
7. $p^\alpha \operatorname{Ext}(A, C[p^\alpha]) = 0$ *for every ordinal α and every compact group A;*
8. *C is a direct factor of a direct product of groups K_α and copies of $\mathbb{J}_p$;*
9. *C belongs to the smallest class $\mathcal{D}$ of compact topological p-groups which contains $\mathbb{Z}/p\mathbb{Z}$ and $\mathbb{J}_p$, is closed under direct factors and direct products such that for every compact group A and ordinal α, $A \in \mathcal{D}$ if and only if both $A[p^\alpha] \in \mathcal{D}$ and $A/A[p^\alpha] \in \mathcal{D}$.*

Dualization of the uniqueness result in Theorem 2.10.2 yields:

Corollary 7.4.14. *Any class $\mathcal{D}$ of compact topological p-groups that contains all simply given topological p-groups, is closed under direct products, and in which the co-Ulm* *invariants distinguish between non-topologically isomorphic groups coincides with the class of simply given topological p-groups.*

Proof. Let $\mathcal{D}$ be a class of groups as in the statement. Then the class $\mathcal{C} = \{\widehat{C} : C \in \mathcal{D}\}$ satisfies the conditions of Theorem 2.10.2, so it coincides with the class of discrete simply presented p-groups. Thus, $\mathcal{D}$ is exactly the class of simply given topological p-groups. □

Exercises.

1. Characterize simply given topological p-groups dualizing the concept of "p-basis" (see Theorem 2.10.1(2)).

7.5 Simply given connected groups

The simply given connected groups are exactly the duals of discrete completely decomposable groups (see Theorems 6.4.19, 3.16.4 and Corollary 3.16.5). To classify them, we dualize the Baer invariants (see Section 1.15). Two 1-dimensional compact connected groups are said to be **equivalent** if they are topologically isomorphic. Following Leopold [96], we call the resulting equivalence classes **patterns** and let $\mathbf{p}(C)$ denote the pattern representing the 1-dimensional compact connected group C. Then Theorem 1.15.3 and Lemma 7.1.1 show that there is a natural one-to-one correspondence between patterns and types of discrete torsion-free rank 1 groups. If $\mathbf{t}$ is a type, then $\hat{\mathbf{t}}$ denotes the corresponding pattern and if $\mathbf{u}$ is a pattern, then $\hat{\mathbf{u}}$ denotes the corresponding type. We define a partial order on the patterns by declaring $\mathbf{u}_1 = \mathbf{p}(C_1) \leq \mathbf{p}(C_2) = \mathbf{u}_2$ if there is a continuous epimorphism $\phi : C_1 \to C_2$. A proper homomorphism is surjective exactly if its adjoint map is injective (Lemma 6.5.1(4)). Therefore, by Proposition 1.15.6(1) ⇔ (3), we obtain

$$\mathbf{u}_1 \leq \mathbf{u}_2 \quad \text{if and only if} \quad \hat{\mathbf{u}}_1 \geq \hat{\mathbf{u}}_2.$$

Let C be a compact connected group. Dually to the typeset of a torsion-free group (see Section 1.15), we define the **pattern set of** C to be $\operatorname{patset}(C) = \{\mathbf{p}(C/H) : H \text{ is a closed connected subgroup of } C \text{ with } \dim(C/H) = 1\}$. Since a closed subgroup of C is connected if and only if its annihilator is pure (see Corollary 6.4.20), we have:

Lemma 7.5.1. *Suppose that C is a compact connected group. Then we have* $\operatorname{patset}(C) = \{\hat{\mathbf{t}} : \mathbf{t} \in \operatorname{typeset}(\widehat{C})\}$, *and* $\operatorname{typeset}(\widehat{C}) = \{\hat{\mathbf{u}} : \mathbf{u} \in \operatorname{patset}(C)\}$.

For a pattern $\mathbf{u}$, we define $C[\mathbf{u}]$ to be the intersection of all closed connected subgroups H of C such that $\dim(C/H) = 1$, and $\mathbf{p}(C/H) \leq \mathbf{u}$, and

$$C^*[\mathbf{u}] = \bigcap_{\mathbf{v}<\mathbf{u}} C[\mathbf{v}].$$

Then the **co-Baer invariants** of C are defined to be the dimensions of the compact factor groups

$$C_{\mathbf{u}} = C^*[\mathbf{u}]/C[\mathbf{u}]$$

for all patterns **u**. The compact groups we defined here are linked to the discrete groups $G(\mathbf{t})$, $G^*(\mathbf{t})$, and $G_{\mathbf{t}}$ (see Section 1.15) as follows:

Lemma 7.5.2 (Leopold [96]). *Suppose that G is a discrete torsion-free group and $\mathbf{t}$ is a type. Then we have:*
1. $(\widehat{G}, G(\mathbf{t})) = \widehat{G}[\hat{\mathbf{t}}]$, *and* $(\widehat{G}, G^*(\mathbf{t})) = \widehat{G}^*[\hat{\mathbf{t}}]$.
2. $\dim \widehat{G_{\hat{\mathbf{t}}}} = r_0(G_{\mathbf{t}})$.

Proof. Lemma 1.10.8 shows that $G(\mathbf{t})$ is the subgroup of G generated by all pure rank 1 subgroups A of G with $\mathbf{t}(A) \geq \mathbf{t}$; thus, $(\widehat{G}, G(\mathbf{t}))$ is the intersection of all closed connected subgroups H of $\widehat{G}$ such that $\dim(\widehat{G}/H) = 1$ and $\mathbf{p}(\widehat{G}/H) \leq \hat{\mathbf{t}}$ (Lemmas 6.4.2, 6.4.4, 6.4.13 and Corollary 6.4.20). It follows that $(\widehat{G}, G(\mathbf{t})) = \widehat{G}[\hat{\mathbf{t}}]$. The proof of the second equation is very similar.

To prove (2), we use Lemma 6.4.12 and (1), obtaining $(G(\mathbf{t})/G^*(\mathbf{t}))\widehat{\ } \cong \widehat{G}^*[\hat{\mathbf{t}}]/\widehat{G}[\hat{\mathbf{t}}]$. Now use Lemma 7.1.1. □

A compact group is called **completely decomposable** if it is a direct product of 1-dimensional compact connected groups. Clearly, we have:

Theorem 7.5.3 (Leopold [95]). *Let C be a compact group. Then C is completely decomposable if and only if $\widehat{C}$ is completely decomposable.*

Then dualization of Baer's Theorem 1.15.4 yields a classification of compact completely decomposable groups in terms of numerical invariants (see Loth [100]):

Corollary 7.5.4. *Let C be a compact completely decomposable group. If $C = \prod_{i\in I} C_i = \prod_{j\in J} D_j$ such that each compact group C_i and D_j is connected and has dimension 1, then there is a bijection $\phi : I \to J$ such that $C_i \cong D_{\phi(i)}$ for all $i \in I$. Two compact completely decomposable groups are topologically isomorphic if and only if they have the same co-Baer invariants.*

Finally, by Theorem 3.16.4 and Corollary 3.16.5, we have:

Corollary 7.5.5. *The class of simply given connected groups coincides with the class of compact completely decomposable groups and is closed under direct factors.*

7.6 Direct factors of simply given groups

Since Warfield groups are generalizations of totally projective p-groups, the following question arises naturally: How can Kiefer's [84] theory be extended to a larger class of compact groups such as the class of compact groups whose duals are Warfield groups? Here we establish a classification theorem and prove a variety of characterizations corresponding to those of Warfield groups as described in Theorem 3.16.1.

First, we construct new invariants using global Stanton invariants (see Section 3.7), which will be added to the co-Ulm invariants. As expected, we need to describe the annihilators of the subgroups $G(M)$ and $G(M^*, p)$ of a discrete group G (see Section 3.6).

Let C be a compact group and $M = [m_{p,i}]_{(p,i)\in\mathbb{P}\times\omega}$ an Ulm matrix. We define

$$C_M = \overline{\sum_{(p,i)\in\mathbb{P}\times\omega} p^i(C[p^{m_{p,i}}])}.$$

Let $X = \{(p,i,j) \in \mathbb{P}\times\omega\times\mathbb{N} : m_{p,i+|j|_p} \neq \infty\}$, and let $\mathcal{F}_M$ be the set of all functions

$$\begin{array}{rccc} e: & \mathbb{N} & \to & X \\ & n & \mapsto & (p_{e_n}, i_{e_n}, j_{e_n}) \end{array}$$

with $j_{e_n} = n$ for all $n \in \mathbb{N}$. For $e \in \mathcal{F}_M$ and $n \in \mathbb{N}$, let $m_{e(n)} = m_{p_{e_n}, i_{e_n}+|n|_{p_{e_n}}}$. Then we define

$$C_{M^*} = \bigcap_{e\in\mathcal{F}_M}\left[\overline{C_M + \sum_{n\in\mathbb{N}} p_{e_n}^{i_{e_n}}(C[p_{e_n}^{m_{e(n)}+1}])}\right] \cap (C_M + C_0),$$

where C_0 denotes the identity component of C.

Notice that $\mathcal{F}_M \neq \emptyset$ exactly if for each $n \in \mathbb{N}$ there exists a $(p,i) \in \mathbb{P}\times\omega$ such that $m_{p,i+|n|_p} \neq \infty$. Thus, $\mathcal{F}_M = \emptyset$ if and only if $M \sim \overline{\infty}$. Since the empty intersection of a family of subgroups is the whole group, this implies that $C_{M^*} = C_M + C_0$ in the case where $M \sim \overline{\infty}$.

For an Ulm sequence $\overline{\alpha} = (\alpha_i)_{i<\omega}$ and a prime p, we define

$$C_{\overline{\alpha},p} = \overline{\sum_{i<\omega} p^i(C[p^{\alpha_i}])}.$$

Letting $I_{\overline{\alpha}}$ denote the set of all $i < \omega$ such that $\alpha_i \neq \infty$, we define

$$C_{\overline{\alpha}^*,p} = \bigcap_{I\subseteq I_{\overline{\alpha}},\, I \text{ infinite}} \left(C_{\overline{\alpha},p} + \overline{\sum_{i\in I} p^i(C[p^{\alpha_i+1}])}\right).$$

It is clear that in the case where $\alpha_i = \infty$ for some i, this intersection is empty, which implies that $C_{\overline{\alpha}^*,p} = C$. Finally, we define

$$C_{M^*,p} = C_M + (C_{M^*} \cap C_{M_p^*,p}).$$

Lemma 7.6.1. *Let G be a discrete group, p a prime, $\overline{\alpha}$ an Ulm sequence, and M an Ulm matrix. Then we have the following:*

1. $\widehat{G}_M = (\widehat{G}, G(M))$.
2. $\widehat{G}_{M^*} = (\widehat{G}, G(M^*))$.
3. $\widehat{G}_{\overline{\alpha},p} = (\widehat{G}, G(\overline{\alpha}, p))$.
4. $\widehat{G}_{\overline{\alpha}^*,p} = (\widehat{G}, G(\overline{\alpha}^*, p))$.
5. $\widehat{G}_{M^*,p} = (\widehat{G}, G(M^*, p))$.

Proof. First, write $M = [m_{p,i}]_{(p,i)\in\mathbb{P}\times\omega}$ and recall the definition $n^{-1}A = \{x \in G : nx \in A\}$ for a subgroup A of G and $n \in \mathbb{N}$ (see Section 6.4). Then $G(M) = \bigcap_{(p,i)\in\mathbb{P}\times\omega} p^{-i}(p^{m_{p,i}}G)$; thus,

$$(\widehat{G}, G(M)) = \Big(\widehat{G}, \bigcap_{(p,i)} p^{-i}(p^{m_{p,i}}G)\Big) = \overline{\sum_{(p,i)} (\widehat{G}, p^{-i}(p^{m_{p,i}}G))} = \overline{\sum_{(p,i)} p^i(\widehat{G}[p^{m_{p,i}}])}$$

by Lemmas 6.4.13, 6.4.14 and 7.1.3. It follows that $(\widehat{G}, G(M)) = \widehat{G}_M$.

To show the second assertion, we recall the notation $m_{e(n)} = m_{p_{e_n}, i_{e_n}+|n|_{p_{e_n}}}$ (where $e \in \mathcal{F}_M, n \in \mathbb{N}$) and write

$$\begin{aligned} G(M^*) &= \langle x \in G(M) : \|x\| \nsim M\rangle + t(G(M)) \\ &= \langle x \in G(M) : nM \ngeq \|x\| \text{ for all } n \in \mathbb{N}\rangle + t(G(M)) \\ &= \langle x \in G(M) : \text{ for every } n \in \mathbb{N} \text{ there is a } (p,i) \in \mathbb{P}\times\omega \text{ such that} \\ &\qquad |p^i x|_p > m_{p,i+|n|_p} \neq \infty\rangle + t(G(M)) \\ &= \Big\langle \bigcup_{e\in\mathcal{F}_M} \Big[G(M) \cap \bigcap_{n\in\mathbb{N}} p_{e_n}^{-i_{e_n}}(p_{e_n}^{m_{e(n)}+1}G)\Big]\Big\rangle + t(G(M)). \end{aligned}$$

Note that the annihilator of tG coincides with $\widehat{G}_0$ (Theorem 6.4.17). Using (1), we obtain

$$\begin{aligned} (\widehat{G}, G(M^*)) &= \bigcap_{e\in F_M} \Big[(\widehat{G}, G(M)) + \Big(\widehat{G}, \bigcap_{n\in\mathbb{N}} p_{e_n}^{-i_{e_n}}(p_{e_n}^{m_{e(n)}+1}G)\Big)\Big] \cap (\widehat{G}, t(G(M))) \\ &= \bigcap_{e\in\mathcal{F}_M} \Big[\overline{\widehat{G}_M + \sum_{n\in\mathbb{N}} p_{e_n}^{i_{e_n}}(\widehat{G}, p_{e_n}^{m_{e(n)}+1}G)}\Big] \cap (\widehat{G}, G(M) \cap tG) \\ &= \bigcap_{e\in\mathcal{F}_M} \Big[\overline{\widehat{G}_M + \sum_{n\in\mathbb{N}} p_{e_n}^{i_{e_n}}(\widehat{G}[p_{e_n}^{m_{e(n)}+1}])}\Big] \cap (\widehat{G}_M + \widehat{G}_0) \\ &= \widehat{G}_{M^*}. \end{aligned}$$

For (3), we write $\overline{\alpha} = (\alpha_i)_{i<\omega}$ and have $G(\overline{\alpha}, p) = \bigcap_{i<\omega} p^{-i}(p^{\alpha_i}G)$; therefore, $(\widehat{G}, G(\overline{\alpha}, p)) = \overline{\sum_{i<\omega}(\widehat{G}, p^{-i}(p^{\alpha_i}G))} = \overline{\sum_{i<\omega} p^i(\widehat{G}[p^{\alpha_i}])} = \widehat{G}_{\overline{\alpha},p}$, as desired.

To show (4), we recall that $I_{\overline{\alpha}}$ denotes the set of all $i < \omega$ with $\alpha_i \neq \infty$. Then

$$G(\overline{\alpha}^*, p) = \Big\langle \bigcup_{I\subseteq I_{\overline{\alpha}}, I \text{ infinite}} \Big(G(\overline{\alpha}, p) \cap \bigcap_{i\in I} p^{-i}(p^{\alpha_i+1}G)\Big)\Big\rangle,$$

which yields

$$\begin{aligned}(\widehat{G}, G(\overline{\alpha}^*, p)) &= \bigcap_{I\subseteq I_{\overline{\alpha}}, I \text{ infinite}} \left[(\widehat{G}, G(\overline{\alpha}, p)) + \overline{\sum_{i\in I}(\widehat{G}, p^{-i}(p^{\alpha_i+1}G))}\right] \\ &= \bigcap_{I\subseteq I_{\overline{\alpha}}, I \text{ infinite}} \left(\widehat{G}_{\overline{\alpha},p} + \sum_{i\in I} p^i(\widehat{G}[p^{\alpha_i+1}])\right) \\ &= \widehat{G}_{\overline{\alpha}^*,p}.\end{aligned}$$

Finally, we have

$$\begin{aligned}(\widehat{G}, G(M^*, p)) &= (\widehat{G}, G(M) \cap (G(M^*) + G(M_p^*, p))) \\ &= (\widehat{G}, G(M)) + (\widehat{G}, G(M^*) + G(M_p^*, p)) \\ &= \widehat{G}_M + (\widehat{G}_{M^*} \cap \widehat{G}_{M_p^*,p}) \\ &= \widehat{G}_{M^*,p}.\end{aligned}$$

□

For a compatibility class c of Ulm matrices, prime p, and equivalence class e of Ulm sequences, the **co-Stanton invariant** of a compact group G is defined as

$$s^*(c,p,e,G) = \sup\{\dim_{p,0}(G_{M^*,p}/G_M) : M \in c \text{ and } M_p \in e\},$$

where $\dim_{p,0}$ denotes the sum of the p-dimension $\dim_p$ and the dimension dim (see Section 7.1). As expected, co-Stanton invariants are the dual counterparts of Stanton invariants (see Section 3.7):

Proposition 7.6.2. *Let G be a discrete group, c a compatibility class of Ulm matrices, p a prime, and e an equivalence class of Ulm sequences. Then $s(c,p,e,G) = s^*(c,p,e,\widehat{G})$.*

Proof. For any $M \in c$ with $M_p \in e$, we have

$$(G(M)/G(M^*,p))\widehat{\ } \cong (\widehat{G}, G(M^*,p))/(\widehat{G}, G(M)) = \widehat{G}_{M^*,p}/\widehat{G}_M$$

by Lemmas 6.4.2, 6.4.12, and 7.6.1(1), (5). Now use Lemma 7.1.1. □

In view of Theorems 3.7.3, 3.7.7, 3.16.1(1) ⇔ (3), Corollary 6.4.11, and duality, we have the following classification result for the duals of Warfield groups:

Theorem 7.6.3 (Loth [102]). *Let C and D be direct factors of simply given groups. Then C and D are topologically isomorphic if and only if they have the same co-Ulm invariants u_p^* ($p \in \mathbb{P}$) and the same co-Stanton invariants s^*.*

Next we establish various characterizations of direct factors of simply given groups, dual to those of Warfield groups in Theorem 3.16.1. First, we need to dualize the concept of global niceness (see Section 3.8). A closed subgroup F of a compact group C is called **smart (in C)** if the factor group

$$\frac{F \cap C[p^\alpha]}{F[p^\alpha]}$$

is p-divisible for all primes p and ordinals α.

Proposition 7.6.4. *A subgroup A of a discrete group G is nice in G if and only if $(\widehat{G}, A)$ is smart in $\widehat{G}$.*

Proof. Let p be a prime, α an ordinal, and $\psi : (G/A)\widehat{} \to (\widehat{G}, A)$ the topological isomorphism, as in Lemma 6.4.2. By Lemma 7.4.6, ψ maps $((G/A)\widehat{}, p^\alpha(G/A))$ onto $(\widehat{G}, A)[p^\alpha]$, and maps $((G/A)\widehat{}, (p^\alpha G + A)/A)$ onto $\widehat{G}[p^\alpha] \cap (\widehat{G}, A)$. Letting $H = (p^\alpha(G/A))/(p^\alpha G + A)/A$, we obtain

$$\widehat{H} \cong ((p^\alpha(G/A))\widehat{}, (p^\alpha G + A)/A) \cong \frac{((G/A)\widehat{}, (p^\alpha G + A)/A)}{((G/A)\widehat{}, p^\alpha(G/A))} \cong \frac{\widehat{G}[p^\alpha] \cap (\widehat{G}, A)}{(\widehat{G}, A)[p^\alpha]}$$

by Lemmas 6.4.2 and 6.4.12. Then Lemma 6.4.15 shows that $(\widehat{H}, H[p]) = p\widehat{H}$; thus, $H[p] = 0$ if and only if $p\widehat{H} = \widehat{H}$ (Corollary 6.4.5). In other words, $H[p] = 0$ exactly if $(\widehat{G}[p^\alpha] \cap (\widehat{G}, A))/(\widehat{G}, A)[p^\alpha]$ is p-divisible. □

We say that a compact group C has a **quasi-decomposition** if it has a closed totally disconnected subgroup K such that $A = C/K$ can be written as $A = \prod_{i \in I} A_i$ for some set I, where $A_i = \mathbb{T}$ $(i \in I)$ such that the exact sequence $0 \to K \to C \xrightarrow{\varphi} A \to 0$ induces exact sequences $0 \to K_i \to C \xrightarrow{\pi_i \varphi} A_i \to 0$ ($\pi_i : A \to A_i$ is the i-th projection map), which yield

$$K + C_M = \bigcap_{i \in I} (K_i + C_M)$$

for all Ulm matrices M. If in addition, K is simply given and smart in C, then we say that C has a **simply given quasi-decomposition**. Such groups are exactly the duals of Warfield groups:

Proposition 7.6.5 (Loth [101]). *Let G be a discrete group. Then we have:*

1. *G has a decomposition basis if and only if $\widehat{G}$ has a quasi-decomposition.*
2. *G has a nice decomposition basis X such that $G/\langle X \rangle$ is simply presented if and only if $\widehat{G}$ has a simply given quasi-decomposition.*

Proof. Suppose that G has a decomposition basis $X = \{x_i : i \in I\}$. For each $i \in I$, we have a commutative diagram

$$\begin{array}{ccccccccc}
0 & \to & \langle X \rangle & \to & G & \to & G/\langle X \rangle & \to & 0 \\
 & & \uparrow \subseteq & & \| & & \uparrow & & \\
0 & \to & \langle x_i \rangle & \to & G & \to & G/\langle x_i \rangle & \to & 0
\end{array}$$

with exact rows. Using Lemma 6.4.2, we obtain the dual diagram

$$\begin{array}{ccccccccc}
0 & \leftarrow & \prod_{i \in I} \langle x_i \rangle\widehat{} & \leftarrow & \widehat{G} & \leftarrow & K = (\widehat{G}, \langle X \rangle) & \leftarrow & 0 \\
 & & \downarrow \pi_i & & \| & & \downarrow \subseteq & & \\
0 & \leftarrow & \langle x_i \rangle\widehat{} & \leftarrow & \widehat{G} & \leftarrow & K_i = (\widehat{G}, \langle x_i \rangle) & \leftarrow & 0.
\end{array}$$

Then K is totally disconnected (Theorem 6.4.17), and $\langle x_i\rangle\widehat{\ } \cong \mathbb{T}$ for all $i \in I$. Let M be an Ulm matrix. Then $\langle X\rangle \cap G(M) = \sum_{i\in I}(\langle x_i\rangle \cap G(M))$ implies that

$$(\widehat{G}, \langle X\rangle) + (\widehat{G}, G(M)) = \bigcap_{i\in I}(\widehat{G}, \langle x_i\rangle \cap G(M)) = \bigcap_{i\in I}(\widehat{G}, \langle x_i\rangle) + (\widehat{G}, G(M)),$$

so by Lemma 7.6.1(1), we obtain $K + \widehat{G}_M = \bigcap_{i\in I}(K_i + \widehat{G}_M)$, as desired.

Conversely, assume that $\widehat{G}$ has a quasi-decomposition. Then there is a closed totally disconnected subgroup K of $\widehat{G}$ such that $\widehat{G}/K \cong \prod_{i\in I} A_i$ with $A_i = \mathbb{T}$ for all $i \in I$ yielding commuting diagrams with exact rows

$$\begin{array}{ccccccccc} 0 & \to & K & \to & \widehat{G} & \xrightarrow{\varphi} & \prod_{i\in I} A_i & \to & 0 \\ & & \downarrow \subseteq & & \| & & \downarrow \pi_i & & \\ 0 & \to & K_i & \to & \widehat{G} & \to & A_i & \to & 0, \end{array}$$

satisfying $K + \widehat{G}_M = \bigcap_{i\in I}(K_i + \widehat{G}_M)$. Then we obtain dual diagrams

$$\begin{array}{ccccccccc} 0 & \leftarrow & \widehat{K} & \leftarrow & G & \xleftarrow{\varphi^*} & \bigoplus_{i\in I}(A_i)\widehat{\ } & \leftarrow & 0 \\ & & \uparrow & & \| & & \uparrow \pi_i^* & & \\ 0 & \leftarrow & (K_i)\widehat{\ } & \leftarrow & G & \leftarrow & (A_i)\widehat{\ } & \leftarrow & 0, \end{array}$$

where the dual group of $\widehat{G}$ is identified with G. For each $i \in I$, we have $(A_i)\widehat{\ } \cong \mathbb{Z}$, so we can write $\varphi^*((A_i)\widehat{\ }) = \langle x_i\rangle$ for some $x_i \in G$. Clearly, $X = \{x_i : i \in I\}$ is a set of independent elements of infinite order, and $G/\langle X\rangle \cong \widehat{K}$ is torsion. To show that $\bigoplus_{i\in I}\langle x_i\rangle$ is a valuated coproduct in G, let M be an Ulm matrix. Then $(G, K + \widehat{G}_M) = (G, \ker\varphi) \cap (G, \widehat{G}_M) = \operatorname{im}\varphi^* \cap G(M)$ and $(G, \bigcap_{i\in I}(K_i + \widehat{G}_M)) = \sum_{i\in I}(G, K_i + \widehat{G}_M) = \sum_{i\in I}((G, \ker\pi_i\varphi) \cap (G, \widehat{G}_M)) = \sum_{i\in I}(\operatorname{im}\varphi^*\pi_i^* \cap G(M))$; thus, $\langle X\rangle \cap G(M) = \sum_{i\in I}(\langle x_i\rangle \cap G(M))$. Therefore, X is a decomposition basis of G.

The second statement follows from (1), Corollary 3.8.6, and Propositions 7.3.2, 7.6.4. □

A smart subgroup F of a compact group C is called **$*$-smart (in C)** if for every closed subgroup D of C with $C/D \cong \mathbb{T}$, there is a positive integer n such that

$$F_M + (n^{-1}D \cap F) = (C_M \cap F) + (n^{-1}D \cap F)$$

for each Ulm matrix M.

Proposition 7.6.6 (Loth [101]). *Let A be a subgroup of a discrete group G. Then A is quasi-sequentially nice in G if and only if $(\widehat{G}, A)$ is $*$-smart in $\widehat{G}$.*

Proof. Let $F = (\widehat{G}, A)$. By Proposition 7.6.4, A is nice in G exactly if F is smart in $\widehat{G}$. Let $\psi : (G/A)\widehat{\ } \to (\widehat{G}, A)$ be the topological isomorphism as in Lemma 6.4.2. Suppose that A is quasi-sequentially nice and assume that D is a closed subgroup of $\widehat{G}$ with

$\widehat{G}/D \cong \mathbb{T}$. Then $B = (G, D)$ is an infinite cyclic group since $\widehat{B} \cong \widehat{G}/(\widehat{G}, B) \cong \mathbb{T}$. Therefore, by Lemma 3.9.6(1)⇒(4), there is an $n \in \mathbb{N}$ such that

$$(G/A)(M) \cap (nB + A)/A = (G(M) + A)/A \cap (nB + A)/A$$

for each Ulm matrix M. Let H_1 and H_2 denote the annihilator in $(G/A)\widehat{\ }$ of the left and right side, respectively, of this equation. Then $H_1 = ((G/A)\widehat{\ }, (G/A)(M)) + ((G/A)\widehat{\ }, (nB + A)/A)$, and $H_2 = ((G/A)\widehat{\ }, (G(M) + A)/A) + ((G/A)\widehat{\ }, (nB + A)/A)$; thus, $\psi(H_1) = (\widehat{G}, A)_M + (\widehat{G}, nB + A) = (\widehat{G}, A)_M + (n^{-1}(\widehat{G}, B) \cap (\widehat{G}, A))$, and $\psi(H_2) = (\widehat{G}, G(M) + A) + (\widehat{G}, nB + A) = (\widehat{G}_M \cap (\widehat{G}, A)) + (n^{-1}(\widehat{G}, B) \cap (\widehat{G}, A))$. Consequently, $F = (\widehat{G}, A)$ is $*$-smart in $\widehat{G}$.

Conversely, assume that $F = (\widehat{G}, A)$ is $*$-smart in $\widehat{G}$, and let B be an infinite cyclic subgroup of G. Then $\widehat{G}/D \cong \mathbb{T}$ with $D = (\widehat{G}, B)$; hence, there is an $n \in \mathbb{N}$ such that for each Ulm matrix M, $\psi(H_1) = \psi(H_2)$, where H_1, H_2 are as in the paragraph above. But then the groups $(G/A)(M) \cap (nB + A)/A$ and $(G(M) + A)/A \cap (nB + A)/A$ have the same annihilator, hence they are identical. By Lemma 3.9.6(4)⇒(1), A is quasi-sequentially nice in G. □

Given a discrete group G and a primitive element $x \in G$ (see Section 3.10), we wish to describe the annihilator $(\widehat{G}, \{x\})$. Suppose that A is a closed subgroup of a compact group C such that $C/A \cong \mathbb{T}$. Then A is called **plain** if for every Ulm matrix $M = [m_{q,i}]_{(q,i) \in \mathbb{P} \times \omega}$, prime p, and $n \in \mathbb{N}$, the following is true: If $n(C_{M^*,p}) \subseteq A$, then one of the following conditions holds:

(a) For every $n' \in \mathbb{N}$, there is a prime q and an $i < \omega$ such that $m_{q,i+|n'|_q} \neq \infty$ and $q^i(C[q^{m_{q,i+|n'|_q}+1}]) \subseteq A$,
or
(b) $m_{p,i} \neq \infty$ for all $i < \omega$, and $np^i(C[p^{m_{p,i}+1}]) \subseteq A$ for infinitely many $i < \omega$.

Lemma 7.6.7. *Let G be a discrete group and $x \in G$. Then x is primitive in G if and only if $(\widehat{G}, \{x\})$ is plain in $\widehat{G}$. For every plain subgroup B of $\widehat{G}$, there is a primitive element $y \in G$ such that $B = (\widehat{G}, \{y\})$.*

Proof. Let $A = (\widehat{G}, \{x\})$, $M = [m_{q,i}]$ an Ulm matrix, p a prime, and $n \in \mathbb{N}$.

First, suppose that $x \in G$ is primitive, and assume that $n(\widehat{G}_{M^*,p}) \subseteq A$. Then $(\widehat{G}, n^{-1}(G(M^*, p))) = n(\widehat{G}, G(M^*, p)) = n(\widehat{G}_{M^*,p})$ shows that $x \in n^{-1}(G(M^*, p))$; thus, $nx \in G(M^*, p)$. Since x is primitive, we have $\langle x\rangle \cong \mathbb{Z}$, which yields $\widehat{G}/A \cong \langle x\rangle\widehat{\ } \cong \mathbb{T}$, and one of the following cases holds:

Case I: $\|x\| \nleq M$. Let $n' \in \mathbb{N}$. Since $nx \in G(M)$, we have $n\|x\| \geq M$. Thus, $n'M \ngeq \|x\|$. Therefore, there is a prime q and an $i < \omega$ such that $|q^i x|_q > m_{q,i+|n'|_q} \neq \infty$. Then $x \in q^{-i}(q^{m_{q,i+|n'|_q}+1}G)$, and we obtain $q^i(\widehat{G}[q^{m_{q,i+|n'|_q}+1}]) = q^i(\widehat{G}, q^{m_{q,i+|n'|_q}+1}G) = (\widehat{G}, q^{-i}(q^{m_{q,i+|n'|_q}+1}G)) \subseteq (\widehat{G}, \{x\}) = A$.

Case II: $nx \in G(M_p^*, p)^{\diamond}$. Then $|np^i x|_p > m_{p,i} \neq \infty$ for infinitely many $i < \omega$. But this shows that for infinitely many $i < \omega$, we have $x \in (np^i)^{-1}(p^{m_{p,i}+1}G)$, and therefore, $np^i(\widehat{G}[p^{m_{p,i}+1}]) = (\widehat{G}, (np^i)^{-1}(p^{m_{p,i}+1}G)) \subseteq A$, as desired.

Conversely, suppose that A is plain in $\widehat{G}$, and assume that $nx \in G(M^*,p)$. Then $x \in n^{-1}(G(M^*,p))$; hence, $n(\widehat{G}_{M^*,p}) = n(\widehat{G},G(M^*,p)) = (\widehat{G},n^{-1}(G(M^*,p))) \subseteq A$. In case condition (a) holds, let $n' \in \mathbb{N}$. Then there is a prime q and an $i < \omega$ such that $m_{q,i+|n'|_q} \neq \infty$ and $(\widehat{G},q^{-i}(q^{m_{q,i+|n'|_q}+1}G)) = q^i(\widehat{G}[q^{m_{q,i+|n'|_q}+1}]) \subseteq A = (\widehat{G},\{x\})$; thus, $|q^ix|_q > m_{q,i+|n'|_q}$. It follows that $n'M \ngeq \|x\|$. Since n' was arbitrary, we obtain $\|x\| \nsim M$. Now assume that condition (b) holds. Then $m_{p,i} \neq \infty$ for all $i < \omega$, and $np^i(\widehat{G}[p^{m_{p,i}+1}]) \subseteq A$ for infinitely many $i < \omega$. Consequently, for infinitely many $i < \omega$, we have $x \in (np^i)^{-1}(p^{m_{p,i}+1}G)$, and therefore, $|np^ix|_p > m_{p,i}$. This shows that x is primitive.

To prove the last assertion, let B be any plain subgroup of $\widehat{G}$. Then $H = (G,B)$ satisfies $\widehat{H} \cong \widehat{G}/B \cong \mathbb{T}$. Thus, $H = \langle y\rangle$ for some element y of infinite order. Then $B = (\widehat{G},H) = (\widehat{G},\{y\})$. Hence, y is primitive in G as we just saw. □

Next, we dualize the concept of a knice subgroup (see Section 3.12). A smart subgroup F of a compact group C is called **k-smart (in C)** whenever the following condition holds: If $C/D \cong \mathbb{T}^k$ for some closed subgroup D of C and $k < \omega$, then there is a positive integer r and an exact sequence

$$0 \to B \xrightarrow{\subseteq} C \xrightarrow{\phi} \prod_{i=1}^{n} C_i \to 0$$

($n \geq 1$, $C_1 = C/F$, $C_i = \mathbb{T}$ for all $i = 2,\dots,n$) with $rB \subseteq D$ so that the induced exact sequences

$$0 \to B_i \xrightarrow{\subseteq} C \xrightarrow{\pi_i\phi} C_i \to 0$$

($\pi_i : \prod_{i=1}^n C_i \to C_i$ is the i-th projection map) yield the following:
(i) $B_1 = F$;
(ii) B_i is a plain subgroup of C for $i = 2,\dots,n$;
(iii) $B + K = \bigcap_{i=1}^n (B_i + K)$ for all subgroups K of the form C_M, C_{M^*}, $C_{\overline{\alpha}^*,p}$ or $C_{M^*,p}$.

Proposition 7.6.8 (Loth [101]). *Let A be a subgroup of a discrete group G. Then A is knice in G if and only if $(\widehat{G},A)$ is k-smart in $\widehat{G}$.*

Proof. By Proposition 7.6.4, A is nice in G exactly if $F = (\widehat{G},A)$ is smart in $\widehat{G}$. Suppose that A is knice in G, and let D be a closed subgroup of $\widehat{G}$ such that $\widehat{G}/D \cong \mathbb{T}^k$ for some $k < \omega$. We let $E = (G,D)$ and have $E \cong \widehat{(\widehat{G}/D)} \cong \mathbb{Z}^k$, so E is a finitely generated subgroup of G. Since A is knice, there is a $*$-valuated coproduct $A \oplus \langle x_2\rangle \oplus \cdots \oplus \langle x_n\rangle$ in G containing rE for some $r \in \mathbb{N}$, where $x_2,\dots,x_n$ are primitive elements of G and $n \geq 1$. Let $A_1 = A$, $A_i = \langle x_i\rangle$ for $i = 2,\dots,n$, and define $C_i = \widehat{G}/(\widehat{G},A_i)$ for $i = 1,\dots,n$. Then $B = (\widehat{G},\sum_{i=1}^n A_i)$ is the kernel of the natural map $\widehat{G} \to \prod_{i=1}^n C_i$, and we have $rB = (\widehat{G},r^{-1}(\sum_{i=1}^n A_i)) \subseteq (\widehat{G},E) = D$.

For $i = 1,\dots,n$, we let $B_i = (\widehat{G},A_i)$. Then $B_1 = F$, and condition (ii) is satisfied by Lemma 7.6.7. Since $\bigoplus_{i=1}^n A_i$ is a $*$-valuated coproduct in G, we have $(\widehat{G},\sum_{i=1}^n A_i)+(\widehat{G},H) =$

$(\widehat{G}, (\sum_{i=1}^n A_i) \cap H) = (\widehat{G}, \sum_{i=1}^n (A_i \cap H)) = \bigcap_{i=1}^n ((\widehat{G}, A_i) + (\widehat{G}, H))$ for each group H of the form $G(M)$, $G(\overline{\alpha}^*, p)$, $G(M^*)$ or $G(M^*, p)$. Thus, Lemma 7.6.1 shows that condition (iii) holds as well.

Conversely, suppose that $F = (\widehat{G}, A)$ is k-smart in $\widehat{G}$, and let S be a finite subset of G. Writing $\langle S\rangle = Z \oplus T$ with $Z \cong \mathbb{Z}^k$ $(k < \omega)$ and T finite (Theorem 1.9.3), we define $D = (\widehat{G}, Z)$ and have $\widehat{G}/D \cong \widehat{Z} \cong \mathbb{T}^k$. Then there is an $r \in \mathbb{N}$ and short exact sequences as described in the definition of "k-smart". Therefore, we have commuting diagrams with exact rows

$$\begin{array}{ccccccccc} 0 & \to & B & \to & \widehat{G} & \xrightarrow{\phi} & \prod_i C_i & \to & 0 \\ & & \downarrow & & \| & & \downarrow \pi_i & & \\ 0 & \to & B_i & \to & \widehat{G} & \to & C_i & \to & 0, \end{array}$$

inducing commuting dual diagrams

$$\begin{array}{ccccccccc} 0 & \leftarrow & \widehat{B} & \leftarrow & G & \xleftarrow{\phi^*} & \bigoplus_i (C_i)\widehat{\ } & \leftarrow & 0 \\ & & \uparrow & & \| & & \uparrow \pi_i^* & & \\ 0 & \leftarrow & (B_i)\widehat{\ } & \leftarrow & G & \leftarrow & (C_i)\widehat{\ } & \leftarrow & 0. \end{array}$$

Now write $\operatorname{im}\phi^* = \bigoplus_{i=1}^n A_i \subseteq G$ with $A_i = \phi^* \pi_i^*((C_i)\widehat{\ }) = (G, \ker(\pi_i \phi)) = (G, B_i)$ for $i = 1, \ldots, n$. Then $B_1 = F$, and for all $i = 2, \ldots, n$, the group B_i is plain in $\widehat{G}$; thus, $B_i = (\widehat{G}, \{y_i\})$ for some primitive element y_i of G (Lemma 7.6.7), and we have $A_i = (G, B_i) = \langle y_i\rangle$. By condition (iii) and Lemma 7.6.1, $A \oplus \langle y_2\rangle \oplus \cdots \oplus \langle y_n\rangle$ is a $*$-valuated coproduct in G. Finally, $rB \subseteq D$ implies that $Z = (G, D) \subseteq (G, rB) = r^{-1}(G, B) = r^{-1}(G, \ker\phi) = r^{-1}\operatorname{im}\phi^* = r^{-1}(A \oplus \langle y_2\rangle \oplus \cdots \oplus \langle y_n\rangle)$. Thus, $A \oplus \langle y_2\rangle \oplus \cdots \oplus \langle y_n\rangle$ contains a nonzero multiple of $\langle S\rangle$. Therefore, A is knice in G. □

A short exact sequence $0 \to A \to B \to C \to 0$ of compact groups is called **sequentially cobalanced** if for each Ulm matrix M, the induced sequence

$$0 \to A/A_M \to B/B_M \to C/C_M \to 0$$

is exact. By Lemma 7.6.1(1), we have:

Lemma 7.6.9. *An exact sequence $0 \to A \to B \to C \to 0$ of discrete groups is sequentially pure if and only if its dual sequence $0 \to \widehat{C} \to \widehat{B} \to \widehat{A} \to 0$ is sequentially cobalanced.*

We now have all the tools available to dualize the various characterizations of Warfield groups given in Theorem 3.16.1. Most of the characterizations can be found in Loth [103].

Theorem 7.6.10. *Let C be a compact group. Then the following conditions are equivalent:*

1. *C is the dual of a Warfield group;*
2. *C is a direct factor of a simply given group;*
3. *C has the injective property relative to all sequentially cobalanced sequences of compact groups;*
4. *C has a simply given quasi-decomposition;*
5. *C has an H^*-family of k-smart subgroups;*
6. *C has a G^*-family of k-smart subgroups;*
7. *C has a G^*-family of pure k-smart subgroups;*
8. *C has an F^*-family of pure k-smart subgroups;*
9. *C has a descending composition series of k-smart subgroups;*
10. *C has a descending composition series of $*$-smart subgroups;*
11. *C has a descending composition series $\{F_\alpha\}_{\alpha\le\lambda}$ of k-smart subgroups such that if $F_\alpha/F_{\alpha+1}$ is infinite, there is a plain subgroup D_α of C containing $F_{\alpha+1}$ such that*
    (i) $C/F_{\alpha+1} = F_\alpha/F_{\alpha+1} \bigoplus D_\alpha/F_{\alpha+1}$;
    (ii) $F_{\alpha+1} + K = (F_\alpha + K) \cap (D_\alpha + K)$ *for each subgroup K of the form C_M, C_{M^*}, $C_{\overline{\alpha}^*,p}$, or $C_{M^*,p}$;*
12. *C has a descending composition series $\{F_\alpha\}_{\alpha\le\lambda}$ of smart subgroups such that if $F_\alpha/F_{\alpha+1}$ is infinite, there is a closed subgroup D_α of C containing $F_{\alpha+1}$ such that*
    (i) $C/F_{\alpha+1} = F_\alpha/F_{\alpha+1} \bigoplus D_\alpha/F_{\alpha+1}$;
    (ii) $F_{\alpha+1} + C_M = (F_\alpha + C_M) \cap (D_\alpha + C_M)$ *for all Ulm matrices M;*
13. *C has an H^*-family of smart subgroups and possesses a quasi-decomposition;*
14. *C has a G^*-family of smart subgroups and possesses a quasi-decomposition.*

Finally, dualization of Theorem 3.16.3 yields a uniqueness theorem for the class of direct factors of simply given groups:

Corollary 7.6.11. *Any class of compact groups that contains all simply given groups is closed under finite direct products and direct factors, and in which the co-Ulm invariants and co-Stanton invariants distinguish between non-topologically isomorphic groups, coincides with the class of direct factors of simply given groups.*

Exercises.
1. Let A be a subgroup of a compact group C and $n \in \mathbb{N}$. If A is plain, then so is $n^{-1}A$.
2. Show that the definitions of smart subgroup given in this section and in Section 7.4 coincide whenever C is a compact topological p-group.
3. In the definition of the co-Stanton invariant $s^*(c, p, e, G)$, $\dim_{p,0} = \dim_p$ in the case where $(\infty, \infty, \ldots) \notin e$, and $\dim_{p,0} = \dim$ otherwise.

7.7 Groups with partial quasi-decompositions

In this section, we determine the duals of PDB groups and establish a classification result. Recall that a discrete group G is said to be a PDB group if it has a nonempty system $\mathcal{C}$ of finite decomposition sets in G such that for every $X \in \mathcal{C}$ and $x \in G$, there is a $Y \in \mathcal{C}$ with $X \subseteq Y$ and $nx \in \langle Y \rangle$ for some $n \in \mathbb{N}$ (see Section 5.1). It is easily verified that G is a PDB group exactly if there exists a nonempty system $\mathcal{A}$ of subgroups

$$A = \bigoplus_{i \in I} A_i \subseteq G$$

with I finite and $A_i = 0$ or $A_i \cong \mathbb{Z}$ $(i \in I)$ such that the following properties are satisfied:

(P1) For each $A = \bigoplus_{i \in I} A_i \in \mathcal{A}$, we have $A \cap G(M) = \bigoplus_{i \in I}(A_i \cap G(M))$ for all Ulm matrices M;

(P2) For each $A \in \mathcal{A}$ and homomorphism $\alpha : \mathbb{Z} \to G$, there is a $B = \bigoplus_{j \in J} B_j \in \mathcal{A}$ and a subset J' of J such that $A = \bigoplus_{j \in J'} B_j$ and $\operatorname{im} \alpha \subseteq n^{-1}B$ for some $n \in \mathbb{N}$.

In fact, if G has such a system $\mathcal{A}$ of subgroups, then $\{X : X$ is a decomposition set in G and $\langle X \rangle \in \mathcal{A}\}$ is a partial decomposition basis for G, and if $\mathcal{C}$ is a partial decomposition basis for G, then the system $\{\bigoplus_{x \in X} \langle x \rangle : X \in \mathcal{C}\}$ satisfies conditions (P1) and (P2).

Dually, we say that a compact group C has a **partial quasi-decomposition** if there exists a nonempty system $\mathcal{D}$ of factor groups of C by closed subgroups such that each $D = C/K \in \mathcal{D}$ can be written as $D = \bigoplus_{i \in I} D_i$ with I finite and $D_i = 0$ or $D_i \cong \mathbb{T}$ $(i \in I)$ so that the following properties hold:

(P1*) The exact sequence $0 \to K \to C \to D \to 0$ induces exact sequences

$$0 \to K_i \to C \to D_i \to 0 \quad (i \in I),$$

which yield $K + C_M = \bigcap_{i \in I}(K_i + C_M)$ for all Ulm sequences M;

(P2*) for every $D = C/K \in \mathcal{D}$ and $\delta \in \widehat{C}$, there is an $E = C/L = \bigoplus_{j \in J} E_j \in \mathcal{D}$ and a subset J' of J such that $K = \bigcap_{j \in J'} L_j$ and $nL \subseteq \ker \delta$ for some $n \in \mathbb{N}$.

For instance, every zero-dimensional compact group C has a partial quasi-decomposition. To see this, let $\mathcal{D} = \{0\}$. If $\delta \in \widehat{C}$, then $\operatorname{im} \delta \cong \mathbb{T}$ is impossible since $C/\ker \delta \cong \mathbb{T}$ would contradict $\dim C = 0$. Thus, $\operatorname{im} \delta \cong \mathbb{Z}/n\mathbb{Z}$ for some $n \in \mathbb{N}$ (Corollary 6.4.6), and it follows that $nC \subseteq \ker \delta$.

Proposition 7.7.1. *A discrete group G is a PDB group if and only if $\widehat{G}$ has a partial quasi-decomposition.*

Proof. Suppose that G has a system $\mathcal{A}$ satisfying (P1) and (P2). We let

$$\mathcal{D} = \{\widehat{G}/(\widehat{G}, A) : A \in \mathcal{A}\}$$

and pick $D = \widehat{G}/K \in \mathcal{D}$ with $K = (\widehat{G}, A)$ and $A = \bigoplus_{i \in I} A_i \in \mathcal{A}$. By Lemma 6.4.22, we obtain $\widehat{G}/K = \bigoplus_{i \in I}(\widehat{G}, H_i)/K$ with $H_i = \sum_{j \neq i} A_j$ $(i \in I)$ and exact sequences

$$0 \to (\widehat{G}, A_i) \to \widehat{G} \to (\widehat{G}, H_i)/K \to 0.$$

If M is an Ulm matrix, then we have $K + \widehat{G}_M = (\widehat{G}, A \cap G(M)) = (\widehat{G}, \sum_{i\in I}(A_i \cap G(M))) = \bigcap_{i\in I}(\widehat{G}, A_i \cap G(M)) = \bigcap_{i\in I}[(\widehat{G}, A_i) + \widehat{G}_M]$; thus, (P1*) is satisfied.

Letting $\delta : \widehat{G} \to \mathbb{T}$ be a continuous homomorphism, we write its adjoint map as $\alpha = \delta^* : \mathbb{Z} \to G$. Then by condition (P2), there is a $B = \bigoplus_{j\in J} B_j \in \mathcal{A}$ and a subset J' of J such that $A = \bigoplus_{j\in J'} B_j$ and $\operatorname{im}\alpha \subseteq n^{-1}B$ for some $n \in \mathbb{N}$. Letting $E = \widehat{G}/L$ with $L = (\widehat{G}, B)$, we have $E \in \mathcal{D}$. Using Lemma 6.4.22 again, we obtain

$$\widehat{G}/L = \bigoplus_{j\in J}(\widehat{G}, N_i)/L$$

with $N_i = \sum_{j\neq i} B_j$ $(i \in J)$, and for each $j \in J$, the annihilator $L_j = (\widehat{G}, B_j)$ is the kernel of the natural map $\widehat{G} \to (\widehat{G}, N_j)/L$. Since $K = (\widehat{G}, A) = (\widehat{G}, \sum_{j\in J'} B_j) = \bigcap_{j\in J'} L_j$ and $nL = n(\widehat{G}, B) = (\widehat{G}, n^{-1}B) \subseteq (\widehat{G}, \operatorname{im}\alpha) = \ker\delta$, we conclude that (P2*) is satisfied.

Conversely, assume that $\widehat{G}$ has a partial quasi-decomposition with system $\mathcal{D}$ as required. Define $\mathcal{A}$ to be the set of all subgroups (G, K) of G such that K is a closed subgroup of $\widehat{G}$ with $\widehat{G}/K \in \mathcal{D}$. Then use the fact that $K = (\widehat{G}, (G, K))$ and Lemma 6.4.22 to show that the system $\mathcal{A}$ satisfies (P1) and (P2). □

Next, we wish to dualize the concept of partial isomorphism (see Section 4.1). Recall that two discrete groups G and H are partially isomorphic if and only if there is a nonempty system I of isomorphisms between subgroups of G and H satisfying the back-and-forth property, which we rephrase as follows: For every map $f : S \to T$ in I and homomorphism $\alpha : \mathbb{Z} \to G$ (or $\beta : \mathbb{Z} \to H$), there is a map $g : S' \to T'$ in I with $S + \alpha(\mathbb{Z}) \subseteq S'$ and $T \subseteq T'$ (or $S \subseteq S'$ and $T + \beta(\mathbb{Z}) \subseteq T'$) such that the diagram

$$\begin{array}{ccc} S' & \xrightarrow{g} & T' \\ \uparrow i_1 & & \uparrow i_2 \\ S & \xrightarrow{f} & T \end{array}$$

commutes, where i_1, i_2 are the inclusion maps.

Dually, we call two compact groups C and D **$*$-partially isomorphic**, written $C \cong_* D$, if there is a nonempty set J of topological isomorphisms between factor groups of C and D by closed subgroups satisfying the **$*$-back-and-forth property**: For every map $\phi : C/A \to D/B$ in J and $\delta \in \widehat{C}$ ($\eta \in \widehat{D}$, respectively), there is a map $\psi : C/A' \to D/B'$ in J with $A' \subseteq A \cap \ker\delta$ and $B' \subseteq B$ ($A' \subseteq A$ and $B' \subseteq B \cap \ker\eta$, respectively) such that the diagram

$$\begin{array}{ccc} C/A' & \xrightarrow{\psi} & D/B' \\ \downarrow \pi_1 & & \downarrow \pi_2 \\ C/A & \xrightarrow{\phi} & D/B \end{array}$$

commutes, where π_1, π_2 are the natural epimorphisms. If a system J satisfies this condition, then we write $J : C \cong_* D$.

It is clear that $C \cong_* D$ exactly if $D \cong_* C$. Notice that $C \cong D$ implies $C \cong_* D$. The converse holds if for instance, C and D are finite:

Example 7.7.2. If C and D are finite groups with $C \cong_* D$, then $C \cong D$.

Proof. Write $C = \bigoplus_{i=1}^n C_i$ and $D = \bigoplus_{j=1}^m D_j$ with each C_i and D_j cyclic (see Theorem 1.8.1). For each $i = 1, \ldots, n$, let $\alpha_i = \phi_i \pi_i$, where $\pi_i : C \to C_i$ is the projection map, and $\phi_i \in (C_i)\widehat{}$ is injective. Then $\alpha_i \in \widehat{C}$ and $\bigcap_{i=1}^n \ker \alpha_i = 0$. Similarly, there are maps $\beta_j \in \widehat{D}$ such that $\bigcap_{j=1}^m \ker \beta_j = 0$. By the $*$-back-and-forth property, there is a topological isomorphism $C/H \to D/K$ with $H \subseteq \bigcap_{i=1}^n \ker \alpha_i$ and $K \subseteq \bigcap_{j=1}^m \ker \beta_j$; thus, $C \cong D$. □

Lemma 7.7.3. *Let G and H be discrete groups. Then $G \cong_p H$ if and only if $\widehat{G} \cong_* \widehat{H}$.*

Proof. Suppose that $I : G \cong_p H$, and let $f : S \to T$ be a map in I. Letting $\phi_S : \widehat{G}/(\widehat{G}, S) \to \widehat{S}$ and $\phi_T : \widehat{H}/(\widehat{H}, T) \to \widehat{T}$ denote the topological isomorphisms as in Lemma 6.4.12, we obtain the topological isomorphism $\psi_f = \phi_S^{-1} f^* \phi_T$. Now let $\delta \in \widehat{G}$ (or $\eta \in \widehat{H}$). Then we obtain the adjoint map $\delta^* : \mathbb{Z} \to G$ (or $\eta^* : \mathbb{Z} \to H$) and can extend f to a map $g : S' \to T'$ in I such that $S + \delta^*(\mathbb{Z}) \subseteq S'$ (or $T + \eta^*(\mathbb{Z}) \subseteq T'$). The commuting diagram

$$\begin{array}{ccc} S' & \xrightarrow{g} & T' \\ \uparrow & & \uparrow \\ S & \xrightarrow{f} & T \end{array}$$

induces the commuting diagram

$$\begin{array}{ccc} \widehat{G}/(\widehat{G}, S') & \xleftarrow{\psi_g} & \widehat{H}/(\widehat{H}, T') \\ \downarrow & & \downarrow \\ \widehat{G}/(\widehat{G}, S) & \xleftarrow{\psi_f} & \widehat{H}/(\widehat{H}, T) \end{array}$$

with $(\widehat{G}, S') \subseteq (\widehat{G}, S + \delta^*(\mathbb{Z})) = (\widehat{G}, S) \cap \ker \delta$ and $(\widehat{H}, T') \subseteq (\widehat{H}, T)$ [or $(\widehat{G}, S') \subseteq (\widehat{G}, S)$ and $(\widehat{H}, T') \subseteq (\widehat{H}, T) \cap \ker \eta$]. Consequently, the system $J = \{\psi_f : f \in I\}$ satisfies the $*$-back-and-forth property, and it follows that $J : \widehat{G} \cong_* \widehat{H}$. The proof of the converse is similar. □

Since partially isomorphic groups are isomorphic whenever they are countable, Corollary 6.3.3 yields the following:

Corollary 7.7.4. *Let C and D be compact metrizable groups. If $C \cong_* D$, then $C \cong D$.*

To classify groups with quasi-decompositions up to $*$-partial isomorphism, we need to modify the co-Ulm invariants and co-Stanton invariants (see Sections 7.2 and 7.6). Let G be a compact group and p a prime. Then we define

$$\widehat{u_p^*}(\alpha, G) = u_p^*(\alpha, G) \wedge \omega$$

for all $\alpha \in \mathrm{Ord}_\infty$ and

$$\widehat{s^*}(c,p,e,G) = s^*(c,p,e,G) \wedge \omega$$

for each compatibility class c of Ulm matrices and equivalence class e of Ulm sequences. Now we are able to dualize Theorem 5.3.8:

Theorem 7.7.5. *Two groups with partial quasi-decompositions are $*$-partially isomorphic if and only if they have the same modified co-Ulm invariants $\widehat{u_p^*}$ ($p \in \mathbb{P}$) and modified co-Stanton invariants $\widehat{s^*}$.*

We call a compact group k-**smart** if it is a k-smart subgroup of itself. Dualizing Theorems 5.5.5 and 5.8.3, we obtain:

Corollary 7.7.6. *The following are equivalent for a compact group C:*
1. *C has a partial quasi-decomposition;*
2. *C is k-smart;*
3. *There is a simply given group D such that $C \times D \cong_* D$;*
4. *There is a direct factor D of a simply given group such that $C \times D \cong_* D$;*
5. *There is a group D with partial quasi-decomposition such that $C \times D \cong_* D$.*

Since the class of PDB groups is closed under direct summands (Corollary 5.5.6), we have:

Corollary 7.7.7. *The class of groups with partial quasi-decompositions is closed under direct factors.*

7.8 Going further

Kiefer [84] characterized condition (H3*) in the definition of a smart system (see Section 7.4) in terms of pseudometrics. Leopold [95, 96] investigated the duals of Butler groups and in [97, 98], he introduced and studied the compact-open topology on $\mathrm{Ext}(C,A)$ for discrete groups A and C, and characterized those groups A and C for which $\mathrm{Ext}(C,A)$ is Hausdorff (or compact, locally compact). His dissertation [97] also deals with the compact-open topology on $\mathrm{End}\, G$ for discrete groups G. In [100], Loth studied the duals of almost completely decomposable groups.

Problems.
1. Let κ be an infinite cardinal. Fuchs and Hill [30] defined a subgroup A of a p-group G to be κ-**separable**, if for every $x \in G$ there is a subset S of G such that $|S| \le \kappa$ and

   $$\sup\{|x+a|_p : a \in A\} = \sup\{|x+s|_p : s \in S\}.$$

   Then G is said to satisfy **Axiom** $3:\kappa$ if G has a family C of κ-separable subgroups satisfying

(FH1) $0 \in \mathcal{C}$;
(FH2) if $N_i \in \mathcal{C}$ $(i \in I)$, then $\sum_{i \in I} N_i \in \mathcal{C}$;
(FH3) if $N \in \mathcal{C}$ and S is a subgroup of G with $|S| \leq \kappa$, then there is an $M \in \mathcal{C}$ such that $\langle S, N \rangle \subseteq M$ and $|M/N| \leq \kappa$.
Dualize the concept of Axiom $3 : \kappa$.

2. Dualization of the various characterizations of certain classes of abelian groups (for example, totally projective p-groups, Warfield groups) yields a description of the corresponding dual groups. Conversely, how can the topological structure of these compact groups be used to determine the structure of their (discrete) dual groups?
3. Dualize the concept of $L^{\lambda}_{\infty\omega}$-equivalence (see Section 4.2).

Symbols

$A \times B$	direct product of A and B
$A \otimes B$	tensor product of A and B
$A \oplus B$	algebraic direct sum of A and B
$A \bigoplus B$	topological direct sum of A and B
$A + B, \langle A, B \rangle$	submodule generated by A and B
$\alpha \restriction A$	restriction of map α to A
$\beta\alpha$	composite of maps α and β
bG	set of compact elements of G
$\mathbb{C}$	group of complex numbers
$C[p^\alpha]$	set of elements in C of p-order $\leq \alpha$
$\chi_G(x)$	$(\lvert x\rvert_{p_i})_{i \in \mathbb{N}}$; characteristic of x
$\operatorname{coker} \alpha$	cokernel of map α
$\mathcal{C}_H^G$	$\{Y \cap H : Y \text{ is a finite } * \text{-decomposition set in } G\}$
$D_{\alpha,p}(A)$	$\{x \in A \cap p^\alpha G : px \in p^{\alpha+2}G\}/(A \cap p^{\alpha+1}G)$
Δ, ∇	diagonal map, codiagonal map
$\dim C$	dimension of C
$\dim_p C$	p-dimension of C
$\dim_{p,0}(C)$	$\dim_p C + \dim C$
$\operatorname{Ext}(C, A)$	group of extensions of A by C
f_p	$f \otimes 1_{\mathbb{Z}_p}$
$\bigoplus G_i$	direct sum of groups G_i
$\prod G_i$	direct product of groups G_i
$G^{\mathfrak{m}}$	direct product of $\mathfrak{m}$ copies of G
G_p	$G \otimes \mathbb{Z}_p$; p-localization of G
$G_{(p)}$	$\{x \in G : \lim_{n\to\infty} p^n x = 0\}$; p-component of G
$G[n]$	$\{x \in G : nx = 0\}$
$G(\mathbf{t})$	$\{x \in G : \mathbf{t}(x) \geq \mathbf{t}\}$
$G^*(\mathbf{t})$	$\langle x \in G : \mathbf{t}(x) > \mathbf{t} \rangle$
$G_{\mathbf{t}}$	$G(\mathbf{t})/G^*(\mathbf{t})$
$G(\overline{\alpha}), G(\overline{\alpha}, p)$	$\{x \in G : \lVert x\rVert_p \geq \overline{\alpha}\}$
$G(\overline{\alpha}^*)^\diamond$	$\{x \in G(\overline{\alpha}) : \lvert p^i x\rvert_p \neq \alpha_i$ for infinitely many $i < \omega$ or x is torsion$\}$
$G(\overline{\alpha}^*)$	$\langle G(\overline{\alpha}^*)^\diamond \rangle$
$G(\overline{\alpha}^*, p)^\diamond$	$\{x \in G(\overline{\alpha}, p) : \lvert p^i x\rvert_p \neq \alpha_i$ for infinitely many $i < \omega\}$
$G(\overline{\alpha}^*, p)$	$\langle G(\overline{\alpha}^*, p)^\diamond \rangle$
$G(M)$	$\{x \in G : \lVert x\rVert \geq M\}$
$G(M^*)^\diamond$	$\{x \in G(M) : \lVert x\rVert \nsim M$ or x is torsion$\}$

https://doi.org/10.1515/9783110427684-008

$G(M^*)$	$\langle G(M^*)^\diamond\rangle$
$G(M^*,p)^\diamond$	$G(M^*)^\diamond \cup G(M_p^*,p)^\diamond$
$G(M^*,p)$	$G(M) \cap \langle G(M^*,p)^\diamond\rangle$
$G_\alpha(M)$	$\{x \in G : \|x\| \geq_\alpha M\}$
$G_\alpha(M^*)^\diamond$	$\{x \in G_\alpha(M) : \|x\| \not\sim_\alpha M$ or x is torsion$\}$
$G_\alpha(M_p^*,p)^\diamond$	$\{x \in G : \|x\|_p \geq_\alpha M_p$ and $\lvert p^i x\rvert_p \neq_\alpha m_{p,i}$ for infinitely many $i < \omega\}$
$G_\alpha(M^*,p)^\diamond$	$G_\alpha(M^*)^\diamond \cup G_\alpha(M_p^*,p)^\diamond$
$G_\alpha(M^*,p)$	$G_\alpha(M) \cap \langle G_\alpha(M^*,p)^\diamond\rangle$
G_0	identity component of G
G_d	LCA group G with discrete topology
$\widehat{G}$	dual group of G
$(\widehat{G},S)$	annihilator of $S \subseteq G$ in $\widehat{G}$
H_α	generalized Prüfer group of length α
$H_{\omega+1}$	Prüfer group
$\mathrm{Hom}(A,C)$	group of homomorphisms from A to C
$I_{\alpha,p}(A)$	image of $\phi : D_{\alpha,p}(A) \to (p^\alpha G)[p]/(p^{\alpha+1}G)[p]$
$\mathrm{im}\,\alpha$	image of map α
$\cong$	isomorphism, topological isomorphism
$\cong_p$	partial isomorphism
$\cong_*$	$*$-partial isomorphism
$=_\alpha$	equal up to α
$\geq_\alpha$	greater than or equal up to α
$\sim_\alpha$	α-equivalent (Ulm sequences), α-compatible (Ulm matrices)
$\mathbb{J}_p$	group of p-adic integers
$\ker\alpha$	kernel of map α
$L_{\infty\omega}^\lambda$	class of formulas with quantifier rank $\leq \lambda$
$L_{\infty\omega}$	$\bigcup_{\lambda\in\mathrm{Ord}} L_{\infty\omega}^\lambda$
$\equiv_\lambda$	equivalence in $L_{\infty\omega}^\lambda$
$\equiv_\infty$	equivalence in $L_{\infty\omega}$
$M = [m_{p,i}]$	Ulm matrix
M_p	p-th row of Ulm matrix M
$\mathbb{N}$	set of natural numbers
nG	$\{nx : x \in G\}$
$n^{-1}A$	$\{x \in G : nx \in A\}$ for a subgroup A of G
$o(x)$	order of x
$o_p(x)$	p-order of x
0_G	zero map on G
1_G	identity map on G
ω	smallest infinite ordinal

Ord	class of ordinals
Ord_∞	class of ordinals together with the symbol ∞
$\mathbb{P}$	set of primes in $\mathbb{Z}$
P_α	Walker group
$p^\alpha G$	set of elements in G of p-height $\geq \alpha$
$\mathrm{patset}(C)$	pattern set of C
$\mathbf{p}(C)$	pattern of C
ϕ^*	adjoint map of ϕ
$P(F, \varepsilon)$	$\{\chi \in \widehat{G} : \chi(F) \subseteq \mathbb{T}_\varepsilon\}$
$\mathbb{Q}$	group of rational numbers
$qr(\varphi)$	quantifier rank of φ
$\mathbb{R}$	group of real numbers
$r(G)$	rank of G
$r_0(G)$	torsion-free rank of G
$r_p(G)$	p-rank of G
$r_{p,0}(G)$	$r_p(G) + r_0(G)$
S^0	$\{x \in G : rx \in S$ for some nonzero $r \in R\}$ for $S \subseteq G$, G an R-module
$\langle S\rangle$	subgroup (or submodule) generated by S
$\overline{S}$	closure of $S \subseteq G$ in G
$\lvert S\rvert$	cardinality of S
$\sum A_i$	subgroup generated by subgroups A_i
$s(e, G)$	e-th Stanton invariant of G
$s(c, p, e, G)$	(global) Stanton invariant of G
$\hat{s}$	modified Stanton invariant
s^*	co-Stanton invariant
$\widehat{s^*}$	modified co-Stanton invariant
$\mathbb{T} \cong \mathbb{R}/\mathbb{Z}$	circle group
$\mathbb{T}_\varepsilon$	$\{z \in \mathbb{T} : \lvert z - 1\rvert < \varepsilon\}$
tG	torsion part of G
t_pG	p-torsion part of G
$\mathbf{t}(x)$	type of element x
$\mathrm{typeset}(G)$	$\{\mathbf{t}(x) : 0 \neq x \in G\}$
$u_p(\alpha, G)$	α-th Ulm invariant of G
$u_{\alpha,p}(G, A)$	α-th Ulm invariant of G relative to the subgroup A
$\hat{u}$	modified Ulm invariant
u^*	co-Ulm invariant
$\widehat{u^*}$	modified co-Ulm invariant
v_p	p-valuation

$w(e, G)$	e-th Warfield invariants of G
$w(c, p, e, G)$	(global) Warfield invariant of G
$\hat{w}$	modified Warfield invariant
w^*	co-Warfield invariant
$\lvert x\rvert, \lvert x\rvert^G$	(p-)height of x, computed in G
$\lvert x\rvert_p, \lvert x\rvert_p^G$	p-height of x, computed in G
$\lVert x\rVert, \lVert x\rVert_G$	Ulm matrix of x, computed in G
$\lVert x\rVert_p, \lVert x\rVert_p^G$	p-Ulm sequence of x, computed in G
$\mathbb{Z}$	group or ring of integers
$\mathbb{Z}/n\mathbb{Z}$	cyclic group of order n
$\mathbb{Z}_p$	$\{\frac{m}{n} : m, n \in \mathbb{Z}$ and $p \nmid n\}$; p-localization of $\mathbb{Z}$
$\mathbb{Z}(p^\infty)$	quasicyclic p-group

Bibliography

[1] P. R. Ahern and R. I. Jewett, *Factorization of locally compact abelian groups*, Ill. J. Math. **9** (1965), 230–5.

[2] D. L. Armacost, *On pure subgroups of LCA groups*, Proc. Amer. Math. Soc. **45** (1974), 414–8.

[3] D. L. Armacost, *The Structure of Locally Compact Abelian Groups*, Marcel Dekker Inc., New York (1981).

[4] D. M. Arnold, *Finite Rank Torsion Free Abelian Groups and Rings*, Lecture Notes in Math., Vol. 931, Springer, Berlin (1982).

[5] D. M. Arnold, R. Hunter and E. A. Walker, *Direct sums of cyclic valuated groups*, Sympos. Math. **23** (1979), 77–84.

[6] R. Baer, *Abelian groups without elements of finite order*, Duke Math. J. **3** (1937) 68–122.

[7] R. Baer, *Abelian groups that are direct summands of every containing abelian group*, Bull. Amer. Math. Soc. **46** (1940), 800–6.

[8] J. Barwise, *Back and forth through infinitary logic*, in Studies in Model Theory, M. D. Morley (ed.), Mathematical Association of America, Studies in Mathematics, Vol. 8 (1973), pp. 5–34.

[9] J. Barwise and P. Eklof, *Infinitary properties of abelian torsion groups*, Annals of Math. Logic **2** (1970), 25–68.

[10] J. Braconnier, *Sue les groupes topologiques localement compacts*, J. Math. Pures Appl., N.S. **27** (1948), 1–85.

[11] M. C. R. Butler, *A class of torsion-free abelian groups of finite rank*, Proc. London Math. Soc. **15** (1965), 680–98.

[12] H. Cartan and S. Eilenberg, *Homological Algebra*, Princeton University Press, Princeton, New Jersey (1956).

[13] R. Charles, *Étude des groupes abéliens primaires de type* $\leq \omega$, Ann. Univ. Saraviensis **4** (1955), 184–99.

[14] P. Crawley and A. W. Hales, *The structure of torsion abelian groups given by presentations*, Bull. Amer. Math. Soc. **74** (1968), 954–6.

[15] P. Crawley and A. W. Hales, *The structure of abelian p-groups given by certain presentations*, J. Algebra **12** (1969), 10–23.

[16] P. Crawley and A. W. Hales, *The structure of torsion abelian groups given by presentations. II*, J. Algebra **18** (1971), 264–8.

[17] J. Dieudonné, *Sur les produits tensoriels*, Ann. Sci. École Norm. Sup. **64** (1948), 101–17.

[18] D. N. Dikranjan, *Introduction to Topological Groups*, https://archive.org/details/Dikran_Dikranjan__Introduction_to_Topological_Groups.

[19] D. N. Dikranjan, I. R. Prodanov and L. N. Stoyanov, *Topological Groups: Dualities and Minimal Group Topologies*, Pure and Applied Mathematics, Vol. 130, Marcel Dekker Inc., New York-Basel (1989).

[20] R. Downey and A. Montalbán, *The isomorphism problem for torsion-free abelian groups is analytic complete*, J. Algebra **320** (2008), 2291–300.

[21] R. Engelking, *Outline of General Topology*, North-Holland, Amsterdam (1968).

[22] S. T. Files, *Endomorphisms of local Warfield groups*, in Abelian Group Theory and Related Topics, Contemp. Math., Vol. 171, American Mathematical Society, Providence, RI (1994), pp. 99–107.

[23] S. V. Fomin, *Über periodische Untergruppen der unendlichen abelschen Gruppen*, Mat. Sb. **2** (1937), 1007–9.

[24] L. Fuchs, *Abelian Groups*, Publishing House of the Hungarian Academy of Science, Budapest (1958).

https://doi.org/10.1515/9783110427684-009

[25] L. Fuchs, *Notes on abelian groups. I*, Ann. Univ. Sci. Budapest **2** (1959), 5–23; *II, Acta Math. Acad. Sci. Hungar. 11 (1960), 117-125.*

[26] L. Fuchs, *Infinite Abelian Groups*, Volume I, Pure Applied Mathematics, Vol. 36, Academic Press, New York (1970).

[27] L. Fuchs, *Infinite Abelian Groups*, Volume II, Pure Applied Mathematics, Vol. 36, Academic Press, New York (1973).

[28] L. Fuchs, *Abelian p-Groups and Mixed Groups*, University of Montreal Press, Montreal (1980).

[29] L. Fuchs, *Abelian Groups*, Springer Monographs in Mathematics, Springer, Cham (2015).

[30] L. Fuchs and P. Hill, *The balanced-projective dimension of abelian p-groups*, Trans. Amer. Math. Soc. **293** (1986), 99–112.

[31] L. Fuchs and K. H. Hofmann, *Extensions of compact abelian groups by discrete ones and their duality theory, I*, J. Algebra **196** (1997), 578–94.

[32] L. Fuchs and K. H. Hofmann, *Extensions of compact abelian groups by discrete ones and their duality theory, II*, in Abelian Groups, Module Theory, and Topology, Proc. in Honour of Adalberto Orsatti, Marcel Dekker, New York (1998), pp. 205–25.

[33] R. O. Fulp, *Splitting locally compact abelian groups*, Michigan Math. J. **19** (1972), 47–55.

[34] R. O. Fulp, *Homological study of purity in locally compact abelian groups*, Proc. London Soc. **21** (1979), 502–12.

[35] R. O. Fulp and P. Griffith, *Extensions of locally compact abelian groups I*, Trans. Amer. Math. Soc. **154** (1971), 341–56.

[36] R. O. Fulp and P. Griffith, *Extensions of locally compact abelian groups II*, Trans. Amer. Math. Soc. **154** (1971), 357–63.

[37] R. Göbel, K. Leistner, P. Loth and L. Strüngmann, *Infinitary equivalence of $\mathbb{Z}_p$-modules with nice decomposition bases*, J. Comm. Algebra **3** (2011), 321–48.

[38] P. Griffith, *Infinite Abelian Group Theory*, University of Chicago Press, Chicago (1970).

[39] H. J. B. Gutiérrez, *Extensions of Topological Abelian Groups*, PhD dissertation, University of Navarra (2016).

[40] S. Hartman and A. Hulanicki, *Les sous-groupes purs et leurs duals*, Fund. Math. **45** (1957), 71–7.

[41] E. Hewitt and K. Ross, *Abstract Harmonic Analysis*, Vol. I, Second Edition, Springer, Berlin (1979).

[42] P. Hill, *Isotype subgroups of totally projective groups*, in Lecture Notes in Math., Vol. 874, Springer, Berlin (1981), pp. 305–21.

[43] P. Hill, *The classification problem*, in Abelian Groups and Modules, CISM Courses and Lectures, Springer, Wien and New York (1984), pp. 1–15.

[44] P. Hill, *Toward the classification of mixed groups*, Arch. Math. **42** (1984), 481–8.

[45] P. Hill, *On the structure of abelian p-groups*, Trans. Amer. Math. Soc. **288** (1985), 505–25.

[46] P. Hill, *Another characterization of totally projective groups*, Houston J. Math. **23** (1997), 21–7.

[47] P. Hill, *On the classification of abelian groups*, Period. Math. Hung. **69** (2014), 41–52.

[48] P. Hill and C. Megibben, *On the theory and classification of of abelian p-groups*, Math. Z. **190** (1985), 17–38.

[49] P. Hill and C. Megibben, *Axiom 3 modules*, Trans. Amer. Math. Soc. **295** (1986), 715–34.

[50] P. Hill and C. Megibben, *Knice subgroups of mixed groups*, in Abelian Group Theory, Gordon and Breach Science Publishers, New York (1987), pp. 89–109.

[51] P. Hill and C. Megibben, *Pure subgroups of torsion-free groups*, Trans. Amer. Math. Soc. **303** (1987), 765–78.

[52] P. Hill and C. Megibben, *The local equivalence problem*, in Abelian Group Theory, Contemp. Math., Vol. 87, American Mathematical Society, Providence, RI (1989), pp. 201–19.

[53] P. Hill and C. Megibben, *Mixed groups*, Trans. Amer. Math. Soc. **334** (1992), 121–42.

[54] P. Hill and W. Ullery, *Isotype separable subgroups of totally projective groups*, in Abelian Groups and Modules, Kluwer Academic Publishers, Dordrecht (1995), pp. 291–300.

[55] P. Hill and W. Ullery, *Almost totally projective groups*, Czech. Math. J. **46** (1996), 249–58.

[56] G. Hjorth, *The isomorphism relation on torsion-free abelian groups*, Fund. Math. **175** (2002), 241–57.

[57] W. Hodges, *Model Theory*, Encyclopedia of Mathematics and Its Applications, Vol. 42, Cambridge University Press (1993).

[58] K. H. Hofmann, *Tensorprodukte lokal kompakter abelscher Gruppen*, J. Reine Angew. Math. **216** (1964), 134–49.

[59] K. H. Hofmann and S. A. Morris, *The Structure of Compact Groups*, de Gruyter, Berlin-New York (1998).

[60] R. Hunter, *Balanced subgroups of abelian groups*, Trans. Amer. Math. Soc. **215** (1976), 81–98.

[61] R. Hunter and F. Richman, *Global Warfield groups*, Trans. Amer. Math. Soc. **266** (1981), 555–72.

[62] R. Hunter and E. A. Walker, *S-groups revisited*, Proc. Amer. Math. Soc. **82** (1981), 13–8.

[63] R. Hunter and E. A. Walker, *Valuated p-Groups*, Lecture Notes in Math., Vol. 874, Springer, Berlin (1981), pp. 350–73.

[64] R. Hunter, F. Richman and E. A. Walker, *Simply presented valuated p-groups*, J. Algebra **49** (1977), 125–33.

[65] R. Hunter, F. Richman and E. A. Walker, *Warfield modules*, in Abelian Group Theory, Lecture Notes in Math., Vol. 616, Springer, Berlin (1977), pp. 87–123.

[66] R. Hunter, F. Richman and E. A. Walker, *Existence theorems for Warfield groups*, Trans. Amer. Math. Soc. **235** (1978), 345–62.

[67] J. M. Irwin and E. A. Walker, *On isotype subgroups of abelian groups*, Bull. Soc. Math. France **89** (1961), 451–60.

[68] C. Jacoby, *The Classification in $L_{\infty\omega}$ of Groups with Partial Decomposition Bases*, PhD dissertation, University of California, Irvine (1980), revised version (2010).

[69] C. Jacoby, *Undefinability of local Warfield groups in $L_{\infty\omega}$*, in Groups and Model Theory: A Conference in Honor of Rüdiger Göbel's 70th Birthday, Contemp. Math., Vol. 576, Amer. Math. Soc., Providence, RI (2012), pp. 151–62.

[70] C. Jacoby and P. Loth, *Abelian groups with partial decomposition bases in $L^{\delta}_{\infty\omega}$, Part II*, in Groups and Model Theory: A Conference in Honor of Rüdiger Göbel's 70th Birthday, Contemp. Math., Vol. 576, Amer. Math. Soc., Providence, RI (2012), pp. 177–85.

[71] C. Jacoby and P. Loth, *Partial decomposition bases and Warfield modules*, Comm. Algebra **42** (2014), 4333–49.

[72] C. Jacoby and P. Loth, *$\mathbb{Z}_p$-modules with partial decomposition bases in $L^{\delta}_{\infty\omega}$*, Houston J. Math. **40** (2014), 1007–19.

[73] C. Jacoby and P. Loth, *Partial decomposition bases and global Warfield groups*, Comm. Algebra **44** (2016), 3262–77.

[74] C. Jacoby and P. Loth, *The classification of $\mathbb{Z}_p$-modules with partial decomposition bases in $L_{\infty\omega}$*, Arch. Math. Logic **55** (2016), 939–54.

[75] C. Jacoby and P. Loth, *The classification of infinite abelian groups with partial decomposition bases in $L_{\infty\omega}$*, Rocky Mountain J. Math. **47** (2017), 463–77.

[76] C. Jacoby, K. Leistner, P. Loth and L. Strüngmann, *Abelian groups with partial decomposition bases in $L^{\delta}_{\infty\omega}$, Part I*, in Groups and Model Theory: A Conference in Honor of Rüdiger Göbel's 70th Birthday, Contemp. Math., Vol. 576, Amer. Math. Soc., Providence, RI (2012), pp. 163–75.

[77] E. R. van Kampen, *Locally bicompact abelian groups and their character groups*, Ann. Math. (2) **36** (1935), 448–63.

[78] I. Kaplansky, *Infinite Abelian Groups*, University of Michigan Press, Ann Arbor (1954, 1968).

[79] I. Kaplansky, *Projective modules*, Ann. Math. **68** (1958), 372–7.
[80] C. Karp, *Finite quantification equivalence*, in The Theory of Models, J. W. Addison, L. Henkin and A. Tarski (eds.), North-Holland, Amsterdam (1965), pp. 407–12.
[81] J. A. Khan, *The finite torsion subgroup of an LCA group need not split*, Period. Math. Hungar. **31** (1995), 43–4.
[82] M. A. Khan, *A notion of purity in locally compact groups*, J. Sci. Univ. Karachi **2** (1973), 128–30.
[83] F. Kiefer, *Die Dualgruppen der total projektiven Gruppen*, Diplomarbeit, Freie Universität Berlin (1980).
[84] F. Kiefer, *The duals of totally projective groups*, in Abelian Group Theory, Lecture Notes in Math., Vol. 874, Springer, Berlin (1981), pp. 297–304.
[85] N. I. Kryuchkov, *On the abelian extensions of locally compact abelian groups*, Math. USSR Sbornik **41** (1982), 511–22.
[86] N. I. Kryuchkov, *Pure injective discrete abelian groups in LCA*, preprint (2004).
[87] N. I. Kryuchkov, *Generalized cotorsion locally compact abelian groups*, Rend. Sem. Mat. Univ. Padova **127** (2012), 107–20.
[88] L. Ya. Kulikov, *Zur Theorie der abelschen Gruppen von beliebiger Mächtigkeit* [Russian], Mat. Sbornik **9** (1941), 165–82.
[89] L. Ya. Kulikov, *On the theory of abelian groups of arbitrary power* [Russian], Mat. Sbornik **16** (1945), 129–62.
[90] L. Ya. Kulikov, *On direct decompositions of groups* [Russian], Ukrain. Mat. Zh. **4** (1952), 230–75.
[91] E. L. Lady, *Almost completely decomposable torsion-free abelian groups*, Proc. Amer. Math. Soc. **45** (1974), 41–7.
[92] M. Lane, *On p-local abelian groups with decomposition bases*, Math. Z. **202** (1989), 129–41.
[93] M. Lane and C. Megibben, *Balanced projectives and Axiom 3*, J. Algebra **111** (1987), 457–74.
[94] K. Leistner, *Partial Isomorphism*, PhD dissertation, University of Duisburg-Essen (2014).
[95] C. Leopold, *Butlergruppen und die Pontrjagin - Dualität*, Diplomarbeit, University of Erlangen-Nürnberg (1995).
[96] C. Leopold, *Butler groups and Lord groups*, Acta Sci. Math. (Szeged) **63** (1997), 397–404.
[97] C. Leopold, *Die kompakt-offene Topologie auf Ext und End*, PhD dissertation, University of Erlangen-Nürnberg (1999).
[98] C. Leopold, *On the compact-open topology of* Ext(C, A), in Abelian Groups and Modules, Trends Math., Birkhäuser, Basel (1999), pp. 293–9.
[99] P. Loth, *Splitting of the identity component in locally compact abelian groups*, Rend. Sem. Mat. Univ. Padova **88** (1992), 139–43.
[100] P. Loth, *The duals of almost completely decomposable groups*, Arch. Math. **68** (1997), 353–8.
[101] P. Loth, *The duals of Warfield groups*, Pac. J. Math. **181** (1997), 333–56.
[102] P. Loth, *Characterizations of Warfield groups*, J. Algebra **204** (1998), 32–41.
[103] P. Loth, *Classifications of Abelian Groups and Pontrjagin Duality*, Algebra, Logic and Applications, Vol. 10, Gordon and Breach, Amsterdam (1998).
[104] P. Loth, *Splitting in locally compact abelian groups*, Rend. Sem. Mat. Univ. Padova **99** (1998), 187–96.
[105] P. Loth, *Direct decompositions of LCA groups*, in Abelian Groups and Modules, Trends Math., Birkhäuser, Basel (1999), pp. 301–7.
[106] P. Loth, *On pure subgroups of locally compact abelian groups*, Arch. Math. **81** (2003), 255–7.
[107] P. Loth, *Compact topologically torsion elements of topological abelian groups*, Rend. Sem. Mat. Univ. Padova **113** (2005), 117–23.
[108] P. Loth, *On t-pure and almost pure exact sequences of LCA groups*, J. Group Theory **9** (2006), 799–808.

[109] P. Loth, *Pure extensions of locally compact abelian groups*, Rend. Sem. Mat. Univ. Padova **116** (2006), 31–40.

[110] P. Loth, *Pure injective and ∗-pure injective LCA groups*, Rend. Sem. Mat. Univ. Padova **133** (2015), 91–102.

[111] S. Mac Lane, *Homology*, Springer, New York (1963).

[112] A. Mader, *Almost Completely Decomposable Groups*, Algebra, Logic and Applications, Vol. 13, Gordon and Breach, Amsterdam (2000).

[113] C. Mathews, *Mixed Groups with Decomposition Bases and Global k-Groups*, Master's Thesis, Auburn University, Alabama (2006).

[114] C. Megibben, *On mixed groups of torsion-free rank one*, Illinois J. Math. **11** (1967), 134–44.

[115] C. Megibben and W. Ullery, *Isotype subgroups of mixed groups*, Comment. Math. Univ. Carol., **42** (2001), No. 3, 421–42.

[116] C. Megibben and W. Ullery, *Isotype Warfield subgroups of global Warfield groups*, Rocky Mountain J. Math. **32** (2002), 1523–42.

[117] C. Megibben and W. Ullery, *On global abelian k-groups*, Houston J. Math., **31** (2005), 675–92.

[118] A. Mekler and J. Oikkonen, *Abelian p-groups with no invariants*, J. Pure Appl. Algebra **87** (1993), 51–9.

[119] J. H. Moore, *Nice Subgroups of Valuated Groups*, Lecture Notes in Math., Vol. 874, Springer, Berlin (1981), pp. 405–16.

[120] J. H. Moore, *A characterization of Warfield groups*, Proc. Amer. Math. Soc. **87** (1983), 617–20.

[121] S. A. Morris, *Pontryagin Duality and the Structure of Locally Compact Abelian Groups*, Cambridge University Press, Cambridge (1972).

[122] M. Moskowitz, *Homological algebra in locally compact abelian groups*, Trans. Amer. Math. Soc. **127** (1967), 361–404.

[123] R. J. Nunke, *Purity and subfunctors of the identity*, in Topics in Abelian Groups, Chicago (1963), pp. 121–71.

[124] R. J. Nunke, *Homology and direct sums of countable abelian groups*, Math. Z. **101** (1967) 182–212.

[125] L. D. Parker and E. A. Walker, *An extension of the Ulm-Kolettis theorems*, in Studies on Abelian Groups, Paris (1968), pp. 309–25.

[126] L. S. Pontrjagin, *The theory of topological commutative groups*, Ann. of Math. **35** (1934), 361–88.

[127] L. S. Pontrjagin, *Selected Works: Topological Groups, Vol. 2* (English translation by A. Brown), Gordon and Breach, New York, 1986.

[128] G. C. Preston, *On locally compact totally disconnected Abelian groups and their character groups*, Pac. J. Math. **6** (1956), 121–34.

[129] H. Prüfer, *Untersuchungen über die Zerlegbarkeit der abzählbaren primären abelschen Gruppen*, Math. Z. **17** (1923), 35–61.

[130] F. Richman, *Mixed local groups*, in Abelian Group Theory, Lecture Notes in Math., Vol. 874, Springer, Berlin (1981), pp 374–404.

[131] F. Richman, *Mixed groups*, in Abelian Group Theory, Lecture Notes in Math., Vol. 1006, Springer, Berlin (1983), 445–70.

[132] F. Richman, *The constructive theory of countably generated Warfield modules*, in Abelian Groups and Noncommutative Rings: A Collection of Papers in Memory of Robert B. Warfield, Jr., Contemp. Math., Vol. 130, American Mathematical Society, Providence, RI (1992), pp. 371–83.

[133] F. Richman, unpublished note (2015).

[134] F. Richman and E. A. Walker, *Valuated Groups*, J. Algebra **65** (1979), 145–67.

[135] H. Sahleh and A. A. Alijani, *On vanishing of pure extensions of torsion-free locally compact*

abelian groups, Int. J. Stat. Math. **6** (2013), 35–7.

[136] H. Sahleh and A. A. Alijani, *The pure injectives and pure projectives in the category of totally disconnected, locallly compact abelian groups*, Math. Sci., Springer **7**, Paper No. 49, 3 p. (2013).

[137] H. Sahleh and A. A. Alijani, *Splitting of extensions in the category of locally compact abelian groups*, Int. J. Group Theory **3** (2014), 39–45.

[138] H. Sahleh and A. A. Alijani, *On component extensions of locally compact abelian groups*, J. Algebra Relat. Top. **4** (2016), 1–11.

[139] R. O. Stanton, *An invariant for modules over a discrete valuation ring*, Proc. Amer. Math. Soc. **49** (1975), 51–5.

[140] R. O. Stanton, *Decomposition bases and Ulm's theorem*, in Abelian Group Theory, Lecture Notes in Math., Vol. 616, Springer, Berlin (1977), 39–56.

[141] R. O. Stanton, *Relative S-invariants*, Proc. Amer. Math. Soc. **65** (1977), 221–4.

[142] R. O. Stanton, *Almost-affable abelian groups*, J. Pure Appl. Algebra **15** (1979), 41–52.

[143] M. Stroppel, *Locally Compact Groups*, EMS Textbooks in Mathematics, European Math. Soc. Publishing House, Zürich (2006).

[144] S. Thomas, *The classification problem for torsion-free abelian groups of finite rank*, J. Amer. Math. Soc. **16** (2003), 233–58.

[145] W. Ullery, *Almost Warfield groups*, Rocky Mountain J. Math. **41** (2011), 2045–55.

[146] H. Ulm, *Zur Theorie der abzählbar-unendlichen abelschen Gruppen*, Math. Ann. **107** (1933), 774–803.

[147] N. Vilenkin, *Direct decompositions of topological groups I*, Mat. Sb., N.S. **19** (1946), 85–154. [English translation from the Russian by E. Hewitt in A.M.S. Translations, Series 1, Volume 23, Providence, Rhode Island (1950).].

[148] E. A. Walker, *Ulm's theorem for totally projective groups*, Proc. Amer. Math. Soc. **37** (1973), 387–92.

[149] E. A. Walker, *The groups P_β*, Symposia Math. **13** (1974), 245–55.

[150] K. D. Wallace, *On mixed groups of torsion-free rank one with totally projective primary components*, J. Algebra **17** (1971), 482–8.

[151] R. B. Warfield, *Classification theorems for p-groups and modules over a discrete valuation ring*, Bull. Amer. Math. Soc. **78** (1972), 88–92.

[152] R. B. Warfield, *Simply presented groups*, in Proceedings of the Special Semester on Abelian Groups, University of Arizona, Tucson (1972).

[153] R. B. Warfield, *A classification theorem for abelian p-groups*, Trans. Amer. Math. Soc. **210** (1975), 149–68.

[154] R. B. Warfield, *Classification theory of abelian groups I: Balanced projectives*, Trans. Amer. Math. Soc. **222** (1976), 33–63.

[155] R. B. Warfield, *Classification theory of abelian groups II: Local theory*, in Abelian Group Theory, Lecture Notes in Math., Vol. 874, Springer, Berlin (1981), pp 322–49.

[156] A. Weil, *L'Intégration dans les Groupes Topologiques et ses Applications*, Hermann & Cie., Paris (1941, 1951).

[157] L. Zippin, *Countable torsion groups*, Ann. Math. **36** (1935), 86–99.

Index

De Gruyter Studies in Mathematics

Volume 72
Francesco Aldo Costabile
Modern Umbral Calculus: An Elementary Introduction with Applications to Linear Interpolation and Operator Approximation Theory, 2019
ISBN 978-3-11-064996-3, e-ISBN 978-3-11-065292-5, e-ISBN (ePUB) 978-3-11-065009-9

Volume 34/1
József Lörinczi, Fumio Hiroshima, Volker Betz
Feynman-Kac-Type Theorems and Gibbs Measures: Volume 1, 2019
ISBN 978-3-11-033004-5, e-ISBN 978-3-11-033039-7, e-ISBN (ePUB) 978-3-11-038993-7

Volume 34/2
József Lörinczi, Fumio Hiroshima
Applications in Rigorous Quantum Field Theory: Volume 2, 2019
ISBN 978-3-11-040350-3, e-ISBN 978-3-11-040354-1, e-ISBN (ePUB) 978-3-11-040360-2

Volume 71
Wolfgang Herfort, Karl H. Hofmann, Francesco G. Russo
Periodic Locally Compact Groups: A Study of a Class of Totally Disconnected Topological Groups, 2018
ISBN 978-3-11-059847-6, e-ISBN 978-3-11-059919-0, e-ISBN (ePUB) 978-3-11-059908-4

Volume 70
Christian Remling
Spectral Theory of Canonical Systems, 2018
ISBN 978-3-11-056202-6, e-ISBN 978-3-11-056323-8, e-ISBN (ePUB) 978-3-11-056228-6

Volume 69
Derek K. Thomas, Nikola Tuneski, Allu Vasudevarao
Univalent Functions. A Primer
ISBN 978-3-11-056009-1, e-ISBN 978-3-11-056096-1, e-ISBN (ePUB) 978-3-11-056012-1